56th International Conference on Experimental Stress Analysis (EAN 2018)

Harrachov, Czech Republic
5-7 June 2018

Editors:

Ales Lufinka
Lubomir Sivcak
Iva Petrikova

ISBN: 978-1-5108-6534-1

Printed from e-media with permission by:

Curran Associates, Inc.
57 Morehouse Lane
Red Hook, NY 12571

Some format issues inherent in the e-media version may also appear in this print version.

Printed by Curran Associates, Inc. (2018)

For permission requests, please contact Czech Society for Mechanics (CSM)
at the address below.

Czech Society for Mechanics (CSM)
Dolejskova 5
Praha 8
182 00
Czech Republic

Phone: +420 266 053 045
Fax: +420 286 587 784

csm@csm.cz

Additional copies of this publication are available from:

Curran Associates, Inc.
57 Morehouse Lane
Red Hook, NY 12571 USA
Phone: 845-758-0400
Fax: 845-758-2633
Email: curran@proceedings.com
Web: www.proceedings.com

Contents

EXPERIMENTAL STRESS ANALYSIS

56th International Scientific Conference

June 5th - June 7th, 2018, Harrachov, Czech Republic

EAN 2018
56th conference on experimental stress analysis

CONFERENCE PROCEEDINGS

Edited by

Aleš Lufinka

Iva Petríková

Lubomír Sivčák

Czech Society for Mechanics

Technical University of Liberec – Faculty of Mechanical Engineering

Title: Experimental Stress Analysis 2018
Conference Proceedings

Editors: Aleš Lufinka
Iva Petríková
Lubomír Sivčák

Published by: Czech Society for Mechanics

Printed by: ReproArt Liberec

The complete set of EAN 2018 Conference Proceedings consist of two parts:

- Book of Extended Abstracts
- CD-ROM (full text of papers)

ISBN 978-80-270-4061-2 (printed, Book of Extended Abstracts)

ISBN 978-80-270-4062-9 (electronic, Conference Proceedings)

Program Committee

Chairman

- doc. Ing. Iva Petríková, Ph.D. (Technical University of Liberec, Czech Republic)

Members

- Ing. Karel Doubrava, PhD. (Czech Technical University in Prague, Czech Republic(
- prof. Ing. Nikolaj Ganev, CSc. (Czech Technical University in Prague, Czech Republic)
- Doc. Ing. Radim Halama, Ph.D. (Technical University of Ostrava, Czech Republic)
- prof. Ing. Stanislav Holý, CSc. (Czech Technical University in Prague, Czech Republic)
- Ing. Lubomír Houfek, PhD. (Brno University of Technology, Czech Republic)
- prof. RNDr. Miroslav Hrabovský, DrSc. (PU and CAS, Olomouc, Czech Republic)
- prof. Ing. Josef Jíra, CSc. (Czech Technical University in Prague, Czech Republic)
- Ing. Aleš Lufinka, PhD. (Technical University of Liberec, Czech Republic)
- prof. Ing. Bohdana Marvalová, CSc. (Technical University of Liberec, Czech Republic)
- doc. Ing. Tomáš Návrat, PhD. (Brno University of Technology, Czech Republic)
- Ing. Pavel Padevět, PhD. (Czech Technical University in Prague, Czech Republic)
- prof. Ing. František Plánička, CSc. (University of West Bohemia, Czech Republic)
- prof. Ing. Milan Růžička, CSc. (Czech Technical University in Prague, Czech Republic)
- Dr.h.c. mult. prof. Ing. František Trebuňa, CSc. (Technical University of Košice, Slovakia)
- Ing. Jaroslav Václavik (Research and Testing Institute Plzen Ltd., Czech Republic)
- doc. Ing. Miloš Vlk, CSc. (Brno University of Technology, Czech Republic)
- Ing. Robert Zemčík, Ph.D. (University of West Bohemia, Czech Republic)

Organizing Committee

Chairman

- doc. Ing. Iva Petríková, Ph.D. (Technical University of Liberec, Czech Republic)

Members

- Ing. Aleš Lufinka, Ph.D. (Technical University of Liberec, Czech Republic)
- Lubomír Sivčák (Technical University of Liberec, Czech Republic)
- Veronika Trnková (Technical University of Liberec, Czech Republic)

DEPARTMENT OF APPLIED MECHANICS

Research and innovation

Mechanics

- Kinematics and dynamics of mechanical systems
- Vibrations diagnostic
- Dynamics analysis and optimization of the systems in terms of vibration reduction with applications in the automotive and mining industries
- Synthesis of controlled mechanical systems, mechatronics-related problems
- Numerical simulation of the dynamics of mechanical systems – multibody dynamics

Elasticity and strength and biomechanics

- Theoretical and experimental research in the response to mechanical loading of components and structures from composite materials capable of large deformations
- Research in the thermomechanical behavior of smart composites – magneto-sensitive elastomers with textile SMA reinforcement
- Experimental research in the thermomechanical behavior of skin and tissues and mechanical response of bones
- Development of 3D models in medicinal applications

- Research in the fatigue of thin structures from smart materials
- Numerical simulation of the response of components and structures to thermomechanical loading – geocomposite materials, PU foams, rubbers reinforced with textile cords or textiles

Contract research

- Stress analysis of machines and structures
- Calculations of structures as regards the fatigue and life
- Mechanical properties of materials
- Kinematics and dynamics of machinery and other mechanical systems
- Vibration analysis and vibroisolation
- Biomechanics

Service offer

- Calculations
- Measuring
- Experiment
- Advisory and consultation service
- Training

DEPARTMENT OF THE DESIGN OF MACHINE ELEMENTS AND MECHANISM

- research, testing, design and optimization of components and mechanisms for the industrial sphere
- research and development for nanofibers production lines
- research and testing of components and complete car seats
- Industrial vibration damping
- CAD training centre

DEPARTMENT OF MACHINE DESIGN CXI

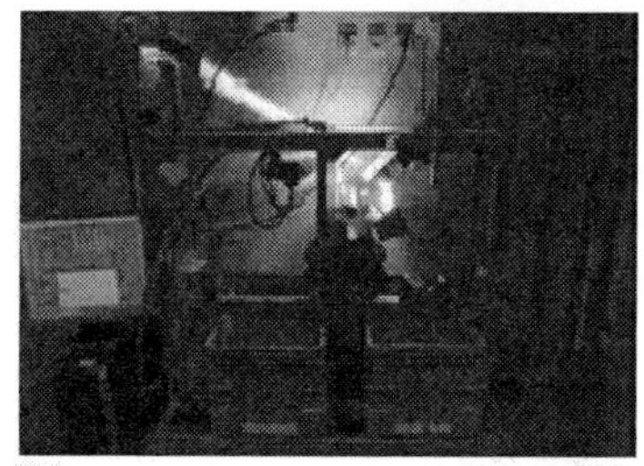

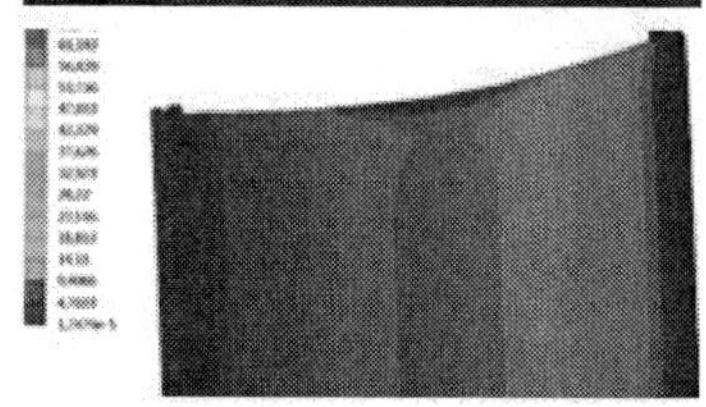

- development of constructions for the innovation of industrial production
- comprehensive solutions to specific problems in the field of engineering technology
- introduction of advanced technologies into the development of production machines
- application of new knowledge of material research into advanced industrial constructions
- examining the mechanical behaviour of the systems using mathematical models

LABORATORY OF APPLIED MECHANICS

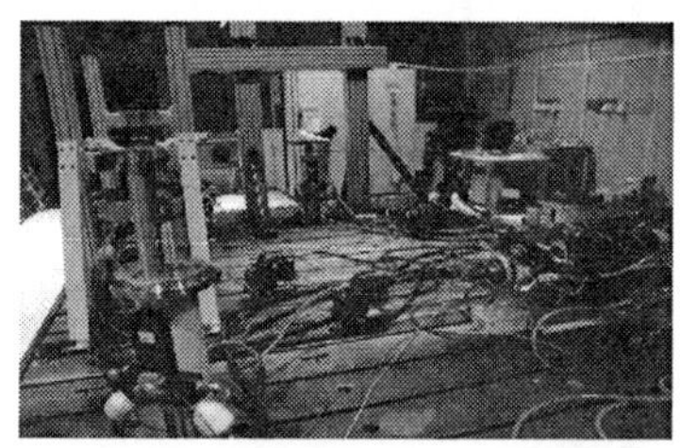

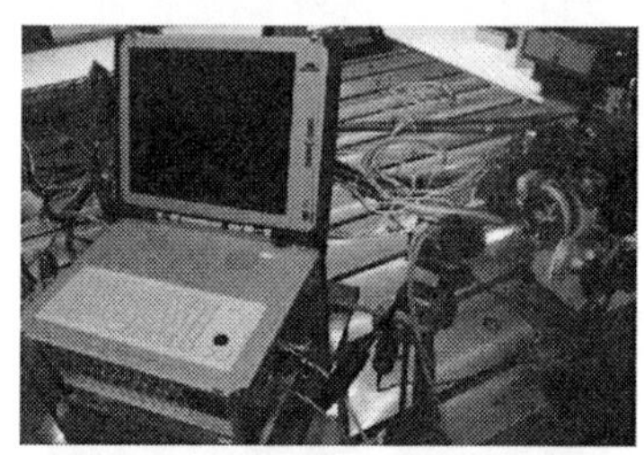

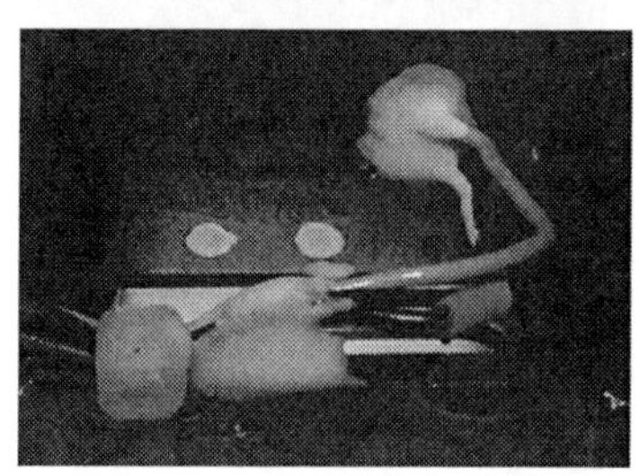

- **testing of mechanical parts and assemblies in the field of solid-state mechanics**
  - one-axis loading up to 250kN
  - multi-axis loading up to 25kN
  - crash tests up to 15m/s
  - vibration tests up to 800Hz
  - hexapod – 6 degrees of freedom platform
  - climatic chamber – the above tests at temperature range -40°C to 140°C and controlled humidity
- **application of foil and semiconductor strain gauges**
- **measurement in and out of the lab**
  - DEWE 5000 – synchronous data and image record
    16 channels, synchronous sampling 500kHz
    camera VGA 600f/s
  - SIRIUS R2DB measurement of noise and vibration, spatial analysis of oscillations
    8 channels, synchronous sampling 200kHz
    camera VGA 600f/s
- **development and production of control and measurement systems and Labview applications**

EAN 2018
56th conference on experimental stress analysis

Measurement of Mechanical Properties of Lead Free Solder

ANTOŠ Daniel[1,a], HALAMA Radim[1,b] and BARTECKÝ Matěj[1,c]

[1] Department of Applied Mechanics, Faculty of Mechanical Engineering, VŠB – Technical University of Ostrava, 17. listopadu 15, 708 33 Ostrava, Czech Republic

[a]daniel.antos.st@vsb.cz, [b]radim.halama@vsb.cz, [c]matej.bartecky@continental-corporation.com

Keywords: solder paste, Sn-3.5Ag-0.75Cu, measurement, cyclic plasticity.

Abstract. This paper describes main results of research focused on mechanical testing of Sn-3.5Ag-0.75Cu solder. Creep, tensile and torsional properties were investigated. The plastic behavior of the material can be studied in the most effective way under pure torsion. The viscoplastic behavior with high sensitivity on strain rate as well as temperature was observed. The results will be subsequently used for the calibration of appropriate constitutive law of the lead free solder.

Introduction

Nowadays almost everyone on the planet Earth daily uses highly advanced and complex electronics such as smart phones, smart televisions, computers etc. These electronics would not work without its intelligent core – printed circuit boards. The board consists of electrically conductive copper traces and non-conductive substrate. Copper traces are sophistically laid on this board in order to connect various components such as resistors, capacitors, semiconductors, etc. In order to ensure right electrical and also mechanical connection between components and copper traces on printed circuit board, the solder joint connection is widely used.

Soldering process dates back to the 5000 years when tin combined with lead was used. Several years ago it was found that lead highly pollute the ground water, which may cause environmental problems and therefore its use was forbidden. From that time, the lead free solder alloys are used for soldering. These solder pastes consist mainly from tin, silver and other chemical elements such as copper, nickel or antimony [1]. In minor representation there are also fluxes and rosin.

Failure usually occur in this tiny solder joint connection when printed circuit board is exposed either to high temperatures or to the alternating temperatures which result in malfunction of electronic device itself. During that, solder joints are usually exposed to multiaxial loading. Finding the cause or even predict malfunction using the finite element method is highly demanded from customers. To be able to do that we firstly need to know material properties of these solder joints.

Design of specimen

Because of the size of the real product, where some components could have less than 5mm, it is really difficult to prepare specimens for measurement of mechanical properties. Also for the right properties of the soldered joint, the temperature time during soldering in solder reflow oven must be respected, see figure 1 [2]. For measurement of mechanical properties of solder joints special samples have to be prepared.

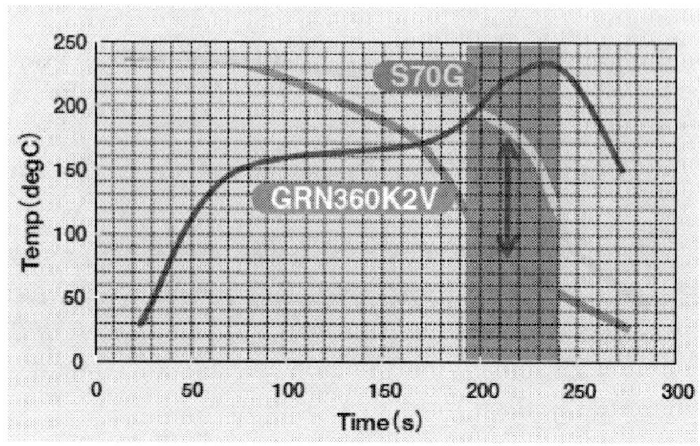

Fig.1 Soldering temperature profile [2]

During our investigations we studied several articles about mechanical properties of solder joints and its measurement ([1], [3], [4], [5]). Interesting and inspired study was the report made by engineers from Royal Instate of Technology KTG, Stockholm [6]. They measured tensile test using the special specimen, which consists of two copper flat bars between which small amounts of solder paste was applied. Because our aim was to measure not only tensile properties but also torsion or even multiaxial mechanical properties, we need to use axisymmetric shape of final specimen. On other hand engineers from Akita Prefectural University used for their testing round shape specimen prepared from ingot casting [7]. This kind of specimen is not suitable for our measurements, because it does not have fluxes and rosin, which are in real solder paste and influence behavior of solder joint. Therefore, we need to use real solder paste and create a special specimen from it. Finally, we combine these two approaches [6], [7].

Our measurement sample consists of copper dog bone round cross-section separated into two pieces between which small amount of solder paste Sn-3.5Ag-0.75Cu is applied. Thickness of solder paste was setup to 0.5mm with diameter of 7mm, see figure 2.

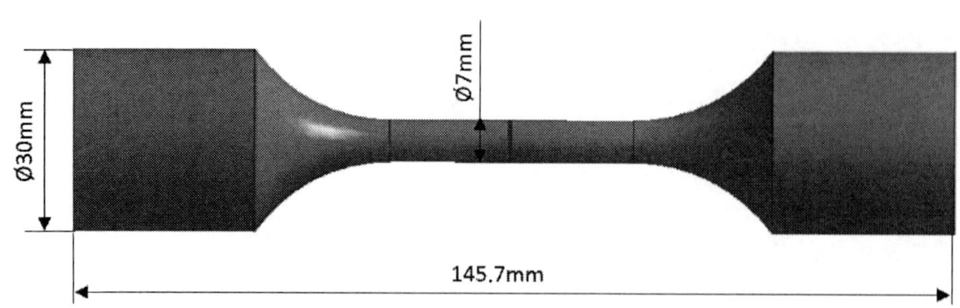

Fig.2 Solder paste specimen

To ensure right dimension of applied solder and also cohesiveness, special soldering tool needed to be designed and manufactured. Important thing was to choose right material for soldering tool with resisting adhesion for applied solder paste. For this purpose, we used S235 steel. Ready-made soldering tool can be seen in the figure 3.

Fig.3 Soldering tool with copper parts of final specimen

Production of specimen

Soldering was done as follows. Firstly, the copper parts of the specimen were placed into soldering tool. Into the created gap, the solder paste was applied. Then such prepared soldering tool was placed into the temperature chamber and heated up to 250°C, which is soldering temperature. After the soldering temperature was reached, the soldering tool was withdrawn from temperature chamber and cooled down to room temperature, see figure 4. Then the specimen was taken out from the soldering tool. Solder area need to be machined to have ideally rounded shape, see figure 5. Such sample was then used for mechanical testing.

Fig.4 Soldering at temperature chamber on the left
and cooling after soldering on the right

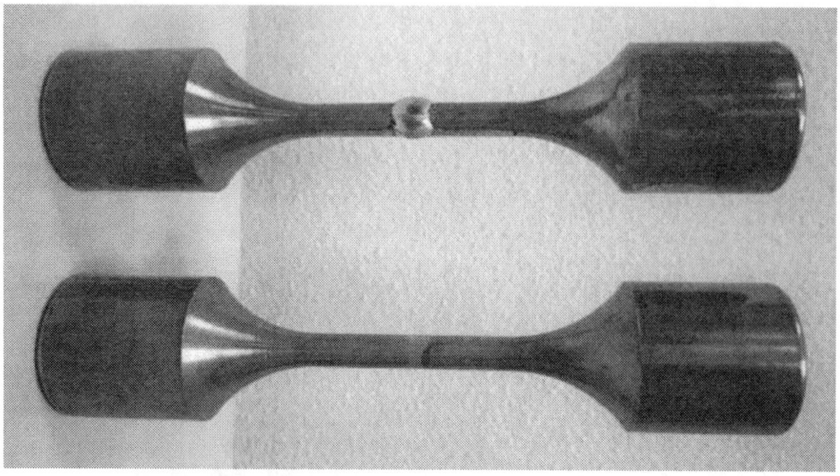

Fig.5 Specimen before and after machining

During specimen creation, a problem with bubbles inside soldering area raised, see figure 6. It is most probable, that these bubbles are either air bubbles trapped inside the solder paste due to imperfect application between copper parts or vacuum bubbles caused by evaporation of fluxes or rosin, which are inside solder paste. Either way after long examination we discovered that bubbles inside solder area are caused by soldering process and non-observance of soldering time. Heating of soldering tool with copper parts and applied solder paste takes more than one hour before soldering temperature was achieved. This approach may also cause degradation of solder paste material. In order to obey the soldering time, firstly the soldering tool with copper parts was pre-heated to soldering temperature. After soldering temperature was reached on copper parts soldering tool was taken out from temperature chamber and solder paste was quickly applied into the gap between copper parts. By this the presence of big bubbles was eliminated. Some amount of tiny bubbles are still present even after soldering process was changed. But these bubbles are also in solder joint area of real product, see figure 7.

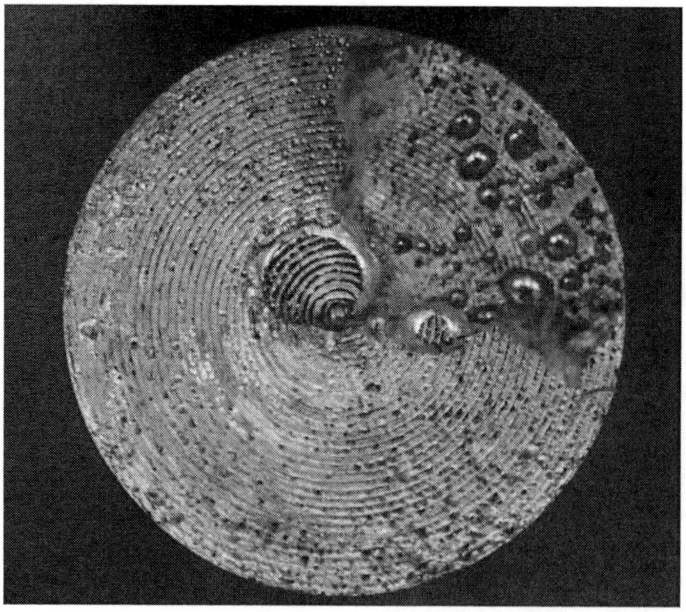

Fig.6 Bubbles inside solder area

Fig.7 Bubbles inside solder joint

Experimental Results

Soldered and machined specimens were then used for tensile, torsion and also creep tests. Different temperatures and strain rates were applied during measurements. Due to the small thickness of solder area, the only results from the tests are elongation and force from jaws. Mechanical extensometer does not have such small grip and optical extensometer was not able to catch whole test due to the low resolution. Therefore the measurements are effected by presence of copper parts. Also elongation before failure is very small and presence of micro residual bubbles could influence measurement. Mechanical properties of solder joint need to be reverse engineered by using finite element method analyses.

Fig.8 Fixture of specimen during tension test (left) and torsion test (right)

Uniaxial tensile tests were driven by deformation. Tests were done for crosshead speed 0.1mm/min, 1mm/min and 10mm/min and for room temperature, 80 °C and 125 °C. Results are presented in the figure 9.

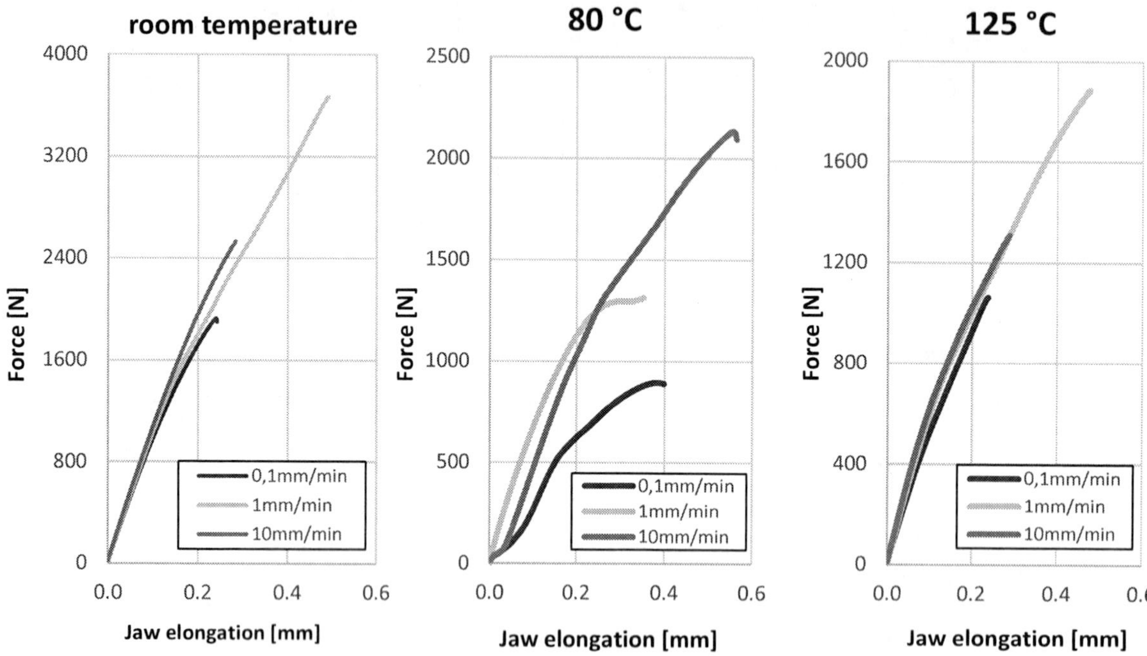

Fig.9 tensile tests records

Torsion tests were driven by shear deformation. These tests were done only for room temperature and for three crosshead speeds: 0.1°/s, 0.5°/s and 1°/s. Records from these tests are presented in the figure 10.

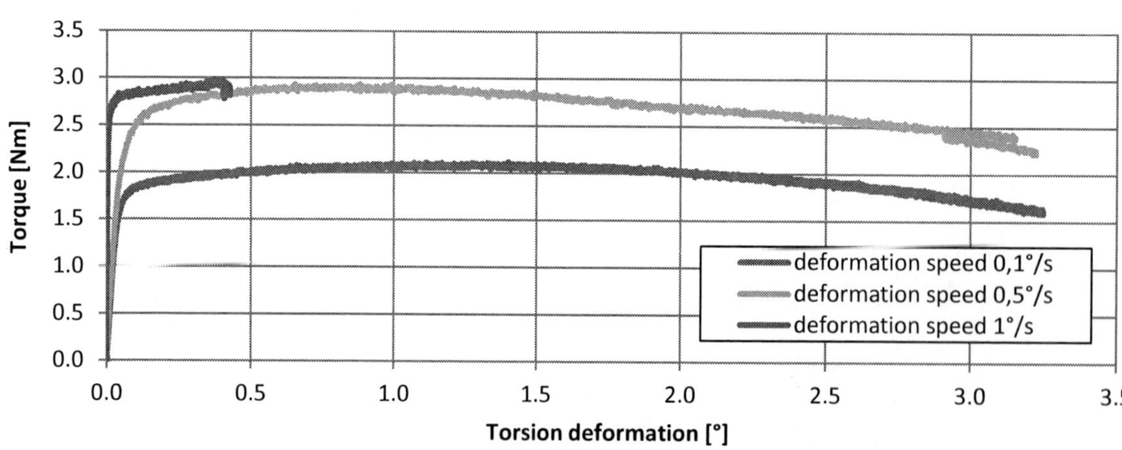

Fig.10 torsion tests records

In the case of creep testing, the loading force was set to 700N. This test was performed for two different temperatures 80°C and 125°C, see figure 11.

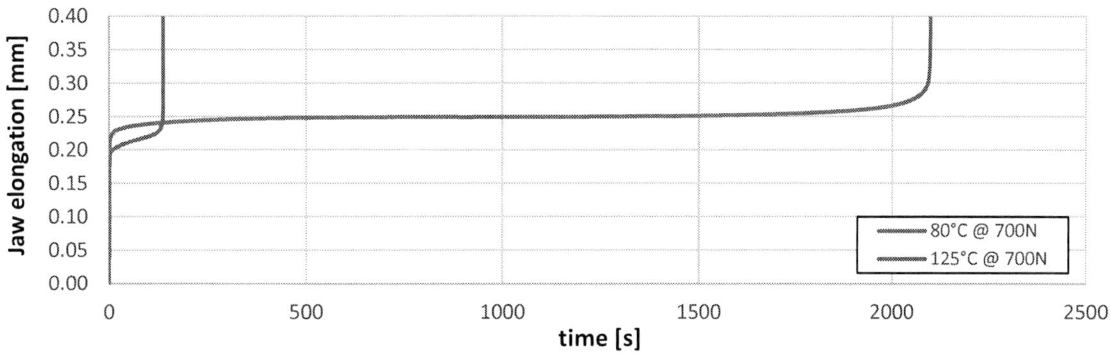

Fig.11 creep tests records

Conclusions

Measurements in tension, torsion and creep shows that behaviour of Sn-3.5Ag-0.75Cu solder joint is rate and temperature dependent. In case of tension tests rate dependency has much smaller effect than the temperature influence. On the other hand, torsion tests show us great influence on crosshead speed, mainly between 0.1°/s and 0.5°/s. During creep, the solder joint behavior is again temperature dependent. These measurements are effected by presence of copper parts and also could be effected by micro bubbles inside of the solder joint. These bubbles are natural and influence the both material strength and number of cycles to failure in the case of cyclic loading. Study of cyclic loading will be published later after deep investigations. Data from these measurements will be used for calibration of a visco-plastic material model, which has been implemented into a finite element code covering strain rate influence as well as creep behaviour of lead free solder pastes.

Acknowledgement

This work was supported by The Ministry of Education, Youth and Sports from the Specific Research Project (SP2018/63). The support of Continental Automotive company is also acknowledged.

References

[1] Yuichiro Toyama, Ikuo Shohji, Effect of Strain Rate on Tensile Properties of Miniature Size Lead-Free Alloys. In: Proceedings of 35th IEEE/CPMT International Electronics Manufacturing Technology Symposium (IEMT), Ipoh, Malaysia; 6 November through 8 November 2012.

[2] Senju Metal Industry Co.LTD, Eco solder® Environment-friendly lead-free solder, Japan. 8 (2013).

[3] Masataka Nishiura, Akihiro Nakayama, Sigeaki Sakatani, Yasuhiro Kohora, Keisure Uenishi and Kojiro F. Kobayashi, Mechanical strength and microstructure of BGA joints using lead free solders, Materials Transactions Vol.43 No.8 (2002) 1802-1807.

[4] Martin Wickham, Jaspal Nottay and Christopher Hunt, A review of mechanical test method standards for lead free solders, National Physical Laboratory, UK. 15 (2001).

[5] Vasu Vasudevan and Xuejun Fan, An acceleration model for lead-free (SAC) solder joint reability under thermal cycling, Electronic Components and Technology Conference (2008). 139-145

[6] Susanne Lewin, Mechanical Reliability of Aged Lead-Free Solders, Royal Institute of Technology, Stockholm. 19 (2012).

[7] Ikuo Shohji, Tomohiro Yoshida, Takehiko Takahashi and Susumu Hioki, Tensile properties of Sn-3.5Ag and Sn-3.5Ag-0.75Cu Lead free solders, Materials Transactions Vol.43 No.8 (2002) 1854-1857.

EAN 2018
56th conference on experimental stress analysis

System of Axle-Box Force Measurement for Experimental Railway Bogie

BAUER Petr[1,a] and KALIVODA Jan[1,b]

[1] Czech Technical University in Prague, Faculty of Mechanical Engineering, Department of Automotive, Combustion Engine and Railway Engineering, Technická 4, 166 07, Prague 6, Czech Republic

[a]Petr.Bauer@fs.cvut.cz, [b]Jan.Kalivoda@fs.cvut.cz

Keywords: Railway bogie, wheel-rail forces, axle-box forces.

Abstract. The article describes the newly developed system for measuring axle-box forces, i.e. forces acting between a bogie-frame and axle-boxes of an experimental railway bogie. One component of the wheelset guidance system was specifically designed was designed with the aim to serve as a three axis load cell. The measurement system was designed to measure x, y and z components of axle-box force independently of each other and independently of the passive resistances in the wheelset guidance mechanism. It consists of 36 strain gauges connected into the three Wheatstone bridges. The load cell design and calibration results are described in detail.

Introduction

Force interaction between rails and railway wheels is undoubtedly one of the most important issues in the development and qualitative evaluation of each railway vehicle. The permitted maximum values of forces applied by the wheel on the rail are defined in the relevant standards. They are based on safety criteria against derailment and the load capacity of the railway infrastructure. Today's effort to build an economic and environment friendly railroad brings a general and sustained demand to reduce wheel-rail contact forces below these legislative limits as much as possible. Reduction of the wheel-rail forces causes not only decrease of vehicle and track wear, it also reduces noise emissions, energy consumption, etc. Therefore, the reduction of wheel-rail contact forces is one of the most important objectives of a long term research activities at the Department of Automotive, Combustion Engine and Railway Engineering CTU in Prague. In order to verify and demonstrate results of mathematical simulations an experimental facility in the form of a roller rig and 1:3.5 scaled experimental bogie [1, 2] is utilized, see Fig. 1.

Fig. 1 The experimental railway bogie at the CTU roller rig

Measurement of contact forces directly in the wheel-rail contact point is virtually impossible. In practice, indirect detection of these forces is used. It is based on the measurement of the strain on the appropriate vehicle parts [4, 5] or the measurement of the deformations of the suspension parts. With respect to the inertia effects of the vehicle components, it is desirable to carry out this measurement as close as possible to the wheel-rail contact point. During rail vehicle homologation tests, the measurement of deformations of wheel disks, wheelset axles, or specially instrumented track sections are used.

Measurement of wheel-rail contact forces at the CTU roller rig

In the case of experimental roller rig bogie, it is possible to use similar approach. However, it is necessary to take into the account also its specificity. At the CTU experimental bogie, the small diameter and high stiffness of the wheel discs as well as frequent need of mounting intervention to the wheelset setup, (3 types of wheelset and several wheel treads with different taper grade are available) would significantly complicate the use of the wheelset for wheel-rail force measurement. Therefore, a system for measuring the lateral component of wheel-rail forces, utilizing transverse roller disk deformation measurement, was developed and applied on all 4 rollers [3]. Although it was not possible to tune the shape of the rollers, hereby formed wheel-roller contact force measurement has a sufficient sensitivity in the lateral direction. By the other hand, the rollers with the current shape and stiffness of the disk are hardly applicable for measurements of the vertical component of wheel-rail force.

Design of a system for axle-box forces measurement

Information on the magnitude of forces between the axle-boxes and the bogie frame are not regarded as an adequate substitute for information about the value of wheel-rail contact forces. The resultant of axle-box forces acting on one wheelset and corresponding resultant wheel-rail forces differ due to the inertia effects of the wheelset. Moreover, from the measured values of axle-box forces it is not possible to determine the distribution of wheel rail forces on the individual wheels. Nevertheless, the axle-box force measurement has significantly meaningful value. The advantage of the measurement of axle-box forces is its relatively easy realization on the vehicle without need to transmit the signal from the rotating parts. Although further development of wheel-rail forces measurement by measurement of roller disc deformation is still planned, it was decided to implement also axle-box forces measurement to the wheelset guidance of the newly developed test bogie. The goal is to measure forces transmitted at each axle-box to the bogie frame in 3 directions:

- F_x – longitudinal component, based on F_x it is possible to calculate action torque of the wheelset steering mechanism,
- F_y - lateral component, F_y provides information on the sum of lateral component of wheel-rail forces.
- F_z - vertical component, F_z provides information on the vertical wheel-rail forces

Load cell design. Considering several possible solutions, the development of a newly designed 3 axis load cell was finally chosen. The main advantage of this solution compared to implementation of industrially produced 3-axis load-cells, is the possibility to adapt certain components of wheelset guidance for axle-box force measurement. This significantly simplifies the design of wheelset guidance itself. In the final design solution (Fig. 2), the axle-box (item 9) is guided by stirrup (item 8), which is on both ends connected by spherical joints (items 6a, b) to the carriages of linear guideway (items 7a, b). Connection of the axle-box and the stirrup is provided via a cylindrical joint (bearings 11a, b). The axle-box can thus rotate around vertical axis towards the stirrup. The end shanks of the stirrup are aimed for the strain gauges placement. The longitudinal axis of the end shanks is at the height of the wheelset axis and also the axis of the control linkages joints is at the same height. Thus, in terms of measurement the stirrup is a beam supported by two supports and loaded by individual components of the axle-box force.

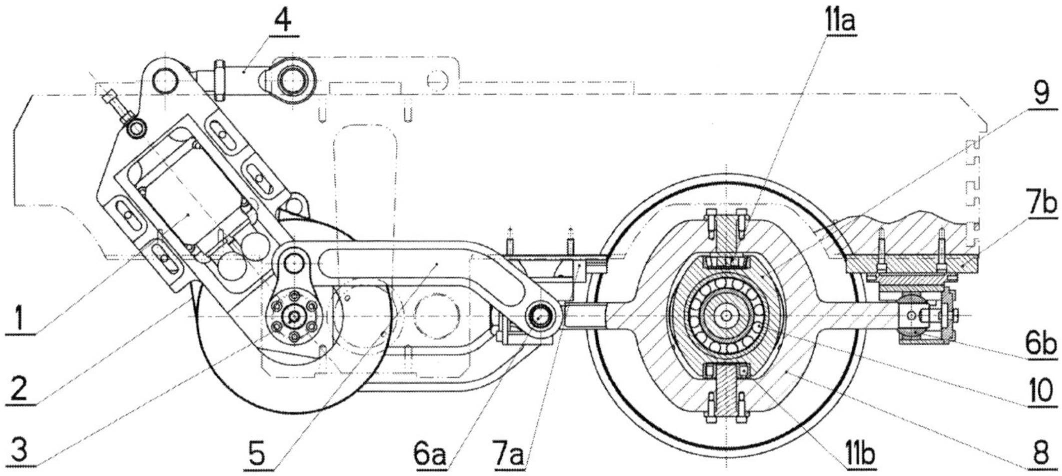

Fig 2 Design of the wheelset steering mechanism and the connection between axle-box and bogie frame

The overall concept of the test bogie expects vertical loading in order to simulate the weight of the vehicle body. For dimensioning of the stirrup in terms of loads F_y and F_z, the maximal vertical load acting upon the test bogie is taken into the account. The magnitude of maximal longitudinal load F_x is given by the maximal yaw torque applied upon a wheelset by the wheelset steering mechanism. Based on these assumptions, the maximal loads of single axle-box load cell are:

- Fx … 2.5 kN
- Fy … 5 kN
- Fz … 5 kN

The stirrup is made of heat-treated C45 steel. Its design was subjected to FEM calculation and shape optimization (Fig 3), in order to achieve as high as possible sensitivity on one hand and not to exceed 500 MPa von Mises stress on the other.

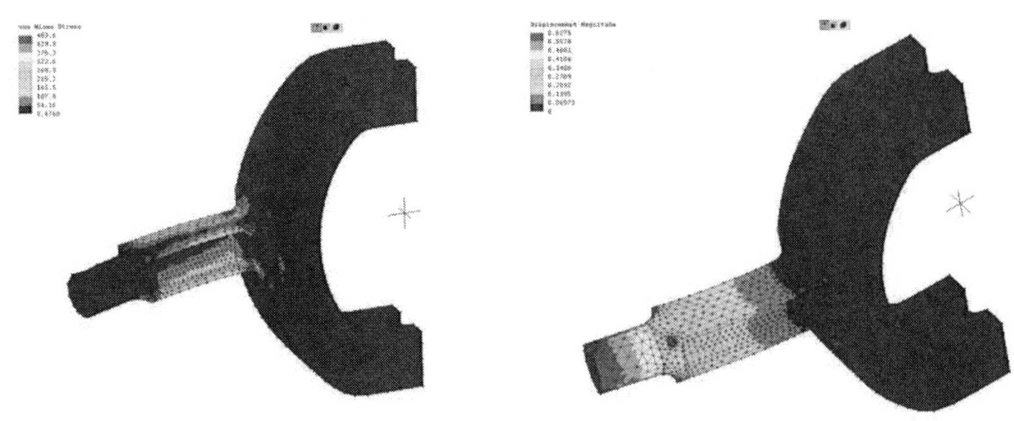

Fig. 3 Sample results of the stirrup FEM analysis. Von Mises stress upon maximum load in all three axes all at once (left). Displacement upon lateral load only (right).

F_x – measurement of the longitudinal component of the axle-box force. The longitudinal position of the axle-box is driven by means of control linkages connected to one of the carriages of a ball linear guide. The force transmitted by linkages F_{CL} (see Fig 4) is the sum of the longitudinal axle-box force F_x and the resistance forces of the linear guides F_{R1} and F_{R2}. In the whole x-section of the shank, closer to control linkages, the normal stress, which is measured by strain gauges T_{x1i} and T_{x1e}, is proportional to sum of the axle-box force F_x and the resistant force of the more distant carriage F_{R2}. Whereas the normal stress at second shank, which is measured by strain gauges T_{x2i} and T_{x2e}, is proportional to resistant force of the more distant linear carriage F_{R2} only. Strain gauges for the measurement of F_x are placed on both sides of the shanks on their neutral axis. Connection of the strain gauges in Wheatstone bridge will automatically compensate resistant force of linear guides from the measured signal (Fig. 4). Also the influence of bending moment originating from lateral axle-box force F_y will be automatically compensated. Due to the placement of strain gauges to the neutral axis, the output signal should be also insensitive to the load caused by the vertical axle-box force F_z.

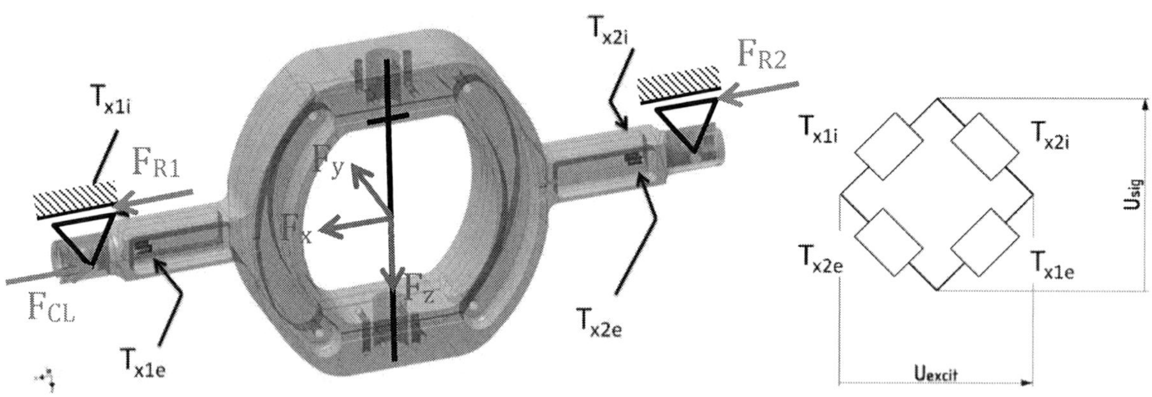

Fig. 4 System for the measurement of the longitudinal component F_x of the axle-box force. Strain gauges position (left) and wiring diagram (right).
$T_{x..}$ – strain gauge designation, F_{CL} – force transmitted by the control linkages, $F_{R.}$ – resistant forces of the linear guides, U_{excit} - power supply of the Wheatstone bridge, U_{sig} – measured response.

F_y – measurement of the lateral component of the axle-box force. Lateral axle-box force F_y bends both the shanks in the horizontal plane and causes strain difference on the inner and outer sides of the shank. Assuming that the stirrup is supported by ideal spherical joints, four strain gauges would be sufficient for measurement of F_y. However, the joints resistance is not negligible. Due to the internal friction, both of the end joints will act by a certain moment, which could influence the measured value of lateral force F_y. Therefore, the proposed system is based on 16 strain gauges (Fig. 5). Compensation of the joints resistance is based on the assumption that the joint friction torques cause a constant bending moment along the shanks. By placing the strain gauges into the two sections and subtracting the signals from those two sections a signal proportional to the lateral force F_y load without influence of friction moments is obtained.

F_z – measurement of the vertical component of the axle-box force. The vertical axle-box force F_z again bends both shanks, but this time in the vertical plane. Measurement of force F_z is by principal analogical to the measurement of the lateral force F_y (Fig. 6).

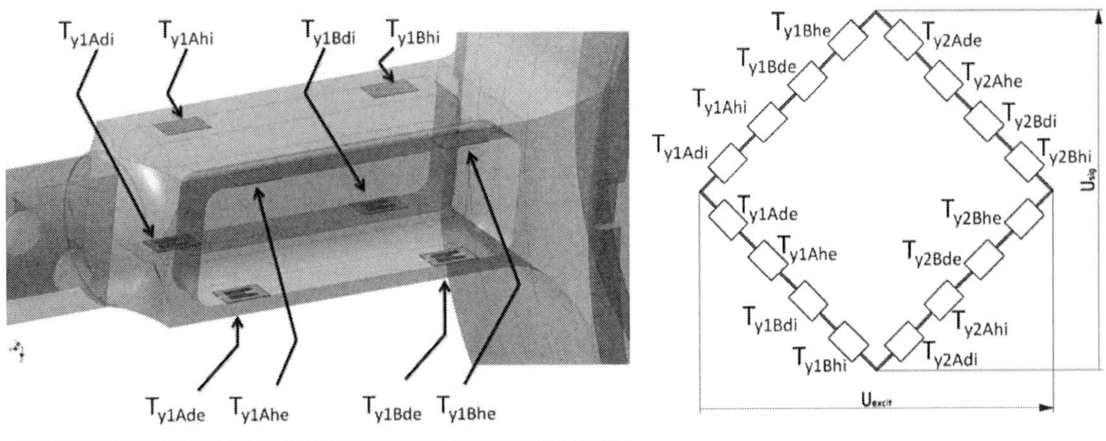

Fig. 5 System for the measurement of the lateral component F_y of the axle-box force.
Strain gauges positions on shank 1 (left), wiring diagram (right).
Strain gauges positions on shank 2 are analogous.

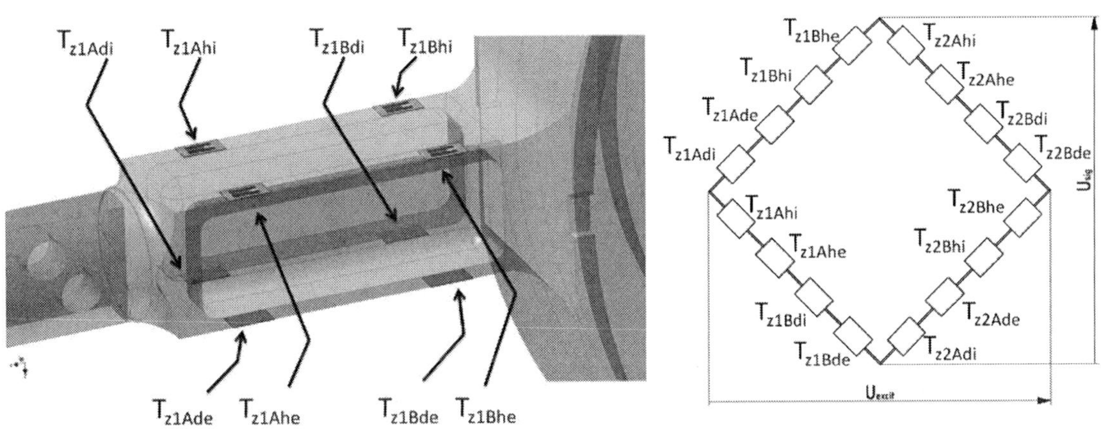

Fig. 6 System for the measurement of the vertical component F_z of the axle-box force.
Strain gauges positions on shank 1 (left), wiring diagram (right).
Strain gauges positions on shank 2 are analogous.

Calibration

The proposed system was implemented on the experimental railway bogie. The stirrup was instrumented by strain gauges and calibrated (Fig. 7). During the calibration, the stirrup was gradually loaded in the x, y and z axes in the range of 0 to 3 kN (see charts on the left on Fig 8, 9, 10). The response of the strain gauge bridges was monitored and compared to the response of calibration load cell. The charts on the right on Fig 8, 9, 10 shows the difference between directly measured stirrup load and the response of stirrup. The difference did not exceed 30 N, which is less than 1% of the measured range. At the same time, the ability of the system to measure independently the x, y and z components of the axle-box forces was confirmed. The mutual influence of the load components is less than 15 N, see the response in unloaded directions on Fig 8, 9, 10.

Fig. 7 Stirrup calibration in x (left), y (middle) and z (right) direction

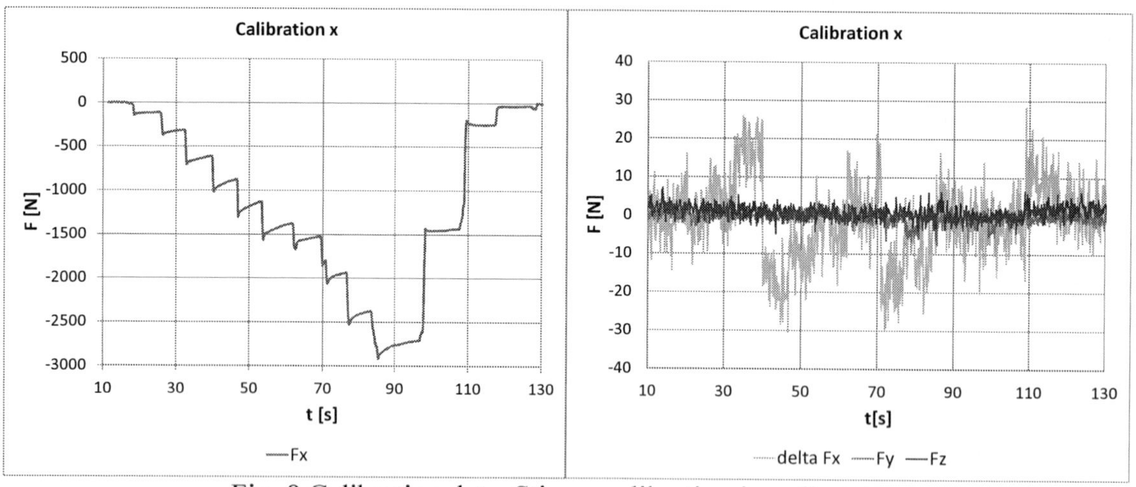

Fig. 8 Calibration data. Stirrup calibration in x direction

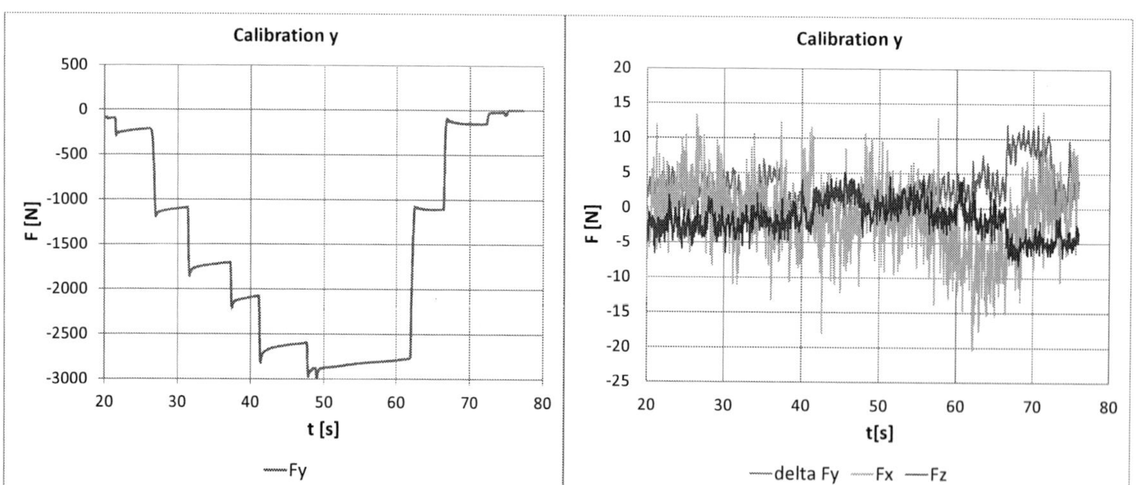

Fig. 9 Calibration data. Stirrup calibration in y direction

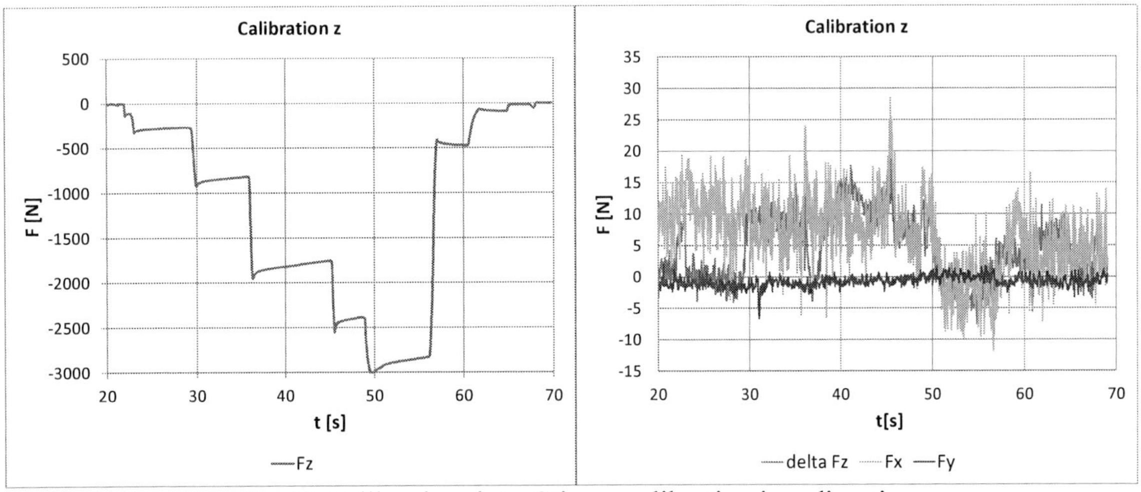

Fig. 10 Calibration data. Stirrup calibration in z direction

Conclusions

The magnitude of wheel-rail contact forces is one of the most important parameters that influences wear of wheels and rails, safety against derailment and many other important parameters of the operation of railway vehicles. Therefore, the reduction of force acting in the wheel-rail contact is still an essential topic in the development of new rolling stock. Because the direct measurement of forces acting in the wheel-rail contact is not possible, it is necessary to determine these forces indirectly, usually by measurement of the deformation of the wheels, rails or other parts of the rail vehicle bogie. One of the parameters, which can be used to determine the magnitude of the wheel-rail contact forces are axle-box forces, i.e. the forces acting between the axle-boxes and the bogie frame. A system capable to measure axle-box forces on the experimental railway bogie was designed. The system is based on stirrup specifically designed component that connects the axle-box and the bogie-frame. The bogie is equipped by four stirrups, one per each axle-box. Each stirrup is instrumented by 36 strain gauges connected to three full bridges. Thus measurement of x, y and z components of a force transmitted between each axle-box and the bogie frame is achieved. The system was successfully calibrated and implemented to the experimental railway bogie. The experimental bogie is thus prepared to conduct experiments focused on reducing the mutual interaction of railway vehicles and the track.

Acknowledgement

This paper was created with the support of Technology Agency of the Czech Republic, project No TE01020038 „Competence Centre of Railway Vehicles“.

References

[1] S. Myamlin, J. Kalivoda and L. Neduzha, Testing of Railway Vehicles Using Roller Rigs [online]. Procedia Engineering. 2017, 2017(187), pp. 688-695. ISSN 1877-7058. doi: 10.1016/j.proeng.2017.04.439.

[2] J. Kalivoda and P. Bauer, Mechatronic Bogie for Roller Rig Tests. In: Six, K., et al., eds. IAVSD 2015 - 24th International Symposium on Dynamics of Vehicles on Roads and Tracks. Graz, 17.08.2015 - 21.08.2015. Boca Raton: CRC Press. 2015, ISBN 9781138028852.

[3] J. Kalivoda and P. Bauer, Scaled Roller Rig Experiments with a Mechatronic Bogie, in: Proceedings of the Second International Conference on Railway Technology: Research, Development and Maintenance. Edinburgh: Civil-Comp Press, 2014, art. no. 317, ISSN 1759-3433. ISBN 978-1-905088-59-1.

[4] M. Dub, J. Kolář, F. Lopot, V. Dynybyl and O. Berka, Experimental analysis of stress state and motion of tram gearbox hinge, in: EAN 2014 - 52nd International Conference on Experimental Stress Analysis, 2014.

[5] S. Myamlin, O. Ten, L. Neduzha, Theoretical and experimental research of strength properties of spine beam of freight cars. in: Science and transport progress. № 3 (51), 2014, pp. 136–145.

EAN 2018
56th conference on experimental stress analysis

Experimental Investigation of Clearance Effect on Single-lap Joint Using Blind Rivet

BEK Lukáš[1,a], KOTTNER Radek[1,b] and LAŠ Vladislav[1,c]

[1] NTIS – New Technologies for the Information Society, Faculty of Applied Sciences, University of West Bohemia, Technická 8, 301 00 Pilsen, Czech Republic

[a]lukasbek@kme.zcu.cz, [b]kottner@kme.zcu.cz, [c]las@kme.zcu.cz

Keywords: Clearance effect, blind rivet, Carbon fibre composite,

Abstract. The experimental investigation of the clearance effect influence on the stiffness and the strength of the riveted single-lap joint using a blind rivet and carbon fibre plates was performed. The investigation for two rivet diameters was performed. For each rivet diameter, samples with four different hole to rivet clearances were prepared while samples were consisting of carbon fibre plates and high strength blind rivets. The samples were loaded by tensile loading up to a failure. Measured force-displacement curves were compared and the average curves for each clearance and the rivet diameter were created.

Introduction

High stiffness and strength in connection with low density predestine composite materials to increasing use of these materials in various applications in sport, automotive, marine, and aerospace industries. Composite structures are usually composed of more composite components or connected to other non-composite parts. Therefore, proper connection must be selected in order to fulfill all requirements of the joint. Widely used joint type is the riveted joint. The possibility of the joining of different materials, low price of rivets, no required fixation of joined components after assembly, and simple creation which includes the drilling of a necessary hole and application of a rivet are the advantages of this joint [1]. However, the strength of the joint can be influenced by the clearance of a hole and a fastener. The creation of a precision hole can be expensive due to the time and a tool needed for this type of operation. Therefore, it is essential to prove whether it is necessary or not.

The clearance effect was investigated in [2], [3], [4], and [5] in case of bolted joint. All these papers concluded that the increasing clearance decreases the stiffness and the strength of the joint. Based on the similarity, this can be concluded also for high strength rivets with solid shank as Hi-Lok, Taper-Lock, Lockbolt, etc. However, the aforementioned joint requires approach from the both sides of the joint for a rivet application. Therefore, blind rivets are widely used due to place restrictions. Due to the fact that the blind rivet shank is reformed during a riveting process, there is a chance that the blind rivet is capable to fill out the clearance and to reduce the negative effect of the clearance effect observed at mentioned high strength rivets with the solid shank.

Therefore, the preparation of test samples and the experimental investigation of the clearance effect on a single-lap joint using the blind rivets were carried out in this work.

Test sample preparation

The single-lap joint test samples were consisting of two carbon fibre components and a high strength TIBULB blind rivet having a diameter of 4.8 mm or 6.4 mm. For the preparation of the carbon fibre components, plates were manufactured by the author in the autoclave ASC Econoclave. The Airbus patented Vacuum Assisted Process (VAP® [6]) was selected as a suitable manufacturing process for the production of the carbon fibre plates. Due to the use of the VAP® membrane, this process brings a benefit of the uniform resin distribution during an infusion resulting in lower material parameter variability and in nearly optimal fibre to resin volume ratio.

The manufactured plates were produced from the non-crimped fabric (NCF) Saertex with the fibres TENAX-J IMS60 E13 24K and from the epoxy resin MSG L285. The plates had the symmetrical layup [90°/0°/+45°/-45°]s. From the manufactured plates, the joint components were cut using a water jet cutter in order to obtain components having the width W=5D, the edge distance E=2D, and the length L=10D, where D denotes the rivet diameter (Fig. 1). To the length of the overall component length, 50 mm for the fixation into a universal testing machine was added. Therefore, components with the dimensions of 107.6 mm x 24.0 mm x 2.4 mm in case of the rivet diameter 4.8 mm and the dimensions of 126.8 mm x 32.0 mm x 2.4 mm in case of the rivet diameter 6.4 mm were prepared.

The hole was done by different reamers in order to produce a perfectly circular hole with the clearance of 0%, 1%, 2%, and 3% compared to the rivet diameter. For each clearance and the rivet diameter, six test samples of the single-lap joint were prepared (Fig. 2).

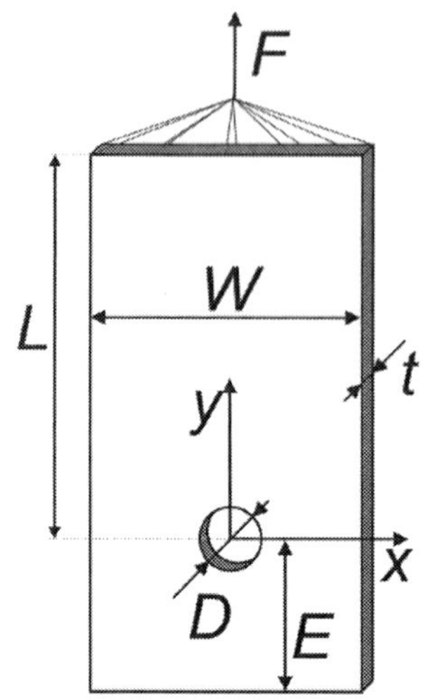

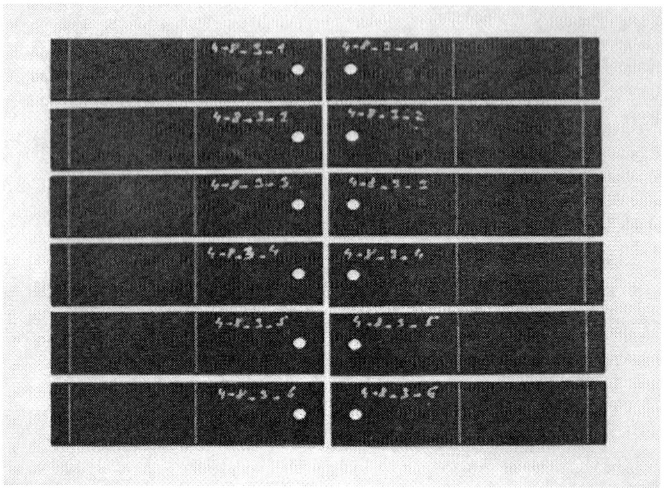

Fig. 1 Geometry of prepared carbon fibre components (left) and example of prepared set with highest hole to rivet clearance of 3% compared to rivet diameter (right)

Fig. 2 Example of prepared test sample

Testing and evaluation

The test samples for each clearance were subjected to a tensile loading using the universal testing machine Zwick/Roell Z050. The displacement controlled loading with the constant head speed of 2 mm/min was used. The load was measured using a load cell and the displacement of the sample was measured using an extensometer. The gage length was set to 60 mm. All samples were loaded up to a failure which was determined as the full failure of the joint or the 80% drop of the loading force compared to the ultimate force. In total, 48 tests were performed.

The results for each hole to rivet clearance were statistically evaluated (see Table 1 and Table 2) and the average force-displacement curves were determined (Fig. 3 thru Fig. 6 and Fig. 8 thru Fig. 11). The average force-displacement curves for each hole to rivet clearance were compared (Fig. 7 and Fig. 12). The stiffness and the strength of the joint were main scopes for the evaluation.

Tab.1 Results for samples with rivet diameter 4.8 mm

	Ultimate force [N]			
Sample number	**Clearance 0%**	**Clearance 1%**	**Clearance 0%**	**Clearance 3%**
1	3605.5	3699.6	3953.8	3742.9
2	3847.1	3814.1	3736.7	3826.1
3	3549.4	3788.4	3774.8	3776.4
4	3731.0	3747.2	3904.6	3569.7
5	3737.6	3765.9	3534.1	3837.5
6	3753.5	3651.4	3865.5	4025.5
Average [N]	**3704.0**	**3744.4**	**3794.9**	**3796.3**
St. dev. [N]	**108.1**	**59.9**	**151.0**	**148.1**
COV [%]	**2.9**	**1.6**	**4.0**	**3.9**
B-basis [N]	**3379.0**	**3564.3**	**3340.7**	**3351.0**

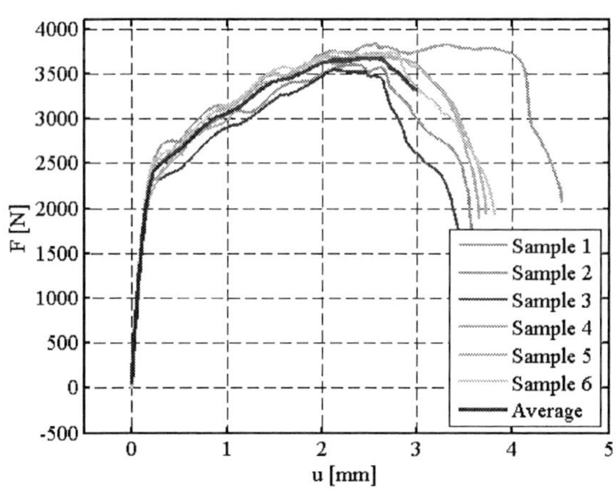

Fig. 3 Force-displacement curves for samples with rivet diameter 4.8 mm and hole to rivet clearance of 0%

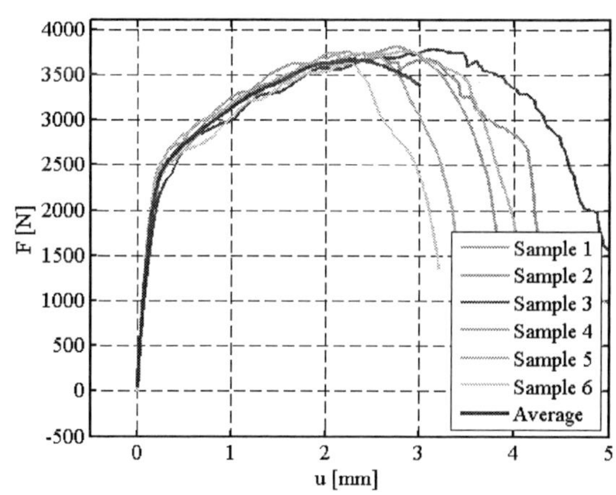

Fig. 4 Force-displacement curves for samples with rivet diameter 4.8 mm and hole to rivet clearance of 1%

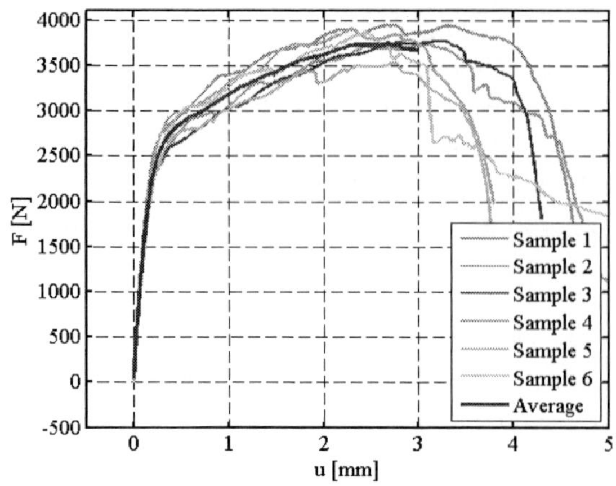

Fig. 5 Force-displacement curves for samples with rivet diameter 4.8 mm and hole to rivet clearance of 2%

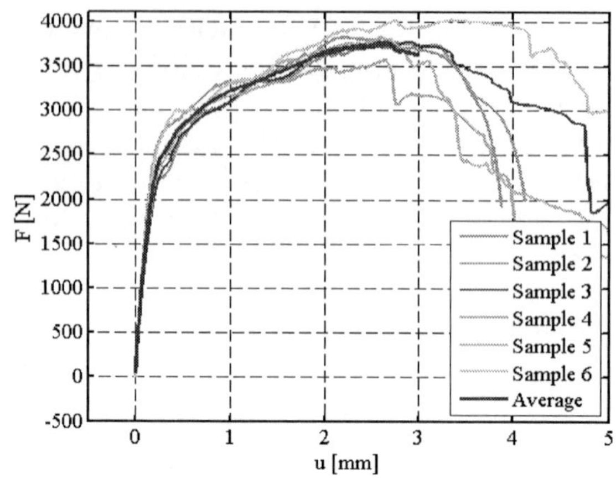

Fig. 6 Force-displacement curves for samples with rivet diameter 4.8 mm and hole to rivet clearance of 3%

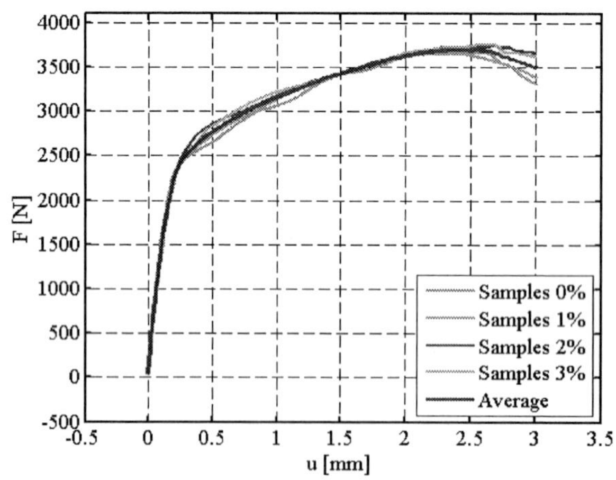

Fig. 7 Average force-displacement curves for samples with rivet diameter 4.8 mm and hole to rivet clearances of 0%, 1%, 2%, and 3%

Tab. 2 Results for samples with rivet diameter 6.4 mm

	Ultimate force [N]			
Sample number	**Clearance 0%**	**Clearance 1%**	**Clearance 2%**	**Clearance 3%**
1	6547.3	6842.1	6883.9	6541.4
2	6659.1	6079.8	6939.0	6471.4
3	6812.8	5948.0	6804.1	6625.5
4	6403.8	6701.7	6134.9	6853.3
5	6519.1	5584.3	6547.5	6876.0
6	6229.0	6417.2	6266.2	6335.4
Average [N]	**6528.5**	**6262.2**	**6595.9**	**6617.2**
St. dev. [N]	201.8	478.5	336.9	214.1
COV [%]	3.1%	7.6%	5.1%	3.2%
B-basis [N]	**5921.7**	**4823.3**	**5582.8**	**5973.4**

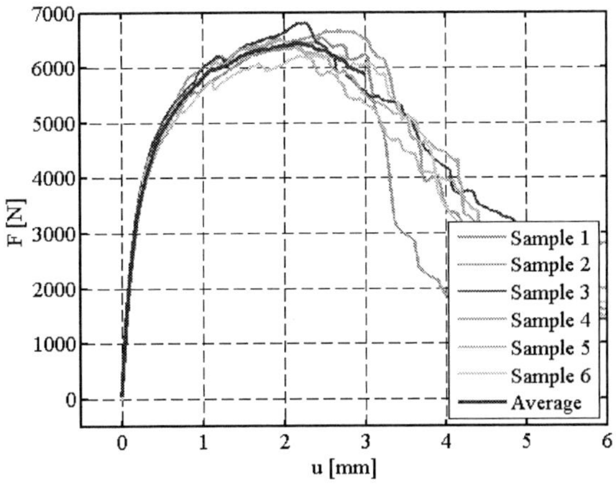

Fig. 8 Force-displacement curves for samples with rivet diameter 6.4 mm and hole to rivet clearance of 0%

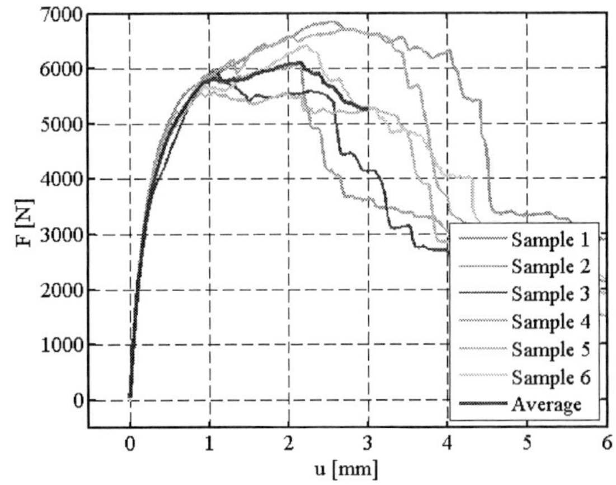

Fig. 9 Force-displacement curves for samples with rivet diameter 6.4 mm and hole to rivet clearance of 1%

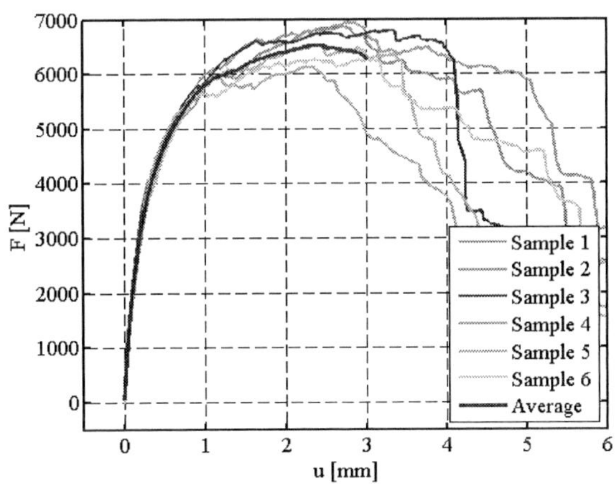

Fig. 10 Force-displacement curves for samples with rivet diameter 6.4 mm and hole to rivet clearance of 2%

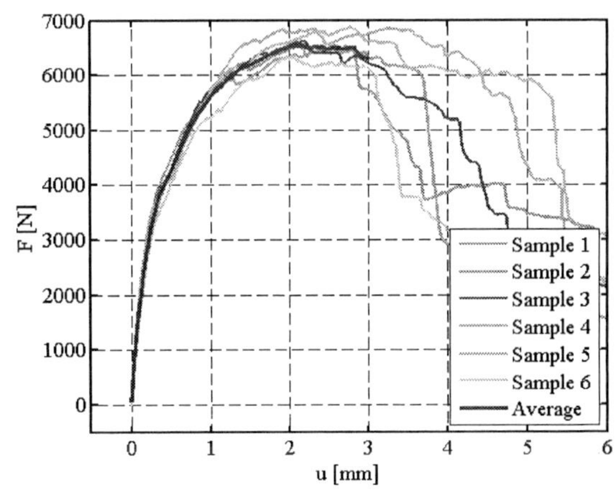

Fig. 11 Force-displacement curves for samples with rivet diameter 6.4 mm and hole to rivet clearance of 3%

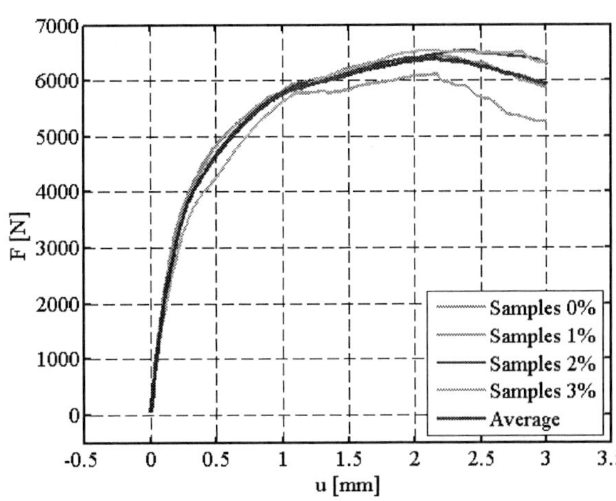

Fig. 12 Average force-displacement curves for samples with rivet diameter 6.4 mm and hole to rivet clearance of 0%, 1%, 2%, and 3%

Result discussion

In case of the samples having the rivet diameter of 4.8 mm, it was observed that the failure occurs in two modes. 15 samples failed due to a rivet failure, 9 samples failed due to the composite bearing. However, the failure mode did not have the influence on the measured ultimate force. The comparison of the average ultimate forces for each hole to rivet clearance (Table 1) proved that there is a negligible difference. The difference between the highest and the lowest average ultimate force was 2.4%.

In case of the samples having the rivet diameter of 6.4 mm, all samples failed due to the composite bearing. The comparison of the average ultimate forces for each hole to rivet clearance (Tab. 2) proved that there is the negligible difference as well. The difference between the highest and the lowest average ultimate force was 5.4%. However, this number is strongly influenced by the pure result of the sample 5 from the set of the samples having the hole to rivet clearance of 1%. Due to the significant difference compared to other samples, it could be determined as an outlier. In case of the removal of this sample, the difference between the highest and the lowest average ultimate force would be 3.3%.

From the comparison of the average force-displacement curves (Fig. 7 and Fig. 12), it can be concluded that the hole to rivet clearance have the influence neither to the strength of the joint, nor to the shape of the force displacement curve.

Conclusion

Series of experiments with the riveted single-lap joint using a blind rivet for joining of carbon fibre plates was performed in order to investigate the influence of the clearance effect on the joint behavior and the tensile strength. Based on the rivet diameter, two failure modes were observed. Based on the comparison of the measured force-displacement curves and the investigated ultimate forces, it was concluded that the clearance effect did not occurred at the blind rivets which is in contradiction with the conclusion made in literature for the high strength rivets with the solid shank.

Acknowledgement

This publication was supported by the project LO1506 of the Czech Ministry of Education, Youth and Sports.

References

[1] H. Schürmann, Konstruieren mit Faser-Kunststoff Verbunden. Springer Verlag, Berlin, Heidelberg, New York, 2007.

[2] C. T. McCarthy, M. A. McCarthy, Three-dimensional finite element analysis of single-bolt, single-lap composite bolted joints: Part I - model development and validation. Composite Structures, Vol. 71, 2005.

[3] M. A. McCarthy, C. T. McCarthy, Three-dimensional Finite element analysis of single-bolt, single-lap composite bolted joints: Part II - effects of bolt-hole clearance. Composite Structures, Vol. 71, 2005.

[4] T. Ireman, Three-dimensional stress analysis of bolted composite single-lap joints. Composite Structures 43, Elsevier, 1998.

[5] M. A. McCarthy, C. T. McCarthy, Finite element analysis of the effect of clearance on single-shear, composite bolted joints. Journal of Plastics, Rubber and Composites, 2003.

[6] Information on: http://www.vap-info.com.

Crack Propagation Prediction in Aluminum Friction Welded Joints

BROŽ Petr[1,a] and DOBIÁŠ Daniel[2,b]

[1]University of West Bohemia, Faculty of Applied Sciences, Univerzitní 8, 306 14 Pilsen, Czech Republic

[2]Czech Technical University in Prague, Klokner Institute, Šolínova 7, 166 08 Prague 6, Czech Republic

[a]broz.petr@tiscali.cz, [b]dobias@klok.cvut.cz

Introduction

The friction welding procedure to achieve lightweight structures proved to be of great importance and the practical prediction of fatigue performance of weldments is a principal element.

In the treatise, in conformity with [1] methodology, an integrated dual boundary element method – finite element method (DBEM – FEM) technique, coupling the welding procedure simulation to the successive crack propagation evaluation, is presented and applied to model multiple crack growth. The friction butt of the precipitation hardened AA 2024 – T3 alloy has been simulated employing a thermo – mechanical FEM model to predict the process caused residual stress field and material softening. The computed stress field is transferred to a DBEM environment and superimposed to the stress field caused by a remote fatigue tensile load applied on a notched specimen. The entire procedure is tried out comparing simulation results with experimental data. The agreement obtained emphasized the predictive of the technique. It has been also analyzed the effect of the residual stress distribution on crack propagation and the interaction between growing cracks.

Owing to wide structural application in the various industries, the aluminum alloy AA 2024 – T3 was chosen as base material. Particularly, the thermo – mechanical FEM model, according to [2] was employed to determine the residual stress state in the butt welded aluminum plates. Further, multiple crack propagation was simulated by the DBEM. The influence of residual stresses on crack growth rates was modeled by the crack growth law in accordance with [3], [4].

DBEM – FEM representation

In thermo – mechanical modeling of friction welding, the most familiar assumption, used to make smaller the computational complexity, is to ignore the material flow in course of welding. This brings about semi – coupled thermo – mechanical models in a Lagrangian formulation where the thermal field is analyzed prior to the mechanical field by separating out the two analyses. The model applied was presented by Sonne at al. [2] The time dependent temperature classification was derived by analyzing the transient heat conduction equation, that is

$$\rho c_p \frac{\partial T}{\partial t} = \frac{\partial T}{\partial x}\left(k\frac{\partial T}{\partial x}\right) + \frac{\partial T}{\partial y}\left(k\frac{\partial T}{\partial y}\right) + \frac{\partial T}{\partial z}\left(k\frac{\partial T}{\partial z}\right) + \dot{Q}'''_{gen} \tag{1}$$

where ρ is the material density, c_ρ is the specific heat capacity, T is the temperature, k is the thermal conductivity and Q_{gen} is the volumetric heat source term. The tool operation is modeled through proper boundary condition applied at the interface between the shoulder and the work piece. A surface heat flux q dependent on the radial place and the local yield stress σ_{yield} , is imposed at the tool shoulder – adjacent material contact area, without modeling the tool sensor, like this

$$q(r,T) = \left(\frac{2\pi n}{60}\right) r \frac{\sigma_{yield}}{\sqrt{3}},$$

(2)

where π is the number of tool revolutions per minute, r is the radial position going from the tool center out, and σ_{yield} is the temperature dependent yield stress.

The progress of microstructure of the Al – alloys 2024 in the T3 temper state in the course of friction welding has been demonstrated to possess a significant influence on the residual stress distribution. In this article, the softening model developed by Myhr and Grong was employed to predict the phase transformation during welding. The same approach was also adopted by Richards et al. The model relates the fraction of dissolved hardening precipitates X_d to the equivalent time of heat treatment $t_{eq} = t/t^*$ (where t is the period of time at a temperature T and t^* is the time for total precipitate dissolution at this temperature) in the following manner:

$$\frac{f}{f_0} = 1 - X_d = 1 - t_{eq}^n = 1 - \sqrt{t_{eq}} \tag{3}$$

$$t_{eq} = \sum_{i=1}^{N_{total}} \frac{\Delta t_i}{t_i^*} = \sum_{i=1}^{N_{total}} \frac{\Delta t_i}{t_{ref}\, exp\left[\frac{Q_{eff}}{R}\left(\frac{1}{T_i} - \frac{1}{T_{ref}}\right)\right]} \tag{4}$$

where t_{ref} is the time for total dissolution at the reference temperature T_{ref}, R is the universal gas constant and Q_{eff} is the effective energy for precipitate dissolution. The small piece of hardening precipitates f/f_0 to the equivalent time via the fraction of dissolved precipitates X_d as demonstrated in Eq. (3), where n is a material constant. The yield stress was predicted through a linear interpolation between the first and entirely dissolved states.

To calculate the transient likewise the residual stress field in the work piece, a normal mechanical model based on the solution of the three static force equilibrium equations is applied, that is

$$\sigma_{ij,j} + p_j = 0 \tag{5}$$

where p_j is the body force at any point within the plate and σ_{ij} the stress tensor. Hooke´s generalized law and linear decomposition of the strain tensor, as well as small strain theory, were applied together with the expression for the thermal strain. The plastic strain evolution is based on the standard J_2 flow theory with a temperature dependent von Mises yield surface. According to [2], isotropic hardening behavior is the most suitable for modeling the mechanical behavior of AA2024 – T3 when combined with a softening model. The yield stress at the instantaneous temperature was determined by interpolating between the upper and lower bound yield stress curves relative to X_d , ie:

$$\sigma_y = \left(\sigma_{y_{max}} - \sigma_{y_{min}}\right)\frac{f}{f_0} + \sigma_{y_{min}} \tag{6}$$

where σ_{max} is the yield stress of the material in the original T3 condition and σ_{min} the yield stress of the in full dissolved material. The upper and lower yield stress curves for AA2024 – T3 are available in [2]. The Vickers microhardness (HV) distribution was calculated by acquiring the same model as in Eq. (6), evidently substituting yield stress with microhardness values, as follows

$$HV = (HV_{max} - HV_{min})\frac{f}{f_0} + HV_{min} \tag{7}$$

using the minimum and maximum hardness values calibrated on the experimental data.

Crack growth model using DBEM

The residual stresses put on in the main and lateral crack surfaces were imported in the commercial kit BEASY for the DBEM growth analysis. An interface procedure already available in the BEASY environment was used for this purpose. The propagation was driven by the remote fatigue load and by the superimposed residual stress field (applied on the crack faces). At each step of propagation, new meshed surfaces were automatically added to the previous crack configuration, consistent with the outcomes of the implemented crack growth law. Again the interface routine scanned the .odb file to extract the residual stress to be applied in addition to the residual stresses already existing on the previous crack surface. Then the remote load was applied again and a new LEFM analysis was performed in an interactive procedure that stopped when a predefined instability crack condition was encountered, i.e. when the stress intensity factor at the crack front reaches a critical value. There is demonstrated in Fig. 1 a block diagram clarified the general routine.

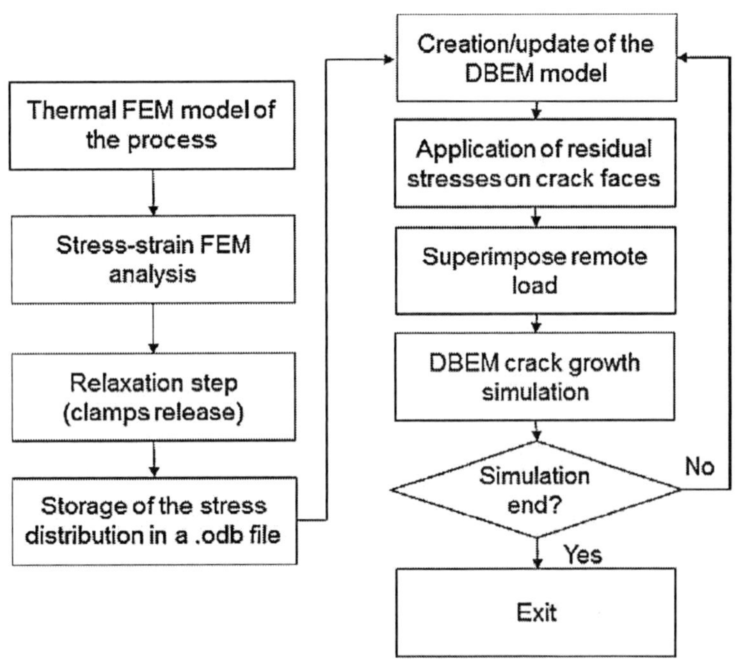

Fig. 1, Block diagram of the executed DBEM – FEM model [1]

Residual stresses influence crack growth because they modify the effective value of the total stress intensity factor at the crack tip, with both the minimum (K_{min}) and the maximum (K_{max}) values in general being alike influenced, going away the parameter $\Delta K = K_{max} - K_{min}$ unchanged. Consequently, the primary effects of residual stresses on crack growth rates are related to the K_{max} variations rather than to the ΔK variations. This is accounted for by a two – parameter approach. According to this, fatigue crack growth is promoted by two driving forces, K_{max} and ΔK. In addition, the procedure assumes that there are two fatigue thresholds $K^*_{max,th}$ and ΔK^*_{th} corresponding to the two driving forces both of which must be simultaneously larger than the relative thresholds for fatigue crack growth to occur. The crack propagation law is expressed by the ensuing manner:

$$\frac{da}{dN} = A(\Delta K - \Delta K^*_{th})^n \left(K_{max} - K^*_{max,th}\right)^m \tag{8}$$

At the same time eq (8) was stamped by best fitting the material parameters *A, n, m* based on literature data. The following values were used for the parameters: ΔK^*_{th} = 1,834,121 N/m$^{3/2}$, $K^*_{max,th}$ = 3,352,014 N/m$^{3/2}$, A = 6.745E – 23 m$^{1.5(n+m)+1}$/N^{n+m}, n = 1.65, m = 0.56 . These parameters are valid for each positive *R* ratio ($\sigma_{min}/\sigma_{max} = K_{min}/K_{max} > 0$). The effects of friction stir welding were emulated by taking into account the residual stress influence on the driving parameters ΔK and K_{max}. Hence, the SIFs used in Eq. (8) were computed as the sum of the SIF corresponding to the remote load and that of the SIF corresponding to process induced residual stresses. In the DBEM analysis, residual stresses were simulated by employing a distribution of tractions on the crack faces checked with the residual stresses caused by the welding method.

Experiments

AA2024-T3 aluminum rolled sheets were connected by friction welds applying a non – consumable Cr – Mo steel tool, adopting an angular tool velocity of 1400 rpm and a welding speed of 70 mm/min. The geometry was characterized by a flat shoulder (20 mm in diameter) with an unthreaded conical pin (6.2 mm in major diameter, a cone angle of 30°and a height of 3.8 mm). The forging action of the tool shoulder was enhanced by imposing a tilt angle of 2°. The dimensions of the adjoining sheets were 200 mm (length), 30 mm (width), and 4 mm (thickness). The welding configuration is indicated in Fig. 2.

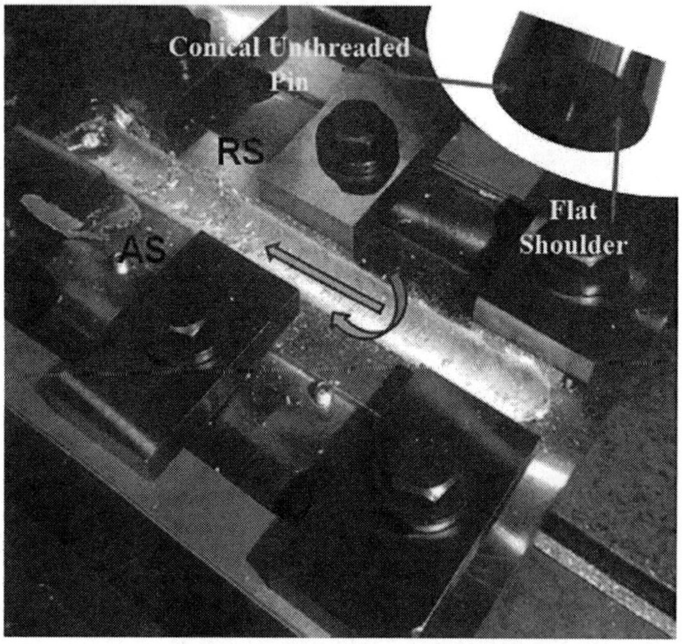

Fig. 2, Welding arrangement

The longitudinal residual stress classification in the not specific section of the welded joint was gained by dint of the contour method. The resulting stress field was already used for crack propagation assessment by some authors, assuming the repetition of the stress field as computed at the mid – length in each cross section orthogonal to the weld line. However, the outlined hybrid (numerical – experimental) method is obviously quite expensive and not generalizable, since the stress analysis is case specific, calling for a repetition of the procedure for each variation of processing parameters and conditions. What is more, accounting for transient effects and stress relaxation in the proximity of the free edges of the sheets was not provided in the crack growth simulations. In that sense, the model in this work stands for a generalization of the algorithm because the residual stresses are predicted numerically taking the different boundary conditions likewise stress relaxation into consideration.

The sensitivity of the model with respect to variations in material properties was assessed by comparing the numerically estimated and experimentally measured microhardness along a line orthogonal to the weld line at the mid – thickness of the specimen. Vickers microhardness was measured using an automatic device applying the following test parameters: indentation load 100gf (0.98 N), loading time 15 s, and indentation speed 60μm/s. the distance between two consecutive indentations was defined as 1 mm. The analysis was performed after 60days of (post welding) natural aging to ensure the establishment of a stable microstructure.

Model validity

Experimental data are got by the contour method. The first part of an appropriate thermo – mechanical model is to possess a realistic account of the temperature field before the follow – up mechanical analysis. Certainly, temperature gradients are both the direct and indirect sources for stresses in view of thermal expansion and metallurgical changes in the aluminum. The transient temperature field during welding (Fig. 3a), stipulated by the thermal computation, is employed in the subsequent stress analysis to get the stress state progress in the welded plate. In Fig. 3b the typical lines in compression at the sides of the moving heat source and in tension then behind the moving heat source can be studied as an outcome of the self – constraining influence acted by the colder neighbouring material.

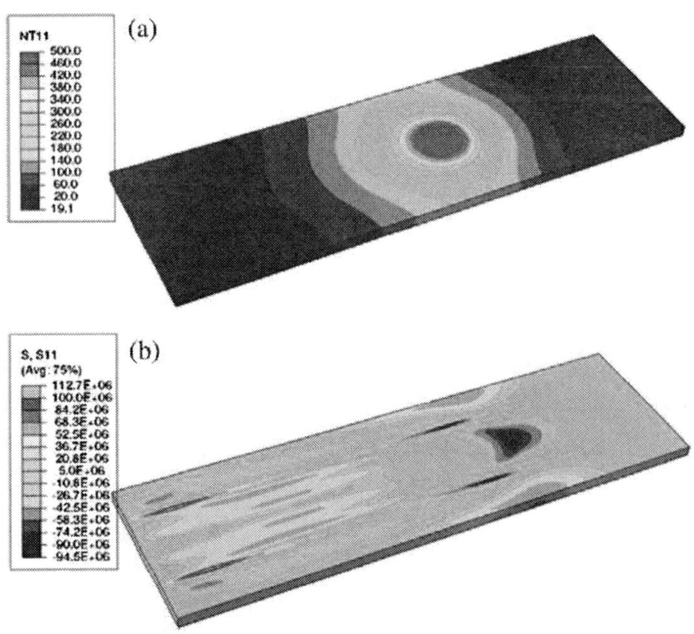

Fig. 3 (a) Temperature field (°C) from the heat transfer study of welding at $t = 100$ s.
(b) The resultant longitudinal stress constituent (N/m^2) from the stress study at the same step [1]

In the friction welding method, heat is set up by the friction between the tool shoulder and the work piece surface, and by the plastic deformation induced in the work piece. The material under and right next to the tool is heated up resulting in expansion; however it is partly constrained by the relatively colder material surrounding this region. Subsequently, the material starts yielding in compression and plastic deformation starts to develop. Besides, the thermal gradients due to non – uniform heat generation and the mechanical boundary conditions (i.e. clamping, contact conditions between work piece and the anvil etc.) also play a significant role for promoting plastic strains and consequently considerable role for publicized plastic strains and accordingly residual stresses.

Crack growth

A multiple crack growth was demonstrated by tests and it was marked by the appearance of the aforesaid lateral crack created from a notch insincerely introduced after the welding process and of a central semi - elliptic crack, it itself nucleated in the weld crown owing to the surface marks left by the tool shoulder. The basic hypothesis, to be verified by the simulations, is that, in the initial phase of the fatigue test, the central crack, due to its reduced size and to the distance from the lateral crack, does not significantly influence the propagation of the lateral crack itself. In this link it should be remarked that in the experiments, the lateral crack, starting from the notch tip, was considered initiated after a pre – cracking phase lasting a number of cycles sufficient to prolong the notch by nucleation of a crack whose superficial length is equal to 0.25 mm. Hence, the initial single crack simulated scenario was based on a lateral through crack with an overall length (2.25 mm) equal to the sum of notch length (2 mm) and pre – crack length (0.25 mm). Taking into consideration that the nucleation time and dimension of central crack were not known a priori and in view of the absence of experimental indication, up to this period, any interaction between the two cracks, the central crack was presupposed small and accordingly not modeled at all in the thought over original crack growth stage.

Eg., to demonstrate the matter investigated, the lateral crack was extended to simulate in a single crack growth, up to 2.5 mm length, before brought in the second central crack in the model. The timing for the introduction of the central crack in the numerical model was prescribed by the experimental watching, as follows: when the lateral crack achieved 2.5 mm in length it is subject to sudden acceleration uncovered and shown by the applied gauges.

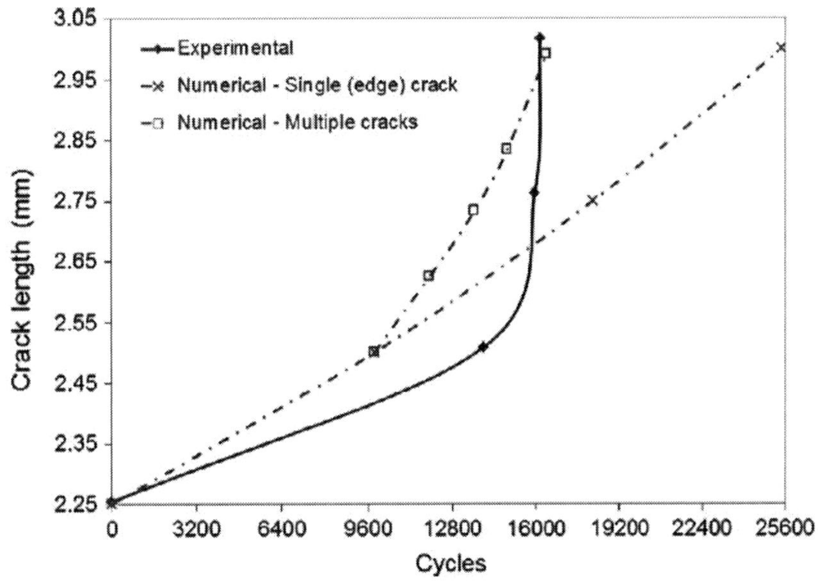

Fig. 4, Lateral crack length experimentally checked and calculated by DBE supposing single crack growth (without central crack) and multiple crack growth (introducing an additional central crack after nearly 10 000 cycles)

The central crack growth proved to be accountable for specimen failure, with connected K_I values (mote II and III are nearly inconsiderable) approaching the material fracture toughness. On the other hand, SIFs along the lateral crack front, according to the J – integral were not adequately high in comparison with the threshold ($K_{max,\ th}$ = 3.352 Pa.m$^{0.5}$) to produce perceptible crack forward movements (Fig. 5).

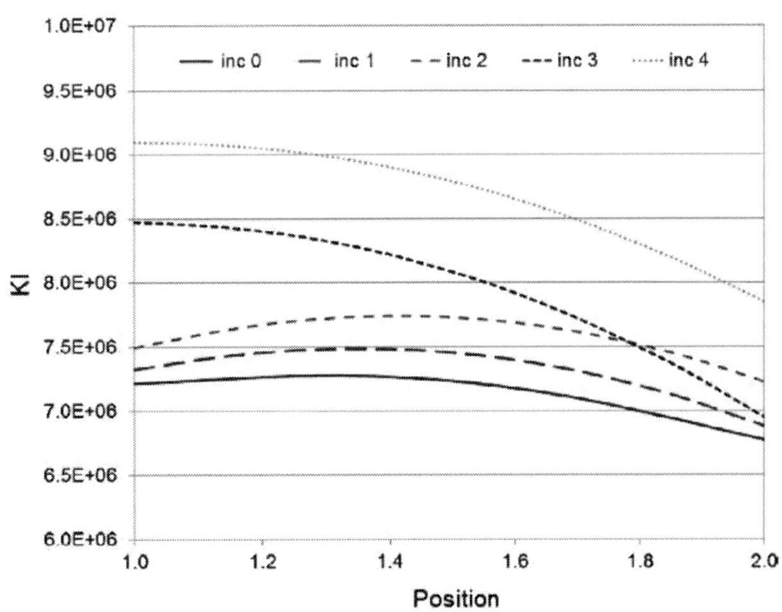

Fig. 5, SIF values (Pa m$^{0.5}$) alongside the crack fronts of the two cracks for each simulation increment (inc 0-9850 cycles; inc 1-1,920 cycles; inc 2-13,660 cycles; inc 3-14,920 cycles; inc 4-16,400 cycles); of crack growth: the coordinate from 1 to 2 relates to crack 1 [1]

Conclusions

Some deductions can be emphasized:

- the realized DBEM – FEM way of tackling proof to be able to efficiently predict multiple crack propagation in the presence of residual stresses
- the crack growth stipulated by the experiment, as conceived by post examination metallographic analyses, was worked by a nearly pure mode I evolution and nearly symmetric form. This finding served as a qualitative corroboration of the residual stress field determined by numerical simulation.
- if the initial crack sets out from the weld line, the process induces opening stresses having an accelerating influence on the crack growth

Acknowledgment

The authors gratefully acknowledge the financial support of the presented research by the University of West Bohemia in Pilsen. This article was prepared with support of the Project SG S – 2016 – 038.

References

[1] R. Citerella, P. Carlone, M. Lepore, GS. Palazzo, Numerical – experimental crack growth analysis in AA2024-T3 FSWed butt joints. Adv Eng Softw 2015; 80: 47-57.

[2] M.R. Sonne, C.C. Tutum, J.H. Hattel, A. Simar, B. deMeester, The effect hardening laws and thermal softening on modeling residual stresses in FSW of aluminium alloy 2024-T3, J Mater Process Technol 2013; 213:477-86.

[3] P. Carlone, R. Citarella, M. Lepore, G.S. Palazzo, A FEM-DBEM investigation of the influence of process parameters on crack growth in aluminium friction stir welded butt joints, Int J Mater Form 2015; 8(4):591-9.

[4] P. Brož, D. Dobiáš, Important Aspects of Testing Welds, EAN 2017, Nový Smokovec (2017), pp 185-186.

EAN 2018
56th conference on experimental stress analysis

Experimental Investigation of Friction Relations in Threaded Joints

CÉZOVÁ Eliška[1,a] and KANAVAL Jan[1,b]

Czech Technical University in Prague, Faculty of Mechanical Engineering, Department of Designing and Machine Components, Czech Republic

[a]eliska.cezova@fs.cvut.cz, [b]jan.kanaval@fs.cvut.cz

Keywords: thread, nuts, friction coefficient, axial load, coating, tightening moment .

Abstract. This paper shows some of the results of the measurements of friction coefficients in stainless steel threads. Three different coatings of the used nuts are considered and tested under various axial loads. The friction conditions for coated standard nuts of M12 and M16 size were measured on a simple test stand. The resulting friction coefficients are plotted with respect to applied load for and pressure in the thread. The results are discussed in detail.

Introduction

The nowadays common use of stainless steel bolts and nuts brought some advantages as well as some disadvantages. The advantage evidently is the corrosion resistance of the connecting parts, which significantly reduces the necessary maintenance costs and prolongates the life cycle of the threaded joints. Besides of the higher price of stainless steel parts (compared to the standard steel), one of the notable disadvantages is in the worsening of friction conditions in the threaded joint. In general higher friction coefficients are typical for stainless steel parts rather than for the parts made of classical steel. This may cause higher tightening moments (up to seizing of the joint) resulting into a reduction of the exerted axial force in the threaded joints, leading to failure of the joint.

In order to overcome this issue, various lubricants or coatings can be applied to the contact surfaces of bolts and nuts ([1], [2], [3]). It was shown, that the use of liquid lubricant, various types of grease or solid lubricants can have positive effect on reduction of friction forces (coefficients) as well as to improve the corrosion resistance of the connecting parts (bolts and nuts). The application of such lubricants on the sliding parts is however connected with certain difficulties especially in outdoor conditions. Also the efficiency of such lubrication degrades quite quickly in time under extreme weather conditions. This is why often the use of solid coatings is recommended in such situations. There exist many different types of suitable coatings. Traditionally they were often formed by a thin layer (single or multiple) of metal (zinc, nickel, chrome, tin, silver, etc.). More recently the modern plastic materials (fluorplastics, like e.g polytetrafluoretylene PTFE, fluorinated ethylene propylene FEP, polyetylene PE, etc.) are able to form a coating layers providing sufficient protection against corrosion and to guarantee optimal friction properties at the same time. Alternatively a special surface treatment can be applied to friction surfaces, based on kind of dry paints, in which solid lubricant particles are dispersed. From the tribological point of view, all these coatings help to prevent the direct contact of the mechanical parts and to avoid dry friction between them.

This kind of coating treatment of connecting parts is of interest e.g. in pipeline repairs using clamps that are fixed using nuts and bolts. These are often made of stainless steel due to its high corrosion resistance needed for the pipes placed under the ground. The difficult conditions of outdoor repairs, including the montage and tightening of clamps, justifies the use of coated connecting parts [5].

The aim of this contribution is to bring some insight into the evaluation of the friction properties of these coatings in threaded joints, with special focus on their degradation by excessive wear under high axial loads.

Experimental setup

The goal of the experiments was to test several bolt/nut couples, subject to different surface treatment, under static axial load(s) during the tightening/loosening of the joint. The test setup consisted from a fixed nut (treated with coating), through the which a bolt (stainless, uncoated) is screwed. The applied moment is measured using the T20WN/20 Nm sensor, from the Hottinger Baldwin Messtechnik GmbH company, calibrated using the torsion moment wrench TMK 200. During the measurement, the bolt was turned continuously, slowly until the value of the moment reached a stable value. The measuring sensor was connected to a computer, recording the time evolution (sequence) of the moment. Based on this record, the stabilized mean value of the moment was determined (using the LABVIEW software package) for each specific case.

During the measurement the bolt was oriented vertically (head up), with the load attached to its free end below the nut. Series of weights was attached to the bolt to study the friction conditions and parameters under different loads. The mass of the applied loads was approximately 83, 162, 202 and 246 kilograms.

The bolt was screwed first down, by turning it to the right, through the (new, coated) nut, while the load weight increased sequentially (4 different weights were attached), and then back up (turning it to the left), taking the weights off again from the end of the bolt. In order to evaluate the effects of the wear of the various coatings, the same experiment was repeated with the nuts that have been previously used. The whole series of measurements was performed independently for M12 and M16 nuts/bolts. See e.g. [4], [5] for details.

Experimental results

Three different coatings of nuts were tested in a series of the above described experiments. The Coating A is zinc, typically used to increase the corrosion resistance of nuts made of standard steel. Coating B is based on PTFE (Polytetrafluoroethylene), while Coating C is a dry sliding paint with molybden suplhide additive. The types of studied coatings, including some of their essential properties are summarized in the table 1.

Tab. 1 Types of applied coatings and their declared properties

Coating	*Type*	*Thickness [μm]*	*Friction coef. [1]*
A	zinc based *(Delta Tone)*	4-10	0.09-0.14
B	PTFE based (Polytetrafluoroethylene) *(Xylan)*	20-35	0.02-0.2
C	dry sliding paint with molybden suplhide (MoS_2) additive *(Gleitmo)*	3-5	0.05-0.1

As mentioned above, the moment needed to turn the bolt was measured for different axial loads (attached weights), different directions of the bolt movement (up/down) and different thread sizes (M12/M16), always for new and reused nuts. Based on the measured moment, the friction coefficient was evaluated for each case using the formula

$$f = \left(\frac{2M_k}{Qd_2} - tan\gamma\right) cos\beta \qquad (1)$$

where Q is axial load, d_2 is pitch diameter of nut thread, M_k is the applied moment, γ is the thread lead angle and β denotes the (half) angle of the thread profile. From the below presented results, i.e. the friction coefficients shown in figures 1-3, is evident that the reused nuts (marked by dashed lines) typically exhibit much higher friction when compared to the new ones (marked by a solid line). This trend is obvious, although its intensity is not the same for all the three studied coatings. This observed phenomena might be caused by the wear of the coating, due to an excessive load in the thread. However this hypothesis has to be verified, for which the presented results should serve as a starting point.

In general the friction forces are proportional to applied loading force, thus this load naturally appears as the independent variable in the graphs presenting the obtained friction coefficients. Because the same load weights were used for all coating and both sizes of the nuts, the horizontal scale is the same for both M12 and M16 nuts. This form of visualization is shown in the left part of the figures 1-3.

However considering the expected wear of the coatings in the threads of the nuts, the contact pressure in the threads can play an important role. Thus, in the right pat of the figures, the same dependence of friction coefficients is always plotted with respect to the corresponding pressure in the thread. The pressure in the thread is computed from

$$p = \frac{Q}{\pi d_2 H_1 z}, \qquad (2)$$

where p is pressure in the thread, H_1 is height of the thread profile with rounded crests and roots while z is the number of support thread. See e.g. [6], [7], [8] for more details.

Because the M16 threads have larger contact surface, the resulting contact pressure is lower (for the same force Q) than while using the smaller M12 nuts. This is why the corresponding range of contact pressures is different for M12 and M16 threads, despite of using the same axial loads. The M16 curves seem now like shifted towards the lower pressures.

The results obtained for the classical zinc based coating are drawn in the figure 1, showing the friction coefficients dependence on both, the axial loading force Q and the resulting contact pressure p.

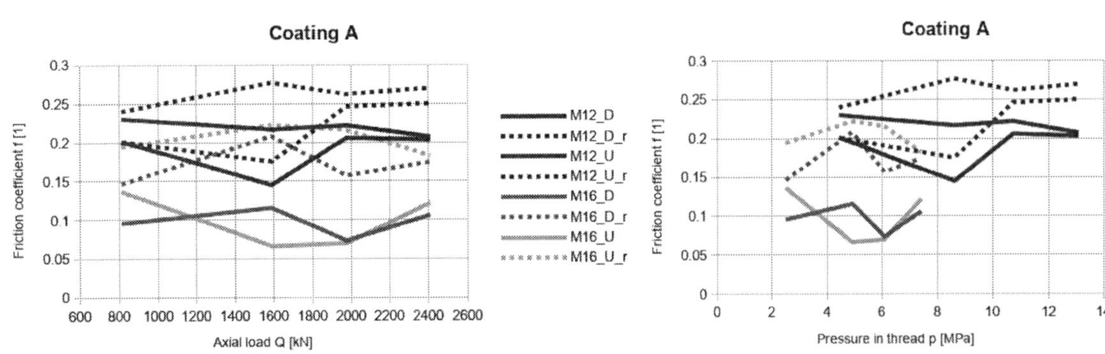

Fig 3 Friction dependence on axial load and pressure in the thread for coating A

In the figure 1, for the zinc coating A, it is obvious that not only the friction coefficients are higher for the reused nuts, but also that the friction coefficients are significantly lower in the larger M16 threads where the range of contact pressures is lower compared with the smaller M12 thread.

The results in the figure 2 for the plastic, PTFE based coating B show similar trend for new versus reused nuts, with higher overall friction in the reused nuts. On the other hand, the friction coefficients for this kind of coating seem to be much less dependent on the pressure in the thread. It should be noted however that the average friction coefficient is visibly higher for this coating in comparison with the previous type A coating based on zinc.

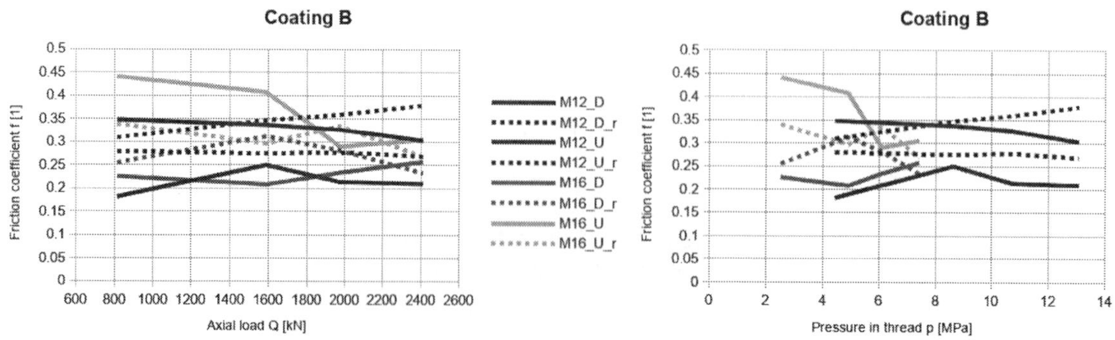

Fig 4 Friction dependence on axial load and pressure in the thread for coating B

The sliding paint, marked as coating C, exhibits yet another kind of behavior. As it can be seen in the figure 3, the friction coefficients are in this case quite insensitive to the variation of the pressure in the thread, however the difference between the new and reused nuts is significant. The friction coefficients in reused nuts are about 2-3 times higher than in the new nuts.

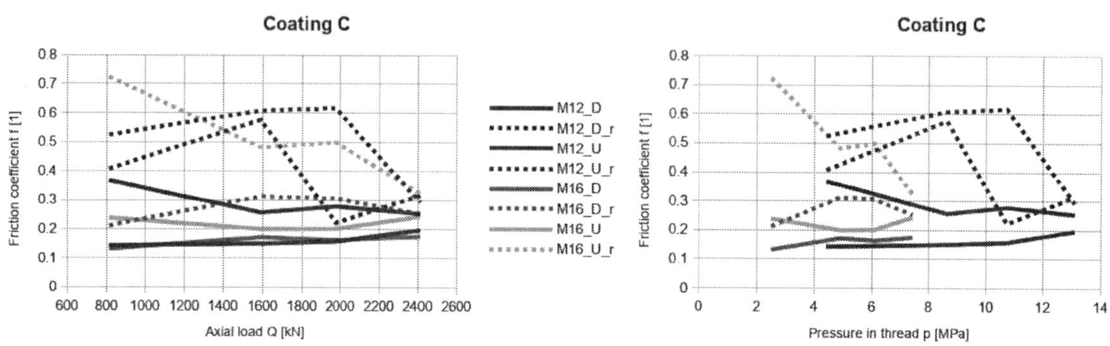

Fig 5 Friction dependence on axial load and pressure in the thread for coating C

Due to this elevated friction in the reused nuts, the overall friction coefficients values are higher than for the previous two coatings A and B. This however strongly depends on the previous use of the nuts.

Conclusions

Our experimental investigation has confirmed that the use of coating of nuts has significant impact on the friction coefficients in thread. The measured friction coefficients for the new nuts were found in (or close to) the range declared by the producers of the coating (see table 1). The reused nuts however show significantly higher friction coefficients close to those found for untreated, raw stainless steel dry surface contact.

The reduction of friction can lead to lower tightening moments needed to ensure the required axial force in the joint. The recommendation of the use of specific surface treatment should however take into account the repeated assembly/disassembly of the joint as well as the level of the contact pressure in the thread, that can lead to excessive wear of the coating and the degradation of its friction properties. The dependence of the coating wear on the contact pressure should further be studied in detail to confirm and possibly better explain the physical mechanisms being responsible for it. The preliminary evaluation of available microphotographs of the sample nuts shows significant reduction of the thickness of the coating layers on the sliding surface of the threads. Thus the thickness of different coatings (shown in the table 1) can be the one of the other crucial factors in prediction of the wear of different types of coatings. This direction of investigation will be followed in our future work.

Acknowledgment. This work was partly funded by the Grant No. SGS16/145/OHK2/2T/12 of the Czech Technical University in Prague.

References

[1] Information on http:/www.fabory.cz, (2016-07-29).

[2] Information on http:/www.fuchslubricants.com, (2016-07-29).

[3] Information on http:/www.technicoat.cz/-data/files/Image/content/xylan, (2016-07-29)

[4] J. Kanaval, F. Starý, E. Cézová: Screw connections complex analysis in the application of advanced coatings and lubricants. In: Book of proceedings of the 57th International Conference of Machine Design Departments. Plzeň: Západočeská universita, 2016, pp. 83-86. ISBN-978-80-261-0609-8.

[5] J. Kanaval, F. Starý, E. Cézová: Remark on repairs of broken pipelines. In: Experimental stress analysis 2017 55th International Scientific Conference. Košice: Faculty of Mechanical Engineering, 2017. ISBN-978-80-553-3167-6, CD-ROM (Full text of paper ISBN 978-80-553-3166-9).

[6] J. Dejl: Konstrukce strojů a zařízení I. Spojovací části strojů. Návrh, výpočet, konstrukce. Ostrava: Montanex, a.s., 2000. ISBN 80-7225-018-3.

[7] V. Švec: Části a mechanism strojů. Spoje a části spojovací. Praha: ČVUT, (2008)

[8] ČSN EN ISO 4762:2004 Šrouby s válcovou hlavou s vnitřním šestihranem (ČSN 02 1143, DIN 912)

EAN 2018
56th conference on experimental stress analysis

3D Experimental Vibration Analysis of Washer Extractor Using Planar Laser Sensors

CIENCIALA Jakub[1,a], MAREK Martin[2,b], FUSEK Martin[1,c] and HALAMA Radim[1,d]

[1]Department of Applied mechanics, Faculty of Mechanical Engineering, VŠB-Technical University of Ostrava, 17. Listopadu 15/2172 Ostrava 708 33, Czech Republic

[2]National Supercomputing Centre IT4Innovations, VŠB-Technical University of Ostrava, 17. Listopadu 15/2172 Ostrava 708 33, Czech Republic

[a]jakub.cienciala@vsb.cz, [b]martin.marek@vsb.cz, [c]martin.fusek@vsb.cz, [d]radim.halama@vsb.cz

Keywords: spatial motion, measurement, washer extractor, 2D laser sensor, vibration

Abstract: This paper deal with measuring and evaluating the spatial motion of industrial washer extractor in two working modes. The article also describes 3D motion measurement device, which is unique by using 2D laser sensors. Washer extractor, on which the measurement was performed, is produced by Alliance Laundry Systems in Příbor. The patented measuring equipment, including software, is characterized by speed and simplicity of the measured data evaluation as well as easy application to the measurement device. This paper is primary focused on measuring the spatial movement of the rotating part of the industrial washer extractor. However, the presented method can be used to measure any object with a known stiffness.

Introduction

Following long-term development of the measuring system, which is evident in contributions [1, 2], the spatial movement analysis of the FX135 washer extractor drum was performed. Measurements were performed for two different wash cycles - rinsing and spinning. For both wash cycles, ordinary textile filling of the drum was used. The measurement aim was to determine the spatial movement of the four selected points of the analyzed body. In the context of the washer extractor manufacturer requirements, the 3D motion analysis of the washing unit drum is necessary especially in case of engineering design. The advantage is the possibility of further use the measured data for example to determine the critical speed, optimize the flexible placing or as the input data of the MPK software.

Since the method of measurement is the content of the patent, only the basic information about the measuring device will be given in subsequent chapters. With respect to paper scope, the evaluation software will not be discussed

Measurement apparatus description

The measurement equipment has undergone several modifications during its development. Current hardware version includes two 2D laser displacement sensors and one axis displacement sensor. Measurement can be complemented by a speed sensor of the washing unit drum. All the sensors used are synchronized together using a separate measuring card. For easy and fast measurement preparation, displacement sensors are mounted on an adjustable measurement stand. This stand allows precise adjustment of the distance and angle of the laser sensors relative to the auxiliary jig. The auxiliary jig is firmly attached by magnets on the movable part of the measured device. The hardware also includes other features that provide power to the sensors and transfer the measured data to the computer.

The above measuring device configuration allows measurement of the spatial movement of any vibrating object. Any complicated preparation or modification of the measured object is not required for the measurement realization.

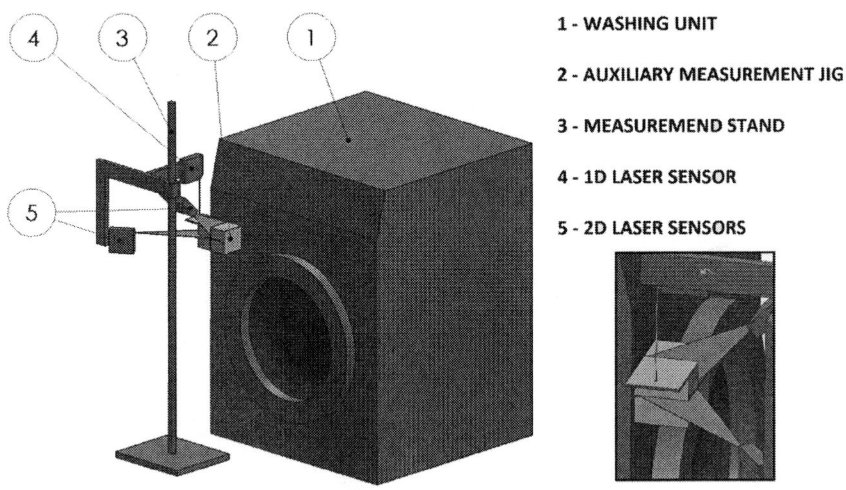

Fig. 13 Measurement apparatus for spatial motion evaluation

The second major part of the apparatus is software, including a measurement module and a data evaluation module. The measurement software allows precise adjustment of individual laser sensors. Another feature of measurement software is the possibility to store measured data to a file. The evaluation part of the software analyzes the measured data and converts it into graphical form. The software allows direct monitoring of measured data over time. It also provides a complete spatial motion analysis in both 2D and 3D visualizations. The spatial motion can be analyzed for any number of points of the measured object.

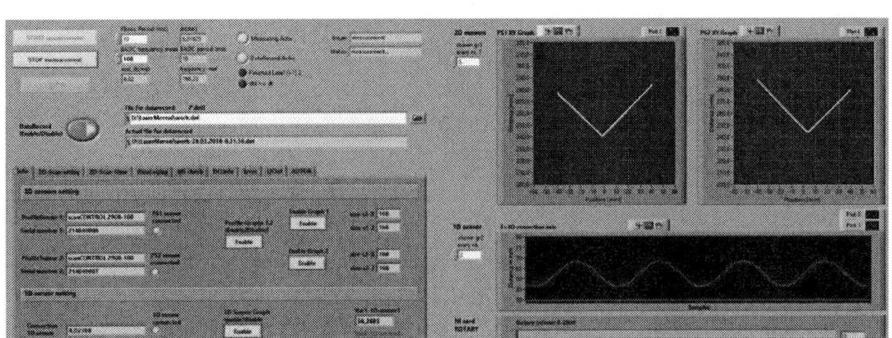

Fig. 14 Measurement software

Measurement analysis

Besides the drum, there are many other members inside the industrial washer extractor, such as control systems, driving unit, water and detergent inlets or drainages. It is necessary to take care to avoid a collision with those members in the design of the washer extractor unit. For the development of industrial washing machines, it is necessary to measure the spatial motion of the washing drum in testing laboratories.

Four points were selected for the washing machine spatial motion analysis. The first point (labeled 0) is also the reference point of the auxiliary measurement jig. The second point (labeled 1) corresponds to the washing drum center at the front of the washing unit. The next two points (labeled 2 and 3) are located at the top and bottom right corners of the industrial washing machine. The layout diagram is shown in figure 3 (b) below. Figure 3 (a) shows photo of practical measurement.

a) b)

Fig. 15 a) real measurement photo, b) measurement points placement scheme

Spatial movement in the rinse program was measured first. The length of this program is 13 minutes. The speed of the washing drum is relatively low in the rinsing mode. During the rinsing cycle, the washing drum is filled by water, which causes drum weight increase and the reference point static displacement. Second, the washing with spinning was measured. The drum speed reaches maximum values during spinning.

The movement in 3D and in XY, ZY, ZX planes was evaluated for individual measurement points. The XY plane corresponds the side view, the ZY plane corresponds the top view, and the ZX plane corresponds the view from the front. The individual measured displacements were also plotted in time for each point.

Table 3 - Measurement results

Washing cycle	Measurement point	0	1	2	3
Rinsing	Δx [mm]	7,4	5,5	15,6	13,9
	Δy [mm]	19,8	53,9	78,8	84,0
	Δz [mm]	15,4	13,1	34,2	49,7
Spinning	Δx [mm]	17,5	46,3	76,3	71,1
	Δy [mm]	15,4	24,7	34,6	35,0
	Δz [mm]	19,9	12,7	20,2	19,2

Graphical results will be presented here only for reference point 0. The following figures show the deflections of point 0 for both measured cycles. The top part of figure 4 shows deviations in the individual planes including the 3D image for the rinsing cycle. The lower part of figure 4 shows the time record of spatial motion in individual coordinate components.

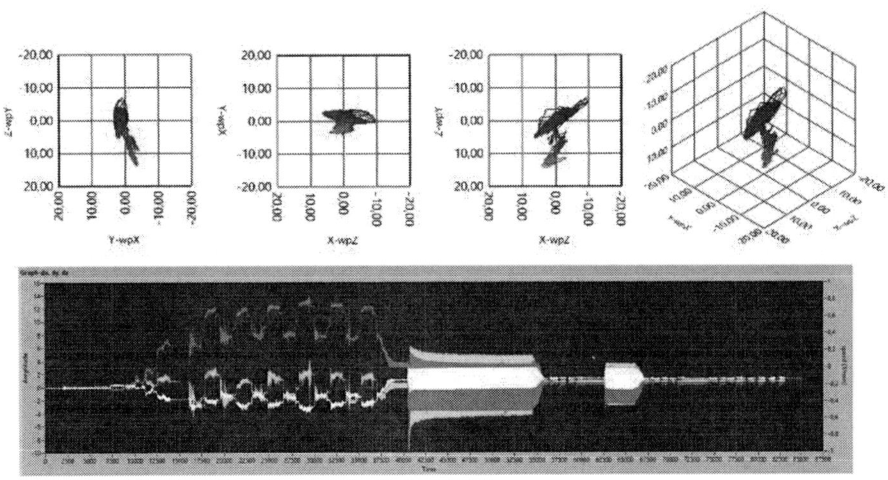

Fig. 16 Motion during the whole rinsing cycle (point 0)

Figure 5 shows the spatial motion of the reference point 0 for the spinning cycle. In its bottom part, the time record of individual deviations throughout the measurement is again shown.

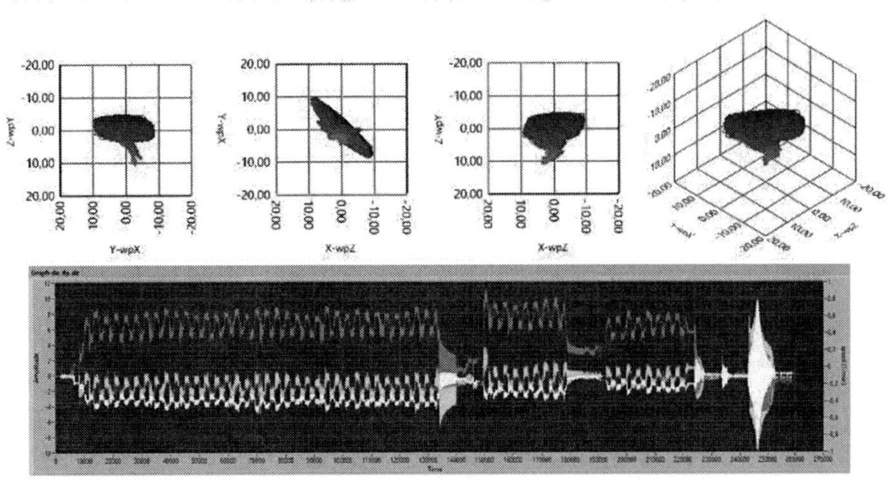

Fig. 17 Motion during the whole spinning cycle (point 0)

Conclusion

This contribution presents a device for accurate measurement and evaluation of spatial movement of rigid structures. The subject of the measurement is the industrial washer FX135. Two washing cycles - rinsing and spinning were selected to measure spatial motion. The data were evaluated at four points in each of the wash cycles. The resulting displacements in individual points are shown in table 1 above.

The washer extractor drum was filled with conventional textile filling during the measurement. It can be assumed considerable variability of the imbalance distribution of the drum in case of repeated measurements. In the case of evaluating a higher number of measurements, it is recommended to use stochastic approaches for evaluating the results.

The measurement result is important statement used by the washer extractor manufacturer to design and develop new equipment.

References

[1] M. Fusek, R. Halama, M. Marek, F. Fojtík, Z. Poruba, Measurement and Evaluation of Spatial Motion of Washer Extractor, In: Proc. of 54th Annual Conference on Experimental Stress Analysis 2016, Department of Mechanics, Faculty of Applied Sciences, University of West Bohemia, 2016, ISBN 978-80-261-0624-1

[2] R. Halama, M. Fusek, F. Fojtík, T. Kot, Real Time Measurement and Evaluation of Washer Extractor Vibrations, In: Proc. of 50th Annual Conference on Experimental Stress Analysis 2012, Tábor, Czech Republic, 4.6.-7.6.2012. p.129-134. ISBN 978-80-01-05060-6.

EAN 2018
56th conference on experimental stress analysis

Formation of Austenite and Ferrite Preferred Orientation During Cold-rolling of Dual and Single-phase Steels

ČAPEK Jiří[1,a], TROJAN Karel[1,b], NĚMEČEK Jakub[1,c], GANEV Nikolaj[1,d] and KOLAŘÍK Kamil[1,e]

[1]Department of Solid State Engineering, Faculty of Nuclear Sciences and Physical Engineering, Czech Technical University in Prague, Trojanova 13, 120 00 Prague 2, Czech Republic

[a]jiri.capek@fjfi.cvut.cz, [b]karel.trojan@fjfi.cvut.cz, [c]nemecj17@fjfi.cvut.cz, [d]nikolaj.ganev@fjfi.cvut.cz, [e]kamil.kolarik@fjfi.cvut.cz

Keywords: dual-phase steel, texture, X-ray diffraction, cold-rolling.

Abstract. The aim of this contribution was to compare the real structure of single-phase with dual-phase steel after cold-rolling. The preferred orientation was studied as a function of thickness reduction of samples. Type and rate of determined preferred orientation depending on material and thickness reduction were studied by X-ray diffraction. The significant differences between intensities and types of texture components were investigated.

Introduction

Duplex steels are dual-phase steels with approx. 50/50 proportion of ferrite and austenite phases. Because of the combination of properties of both phases in dual-phase microstructure, duplex steels are distinguished by good properties in many environments, where the standard austenitic and ferritic steels are usually used [1]. Due to very frequent presence of preferred orientation (texture) in these materials, analysis and consequent interpretation of the texture is crucial, especially in material engineering. The importance of study of texture resides in the anisotropy of most material properties. Because of different mechanical properties of ferrite and austenite phases, their behaviour during deformation can be different in dual-phase steel in comparison with single-phase steels [2].

Orientation distribution function (ODF) is usually used for texture interpretation. The ODF is usually described using Euler space defined by Euler angles φ_1, Φ, φ_2 in Bunge notation [3]. The ODF denotes the bulk density of crystals with the orientation defined by Euler angles and the multiple of a random distribution unit. Euler angles of materials are presented in the range (φ_1, Φ, φ_2) $\in$ <0°; 90°> for cubic crystal and orthorhombic sample symmetries. The ODF is usually interpreted using texture components and textures with fibre components, shortly fibres. For materials with different crystallographic symmetry or lattice, material, chemical and phase composition, and thermo-mechanical properties, the typical texture components and fibres can be varied [2, 3, 4], see Fig. 1. These fibres are usually interpreted according to rolling (RD), transversal (TD) and normal (ND) directions.

The typical texture components and fibres (α_1, ε, γ) of deformed bcc – body centred cubic and (α, τ, η) of fcc – face centred cubic materials are interpreted in Fig. 1. According to stacking fault energy (SFE), two main types of fcc material textures exist: Brass – with small SFE and Copper – with high SFE. Brass type usually contains Brass (Bs) and Goss (G) texture components [3], see Fig. 1. The potential difference between duplex and austenite steel may be caused by the difference in SFE, which depends on chemical composition.

According to [5], the SFE of austenite steel is around 18 mJ/m^2 and for phase of duplex steel, the SFE is 10 mJ/m^2.

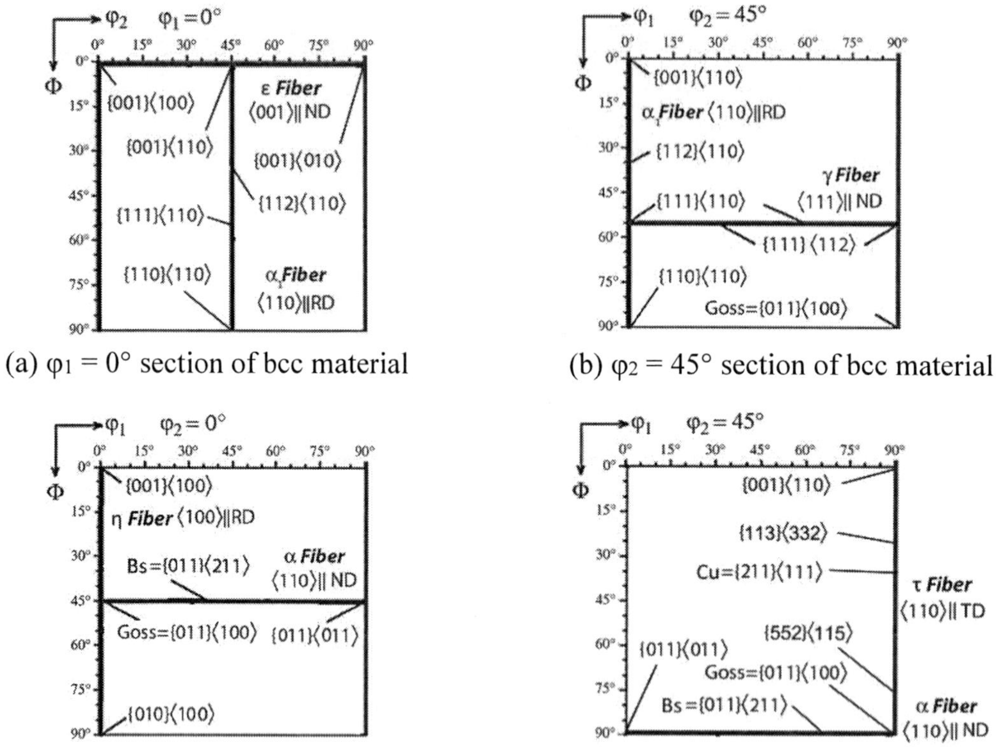

(a) $\varphi_1 = 0°$ section of bcc material (b) $\varphi_2 = 45°$ section of bcc material

(c) $\varphi_2 = 0°$ section of fcc material (d) $\varphi_2 = 45°$ section of fcc material

Fig. 1 Illustration of typical texture components and fibres of deformed cubic materials [3]

Experiment

The plate shape samples of size 19×120 mm^2 and different thicknesses were made of AISI 420 (ferritic), AISI 304 (austenitic), and AISI 318LN (duplex) type of stainless steels. The samples were cold-rolled with 0, 10, 20, 30, 40, and 50 % reduction of thickness so that the final thickness of all samples was 1.5 mm. For this reasons, samples made of ferritic, austenitic and duplex steel were marked as F0–F50, A0–A50 and D0–D50, respectively.

The *X'Pert PRO MPD* diffractometer with cobalt radiation was used for surface analyses of the rolled samples by X-ray diffraction (XRD). The texture analysis was based on ODF calculation from experimental pole figures which were obtained from three diffraction lines *{110}/{220}*, *{200}*, *{211}* of ferrite phase and *{111}/{311}*, *{200}*, *{220}* of austenite phase. The *MATLAB^{TM}* toolbox *MTEX* software was used for processing of all experimental data [6].

Results and discussions

ODF sections of samples with 0, 30 and 50 % deformation (reduction of thickness) are shown in Figs. 3. Because of grinding surface during samples preparation, the initial textures of samples without deformation (F0, A0 and D0) were composed of grinding texture components.

The typical texture components and fibres were observed in the ferritic samples F0–F50, see Figs. 3a, 3c, 3e. The strength of the mentioned texture components and fibres increased with deformation. The dominant α_1 and γ-fibres, which contain *{001}<110>* (rotated cubic),

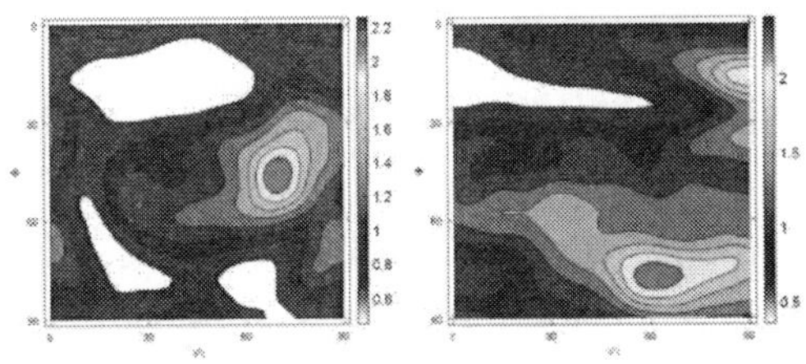

(a) Sample F0

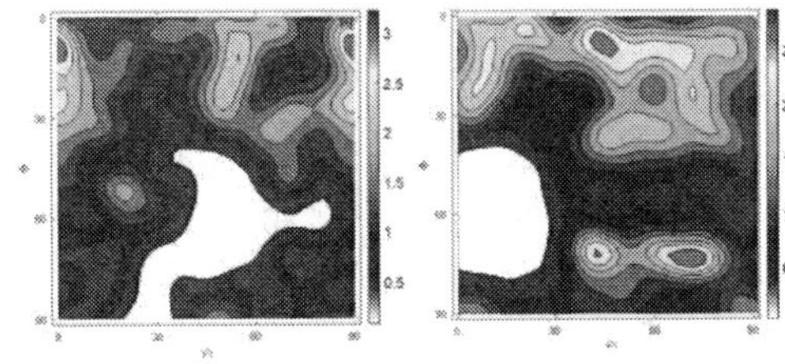

(b) Ferrite phase of sample D0

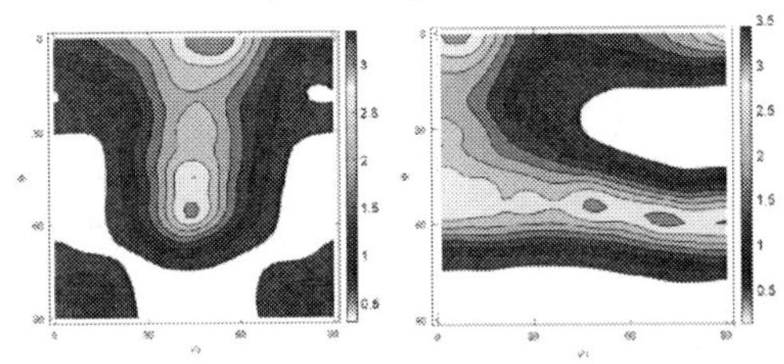

(c) Sample F30

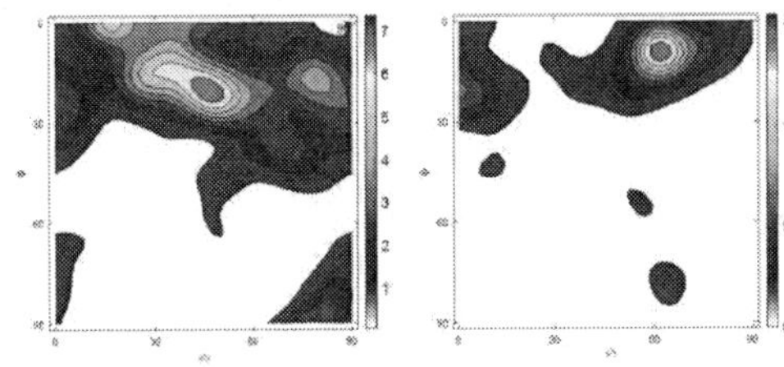

(d) Ferrite phase of sample D30

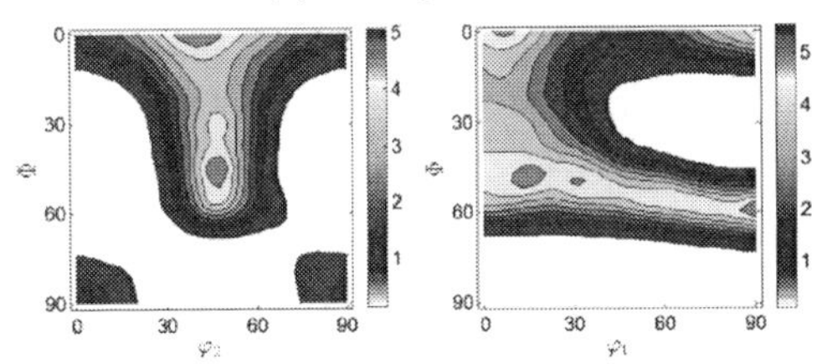

(e) Sample F50

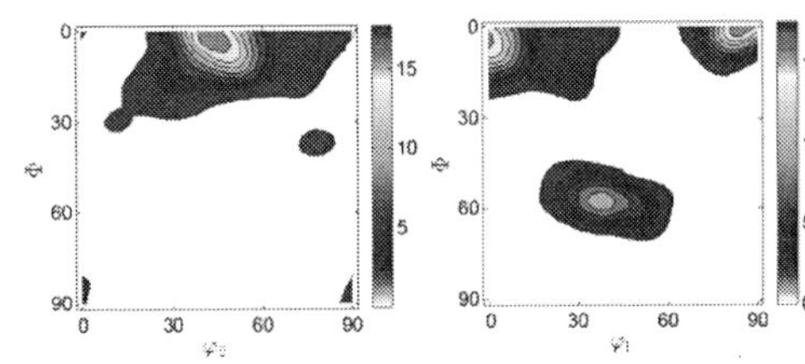

(f) Ferrite phase of sample D50

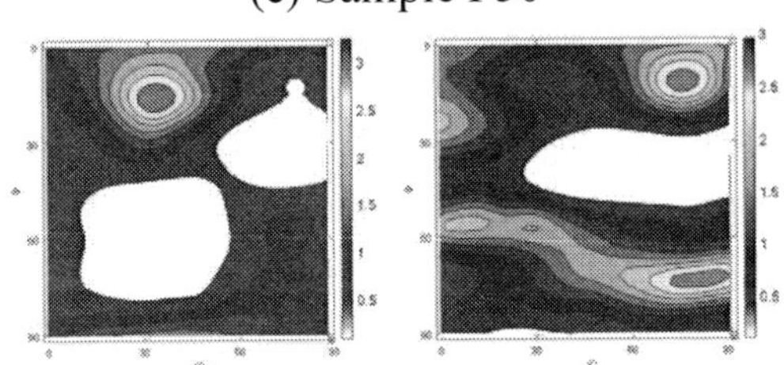

(g) Sample A0

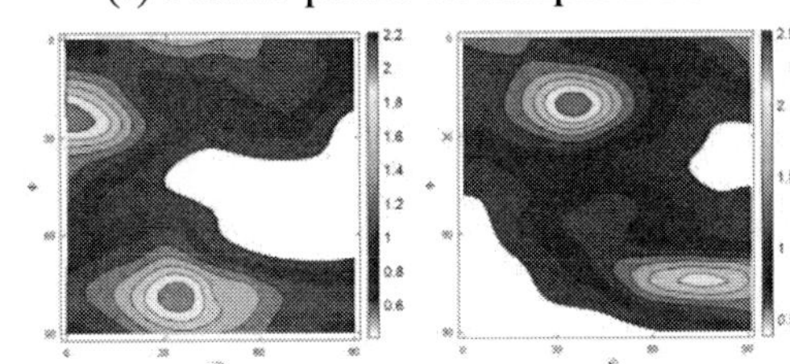

(h) Austenite phase of sample D0

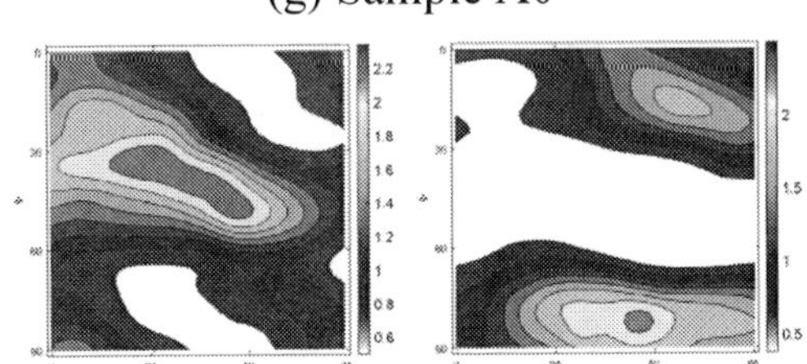

(i) Sample A30

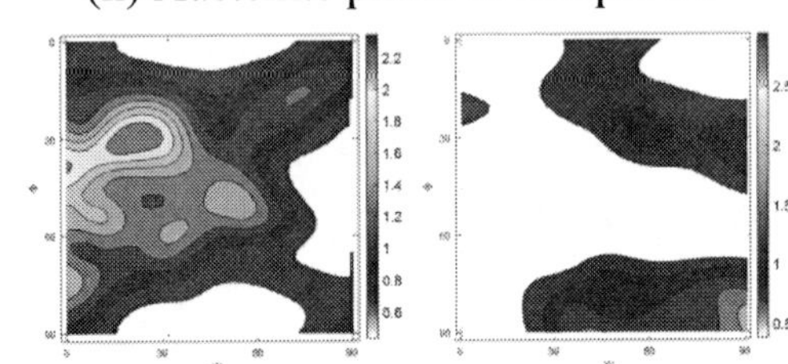

(j) Austenite phase of sample D30

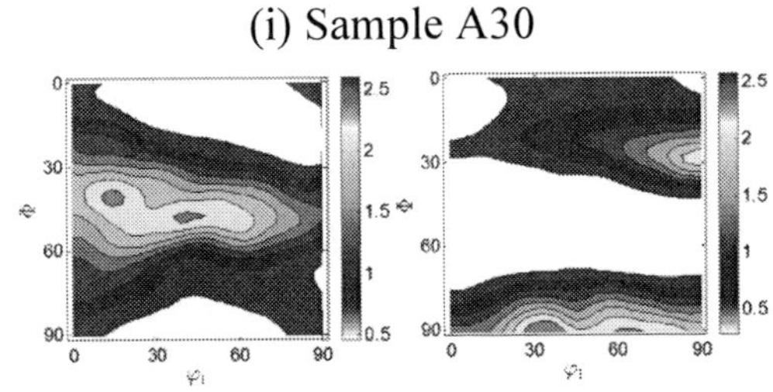

(k) Sample A50

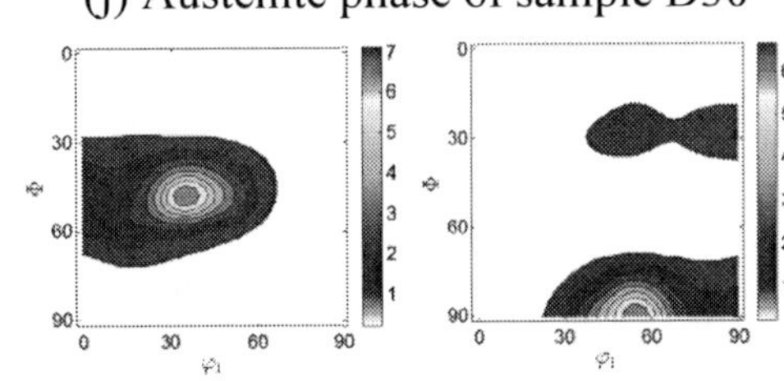

(l) Austenite phase of sample D50

Fig. 3 ODFs in the $\varphi_1 = 0°$ (left) and $\varphi_2 = 45°$ (right) sections for ferrite phases, and $\varphi_2 = 0°$ (left) and $\varphi_2 = 45°$ (right) sections for austenite phases

{112}<110> and *{111}<110>* texture component, were generated during cold-rolling, see Fig. 4a.

Grains transitioned from the rotated cubic *{001}<110>* to the cubic component *{001}<100>* during rolling of the austenitic samples A0–A50, see Figs. 3g, 3i, 3k. The G component *{110}<001>* was generated after 30 % deformation. Rotating these grains around *<100>*||RD axis, the η-fibre was formed. If grains with G orientation rotates around *<110>*||ND axis, the α-fibre and Bs texture component could be observed, see Fig. 4b. The τ-fibre is composed only of G and Cu texture components. However, the Cu texture component is more common for materials with high SFE.

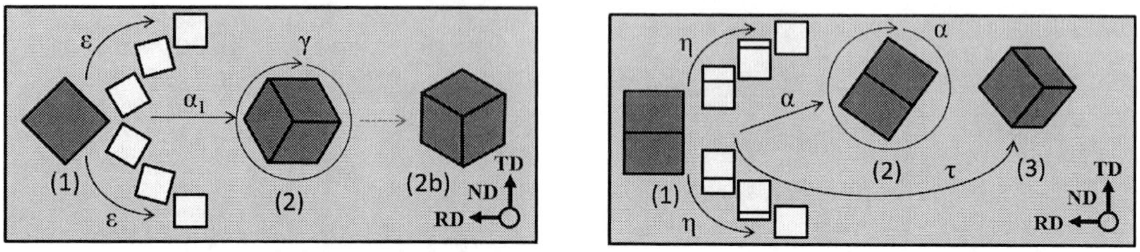

(a) Ferrite grains – *(1)* *{001}<110>* (rotated cubic), *(2)* *{112}<110>* and *(2b)* *{111}<110>* texture component

(b) Austenite grains – *(1)* G, *(2)* Bs and *(3)* Cu texture component

Fig. 4 Typical orientation of bcc (left) and fcc (right) materials during rolling

The mutual influence of ferrite and austenite phases and mainly different mechanical properties of both phases result in limitation of movement of ferrite (Figs. 3b, 3d, 3f) and austenite (Figs. 3h, 3i, 3l) grains during deformation. The harder ferrite phase (ferrite – 400 HV and austenite – 362 HV) keep energy until deformation exceeds a certain limit. This released energy results in sudden ferrite grains rotation to another discrete orientation. From this reason, the observed texture components were very strong and not typical for this deformation of bcc materials or phases. However, typical texture components were generated in the sample with higher deformation.

Texture components of austenite phase of duplex steel (Figs. 3h, 3i, 3l) were stronger in comparison with single-phase steel. With increasing of deformation, Bs and G texture components were observed. In spite of very small SFE (10 mJ/m^2), the unusual Cu texture component was analysed too. The majority of grains had Bs orientation for the sample with 50 % deformation.

Conclusions

The present study showed:

- The type and rate of texture components were different for dual and single-phase steels.
- The higher rate of texture was accompanied by increasing degree of reduction of single-phase samples.
- The typical texture components and fibres were observed for cold-rolled single-phase steels. The generation of Cu texture component of austenitic steel, which is more common for materials with high SFE, is less frequently analysed.
- The textures for both the phases of duplex steel were stronger in comparison with single-phase steels. The presence of fibres was limited and mostly discrete texture components were observed. For high deformation of austenite phase, only Brass component developed texture, which is typical for steels with low SFE.

Acknowledgement

This work was supported by the governmental funding of Technological Agency of Czech Republic – project No. TH02010664 and the Student Grant Competition of the Czech Technical University in Prague, grant No. SGS16/245/OHK4/3T/14.

References

[1] R. Dakhlaoui, C. Braham, A. Baczmański, Mechanical properties of phases in austeno-ferritic duplex stainless steel – Surface stresses studied by X-ray diffraction, Mater. Sci. Eng.: A. 444 (2007) 6-17.

[2] J. Ryś, W. Ratuszek, M Witkowska, Comparison of the Rolling Texture Formation in Duplex Steels with Various Initial Textures, Arch. Metall. Mater. 51 (2006) 495-502.

[3] H.J. Bunge, Texture Analysis in Materials Science, Butterwort, London, 1982.

[4] H. Hu, Texture of metals, Text. 1 (1974) 233-258.

[5] W. Reick, M. Pohl, A.F. Padilha, Determination of stacking fault energy of austenite in a duplex stainless steel, steel res. int. 67 (1996) 253-256.

[6] F. Bachmann, R. Hielscher, H. Schaeben, Texture Analysis with MTEX | Free and Open Source Software Toolbox, Solid State Phenomen. 60 (2010) 63-68.

EAN 2018
56th conference on experimental stress analysis

Experimental Analyses of Limit States, Stress-Strain Distribution, Defects and Failure Mechanisms of High-Pressure Gas Pipe

ČERNÝ Ivo[1,a], KEC Jan[1,b], JERSÁK Miroslav[1,c], POUPA Miloslav[1,d], ČERNÁ Marie[1,e] and PAVELKOVÁ Romana[2,f]

[1]SVÚM a.s., Tovární 2053, 25088 Čelákovice, Czech Republic

[2]NET4GAS s.r.o., Na Hřebenech II 1718/8, 140 21 Praha 4, Czech Republic

[a]Ivo.Cerny@seznam.cz, [b]kec@svum.cz, [c]jersak@svum.cz, [d]poupa@svum.cz, [e]marienkcerna@seznam.cz, [f]romana.pavelkova@net4gas.cz

Keywords: High-Pressure Gas Pipe, Defects, Dents, High Pressure Test, Cyclic Loading.

Abstract. The paper contains results and discussion of a comprehensive experimental programme aimed at evaluation of limit states, stress-strain distribution, defects and failure mechanisms of a section of high-pressure gas pipeline taken from an actual line of the nominal diameter DN700, made of an L290NB steel (API 5L X42 steel), operated at maximum pressure 6.1 MPa, with the aim to confirm the service safety and reliability. The experiments were performed using both laboratory specimens machined from the pipe and full scale pipe section of the length 3820 mm. During the full-scale pressure test, the pipe destruction occurred at pressure 12.7 MPa, which is considerably higher than theoretical failure considering the biaxial loading. The pipeline in these conditions was therefore evaluated as safe with a safety factor more than 2. It should be, however, pointed out that the final destructive crack was not initiated in the area of minimum wall thickness as expected, but at the dent with an evident assistance of the surface small cracks. Therefore, an attention should be paid to such defects in reality.

Introduction

Natural-gas pipeline accidents mostly result in major damage to buildings and other constructions located not only nearby but also quite far away and often put a lot of people in danger of injury or death. Therefore, safety and reliability management of high pressure gas pipelines is one of the most important issues for the operators and big effort is being put to various investigation projects with the aim to minimise probability of unexpected pipeline explosions. One of the research fields is risk assessment of high pressure pipeline explosion and planning of safe distances between the pipeline and buildings on the basis of probability methods like Monte Carlo simulations etc. [1]. The most effective method, however, looks to be a comprehensive theoretical and experimental evaluation of existing overall operated pipeline conditions using different available tools like intelligent pigging, regular internal inspections of corrosion occurrence and its growth, pressure cycle induced fatigue crack growth assessment and other tools [2], followed by a dedicated research programme dealing with a specific technical problem of damage to the pipeline, either local or of a fairly global character.

This contribution contains selected aspects and crucial results of a complex experimental programme aimed at evaluation of limit states, stress-strain distribution, defects and failure mechanisms of a section of high-pressure gas pipeline, taken from an actual pipe line of the nominal diameter DN700, made of an L290NB steel (API 5L X42 steel) commonly used in high pressure gas pipelines in Czechoslovakia in the past, operated at maximum pressure 6.1 MPa, with the aim to confirm the service safety and reliability. The NDT internal inspection of the pipeline section did not indicate any actual defects, only fairly reduced pipe wall thickness with 9.5 mm minimum values. During careful and detailed visual inspection of the pipe section, two dents were found, the first one quite inexpressive of the diameter approximately 100 mm and depth less than 5 mm. The second one was even smaller, almost negligible. The experiments were performed using both laboratory specimens machined from the pipe and full scale pipe section of the length 3820 mm.

Experimental Programme

A comprehensive experimental programme was proposed with the following aims:

- to verify limit states of the pipe integrity, stability of strength and deformation properties and limit pressure values for safe service,
- to evaluate changes of material mechanical properties after loading close to limit states,
- to evaluate effects of dimension irregularities, particularly variable wall thickness, on damage mode and limit states of pipe integrity under biaxial stress,
- to verify whether some creep occurs at stresses corresponding to service loading.

The most important phases of the experimental programme contained:

- evaluation of the material composition, microstructure and mechanical properties, effects of limit loading on changes of mechanical properties,
- visual inspection of all the pipe, detailed non-destructive ultrasonic measurement of pipe wall thickness using mesh 100 x 100 mm,
- experimental stress-strain analysis at numerous important points like minimum and maximum wall thickness, at two inexpressive dents, found during the visual inspection,
- static pressure test to service pressure 6.2 MPa with 100 hours low temperature creep test, when strain values were corrected considering pressure and temperature changes,
- fatigue cycle pressure test to 1000 cycles and final pressure test to destruction,
- fractographical analyses of fracture surface to explain crack initiation mechanisms and failure mode.

Chemical composition was analyzed using optical-emission method, device SPECTROMAXX. Mechanical properties were evaluated using SVÚM a.s. laboratory machines, independently calibrated by the Czech Metrology Institute within the laboratory accreditation. Static tensile tests were performed according to the ČSN EN ISO 6892-1 on the test machine Instron 1185, impact bending tests according to ČSN ISO 148-1 on test hammer PSWO30. Fracture toughness tests were carried out according to ČSN EN ISO 12737 on Instron 1185 machine using three point bend specimens.

Material microstructure was evaluated using optical microscopes. Strain values were measured and recorded using high precision Hottinger Baldwin Messtechnik GmbH device HBM UPM 60 with computer data recording. HBM strain gauges (SG), mostly of 6 mm grid length were used (type 1-LY11-6/120). Fractographical analyses was performed using scanning electron microscope (SEM) Zeiss.

Results and Discussion

Evaluation of Wall Thickness. Thickness of the pipe wall was one of the most important issues as NDT internal inspection indicated areas with reduced thickness. Therefore, this was the primary reason of the experimental programme – to verify, how and whether the reduced thickness affects the limit pressure of the pipe destruction.

As already mentioned, surface network of measurement points with distances 100 x 100 mm was prepared for the ultrasonic measurement. As the pipe section length was 3820 mm and actual circumference almost 2400 mm, there were 38 points in longitudinal direction and 22 points at circumferential direction. Circumferential positions were indicated by alphabetical letters A – W, where position A was on the top line – so called 12 o´clock, longitudinal positions by numerical order 1 – 38. By this method, each position was clearly defined.

According to the literature and codes commonly used for high pressure gas pipelines [3, 4], the minimum wall thickness t is given by following formula:

$$t = P\,D_0 / 2\,F\,E\,T\,S_y \qquad (1),$$

where P is internal service pressure, D_0 is outer pipe diameter, S_y is yield stress and F, E, T are safety coefficients expressing pipeline position in relation with buildings and roads, types of welding joints on the pipeline and temperature conditions, respectively. Temperature coefficient for normal temperatures is T = 1, joint factor for the specific method used also E = 1. Basic design factor for areas outside buildings and roads is F = 0.8. Then minimum pipe thickness for the service pressure P = 6.1 MPa made of steel with minimum yield stress 290 MPa is t = 9.5 mm.

Results of the thickness measurement at individual points in the area of reduced thickness are in Table 1. In addition, circumferential average thickness values for individual longitudinal positions and also longitudinal average values for individual circumferential position were evaluated to obtain simplified, clear information about longitudinal and circumferential distribution of wall thickness reduction – Fig. 1

Tab.1 Actual values of wall thickness at individual measurement points in the area of wall thickness reduction, position of the dents indicated by red circles (G29, negligible dent G25)

	A	B	C	D	E	F	G	H	I	J	K	L	M
17		9.45	9.63	10.76	10.07	9.58	9.70	9.58	9.63				
18	9.72	9.56	9.58	9.16	9.28	8.94	9.09	9.28	9.34	9.68	9.89	10.03	9.77
19	10.13	9.34	9.15	9.31	8.90	8.53	9.25	9.01	8.92	8.53	9.12	9.13	9.28
20	10.68	9.53	9.03	8.41	8.81	8.46	8.97	8.93	8.69	9.01	8.78	9.24	9.58
21	10.55	9.61	8.96	8.54	8.62	8.48	8.94	8.78	8.73	9.09	9.53	9.84	9.36
22	10.12	9.37	9.08	8.75	8.91	8.92	8.99	8.81	9.09	9.09	9.41	9.40	9.32
23	10.27	9.48	8.88	8.64	8.97	8.59	8.67	8.97	9.26	9.40	9.31	9.86	10.08
24	10.33	9.71	9.56	8.75	8.71	8.63	8.92	8.92	9.31	9.25	9.35	9.57	9.34
25	10.34	10.03	9.35	9.04	9.07	8.93	9.33	8.98	9.19	9.21	9.72	9.50	9.62
26	10.35	10.09	9.46	8.85	9.16	9.01	9.28	9.05	9.38	9.60	9.42	9.45	9.56
27		9.62	9.80	9.23	9.41	9.37	9.62	9.48	9.46				
28		9.97	9.86	9.57	9.62	9.58	9.86	9.58	9.38				
29		9.67	9.78	9.61	9.55	9.52	10.01	9.45	9.58				
30		9.31	9.63	9.87	9.90	9.74	9.74	9.47	9.63				
31		9.28	9.89	9.88	9.62	9.25	9.60	9.64	9.49				
32		10.07	9.78	9.60	9.37	9.32	9.53	9.48	9.30				
33		10.06	9.90	9.58	9.19	9.51	9.64	9.58	9.55				
34		9.94	9.90	9.82	9.56	9.09	9.16	9.37	9.28				
35		9.19	9.47	9.85	9.14	9.07	9.32	9.10	9.21				
36	10.09	9.42	9.28	9.08	9.14	8.69	9.23	9.64	9.44	9.36	9.75	9.87	9.71
37	10.07	9.02	9.32	9.30	8.78	8.69	9.26	9.18	9.03	9.35	9.30	9.40	9.59
38	10.37	9.68	9.20	8.94	9.14	8.76	9.25	8.45	9.13	9.37	9.31	9.45	9.56

It follows from Table 1 that there was an area of approximately 800 mm in diameter with wall thickness less than 9 mm, namely around 8.5 mm, the minimum value being 8.41 mm. This thickness is almost by 12 % lower than minimum allowed thickness for this type of pipe. Furthermore an angle dependence of wall thickness follows from Fig. 1, where circumferential differences of the thickness are more than 2 mm, which is a considerably high value. The angle position of minimum and maximum wall thickness is 180° which confirms that the thickness inhomogeneity occurred during pipe manufacture – the pipe was seamless and rolled. The average value from all the measured points was 9.74 mm. Though this thickness is just 0.2 mm above the minimum requested value, the average thickness would be acceptable provided that it was uniform, which was not the actual case.

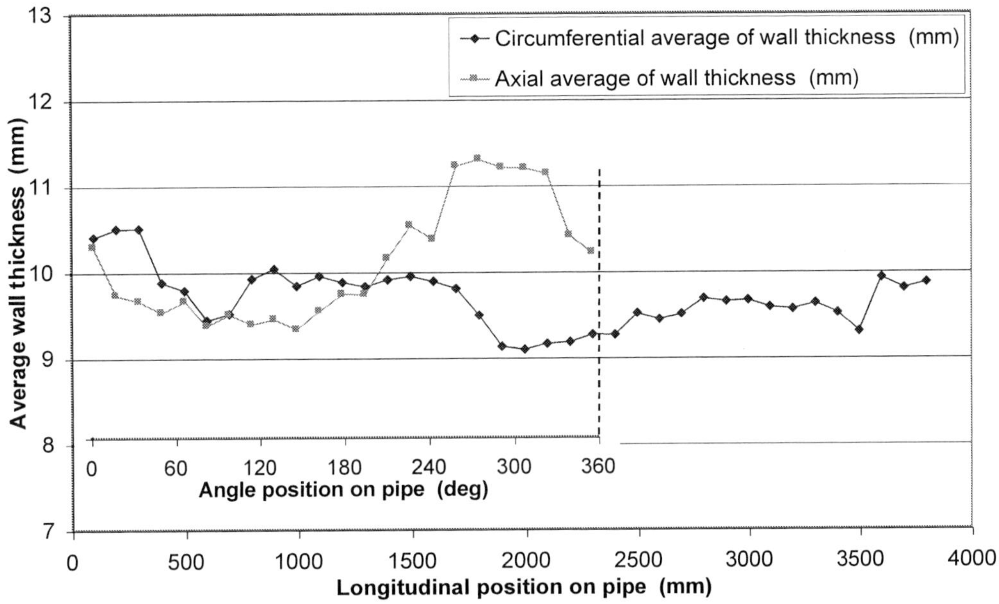

Fig. 1 Longitudinal and wall thickness distribution

Chemical Composition and Mechanical Properties. Content of elements in mass % was the following: C 0.15, Si 0.23, Mn 0.96, P 0.024 and S 0.029. The chemical composition was fully in accordance with the X42 (L290NB) standard.

Results of static tensile tests are in Table 2. Three specimen series were tested in both longitudinal and circumferential directions. The first series and second series were taken from the areas of maximum and minimum wall thickness, respectively. The third series was taken from a ring cut after the 100 hours creep test at internal pressure between 6.1 and 6.2 MPa.

Tab.2 Results of static tensile tests

Specimen direction	**Specimen serries**	**R_{eH} [MPa]**	**R_{eL} [MPa]**	**R_m [MPa]**	**A [%]**	**Z [%]**
longitudinal	1	325	321	496	26.0	44.9
	2	332	-	485	22.7	50.0
	3	327	321	472	22.2	53.0
Specimen direction	**Specimen serries**	**$R_{p0.2}$ [MPa]**	**$R_{t0.5}$ [MPa]**	**R_m [MPa]**	**A [%]**	**Z [%]**
circumferential	1	299	316	494	26.1	46.5
	2	295	303	485	24.6	47.5
	3	297	300	470	22.5	42.2

It can be concluded that mechanical properties satisfied the minimum requirements with the exception of elongation A, where the minimum value is 23 %. Elongation of the material taken from the area of minimum wall thickness was negligibly lower. Even lower was elongation after the 100 hours creep loading. Proof stress in the circumferential direction was close to the minimum requirements and it was not affected either by the position in the pipe or by the creep loading. Maximum strength fully satisfied the requirements, but it was slightly affected by both position in the pipe and creep loading – material from lower wall thickness area had lower maximum strength and material after the creep loading even more low.

Tab.3 Results of impact bending KV_2 and fracture toughness K_{IC} tests

Direction	Series	KV_2 [J]	Fracture toughness K_{IC} [MPa*m^{0,5}]
longitudinal	1	165.3	65.4
	2	45.5	62.2
	3	49.7	65.3
circumferential	1	29.2	61.6
	2	29.8	58.3
	3	32.1	57.0

Results of impact bending and fracture toughness tests are in Table 3. Concerning brittle / ductile properties, the steel is acceptable, but there are distinctly worse results in circumferential direction in terms of both impact bending and fracture toughness. Quite surprising is the very high value of impact bending in longitudinal direction of the material taken from the wall with high thickness – series 1, which is confirmed by fracture toughness, though with much lower difference. In general, fracture toughness in longitudinal direction corresponds quite well to the literature data [5]. Scatter of results, particularly of impact bending indicates that the material homogeneity in macroscopic scale is not optimum.

Metallographic Analysis. Microstructure was evaluated using an optical microscope Neophot II. Microstructure was of ferritic-pearlitic type. Residues of casting structure could be randomly found – Fig. 2, which can be considered as microstructure defects. Ferritic grain size evaluated according to ČSN EN ISO 643 corresponded to 6-7. Concerning material purity, oxide content was low unlike content of sulphides evaluated according to ČSN EN ISO 4967, which was at the end of the scale limit, perhaps even above it. Decarbonization to 0.2 mm depth was visible near both outer and inner surfaces – Fig. 3. Quite frequent were inclined cracks of length up to 0.5 mm, which occurred likely during manufacture – rolling.

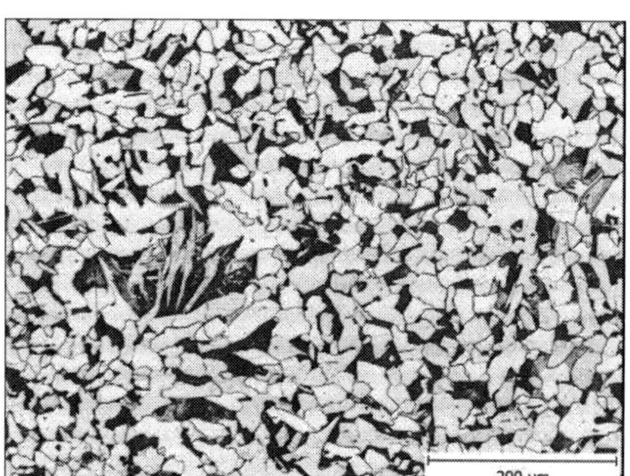

Fig. 2 Ferritc-pearlitic microstructure with residues of casting structure

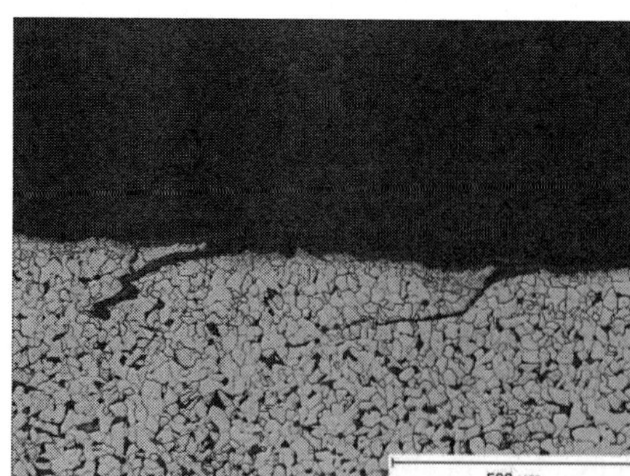

Fig. 3 Decarburized surface with cracks

Stress-Strain Analysis. Total amount of 16 couples of strain gauges (SGs) were glued on the pipe surface, one circumferential and one longitudinal SG at each of the 16 positions with the aim to evaluate:

— strains along the pipe length – on the line B, almost along the top line,
— strains along the circumferential line near the pipe centre – position 25,
— strains at areas of maximum, medium and minimum wall thickness,
— strain in the centre of the dents.

Strain values were recorded during the first stage of the pressure test to service pressure 6.2 MPa. In the following diagram, the theoretical stress was not calculated as an average stress using average thickness value, but it was calculated individually for each SG position considering actual wall thickness corresponding to the measured point.

Total survey of stress-strain dependencies with the exception of strains measured in the area of dents is shown in Fig. 4. The diagram contains record of strains along the line B – B12, B25, B33, circumferential strains at longitudinal position 25 – L25, O25, S 25 and strains at points of minimum and medium wall thickness, respectively – D20 and H16. The diagram contains regression lines of all points of circumferential and longitudinal measurements, respectively, and theoretical stress-strain lines calculated for medium wall thickness considering biaxial stress state.

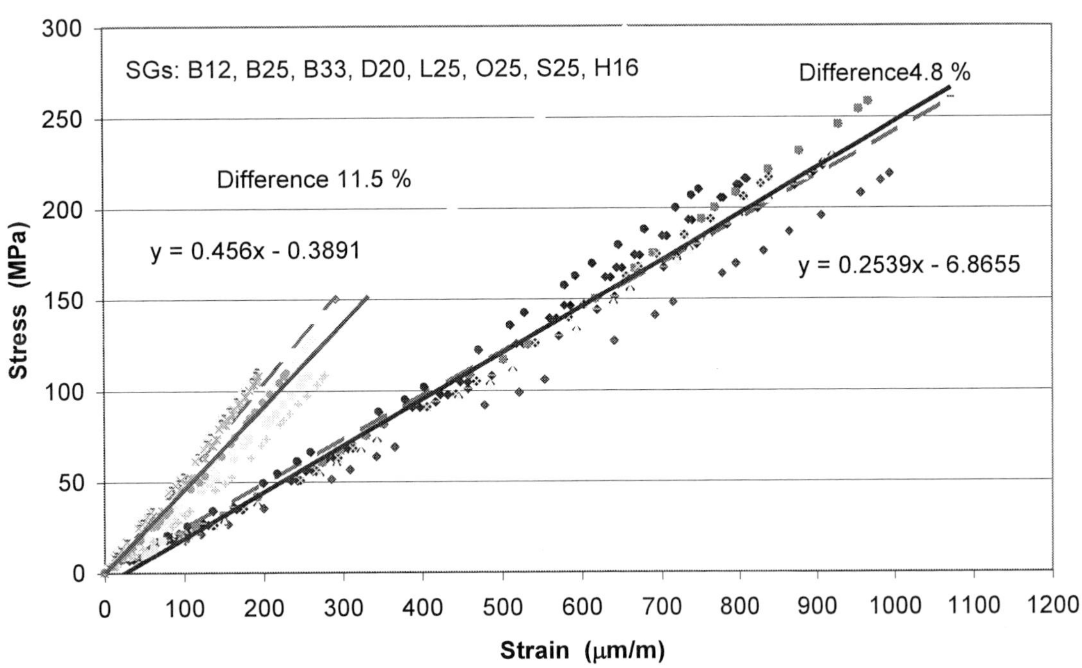

Fig. 4 Total survey of stress-strain dependencies with exclusion of measurement at defects

In Fig. 4, there is quite a good agreement between the regression line corresponding to circumferential measurements, the difference being bellow 5 %. As regards longitudinal measurement, the difference between the regression and theoretical lines is considerably higher, almost 12 %, the actual strains being higher. This fact can be, however, explained. Most of longitudinal measurements were carried out near the line B, i.e. near the top line. The tested pipe was not supported on rollers but just on wooden cross beams near the pipe margins, which partially obstructed free longitudinal motion at the pipe bottom. This eventually resulted in a kind of additional bending, when bottom strains were bellow average and top strains higher than average.

Much bigger differences are those between results measured at individual points and theoretical mean values. The differences are up to ±15% and even up to ±25% for circumferential and longitudinal measurements, respectively. Such differences cannot be explained by different wall thickness only. The likely reason is a more complex stress-strain redistribution in the pipe due to the variable wall thickness.

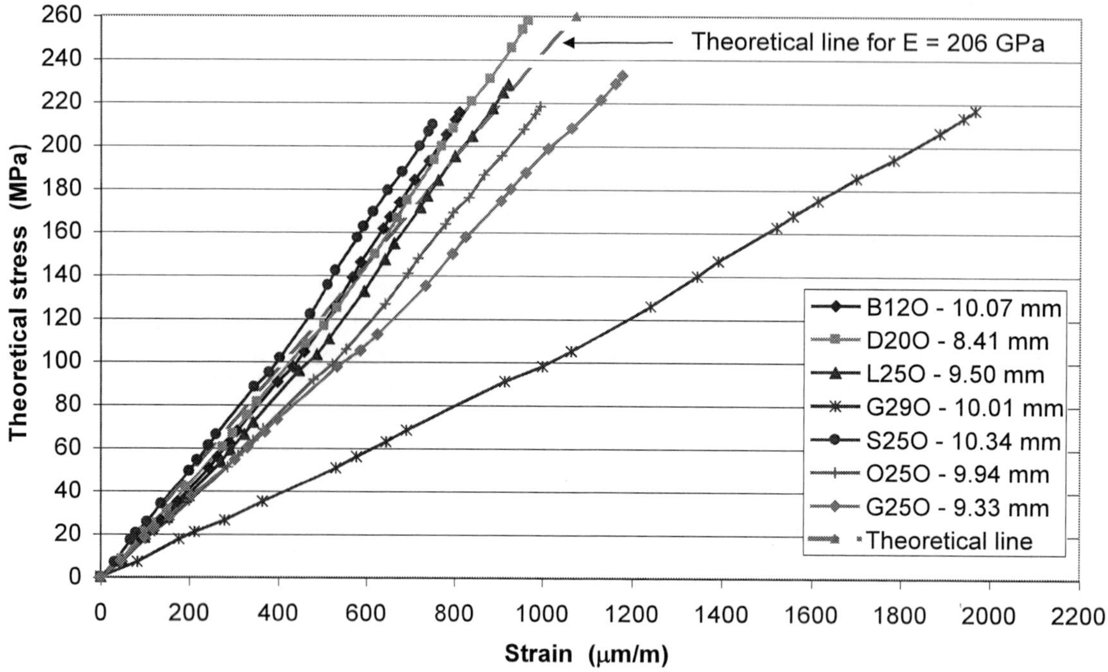

Fig. 5 Stress-strain dependences of selected circumferential measurement points including the two dents

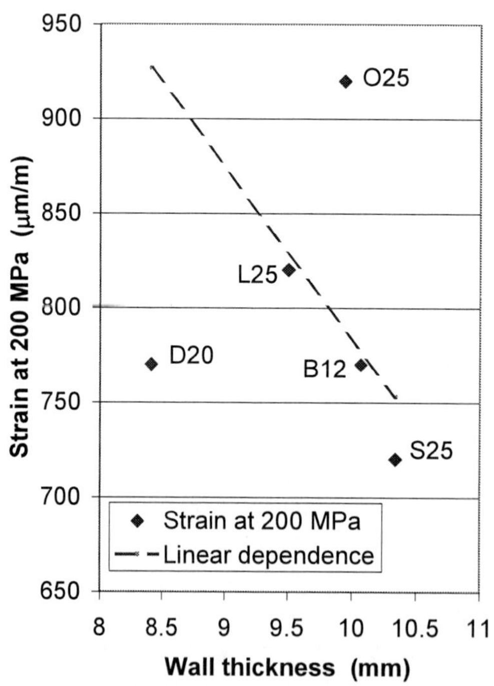

Fig. 6 Strains at 200 MPa at selected measurement points

Stress-strain dependences of selected circumferential measurement points including the two dents are shown in Fig. 5. It is evident that strains at the centre of the larger dent were more than 2.3-times higher in comparison with theoretical values for undamaged wall. Note that the dent actually was shallow, almost invisible, with the depth less than 5 mm and diameter approximately 100 mm. Some increased strains were measured at the centre of the second, quite negligible dent.

Individual values of strains at the stress 200 MPa, except the two dents, are shown in Fig. 6. This figure confirms that actual stress-strain distribution was more complex than it would correspond just to different wall thickness. If the strain only was dependent on the wall thickness at corresponding measurement point, the points would lie at least approximately along the linear line. This concerns only the points L25, S25 and B12. Strain values at points D20 and O25 are fare from this ideal line indicating that the stress-strain distribution is much more complicated.

Evaluation of Pipe Creep. Pipeline creep at room temperature and high pressure is a phenomenon recently studied in several literature contributions [6,7]. In this work, possible creep was measured at internal pressure 6.1 – 6.2 MPa, i.e. at maximum service pressure. This loading corresponded to circumferential stress 222 MPa, if mean wall thickness 9.74 mm was considered, which was 86 % of proof stress under biaxial conditions in the pipe. The measurement was performed for the period of 100 hours. The problem complicating an exact evaluation were temperature dilatations. The pipe could not be placed at any air conditioned room, just in a special uncovered basement in a special hall. Though the temperature changes were not so strong in the basement, they significantly affected the results, though the pipe was full of high pressure water. Strain records at all measurement points were carefully corrected considering material dilatations, but the measurement accuracy was not better than 10 – 15 μm/m. Final strain changes after the test were within this interval, even changes at the centre of the dent. It could be therefore concluded that no measurable creep occurred. Nevertheless, a ring was taken from the pipe for evaluation of mechanical properties after the creep test – Tables 2 and 3.

Pressure Test to Destruction. Loading by the internal pressure to final pipe failure was performed quasistatically with maximum pressure increments 0.27 MPa / min. Pressure values corresponding to proof stress $R_p0.2$, stress $R_t0.5$ and final pipe failure were 7.9 MPa, 8.3 MPa and 12.7 MPa, respectively. If these values are recalculated to stresses using mean pipe wall thickness 9.74 mm, the corresponding stresses are 284 MPa, 298 MPa and 456 MPa, respectively. Note that the pipe is loaded biaxially with $\sigma_2 = \sigma_1 / 2$, σ_1 being circumferential and σ_2 longitudinal stress. Considering this biaxiality and HMH hypothesis, maximum reduced stress should correspond to 0.87 uniaxial stress. In such case the pipe proof stress should be 258 MPa and destruction should occur at approximately at 420 MPa of the circumferential stress. Actual stress values were reduced by the factor 0.95, higher than 0.87 according to the HMH hypothesis. It looks that the HMH hypothesis was slightly too conservative for this case of pipe and the material X42.

Fractographical Analyses of Failure Initiation Area. Final pipe failure occurred at area of the dent, near the point of maximum measured strains, i.e. near the dent centre, but not exactly there. The wall thickness at the failure area was 9.58 mm. The failure did not occur in the area of the minimum wall thickness, which was supposed. This is very important fact indicating that even small, almost negligible dents reduce pipe strength at least slightly. Note that in general, dents caused numerous pipeline failures in the past and recently [8]. Fractographical analysis provided further important information. In the fracture initiation area, intergranular surface cracks were observed, likely the same cracks as shown in Fig. 3. It should be, however, pointed out that both the phenomenona supporting the final failure, namely presence of the dent and surface cracks, reduced final pipe strength just negligibly, because final failure occurred at pressure by several percent higher than according to the HMH hypothesis applied to pipe with constant thickness and without any defects.

Conclusions

The study presented in the paper contains selected aspects and crucial results of a comprehensive experimental programme aimed at evaluation of limit states, stress-strain distribution, defects and failure mechanisms of a section of high-pressure gas pipeline, taken from an actual pipeline line of the nominal diameter DN700, made of an L290NB steel (API 5L X42 steel), operated at maximum pressure 6.1 MPa, Since the NDT internal inspection of the pipeline section indicated reduced pipe wall thickness, the primary aim was to confirm the service safety and reliability. The main results can be summarized as follows:

- Actual wall thickness was even lower than indicated by the NDT method. There was a large area of approximately 800 mm in diameter with wall thickness less than 9 mm, around 8.5 mm, the minimum value was 8.41 mm, almost by 12 % lower than minimum allowed thickness for this type of pipe. Circumferential differences of the thickness were more than 2 mm. The angle position of minimum and maximum wall thickness was 180° which confirmed that the thickness inhomogeneity occurred during rolling of the seamless pipe. The average value from all the measured points was 9.74 mm. Though this thickness was just 0.2 mm above the minimum requested value, it would be acceptable provided that it was uniform, which was not the actual case.
- In general, mechanical properties and material microstructure satisfied the requirements, though in some cases close to limits. Fairly high was content of sulphides, fortunately homogeneously distributed, not in unfavourable chains. Decarbonization was low, up to 0.2 mm near both inner and outer surfaces. The most problematic point was an occurrence of frequent surface inclined cracks of the length up to 0.5 mm, filled with ferrous oxide products.
- Results of the comprehensive experimental stress-strain analyses were in average close to the theoretical values, but individual measurements differed by more than ± 15 % in circumferential direction and even more in longitudinal direction, in the latter case likely affected by fixed wooden supports used. The differences were probably caused by the irregularities in the wall thickness. Strains measured at the dent were more than twice as much higher, though the dent was very small and inexpressive. No creep was observed during 100 hours test at service loading between 6.1 and 6.3 MPa, which corresponded to 86 % of proof stress. No creep occurred even at the dent.
- Final pipe destruction occurred at pressure 12.7 MPa, more than twice as much in comparison with service pressure, at average theoretical stress of 95 % of circumferential material strength, which is considerably higher than theoretical failure considering the biaxial loading. The pipeline in these conditions was therefore evaluated as definitely safe, with a safety factor more than 2. What should be, however, pointed out is that the final destructive crack was not initiated in the area of minimum wall thickness as expected, but at the dent with an evident assistance of the surface small cracks. Therefore, an attention should be definitely paid to such defects in reality.

Acknowledgement

The support of the institutional funding MSM2579700001 is acknowledged.

References

[1] P. Russo, F. Parisi, Risk-targeted safety distance of reinforced concrete buildings from natural-gas transmission pipelines, Reliab. Eng. Syst. Safe. 148 (2016) 57-66.

[2] Proc. of the 27th Pipeline Pigging and Integrity Management Conference, Clarion Technical Conferences, USA (2015) 392 p.

[3] Pipeline design consideration and standards, Ed. Society of Petroleum Engineers – SPE International (2015) 13 p.

[4] ANSI/ASME Standard B31.8, Standard for Gas Transmission and Distribution Piping Systems, New York City, 1999

[5] C. San Marchi, B.P. Somerday, Technical reference of hydrogen compatibility of materials, Sandia Report, SAND2008-1163, Sandia National Laboratories, Albuquerque (2008), 225 p.

[6] D. Nie, J. Zhao, T. Mo, W. Chen, Room temperature creep and its effect on flow stress in a X70 pipeline steel, Materials Letters 62 (2008) 51–53.

[7] Ľ. Gajdoš, M. Šperl, P. Pařízek, Rozvoj plastických deformací v oceli při konstantním zatížení a jejich význam pro plynárenskou praxi, PLYN XCIII (2013) 198–203 [in Czech].

[8] Q. Yang, J. Shuai, Engineering evaluation method for dented pipeline, Acta Petrolei Sinica 31, 4 (2010) 649-653.

EAN 2018
56th conference on
experimental stress analysis

Experimental Determination of Viscoelastic Behavior of Core for Composite Sandwich Beam

ČERNÝ J. Miroslav[1,a] and ŠLAPÁK Pavel[1,a]

[1] Czech Technical University in Prague, Klokner Institute, Solinova 7, 166 08 Prague 6, Czech Republic

[a] cerny@hppro.klok.cvut.cz

Keywords: viscoelasticity, experimental assessment, composite sandwich beam

Abstract. The experimental assessment of viscoelastic properties of the core for sandwich beams is given. Theoretical results are compared with experimental values.

Introduction

Foam-core sandwich panels and beams offer a high stiffness per unit weight, but creep even at room temperature, what is limiting their use in structural applications [1],[2]. In this paper, the experimental determination of viscoelastic behavior of the core for sandwich beams will be given. Theoretical results even for sandwich beam are compared with data from tests [3].

Experiment

For the mechanical long-time tests was supplied the plate is a sandwich structure, which is part of the sandwich of the wagon-body for means of transport, produced by the Czech company VARIEL s.r.o. Test specimen with the shape of a cube with dimensions 50x50x50 mm was inserted into the set-up, manufactured in the Klokner Institute, enabling the long-time loading by a constant force. The longitudinal deformation of the test body were measured by inductive sensors (LVDTs) and resistive strain gauges. Test specimens were 24 h loaded and then 24 h unloaded. (Fig.1 and 2 show a loading of the sandwich beam [3]).

Fig.1 Testing set-up

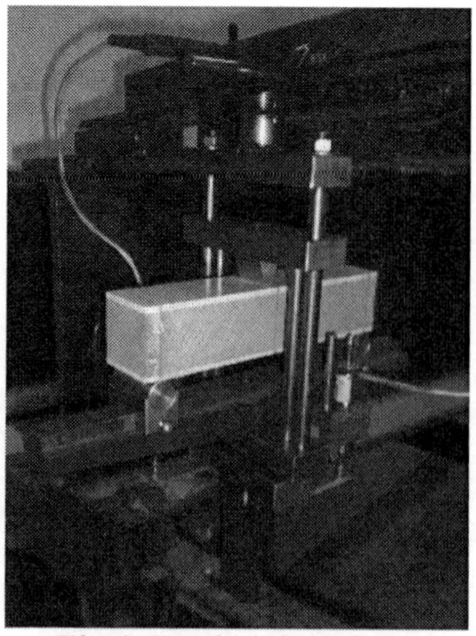

Fig.2 Sandwich beam

Mathematical Approximation of Viscoelastic Behavior

In viscoelastic problems for the evaluation of the time effect on stress and deformation are usually used the rheological models containing elastic and viscous elements operating in parallel or in series, in order to capture the characteristics of rheological models.

Basic of viscoelastic models are two-element models Kelvin-Voigt and Maxwell. Kelvin-Voigt model consists of parallel springs (E) and dampers (K). Maxwell's model consists of serial connection of the springs and dampers. Differential relations for two-element Voigt models and Maxwell can be written directly by a superposition of deformation speeds.

The three-element models containing three material characteristics are often used, in particular Poynting - Thomson model representing the serial connection of spring and Kelvin-Voigt element. The used model is also called a standard visco-elastic material.

$$\frac{1}{E_1}\sigma(x,t) = \varepsilon(x,t) - \frac{E_1}{K}\int_0^t e^{-\frac{E_1+E_2}{K}(t-\tau)}\,\varepsilon(x,\tau)\,d\tau$$

In the inverted physical equation we get thus

$$\varepsilon = E_1^{-1}\left(\sigma + \frac{E_1}{K}\int_0^t e^{-\frac{E_2}{K}(t-\tau)}\,\sigma(\tau)\,d\tau\right)$$

For the constant stress we get for the Poynting-Thomson model the relationship

$$\varepsilon = E_1^{-1}\left(1 + \frac{E_1}{K}\int_0^t e^{-\frac{E_2}{K}(t-\tau)}\,d\tau\right)\sigma_0$$

and after integration

$$\varepsilon = \frac{\sigma_0}{E_1} + \frac{\sigma_0}{E_2}\left(1 - e^{-\frac{E_2}{K}t}\right)$$

Results from experiment

An element subjected to the load constant in time was after 24 hrs load removed and the results of the experiment were compared with viscoelastic behaviour according to the Poynting-Thomson model. An arrangement of the experiment has been similar to Fig.1.

The theoretical course of the strain is dependent from the three material constants E_1, E_2, K.

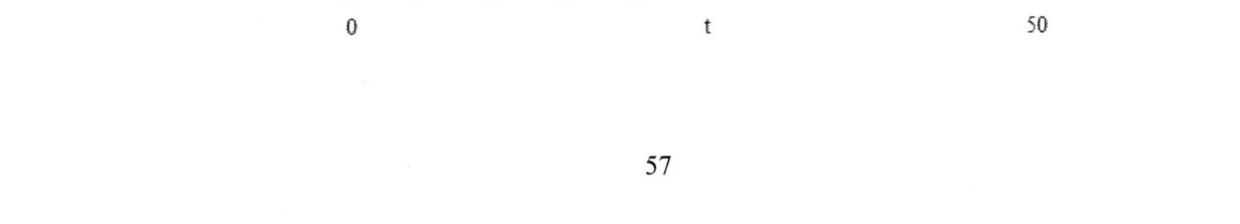

For the estimated standard linear material can be the values of the respective coefficients determined on the basis of carried out experiments. The size of the module E_1 follows from the values $\varepsilon = \varepsilon_0$ at the immediate load at the time $t = 0$ still didn't Kelvin model with the structural formula $E \,/\, K$ and the Poynting-Thomson model is then reduced to an elastic model Hooke. Out of here is $E_1 = \dfrac{\sigma_0}{\varepsilon_0}$. On the other hand, the value of E_2 can be approximated using the estimated values of the deformation in time $t \to \infty$, when the limit deformation of the coming term $\varepsilon_\infty \approx \dfrac{\sigma_0}{E_1} + \dfrac{\sigma_0}{E_2}$. What concerns the value of K, it is necessary to know the value of the deformation for the next time indication, e.g. for $t = t_1$. We obtain the relation $K = K(E_2)$. For the estimated standard linear material can be the values of the respective coefficients determined on the basis of carried out experiments.

The proposed procedures illustrate their use on a numerical example. Let
$\varepsilon_0 = 1{,}625.10^{-3}, \;\; \varepsilon_1 = 2{,}0.10^{-3}, \;\; t_1 = 24h, \;\; \sigma_0 = 0{,}04\,MPa, \;\; t_1 = 24h$.

Then comes the $E_1 = 24{,}6$ MPa, E_2 in the range of 27 to 45 MPa , which correspond to values of K in the range 1970 to 2220 MPa.h.

Conclusion

The experimental determination of viscoelastic behavior of the core for sandwich beams has been given and Poynting- Thomson model used.

If the theoretical curve for creep under mechanical model for the measured values, we have the possibility of changes of parameters or switch to another mathematical model. Formally we can also carry out replacement base e in exponential function by another suitably elected number, for example a. According to whether a <e, or a>e , the new curve is upper or under of the original creep curve, respectively.

References

[1] Cerny M.J., Slapak P., Behavior of Composite Sandwich Beam in Bending, Proceedings, ECCM15, Venice (2012)

[2] Cerny M.J., Slapak P., Visco-elastic Bending of Sandwich Beam with Foam Core, Proceedings, EAN2016, Srni (2016)

[3] Cerny M.J., Slapak P., Viscoelastic Bending of Symmetrical and Nonsymmetrical Composite Sandwich Beam, EAN2017, Novy Smokovec (2017)

EAN 2018
56th conference on experimental stress analysis

Stress Intensity Factor for Corrosion Fatigue on Steel Alloy in Steam Turbines

ČERNÝ Miroslav[1,a]

[1] Czech Technical University in Prague, Klokner Institute, Solinova 7, 166 08 Prague 6, Czech Republic

[a] cerny@hppro.klok.cvut.cz

Keywords: corrosion fatigue, stress intensity factor, steam turbines

Abstract. The corrosion pits in steel alloy for steam turbines can be considered as surface cracks for determining threshold limits for fatigue cracking. The threshold can be defined using linear elastic fracture mechanics (LEFM). Then the stress intensity factor should be used even for evaluating the influence of corrosion pits on fatigue such materials.

Introduction

The corrosion pits in steel alloy for steam turbines can be considered as surface cracks for determination of threshold limits for fatigue cracking. The threshold can be defined using linear elastic fracture mechanics (LEFM). Then the stress intensity factor should be used even for evaluating the influence of corrosion pits on fatigue such materials.

It has been observed, that most pits have roughly a semi-elliptical shape with the width at the surface, $2c$, and the pit depth a. Further, investigations show that the pits can be treated as semi-circular surface cracks with a geometry described above.

Fatigue failure in steam turbine blades is a critical issue since high frequency fatigue loading is induced by unsteady steam flow. The cyclic stresses are low compared to centrifugal forces due to the rotation which leads to high stress ratios.

In the low pressure part of steam turbines, fatigue crack initiation at corrosion pits is a very frequently observed cause of damage. Especially in the last two blading stages – where transition from dry to wet steam takes place – pitting may occur when unexpected operating conditions, for example, due to leaking of the condenser are present.

Stress intensity factor ΔK

Calculation of the threshold stress intensity factor ΔK_{th} for the corrosion pits can be calculated from the relationship:

$$\Delta K_{th} = \alpha.\Delta\sigma_{th}.\sqrt{\pi a}$$

ΔK_{th} is the range of the threshold stress intensity factor, $\Delta\sigma_{th}$ is the threshold fatigue strength, a is the pit depth and α is a geometry factor. For semi-circular surface cracks, α is approximated by 0.67.

For a semi-elliptical defect the threshold stress intensity factor is following:

$$\Delta K_{th} = \frac{\Delta\sigma_{th}\sqrt{\pi a}[1.13 - 0.07(a/c)^{1/2}]}{[1+1.47(a/c)^{1.64}]^{1/2}}$$

where c is half pit width.

Because above equations are based on linear elastic fracture mechanics, ΔK_{th} should be a constant independent of crack size. This is clearly not the case.

The limitation of LEFM is not surprising as the overall 'crack' length including the pit is small; a non-propagating crack formed at the pit may have an overall length which includes the pit but does not have the plastic wake history of a crack of the same length, the crack will be contained within the plastic field of the pit.

In order to account for the breakdown of the LEFM concept for small crack sizes, El Haddad and co-workers proposed an empirical relationship for the stress intensity factor range based on the notion of an effective crack length (a+a0):

$$\Delta K_{th} = \alpha.\Delta\sigma_{th}\sqrt{\pi(a+a_0)}$$

where ΔK is the range of the stress intensity factor, $\Delta\sigma_{th}$ is the threshold fatigue strength and a_0 is a constant.

This relationship has been shown to describe short crack and pit behaviour well for a range of systems, and a_0 provides a link to long cracks. The a_0 value can be evaluated from the limiting condition where the physical crack size, a, approaches zero, i.e. $a_0 \gg a$ and giving

$$a_0 = \frac{1}{\pi}\left(\frac{\Delta K_{th}}{\alpha.\Delta\sigma_0}\right)^2$$

where $\Delta\sigma_0$ is the fatigue limit of a smooth specimen.

Kitagawa- Takahashi Diagram

Since fatigue crack is expected to nucleate at the mouth of the pit (i.e. at the end of the c axis), the dimension c (half pit) will be used to represent the pit or crack size. With this approximation and that the pit size is small in terms of crack behavior, the stress intensity factor of pits ΔK and finally a Kitagawa- Takahashi (K-T) diagram can be created.

The shown diagram is for 403/410 12% Cr martensitic steel and temperature 90°C. The K-T diagram correlates pit-to-crack fatigue data with the use of the approach El Haddad for short cracks (Fig.2). The data are shown for specific values of R (ratio of minimum to maximum stress for combined steady and cyclic stress). As a result, K-T diagrams can be used for assessment of the fatigue limit and life-time of corroded parts of steam turbines.

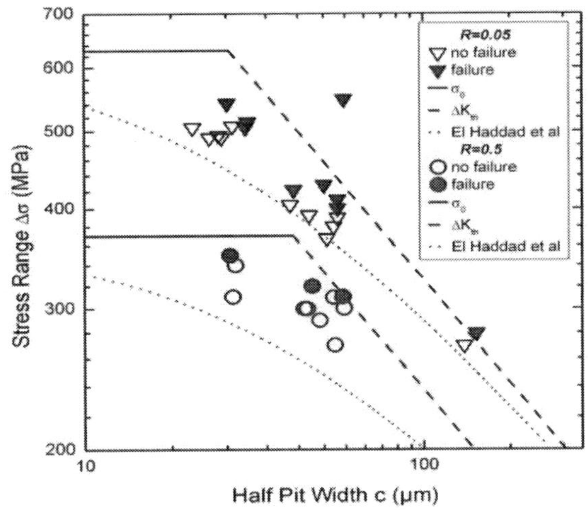

Fig.1 Kitagawa- Takahashi Diagram

True pit geometry

The description of true pit geometry has been published by author earlier [2]. The max. and min. depths of pits and their densities have been found. In [3] are shown the mentioned characteristics and Table 1 in [3] comprises the parameters of corrosion evaluated for part of steam turbine at power plant ETU (alloy X22CrMoV12-1).

A microscopy system Nikon Eclipse ME600 with image analysis software Nikon Elements has been used [2], [3]. The max. and min. depths of pits and their densities have been found for eight areas. Table 1 comprises the parameters of corrosion evaluated for part of turbine packing ETU23 (alloy X22CrMoV12-1).

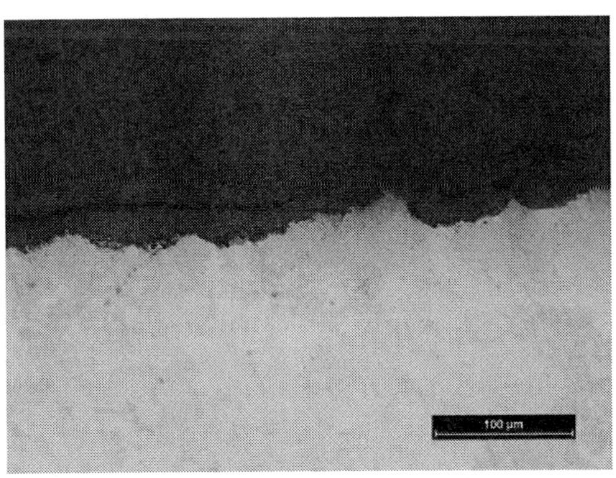

Fig.2 True geometry of corrosion

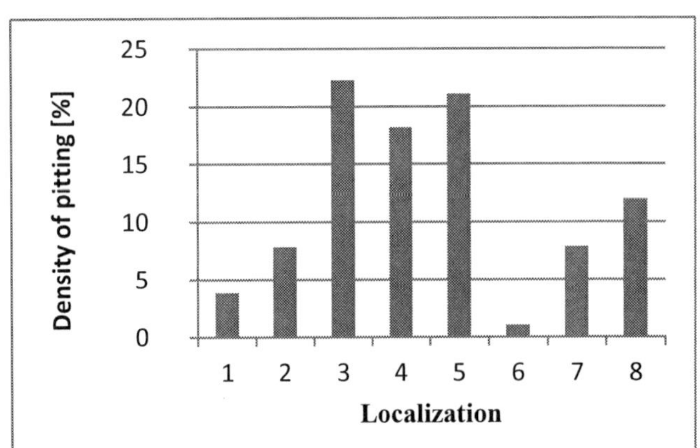

Fig.3 Density of pitting

Stress intensity factor for corrosion fatigue damage

Since fatigue crack is expected to nucleate at the mouth of the pit (i.e. at the end of the c axis), the dimension c is used to represent the pit or crack size. With this approximation and that the pit size is small in terms of crack behavior, the stress intensity factor of pits ΔK for long cracks and cracks that were initiated at corrosion pits can be created.

The calculation for pit depth and different aspect ratio for 403/410 12% Cr martensitic steel and temperature 90°C has shown threshold stress intensity factors (a=20, ΔK_{th}=4,959, a=30, ΔK_{th}=5,693) and corresponding stresses.

Acknowledgement

The presented work has been supported by grant No TE01020068.

Conclusions

The threshold limits for fatigue cracking can be defined using linear elastic fracture mechanics (LEFM). Then the stress intensity factor should be used even for evaluating the influence of corrosion pits on fatigue such materials.

References

[1] Černý M.: Damage and Localized Corrosion in Steam Turbines, Conference EAN 2014, Proceedings (2014)

[2] Černý M.: Localized Corrosion in Alloy Steel of Steam Turbines, Applied Mechanics and Materials, vol.827, Trans Tech Publications (2016) doi:10.428/www.scientific.net/AMM.827.177

[3] Černý M.: Modeling of Corrosion Fatigue Crack Initiation in Steel Alloy for Steam Turbines, Conference EAN 2016, Proceedings (2016)

[4] Černý M.J: Modeling of Corrosion Fatigue Damage in Steel Alloy for Steam Turbines, Conference EAN 2017, Proceedings (2017)

EAN 2018
56th conference on experimental stress analysis

Experimental Verification of Jet Engine Composite Inlet for Bird and Hail Stone Impact Resistance

DOUBRAVA Radek[1,a], OBERTHOR Martin[1,b], BĚLSKÝ Petr[1,c], DVOŘÁK Milan[2,d] and DOUBRAVA Karel[2,e]

[1]VZLÚ (Czech Aerospace Research Centre), Beranových 130, 19905, Prague 9 – Letnany, Czech Republic

[2]Czech Technical University in Prague, Faculty of Mechanical Engineering, Technicka 4, 16607 Prague 6, Czech Republic

[a]doubrava@vzlu.cz, [b]oberthor@vzlu.cz , [c]belsky@vzlu.cz, [d]Milan.Dvorak@fs.cvut, [e]Karel.Doubrava@fs.cvut.cz

Keywords: FEM, Bird, SPH, impact

Abstract. A simultaneous work between experiments and numerical simulations is necessary to well understand the behavior of a bird or hail during an impact and also the reaction of the impacted material and structure. A simulation methodology is developed and validated to certify the bird-strike resistance of composite air ducts designed for a new generation of jet training aircraft. The physical bird and hail impact tests were performed on real composite part. The results form numerical simulation were compared with test results. Numerical simulation was used also for test preparation and test rig design and optimization for simulation influence of surrounding aircraft structure.
The validated modelling procedure allows for the analysis of numerous bird-strike scenarios, improving the optimisation procedures for aircraft component design, and reducing the cost of development by reducing the need to manufacture test prototypes.

Introduction

Bird and hail strike is a major threat for aircraft structure, as it can lead to serious structural damage. The vulnerable parts of planes or helicopters are windshield, nose, fuselage panels, wing and empennage leading edges, rotor blades, fan blades and engines inlets. Therefore, the international certification regulation requires [1] all forward facing components to prove a certain level of bird strike resistance before they can be employed in an aircraft. A bird or hail impact experiment provides a direct method to examine the impact structure resistance.

The testing program was established to assist in the selection of a composite material and optimised lay-up from the point of view of energy absorption from bird and hail impacts on the air ducts of a new generation of jet training aircraft. The approach for development of a composite air duct for a new generation of jet training aircraft, shown in Figure 1.

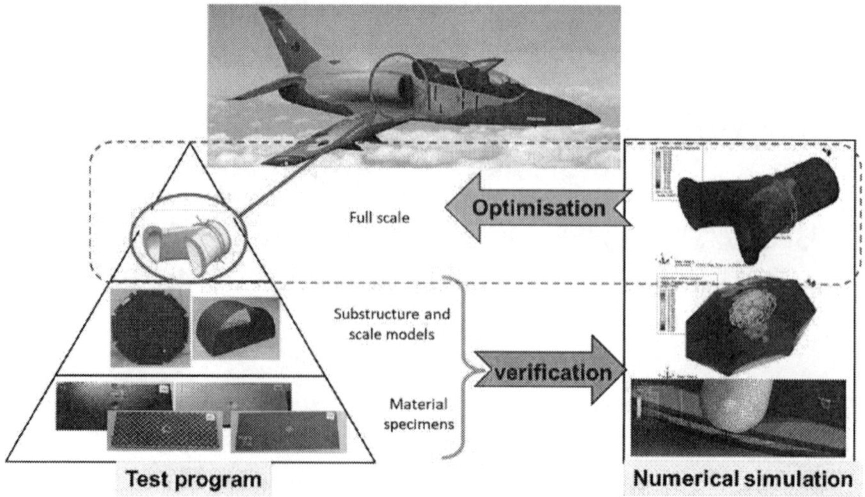

Fig. 1 Building block approach for testing and numerical simulation

The tests and analysis were initiated using low impact energy tests on small test specimens [2] according the procedures given in ASTM D7136M [3] and can be expanded to high-speed impact tests on flat test specimens to verify the manufacture and performance of a complexly shaped part. The flat test specimen used in the high-speed impact resistance verification tests was designed to provide confirmation of the performance of the selected composite material and to assist in the finite element modelling of the global structure. All numerical models were calibrated on the basis of experimental results gathered with an eye toward damage initiation and propagation behaviors. Figure 2 and 3 shows example of bird and hail stone strike experimental and numerical analysis on flat composite test specimens [4].

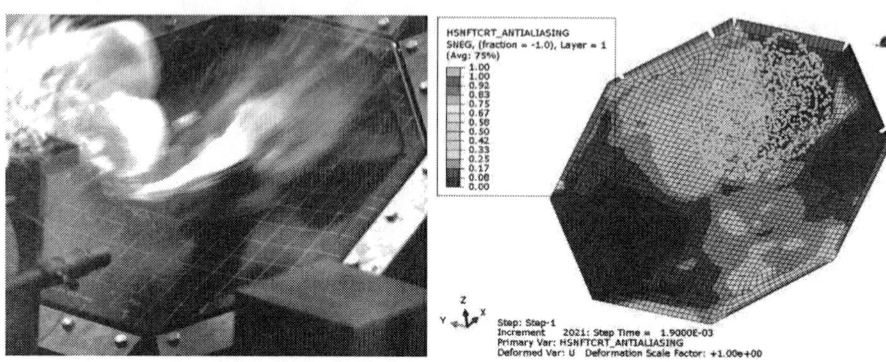

Fig.2 Bird strike test on flat test composite panel (left) and numerical simulation (right) [4]

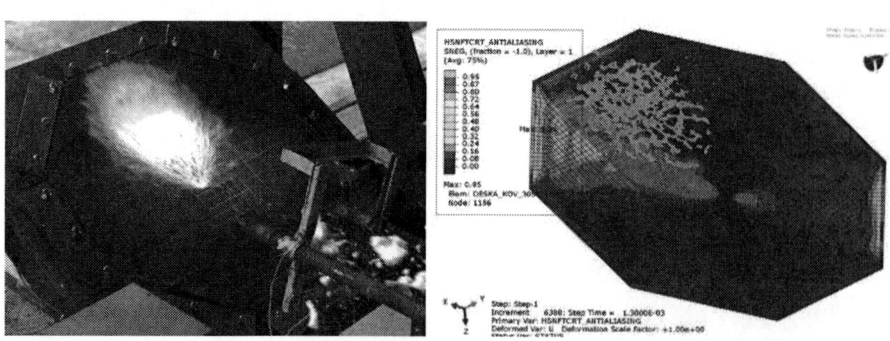

Fig.3 Hail stone impact test on flat test composite panel (left) and numerical simulation (right) [4]

Materials and methods

The physical bird and hail impact tests were performed at the Czech Aerospace Research Centre (VZLÚ) according airworthiness requirements [5] [6]. The required impact velocity was achieved using a properly pressurized air gun-type pressure vessel. The projectiles were accelerated by compressed air through the smooth borehole of the gun barrel up to its required velocity according to the specifications. Figure 4 shows the air gun test facilities.

Fig. 4 VZLU air gun used for high speed impact tests

Projectiles

The projectiles used in the impact tests is 1 kg (nominally) bird [5] and for hail simulation, ice ball 25 mm and 50 mm diameter [6] (see figure 5).

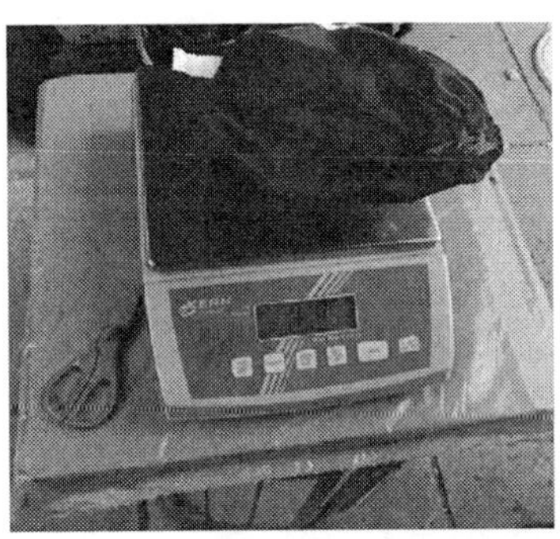

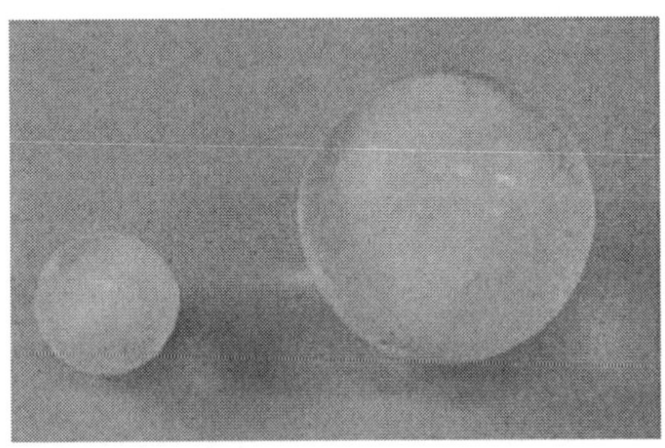

Fig.5 Weighting of bird in textile bag (left) for bird strike test and ice balls (25 mm and 50 mm) for hail strike tests

Test specimen and boundary conditions

The full-scale parts of inlet were manufactured in Aero Vodochody Aerospace company using IIcxply 8552/AGP193-PW prepregs [7]. The test stand was design and numerical verified from view point of real boundary condition simulation (stiffness of fuselage). Material properties for Hexply 8552/AGP193PW-37 [8] provided in the literature [7] were used for numerical analysis and are given in Table 1.

Tab. 1 Elastic material properties for Hexply 8552/AGP193PW-37 [8]

	E1 [MPa]	E2 [MPa]	Nu12 [1]	Nu13 [1]	Nu23 [1]	G12 [MPa]	G13 [MPa]	G23 [MPa]
Hexply 8552/AGP193-PW	60000	60000	0.046	0.32	0.32	5280	5280	3580

Where E is the Young's modulus, Nu is the Poisson's ratio, and G is the shear modulus.

The test stand was design and manufactured in Aero Vodochody Aerospace company. The design was and optimized from view point of real boundary condition (stiffness of fuselage) by numerical simulation in VZLÚ.

Figure 6 shows test stand and test arrangement for impact tests.

Fig.6 Test stand (left) and arrangement of impact tests (right)

Bird strike test procedure

The required impact speed for test of full scale test specimen was 594 km/h (165 m/s) for bird and 770 km/h (214 m/s) for hail stone [9]. The real speed achieved during the tests was 629 km/h for bird and 789 km/h for hail stone.

Figure 7 shows summary of measurement area for verification of structure behavior after the tests.

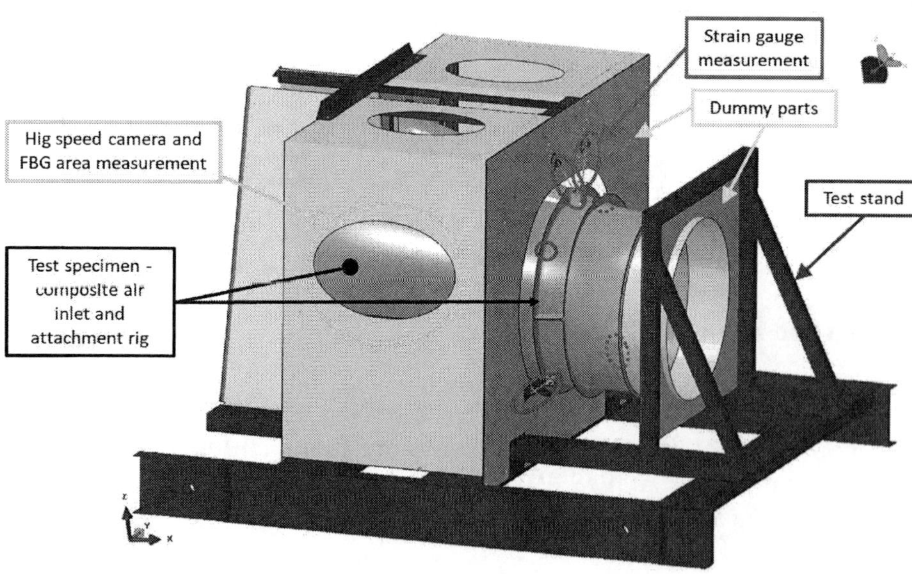

Fig.7 Shema of impact test assembly and measurement

Figure 8 shows example of high speed camera measurement from point of view of displacement in impacted area and analysis of hail stone projectile speed.

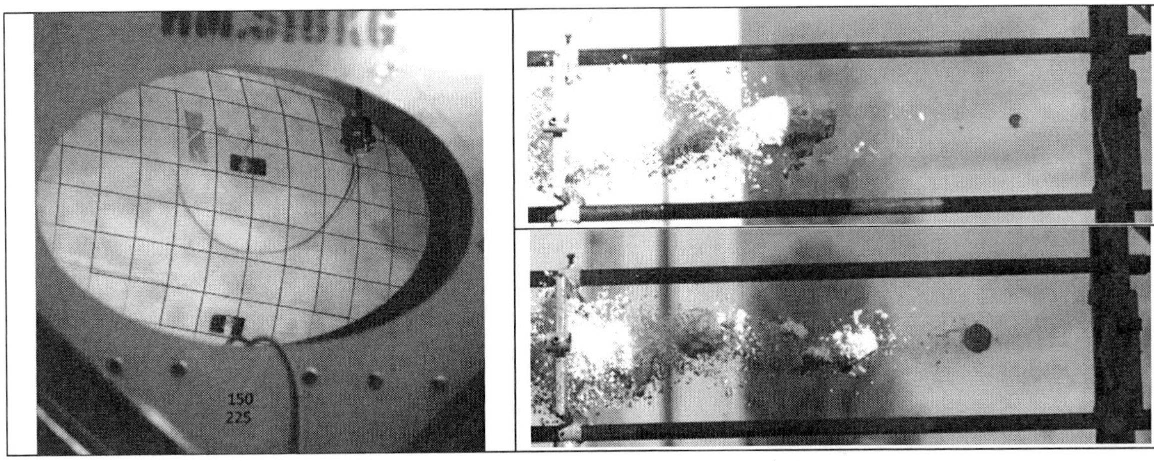

Fig.8 High speed camera pictures for deformation analysis in impacted area (left) and hail stone speed analysis (right upper - 25 mm ice ball; right bottom - 50 mm diameter ice ball)

The measurement by Fiber Bragg Grating (FBG) sensors and strain gauge measurement were performed by Czech technical university (CTU) [10]. Figure 9 shows measurement of strain on composite parts by FBG and attachment rods sensors by strain gauges.

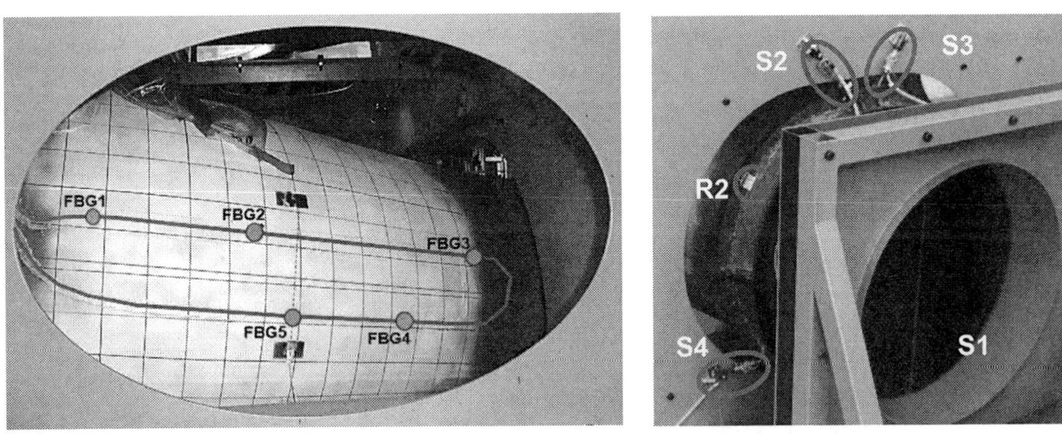

Fig.9 FBG (left) and strain gauge measurement (CTU) [10]

Numerical simulation

The numerical simulation was focused from point of view of high energy impact on birds strike test simulation.

A bird strike is a high-velocity impact in which materials with a huge difference in properties (a bird is a soft impacting material compared to the stiff material of the target aircraft body) come into contact with each other, resulting in nonlinear material behavior, high strain rates, and extremely large deformations. Nonlinear finite element software has the capability to predict the loads and deformations of both the bird projectile and the complex aircraft component being impacted within acceptable levels of accuracy. In high-velocity impacts, the pressure on the bird tissues severely exceeds their limits, causing the bird material to behave like a fluid [3].

Thus, the bird material can be described by the so-called "elastic-plastic hydrodynamic" material model. The hydrodynamic material model is defined in ABAQUS [11] by a tabulated equation of state using Hugoniot curves for water-like homogenized bird materials [12] [13]. The bird nodes were charged with an initial velocity, and a combination of tensile failure and shear failure criteria were used.

The geometry of the projectile (bird) was idealized as a 60-mm-long cylinder with two hemispheric ends having a radius of 60 mm. The bird geometry used in the simulation reflects the general geometry of the bird projectile used in the physical experiments. Figure 10 shows the geometry of the bird model, which was meshed by 10,770, C3D8R 8-node linear brick elements with conversion to particle elements (SPH – Smoothed Particle Hydrodynamics) [11]. The density of the bird material in the model for the defined volume was established to reflect the weight of the birds used in the physical tests.

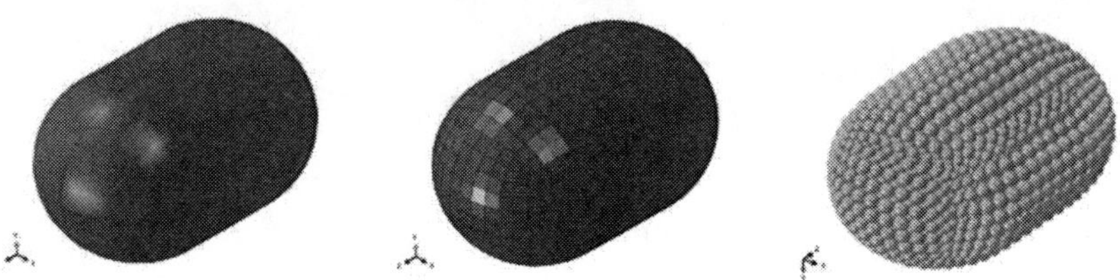

Fig. 10 Geometric model (left), FE mesh (centre), and SPH elements (right), of the idealised bird

To simulate the composite structure, 4-node shell elements (S4R) with a mesh size of 10 mm were used. The test rig was simulated by 4-node shell elements (S4R), truss elements (T3D2) and beam elements (B31).

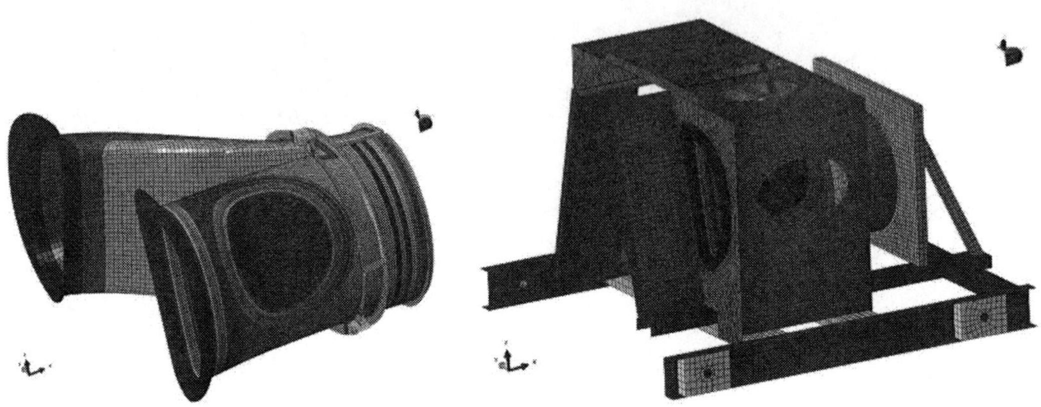

Fig.11 FE model of composite inlet (left) and FE model of test assembly (right)

The FE simulations were performed using the ABAQUS FE software package [11]. An explicit solver with double precision was used for the analysis. The general contact was used for contact analysis. From the point of view of the damage analysis of a composite material, the Hashin's damage material model was used [14].

Figure 11 shows the FE mesh of composite test specimen and full-scale test. Figure 12 shows examples of composite and test rig FE analysis.

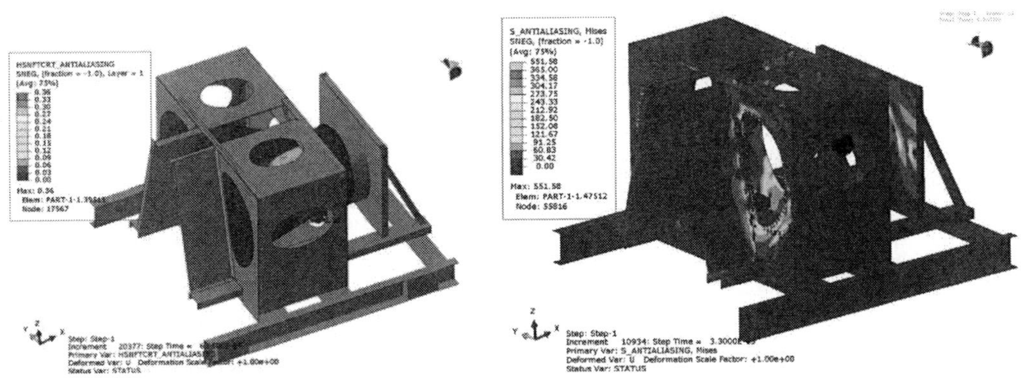

Fig. 12 FE analysis of composite parts (left) and test rig with dummy parts (right)

Results and discussion

Impact test results

The tests were analyzed as follows:

- The displacement in impact area from high-speed camera images.
- The strain from FBG sensors (CTU).
- The forces on rod attachments.
- Non-destructive testing (NDT) before and after impact.

Deformation was analyzed on the basis of the movement of raster lines and scaled on the basis of known dimensions and the camera position. This figure also summarizes the results from the NDT measurements of the demonstrator after testing, to correlate with numerical simulation.

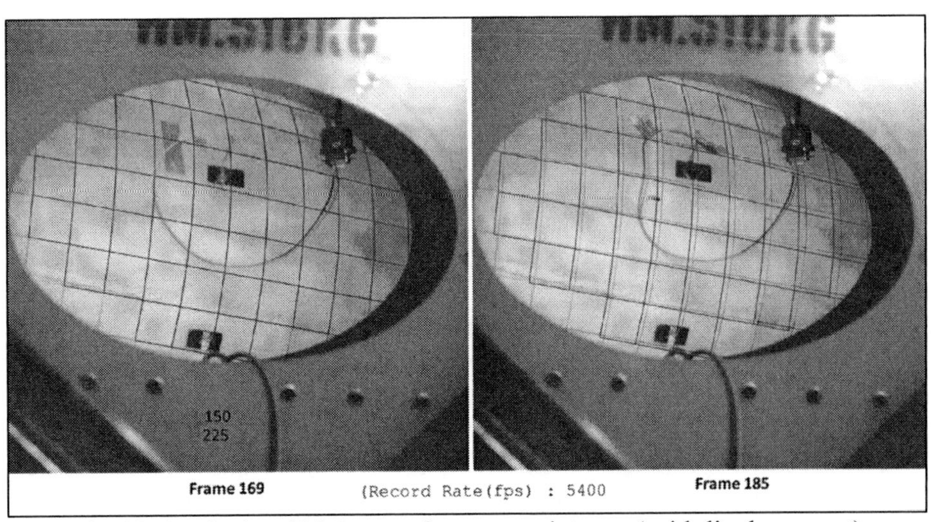

Fig.13 Analysis of high speed camera pictures (grid displacement)

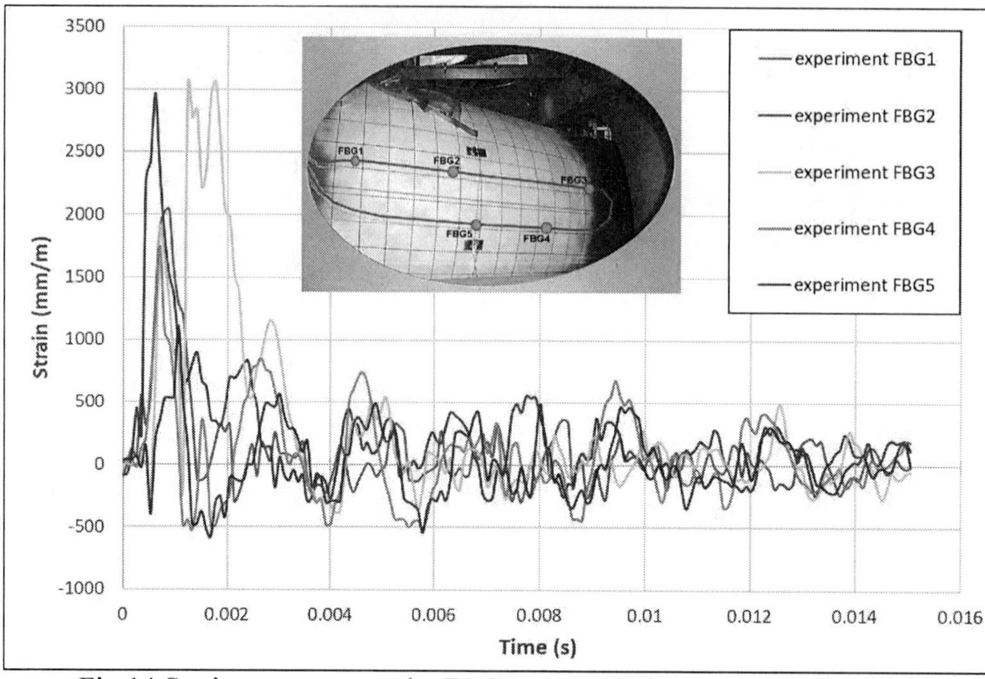

Fig.14 Strain measurement by FBG sensors during impact (CTU) [10]

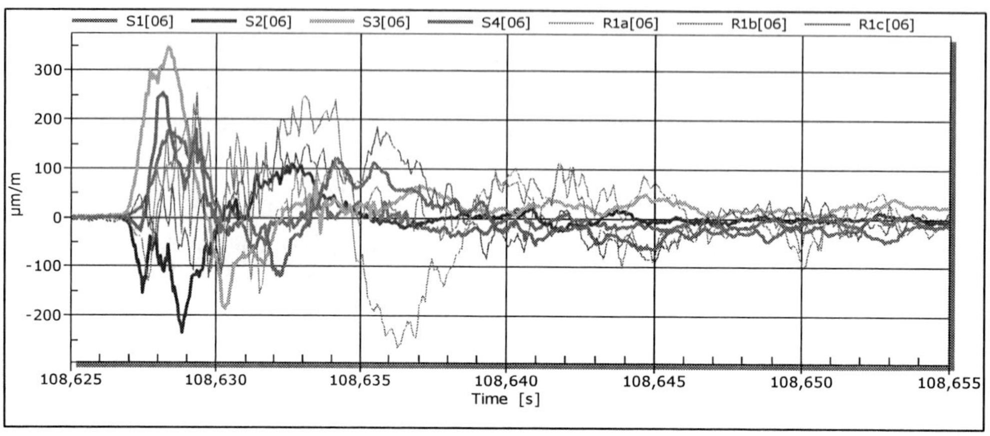

Fig.15 Strain gauge measurement during impact on attachment rods and attachment rig (CTU) (see figure 9) [10]

The whole composite air duct assembly was subjected to detailed non-destructive inspections before and after the impact tests. The NDT activities consisting of visual and ultrasonic inspections were performed by Level 2 qualified NDT staff.

During the inspections, various ultrasound technologies and methods were tested to find the most appropriate inspection procedure (figure 16). Both conventional A-scan inspection using single-element UT probe and B-scan or C-scan inspection using Phased Array technology were performed.

This ultrasonic equipment was used for the activities:

- Flaw detector Omniscan MX2 with UT module OMNI-M-PA16128
- Single-element probe Sonatest PRDT 2550 5MHz
- PA probe 5L64-A2 with contact wedge SA2-0L
- PA probe 2.25L128-I3 with contact wedge SI3-0L
- Manual PA scanner RollerFORM 5MHz

Fig.16 NDT measurement: a) Sonatest PRDT2550 probe; b) PA 5L64 probe; c) PA probe 2.25L128 and d) RolerForm 5 MHz.

Comparison of results

The results from the numerical simulation were compared with the experiment results for bird strike test.

Figure 17 shows the results of the qualitative measurement of the maximum displacement of composite inlet. The measurement was performed for maximal displacement in the measurement are corresponding to the time impact 3 ms.

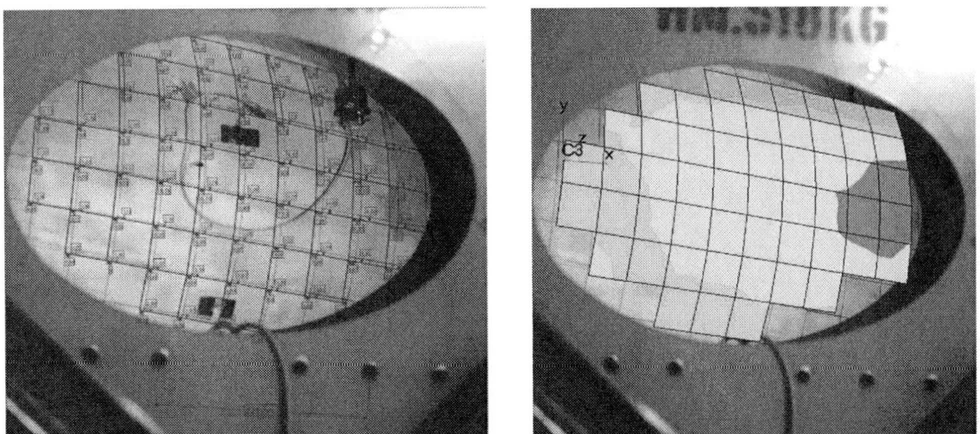

Fig.17 Analysis of displacement on the base of grid movement from high-speed camera pictures

Figure 18 shows comparison of displacement measurement from experiment and result of numerical simulation.

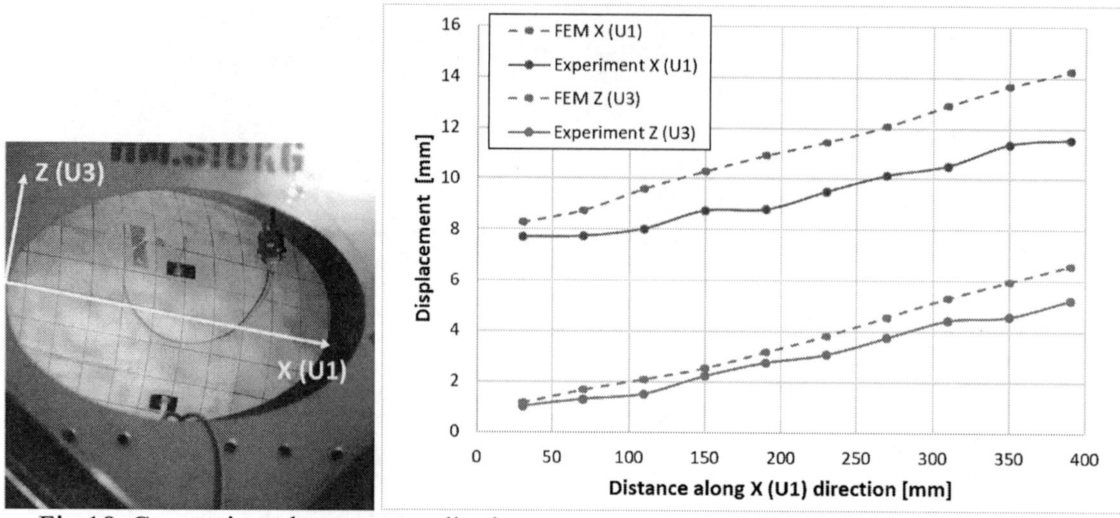

Fig.18 Comparison between qualitative measurement of displacement from high speed camera picture and numerical simulation for time impact 3 ms.

The differences between experiment and simulation is more based resolution of pictures from high speed camera and distortion by perspective.

Figure 19 shows comparison between experiment and numerical simulation on strain gauge measurement on attachment rods.

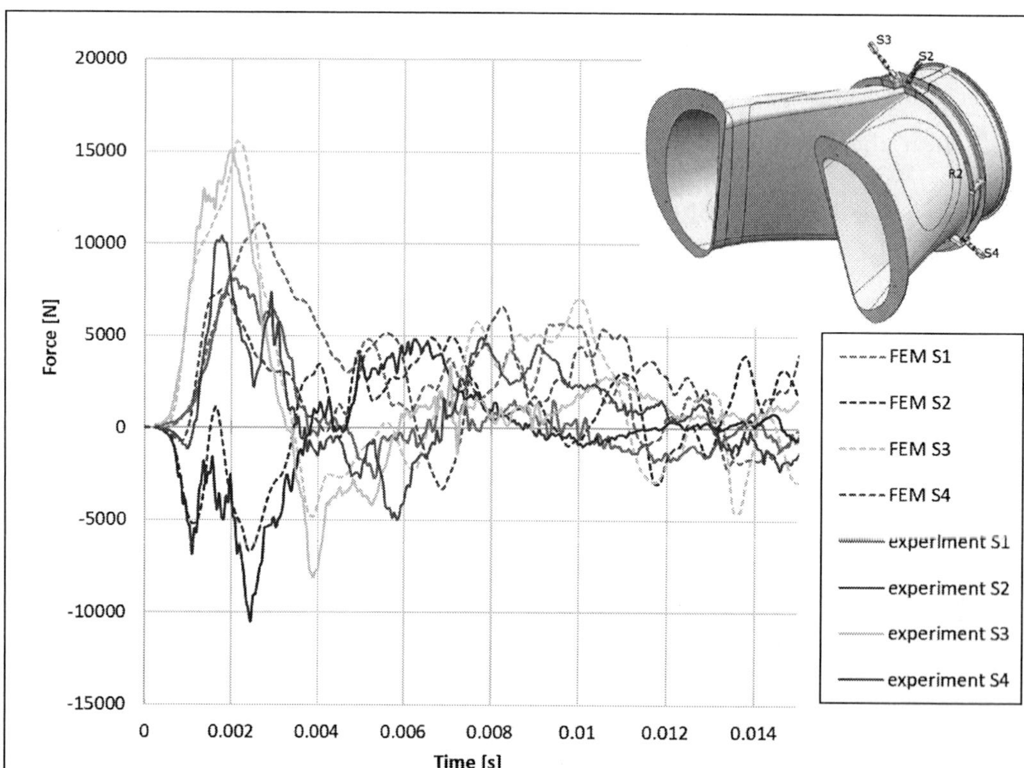

Fig.19 Comparison between experiment measurement of attachment rods load during impact and numerical simulation results. The results were compare for maximal peak of load.

The differences between experiment and simulation form figure 19 is more based on unknown or difficult defined boundary condition e.g. pretension, friction, damping etc.

Figure 20 to 24 shows comparison between results from strain measurement on FBG sensors and numerical simulation. The data used for comparison not filtered.

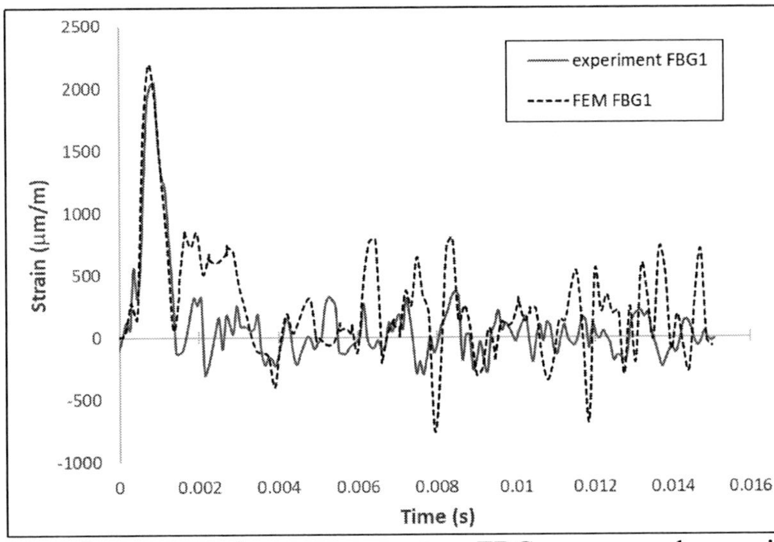

Fig.20 comparison between strain measurement on FBG sensors and numerical simulation for sensors FBG1 (see figure 9)

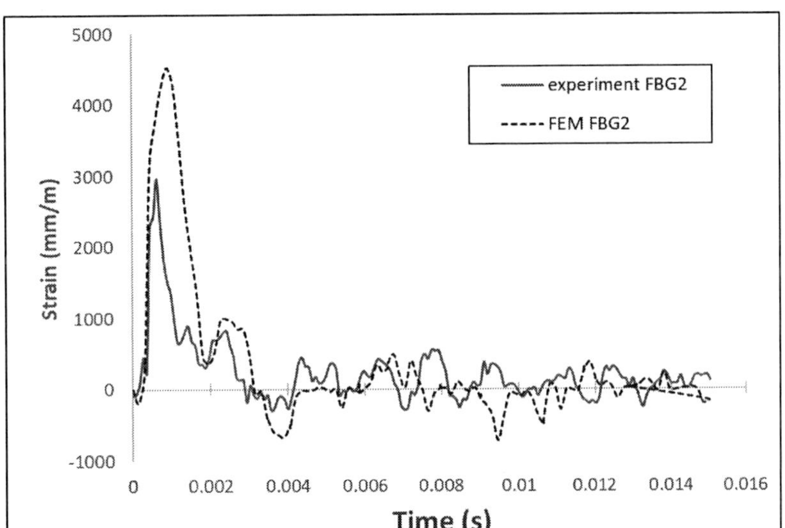

Fig.21 comparison between strain measurement on FBG sensors and numerical simulation for sensors FBG2 (see figure 9)

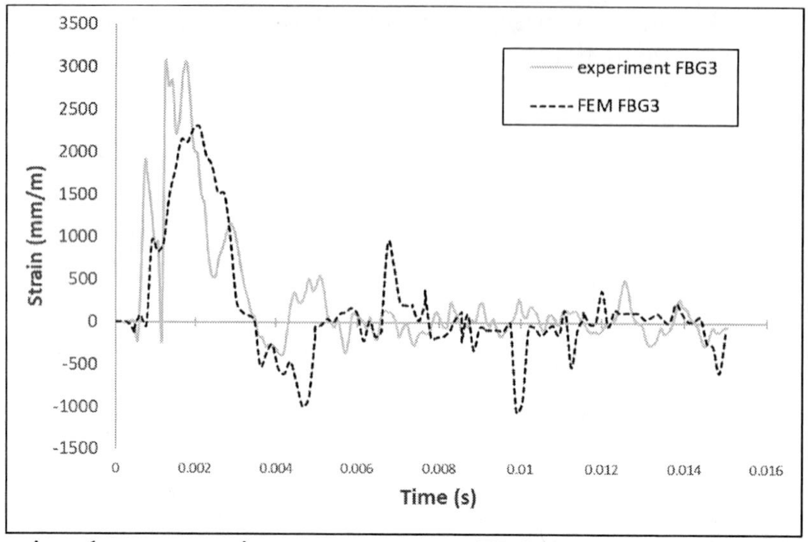

Fig.22 comparison between strain measurement on FBG sensors and numerical simulation for sensors FBG3 (see figure 9)

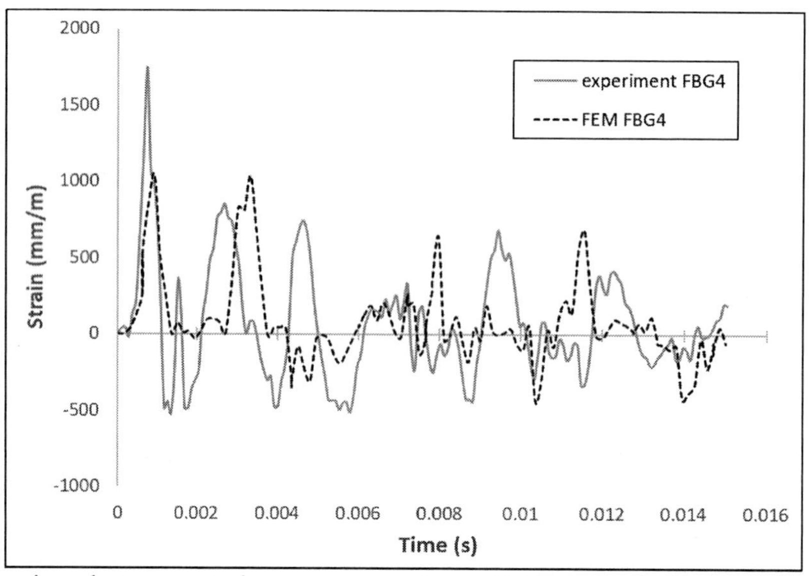

Fig.23 comparison between strain measurement on FBG sensors and numerical simulation for sensors FBG4 (see figure 9)

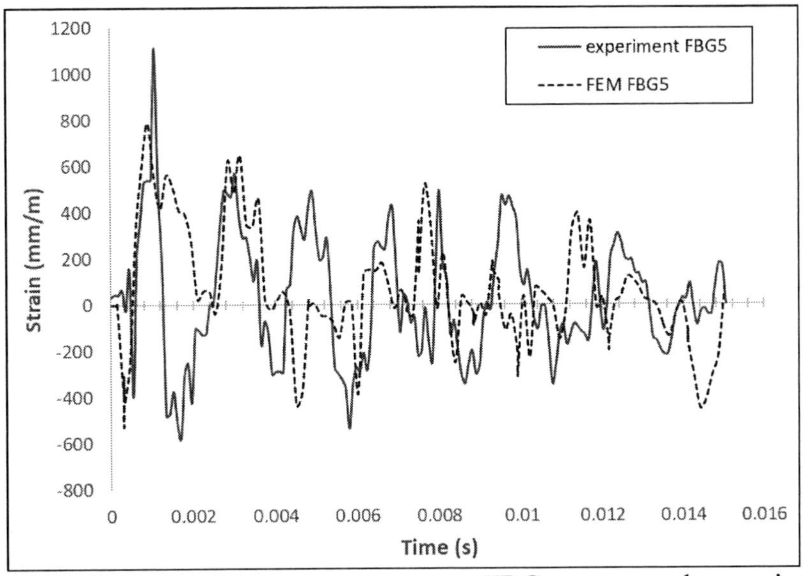

Fig.24 comparison between strain measurement on FBG sensors and numerical simulation for sensors FBG5 (see figure 9)

The results from FBG sensors shows very good correlation with results from numerical simulation especially in sensors near the impact point (sensors (FBG1, FBG2 and FBG5).

The NDT inspection confirm that no damage observed before and after the tests.

Conclusions

The comparison between test and simulation shows good harmony in prediction of structure behavior. The result of tests confirms impact resistance of proposed composite structure design for real service operation according airworthiness requirements.

This work was supported by the Technology Agency of the Czech Republic (TA ČR) TE02000032 / Advanced Aerostructures Research Centre project.

References

[1] Cleary EC, Dolbeer RA, Wright SE. Wildlife Strikes to Civil Aircraft in the United States 1990-2005. Federal Aviation Administration National Wildlife Strike Database, 2005.

[2] Batra RC, et al. Damage and failure in low energy impact of fiber-reinforced polymeric composite laminates. Composite structures, 2012, Vol. 94(2), pp. 540-547

[3] ASTM D7136M – 07 Standard Test Method for Measuring the Damage Resistance of a Fibre-Reinforced Polymer Matrix Composite to a Drop-Weight Impact Event.

[4] Doubrava, R., Raška, J., Oberthor, M: Verification of numerical analysis for L-39NG air inlet development by bird strike tests on flat composite panels, Transfer 29/2017, ISSN 1801 – 9315 (in Czech)

[5] ASTM F330 -10 Standard Test Method for Bird Impact Testing of Aerospace Transparent Enclosures.

[6] ASTM F320-05 Standard Test Method for Hail Impact Resistance of Aerospace Transparent Enclosures.

[7] HexPly® 8552 – Product Data http://www.aerospares.hu/files/hexcel/hexply_8552.pdf, March 2017

[8] Sánchez-Sáez S, Barbero E, Navarro C. Analysis of the dynamic flexural behaviour of composite beams at low temperature. Composites Science and Technology, 2007, vol. 67(11), pp. 2616-2632, https://doi.org/10.1016/j.compscitech.2006.12.002

[9] Defence Standard 00-970 Part 1 Section 3, Design and Airworthiness Requirements for Service Aircraft.

[10] Dvořák, M., Doubrava, K.: Test Results Collection (Demonstrator Test), report CTU 12105/17/29, 2017 (in Czech)

[11] ABAQUS 6.14 theory manual, <http://abaqus.software.polimi.it/v6.14/>

[12] Hedayati R, Sadighi M. Bird Strike: An Experimental, Theoretical and Numerical Investigation. Woodhead Publishing, ISBN: 978-0-08-100093-9, Elsevier 2015

[13] Heimbs S. Computational Methods for Bird Strike Simulations: A Review. Computers & Structures, 2011, Vol. 89(23) pp. 2093-2112.

[14] Hashin Z. Failure criteria for unidirectional fiber composites. ASME Journal of Applied Mechanics, 1981, Vol. 47, pp. 329-334.

EAN 2018
56th conference on experimental stress analysis

Axle Gearbox Hinge as a Diagnostic Member for Rail Vehicle Drive

DUB Martin[1,a], KOLÁŘ Josef[2,b] and LOPOT František[1,c]

[1]Czech Technical University in Prague, Faculty of Mechanical Engineering, Department of Designing and Machine Components, Technická 4, 166 07, Prague 6, Czech republic

[2]Czech Technical University in Prague, Faculty of Mechanical Engineering, Department of Automotive, Combustion Engine and Railway Engineering, Technická 4, 166 07, Prague 6, Czech republic

[a]martin.dub@fs.cvut.cz, [b]josef.kolar@fs.cvut.cz, [c]flopot@seznam.cz

Keywords: rail vehicle; drive; simulation; dynamics; diagnostics; axle gearbox

Abstract. Railway transport offers high travel comfort, shipping speed and efficiency. Due to the high speed and large number of transported persons, it is necessary to ensure the safety in the best possible way. Apart from the legislation and regular service, the continuous monitoring during the operation is carried out. Two basic types of the drive are usually used for the public transport – partly and fully sprung drive. This contribution deals with the partly sprung drive, exactly with the possibilities of the usage of the axle gearbox hinge as a diagnostic member for the bogie group including the drive.

Introduction

Railway transport offers high travel comfort, shipping speed and efficiency. Due to the high speed and large number of transported persons, it is necessary to ensure the safety in the best possible way. Apart from the legislation and regular service, the continuous monitoring during the operation is carried out. Two basic types of the drive are usually used for the public transport – partly and fully sprung drive. This contribution deals with the partly sprung drive, exactly with the possibilities of the usage of the axle gearbox hinge as a diagnostic member for the bogie group including the drive, [1].

Partly sprung drive

Partly sprung drive consists of the traction engine (usually asynchronous motor) and transmission with axle gearbox. The engine is embedded on the bogie frame and can be taken as a suspended mass. The big gear of the axle gearbox is embedded directly on the wheelset axle. The reaction force caused by the input and output torque inequality is captured by means of the hinge, which is connected to the bogie frame. Thus, the axle gearbox becomes the partly suspended mass. Partly sprung drive is simpler and more economical than the fully sprung drive. On the other hand, it is less suitable on rail tracks with worse quality due to the larger amount of non-suspended mass. Thus, it is important to analyse the dynamic behaviour of such drive. Partly sprung drive is usually used in combination with one-stage gearbox, two-stage gearbox or the gearbox with inserted gear. One-stage gearbox has simple design and also small width, the schema can be seen in Fig. 1. It is used for example in Prague metro *M1*, Fig. 2.

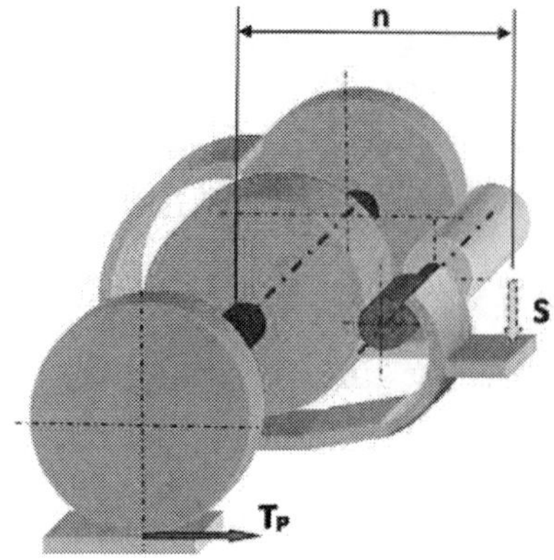

Fig.18 Schema of the partly sprung drive with one-stage gearbox, [1]

Fig.19 Prague metro M1 drive, [2]

The gearbox with inserted gear is used when larger axis distance is required due to the size of the engine. This type can be found in the high-speed train *AGV* drive, Fig. 3. For higher ratios two-stage gearbox is used, e.g. Prague tram *Škoda 14T*, Fig. 4.

Fig.20 AGV high-speed train, [3]

Fig.21 Škoda 14T tram, [4]

The gearbox hinge is very important and interesting part from the dynamic point of view. It is the link between the spatial and torsional dynamics. Previously, it was successfully used for the measurement of various quantities such as the input torque [1, 5, 6]. During these measurements the hinge served as a sensor and an idea to use the hinge as a diagnostic member for rail vehicles with partly sprung drive originated. This diagnostic member should monitor the behaviour of the bogie during the operation of the vehicle and the in-time signal processing should reveal the changes in the bogie group structure, such as damages of the suspension. The hinge could be used also for the predictive service of the vehicle drive. The reaction force in the hinge of the one-stage axle gearbox (Fig. 1) is defined as

$$S = \frac{M(i_c + 1) - J_p\ddot{\psi}}{n}, \tag{1}$$

where M is the input torque, i_c is the gearbox ratio, J_p is the gearbox moment of inertia, ψ is the angle of the gearbox sway and n is the distance between the hinge and wheelset axle.

Dynamic simulation model

To prove and verify above mentioned proposals the dynamic simulation model of the partly sprung drive is assembled. It is a non-linear dynamic model of a single wheelset, which is mounted on the roller test rig [7, 8, 9, 10]. This test rig enables torque loading of the drive and transversal and vertical excitation of the wheelset. The parametric model describes both spatial and torsional dynamics of the system and includes eighteen degrees of freedom.

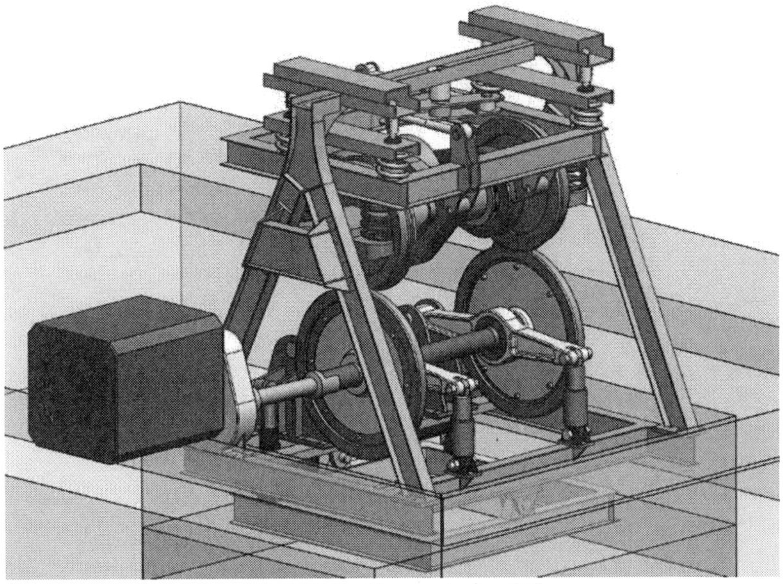

Fig.22 Roller test rig, [1]

Spatial Dynamics. Spatial dynamics of the wheelset and the bogie frame (see Fig. 6) is modeled with consideration of all degrees of freedom and geometrically non-linear. Free body diagrams are defined in general positions, thus all the contributions of forces and moments are included in the equations and higher order expressions are not neglected. Detail description was introduced in [11].

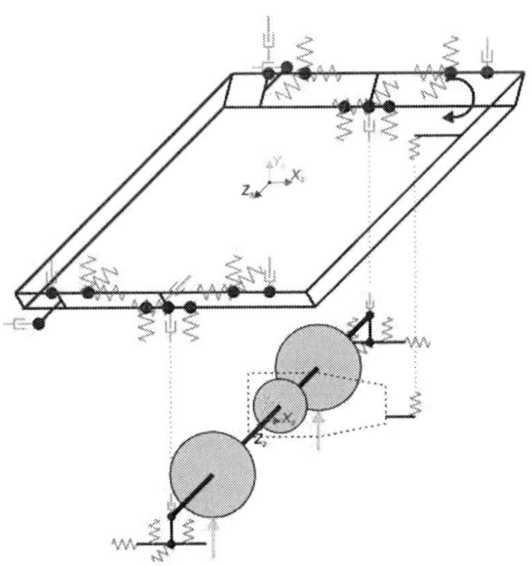

Fig.23 Spatial dynamics – schema [1]

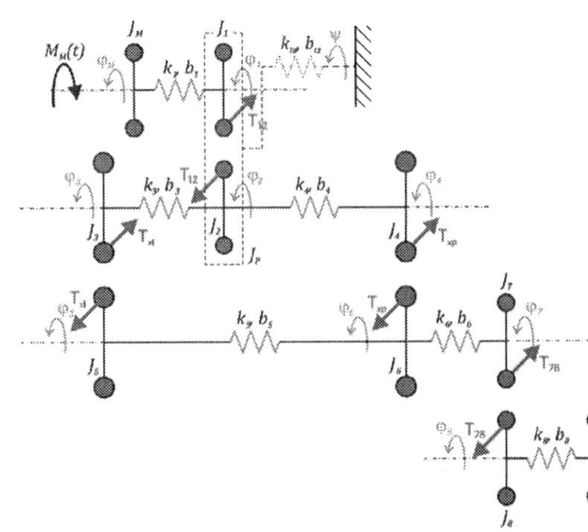

Fig.24 Torsional dynamics – schema, [1]

Torsional Dynamics. The torsional system (see Fig. 7) is consisted of the asynchronous motor, gear coupling and one-stage axle gearbox. The gear ratio is realised by the pinion and big gear, that is mounted to the wheelset. The gearbox is not in the geometrical centre of the wheelset so the stiffness of the left and right part of the axis is different. The driving forces are transmitted from the wheels to the roller shaft, which is coupled through the propeller shaft to the one-stage reduction gearbox and loading motor. The axle gearbox is connected with the bogie frame by means of the hinge, see Fig. 8. Thus the hinge is influenced both by the spatial and torsional dynamics.

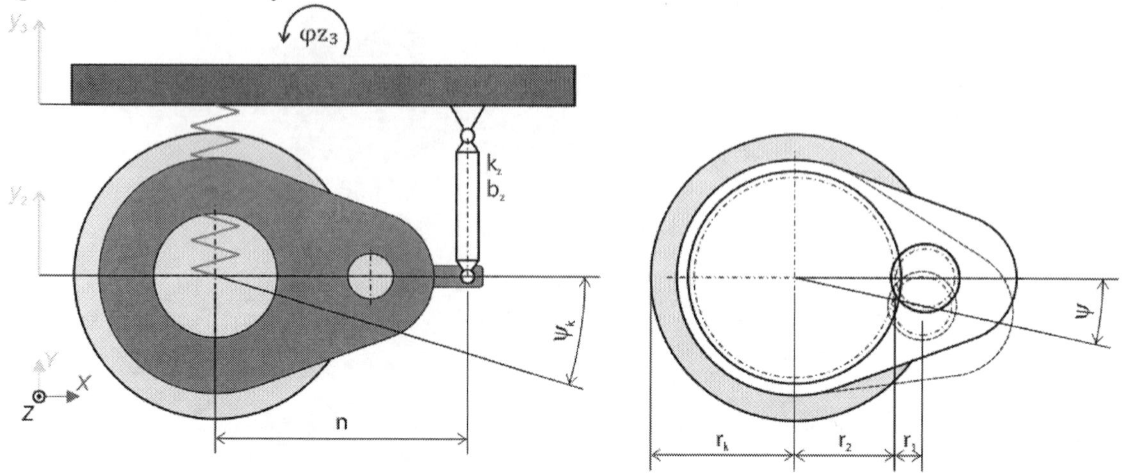

Fig.25 The schematics for the gearbox sway, [1]

The total angle of sway ψ is given as a combination of the sway due to the input torque ψ_m and the sway due to the kinematic excitation ψ_k (see Fig. 8)

$$\psi = \psi_m + \psi_k = \frac{\varphi_1 - \varphi_2 i_c}{1 + i_c} + \frac{y_2 - y_3}{n} - \varphi_{z3}, \tag{2}$$

where φ_1, φ_2 and φ_{z3} are angles of rotation of the pinion, big gear and the bogie frame respectively, y_2 and y_3 are the vertical deflections of the wheelset and the bogie frame. This movement also influences the forces in gear meshing, the tangential force T_{12} can be derived from the dynamic equation of equilibrium for the gearbox

$$T_{12} = \frac{J_P\ddot{\psi} + b_z n^2 \dot{\psi} + k_z n^2 \psi}{r_1 + r_2}, \tag{3}$$

where J_p is the moment of inertia of the axle gearbox, ψ is the angle of rotation, k_z, b_z is the stiffness and the damping of the hinge, r_1, r_2 are radii of the pinion and big gear, n is the distance between the hinge and the axle.

System of equations. The final set of motion equations was assembled in program *Maple* and exported into program *Matlab*. Eighteen second-order differential equations are solved there by means of *Matlab* function for solving moderately stiff ordinary differential equations *ode23t*. The dynamic model is fully parametric and can be used for testing the dynamic response with various parameters of stiffness, damping, dimensions etc.

Testing and results

The dynamic model was tested in various test regimes – torsional loading of the drive, vertical oscillations of the wheelset (simulation of the drive on the rough rail track) and the lateral movement of the wheelset, [1, 11]. For testing diagnostic abilities of the hinge, various test simulations are performed and its response – axial reaction force behaviour is analysed. In this contribution, the changes/damages in primary and secondary suspension are carried out. Exactly, the stiffness of one primary or secondary spring is reduced. The first test case is performed during the drive on the rough rail track. The roughness is simulated by the sum of sine functions

$$h_i(t) = A_1 \sin 2\pi f_1 t + A_2 \sin 2\pi f_2 t + A_3 \sin 2\pi f_3 t, \tag{4}$$

where the amplitudes A_i are 2 mm, 3 mm and 1 mm, the frequencies are dependent on the driving speed corresponding to the wave lengths 50 m, 20 m and 5 m and they are listed in the Table 1.

Tab.1 Frequencies of the rail track roughness for the wheelset excitation

v [km/h]	f_1 [Hz]	f_2 [Hz]	f_3 [Hz]
54	0.3	0.75	3
90	0.5	1.25	5

During the following test, one of four primary springs is taken as damaged, it is simulated with the 10% stiffness. The test is performed with drive speed 54 km/h and 90 km/h. To compare obtained results it is convenient to transform the time behaviour of the hinge force to the frequency domain. The results for both drive speeds can be seen in Fig. 9 and Fig. 10.

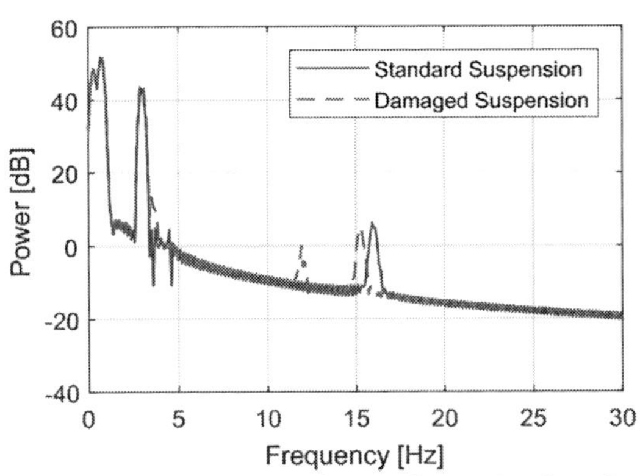

Fig.26 Signal comparison of standard and damaged primary suspension during drive on rough rail track, v = 54 km/h

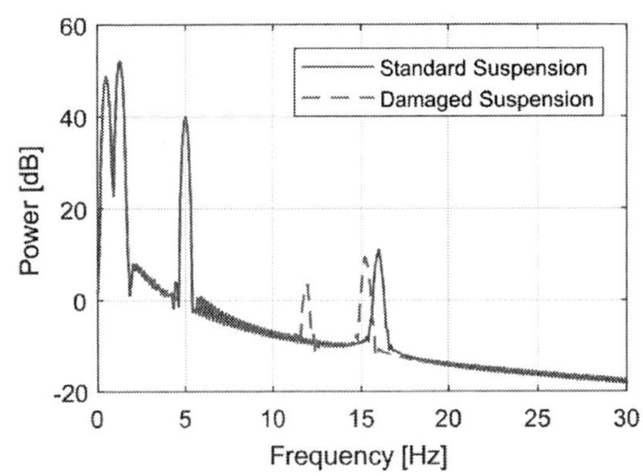

Fig.27 Signal comparison of standard and damaged primary suspension during drive on rough rail track, v = 90 km/h

There is appreciable difference between the spectrum with standard and damaged spring. In both cases, there can be seen three excitation frequencies from the roughness of the rail track. If the spring is damaged, there can be seen the change in the frequency spectrum – one frequency round 16 Hz moves to the lower frequencies and furthermore new one appears at frequency round 12 Hz. Both frequencies correspond to the natural frequencies of the bogie frame. The frequency round 16 Hz is related to the vertical oscillations of the bogie frame and it could be simplified calculated by means of following formula

$$f = \frac{1}{2\pi}\sqrt{\frac{k_p + k_s}{m}}, \tag{5}$$

where k_p and k_s is the stiffness of the primary and secondary suspension respectively and m is the mass of the bogie frame. Similar settings are set during the test with damaged secondary suspension, which is again simulated with 10% stiffness. Results for both drive speeds can be seen in Fig. 11 and Fig. 12. There can be seen equivalent behaviour of the hinge response in frequency domain, even the changes are not so significant.

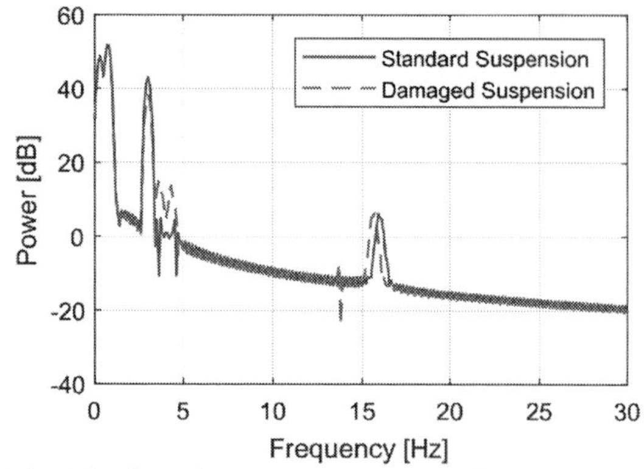

Fig.28 Signal comparison of standard and damaged secondary suspension during drive on rough rail track, v = 54 km/h

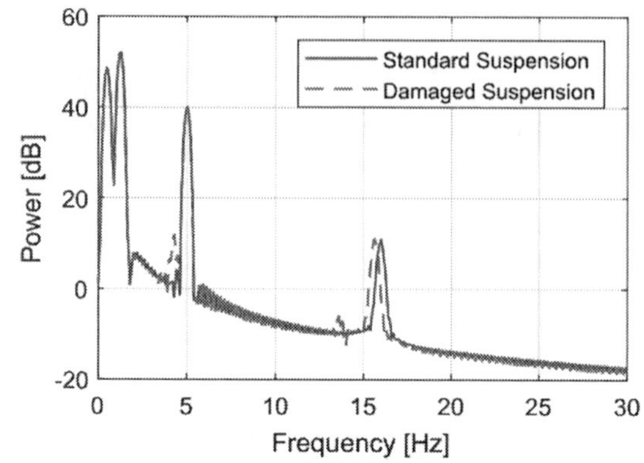

Fig.29 Signal comparison of standard and damaged secondary suspension during drive on rough rail track, v = 90 km/h

In the third test, there is investigated the response during lateral movement of the wheelset during the drive speed 90 km/h. Both primary suspension and secondary suspension spring stiffness is reduced to the 90%. The changes in the stiffness is much smaller than in previous cases. In figures Fig. 13 and Fig. 14, there can be seen significant changes in the response spectra. Both in the frequency and in the power magnitude.

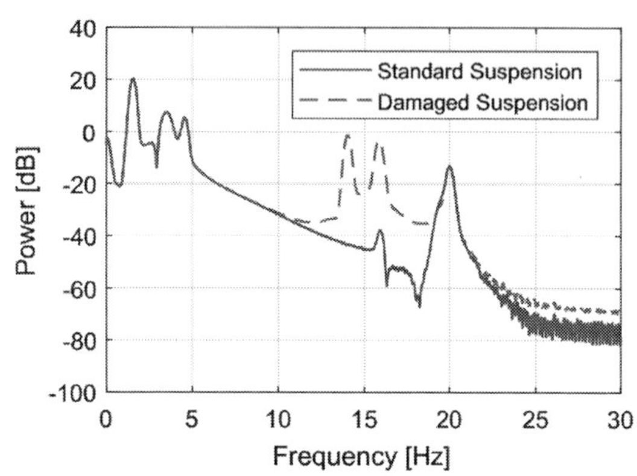

Fig.30 Signal comparison of standard and damaged primary suspension during lateral movement of the wheelset, v = 90 km/h

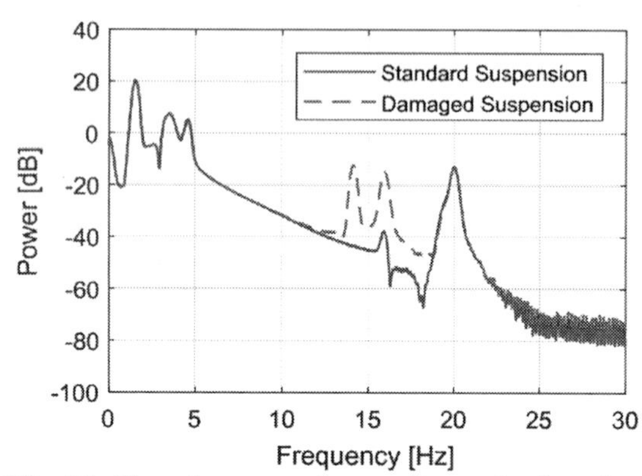

Fig.31 Signal comparison of standard and damaged secondary suspension during lateral movement of the wheelset, v = 90 km/h

Conclusions

In this contribution, there was introduced the problematics of drives especially the partly sprung drive with axle gearbox hinge. The specific partly sprung drive was described by means of the dynamic model. It was used for testing of the drive in various test regimes with different parameters of the suspension. That should simulate possible damages of the suspension springs during the operation of the vehicle and their in-time detection. From above mentioned facts and results follows that the axle gearbox hinge in various partly sprung drives can be conveniently used as and diagnostic component. It is also important aspect that the hinge can be used for diagnostics without considerable changes of its design. The diagnostic hinge could be used also for data acquisition, which is helpful in designing new drives and bogies. The hinge as a diagnostic component should serve also for predictive service. That can significantly help to prevent damages of drives and bogies by means of early warning.

Acknowledgement

This work was supported by *The Technology Agency of the Czech Republic project No: TE01020038 "Competence Centre of Railways Vehicles"*.

References

[1] M. Dub, Design of Autodiagnostic Method for Rail Vehicles Drives, Praha, 2016, Dissertation Thesis, Czech Technical University in Prague, Faculty of Mechanical Engineering.

[2] Information on: www.metroweb.cz

[3] Information on: www.alstom.com

[4] Information on: www.skoda.cz

[5] M. Dub, J. Kolář, F. Lopot, V. Dynybyl and O. Berka, Experimental analysis of stress state and motion of tram gearbox hinge, in: EAN 2014 - 52nd International Conference on Experimental Stress Analysis, 2014.

[6] M. Dub, J. Kolář, F. Lopot, O. Berka and V. Dynybyl, Loading of the Axle Gearbox for Metro During Drive Simulation, in: EAN 2016 - 54nd International Conference on Experimental Stress Analysis, 2016.

[7] K. Petr, V. Dynybyl and J. Chmelař, Test Rig for Gearboxes of Rail Vehicles, in: Modern Methods of Construction Design, 54th International Conference of Machine Design Departments. Hejnice, Cham: Springer International Publishing AG. 2014, p. 169-175, Lecture Notes in Mechanical Engineering, ISSN 2195-4356, ISBN 978-3-319-05203-8.

[8] K. Petr and V. Dynybyl, Methods of testing gearboxes of rail vehicles. in: Applied Mechanics and Materials, 51st Conference on Experimental Stress Analysis, Litoměřice, 11.06.2013 - 13.06.2013, Uetikon-Zurich: Trans Tech Publications, 2014, s. 277-282, ISSN 1660-9336, ISBN 978-3-03785-977-3.

[9] S. Myamlin, J. Kalivoda and L. Neduzha, Testing of Railway Vehicles Using Roller Rigs, Proc. of 10th Intern. Scientific Conf., Transbaltica 2017, Procedia Engineering. ISSN 1877-7058, Vilnius, 2017, Vol. 187, p. 688-695.

[10] S. Myamlin, S. Dailidka and L. Neduzha, Mathematical Modeling of a Cargo Locomotive, Proc. of 16th Intern. Conf., Transport Means. 2012, ISSN 1822-296 X, Kaunas, 2012, p. 310-312.

[11] M. Dub, J. Kolář, F. Lopot and V. Dynybyl, Dynamic Simulation of a Driving Wheelset on the Roller Test Rig, in: The Third International Conference on Railway Technology: Research, Development and Maintenance, Civil-Comp Press, 2016, paper 260.

EAN 2018
56th conference on
experimental stress analysis

Non-destructive Approach to Testing of Concrete Damaged by Fire with the Usage of Fast Fourier Transformation and Wavelet Transformation of Signal Recorded by Impact-Echo Method

DVOŘÁK Richard[1,a], CIKRLE Petr[2,b] and KAREL Ondřej[2,c]

[1] Institute of Physics, Brno University of Technology, Czech Republic

[2] Institute of Building Testing, Brno University of Technology, Czech Republic

[a]dvorak.r1@fce.vutbr.cz, [b]cikrle.p@vutbr.cz, [c]ondrej.karel@vutbr.cz

Keywords: NDT, Non-destructive testing, high-temperature degradation, bridges, concrete

Abstract. This publication presents results of testing of specimens taken from highway bridge, which was damaged by fire originated from car accident. Test specimens were cut from core bores removed from fire damaged areas and areas not affected by fire.

Specimens were tested by non-destructive acoustic impact-echo method and ultrasonic pulse velocity method. The measured values were supplemented by destructive tests.

The intention of this assessment is to compare different interpretation approaches to these acoustic methods. Impact-echo (IE) is method based on exciting the test specimen by mechanical impact [1]. As the mechanical wave spread through the material it is absorbed, reflected or refracted on each interface of materials of different acoustic impedance and by external boundaries. Each interface works as new source of vibrations, which are recorded by attached piezoelectric sensor. The recorded signal is then transformed by Fourier Fast transformation (FFT) [2] to frequency spectrum.

In this publication two different approaches of interpretation of impact-echo measurements are compared. Locating the dominant resonance frequency in frequency spectrum obtained by FFT and by continuous wavelet transformation (CWT) of signal.

Continuous wavelet transformation of signal helps understand which certain frequency occurred at specific point in time of recorded signal [3]. Presented paper focus on comparison of these signal imaging techniques.

Introduction

Concrete is the most used structural material in civil engineering, thus majority diagnostic approaches aim to study and evaluate the condition of concrete structures, wear state of structure, flaws, and defects. These approaches can be divided to destructive, semi-destructive and non-destructive methods of testing [4].

Diagnostics of concrete structure tends to reveal the condition of structure, analyze the measured data and designing a suitable economical retrofitting process, depending on the scale of damage. There is a demand to keep structure intact during the testing and without notable intervention with ongoing operation of construction. Therefore, the non-destructive methods of testing are most wanted for this assessment.

In this publication a diagnostics of highway bridge is being discussed and evaluated. The origin of bridge can't be mentioned due to car accident investigation and ongoing research by Institute of Building Testing. The bridge was built in 2015 so it is a new structure. The damaged area consists of three concrete pier columns, cross beam, and bottom side of bridge superstructure. Fire emerged from car accident caused the damage to concrete elements.

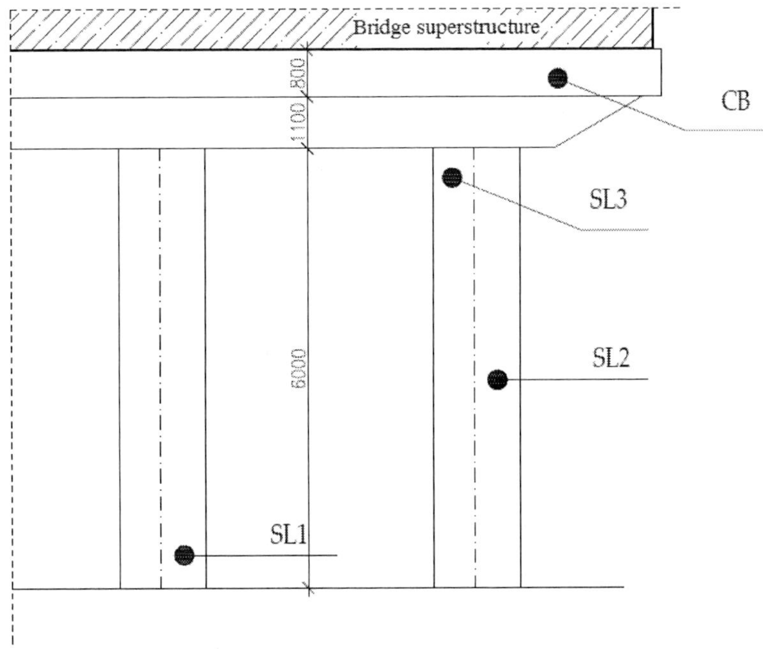

Fig.6 Scheme of damaged bridge and origin of each test specimen

In the presented publication four test specimens removed from bottom (SL1), middle (SL2) and top part of damaged columns (SL3) and from cross beam (CB) right above the fire are assessed. The scheme of bridge is at Fig.6. The four core bores were cut into test beams of parameters stated at Tab.4.

Tab.4 Basic parameters of test specimens

Parameter of test specimens	Name of test specimen			
	SL1[1]	SL2	SL3	CB
Weight [kg]	0.551	0.491	0.687	0.619
Length [mm]	160.1	121.0	167.4	159.2
Width [mm]	39.1	42.2	42.4	41.5
Height [mm]	40.4	42.2	42.4	41.5
Density [$kg \cdot m^{-3}$]	2177	2280	2033	2260
Origin	Footing of column	Middle of column	Head of column	Cross beam

[1] Test specimen SL1 was removed from second column to serve as a reference specimen.

Used methods

To determine physical-mechanical properties of affected areas, a destructive method on removed core bores was conducted to obtain compressive strength [5] of core bores from reference and thermally affected areas.

Non-destructive testing by Impact-Echo (IE) method was done on test specimens removed from bridge columns and cross beam. IE method is complex non-destructive acoustic method, which can provide information about inner structure of tested element and presence of internal flaws, air voids or cracks. This ability is based on skill of observer, who can process the measured data and interpret them. In Impact-Echo measurements are most often used piezoelectric sensors, which produce analog signal. The signal in time domain is recorded by different kinds of recording devices, such as digital oscilloscope. The typical setup for IE measurement is in the Fig.7.

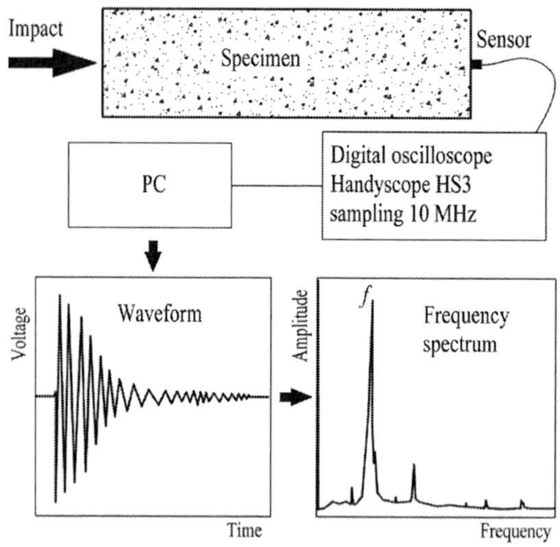

Fig.7 Typical setup for IE measurement [6]

Measurement by Impact echo was conducted on test specimens placed on rubber pads. As exciter of mechanical stress waves a spherical hammer of total mass 25.5 g was used. For recording of vibrations, a piezoelectric sensor MIDI 446s12 was used and for processing the signal was used a digital oscilloscope Handyscope HS3 with sampling speed of 10 MHz. For recording the longitudinal waves, receiver was placed in the front of test beam and exciter hit the test specimen at the opposite side [7].

The most common tool to interpret the measured signal recorded during Impact-Echo test is fast Fourier transformation (FFT) [2]. For example, the Fourier transform of the signal $f(t)$ is:

$$F(\omega) = \frac{1}{\sqrt{2\pi}} \int_{-\infty}^{\infty} f(t)\, e^{-i\omega t} dt \qquad (1)$$

This transform is a time-frequency representation as well as a function of time t_0 and frequency ω and provides an approximation to the frequency content of the signal over a small-time interval about time t_0. This approach is most widely used for signal processing in building vibro-acoustic diagnostics [1], [6], [8] as the dominant resonance frequency is most often assessed. The output of the FFT is frequency spectrum where dominant and harmonic frequencies are evaluated. We can describe a presence of defect or change in physical-mechanical properties by change of dominant frequency.

Continuous wavelet transformation (CWT) can be viewed as a filtering of the signal by a dilated version of mother wavelet $\psi(t)$. The wavelet can be viewed in time window with adjusted width and height, depending on scale – widening for long-scale (low-frequencies) information and narrowing for small-scale (high-frequency) content.

$$\Psi_{s,\tau}(t) = \frac{1}{\sqrt{|s|}}\Psi\left(\frac{t-\tau}{s}\right) \tag{2}$$

The CWT is a time-frequency, or more correctly a time-scale representation and produces a complex-valued surface. As the Gabriella Epasto and all mentioned [8], the CWT method of interpreting the signal can provide the dominant frequency as well as FFT, but in case of noised signal and heavily damaged material a CWT method can provide more parameters and a rapid detection of flaws and defects. In the presented paper a MATLAB R2017 software was used for calculating the spectrograms of IE signal.

Results of testing

Test specimens were tested in longitudinal way and obtained signal was processed by FFT and CWT. In the Fig.8 we can distinguish distribution of dominant frequency for each test specimen. Test specimen SL1 removed from footing of column have its resonance frequency at 8.77 kHz and test specimen shows visible marks of degradation by de-icing salts. The footings of other columns in the row have shown similar level of degradation distinguishable during in-situ inspection. The footings of other columns indicated also presence of salt degradation and freeze-thaw damage.

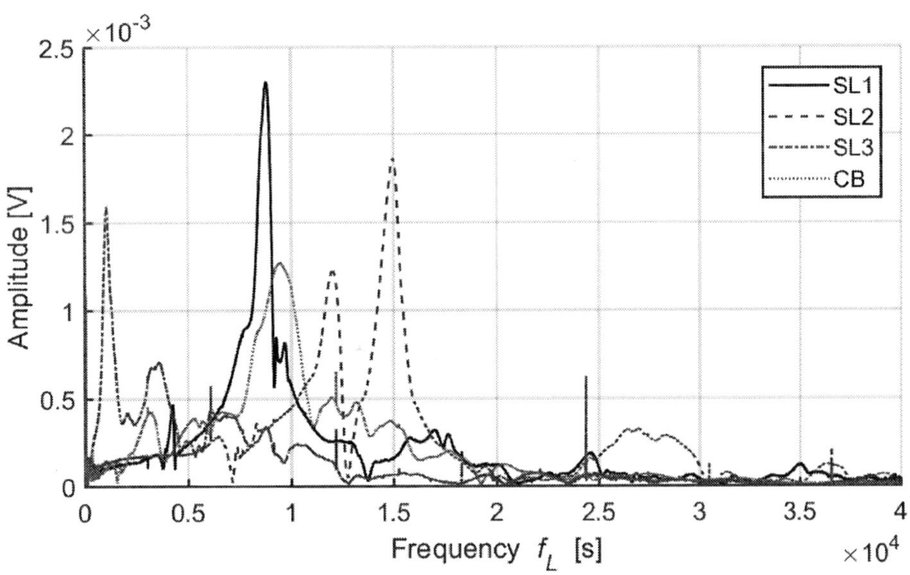

Fig.8 Change of longitudinal resonance frequency for each test specimen

Specimen SL2 removed from the middle of the column at height 2.5 m have resonance frequency 15.01 kHz and was the most preserved test specimen. Surface was black from smoke, but apart other test specimens SL2 remains its physical-mechanical properties at level, that we can describe as reference to other test specimens. This part of the column was spared from de-icing salts and direct flames emerged from car accident.

Both test specimens SL3 removed from head of the column and CB from cross beam were noticeably damaged by fire with presence of spalling [9] and rebars were exposed. Test specimen SL3 had lowest measured resonance frequency 1.02 kHz among all specimens and specimen CB had resonance frequency 9.48 kHz. At each affected surface a destructive compressive strength test was done on removed core bores. The comparison of measured resonance frequencies and compressive strength of concrete is at Fig.9.

In this graph noticeable correlation of cubic compressive strength and resonance frequencies for test specimen SL1, SL2 and SL3 can be seen. The footing of column from which test specimen SL1 was removed had compressive strength 62.1 MPa, the middle of column had 76.9 MPa and the head of column had 60.6 MPa. Cross beam had compressive strength 67.1 MPa and test specimen had resonance frequency 9.48 kHz. This frequency was determined by FFT interpretation of measured signal, and the correlation between strength and resonance frequency occurred for every test specimen SL1, SL2, SL3 and CB.

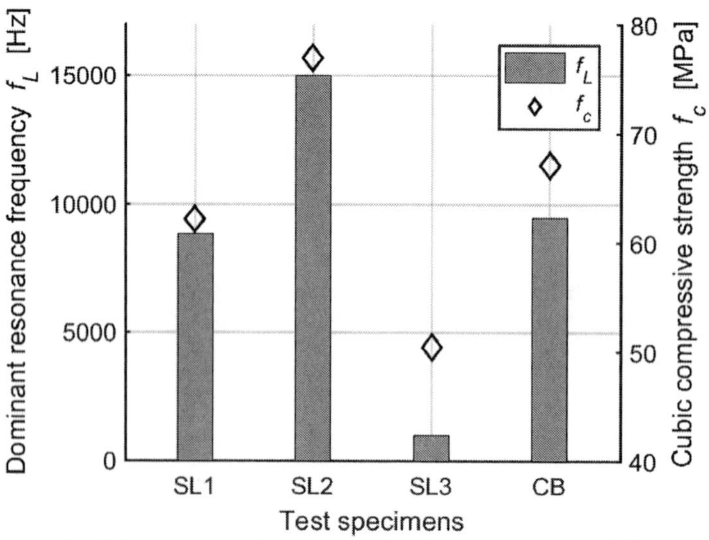

Fig.9 Comparison of cubic compressive strength of core bores and dominant longitudinal frequency from IE measurement of cut specimens

Spectrograms of signals obtained by CWT shows slightly different information about test specimens as can be seen in Fig.10 to Fig.13.

We can distinguish the occurrence of frequencies in time domain and with specific magnitude. At Fig.10 a quite recognizable frequency occurs at 8.67 kHz and keep its partial magnitude from beginning up to 6.7 ms. When we look at Fig.11 a bright resonance frequency and its harmonic frequencies can be recognized at 15 kHz. From this spectrogram can be harmonic frequencies distinguished apart from the spectrograms of other 3 specimens.

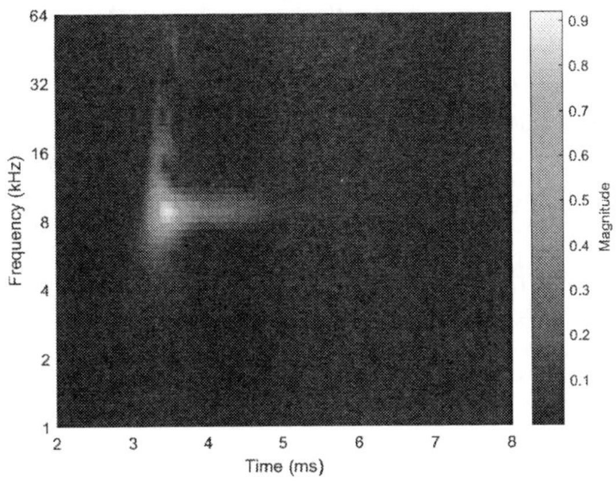

Fig.10 Wavelet transformation of signal from specimen taken from footing of column (SL1)

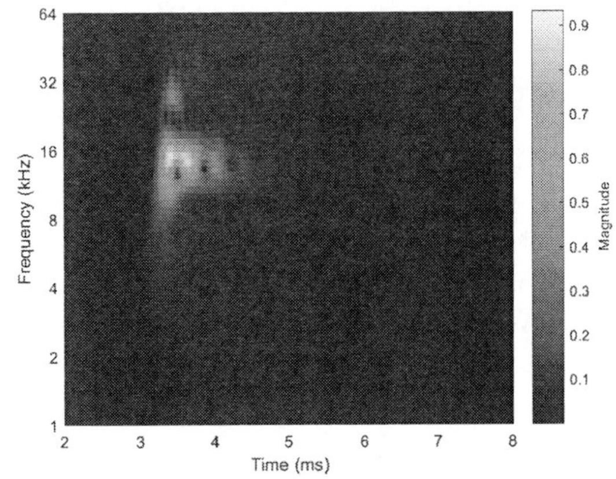

Fig.11 Wavelet transformation of signal from specimen taken from middle of column (SL2)

In Fig.12 a several dominant frequencies can be distinguished at 9.29, 7.55, 3.77 and 1.08 kHz. Specimen SL3 was removed from most damaged part of tested structure, and its resonance frequency obtained by FFT (1.02 kHz) match highlighted frequency 1.08 kHz at bottom of spectrogram. It seems that for the most degraded test specimen SL3 doesn't apply same rule as for the rest of the test specimens, where areas with highest magnitude match its resonance frequency obtained by FFT. At SL3 a center of rather wider area with 62% of maximum magnitude brightness seems to be the resonance frequency.

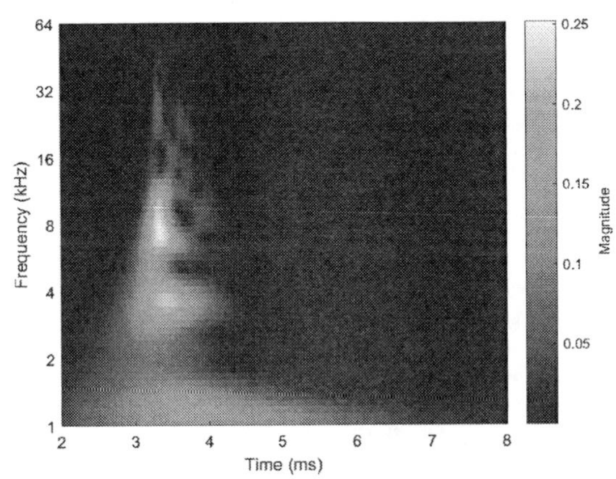

Fig.12 Wavelet transformation of signal from specimen taken from head of column (SL3)

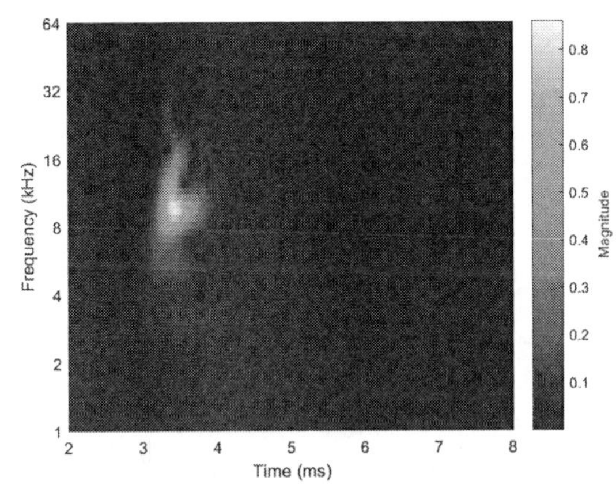

Fig.13 Wavelet transformation of signal from specimen taken from cross beam (CB)

Spectrogram of test specimen removed from cross beam at Fig.13 have brightness magnitude at resonance frequency 9.95 kHz, which is slightly higher than frequency from FFT (9.48 kHz). We can recognize deformed distribution of bright magnitude by similar way as in Fig.10 and Fig.12.

Conclusions

To design a suitable retrofitting process for construction damaged by fire a right diagnostic approach must be set. For this assessment numerous diagnostic procedures have been used over the years.

Presented publication compared two approaches to interpret measured signal from Impact-Echo measurement – a Fast Fourier transformation and Continuous Wavelet Transformation.

The measurement was conducted on test specimens removed from bridge damaged by fire emerged from car accident. Both methods were compared with cubic compressive strength of test specimens removed by core bore in-situ from bridge elements.

It seems that shape and dispersion of low magnitude areas of spectrograms could indicate presence of ongoing degradation by different factors. In such case, resonance frequency in CWT spectrogram can be represented by different magnitude than by the highest magnitude on spectrogram of signal.

Measured resonance frequencies corelated with compressive strength results both for FFT and CWT. Spectrograms provided additional information about dispersion of frequency over the duration of IE signal.

Acknowledgment

This publication was created under the project FAST-J-18-5275 supported by Faculty of Civil Engineering of Brno University of technology.

References

[1] MALHOTRA, V. a Nicholas CARINO. *Handbook on nondestructive testing of concrete*. 2nd ed. Boca Raton, Fla.: CRC Press, 2004. ISBN 08-031-2099-0.

[2] BRACEWELL, R. *The Fourier Transform & Its Applications*. 3. USA: McGraw-Hill Science/Engineering/Math, 1999. ISBN 0073039381.

[3] SADOWSKY, J. The Continuous Wavelet Transform: A Tool for Signal Investigation and Understanding. In: *Johns Hopkins APL Technical Digest - Volume 15, Number 4*. USA, The Johns Hopkins University, 1994, s. 306-318.

[4] HOłA, J., J. BIEń, Ł. SADOWSKI a K. SCHABOWICZ. Non-destructive and semi-destructive diagnostics of concrete structures in assessment of their durability. *Bulletin of the Polish Academy of Sciences Technical Sciences*. 2015, **63**(1), -. DOI: 10.1515/bpasts-2015-0010. ISSN 2300-1917. Dostupné také z:
https://www.degruyter.com/view/j/bpasts.2015.63.issue-1/bpasts-2015-0010/bpasts-2015-0010.xml

[5] ČSN EN 12 390-3. *Zkoušení ztvrdlého betonu - Část 3: Pevnost v tlaku zkušebních těles*. Praha: Český normalizační intitut, 2009.

[6] CARINO, N. J. The Impact-Echo Method: An Overview. In: *Proceedings of the 2001 Structures Congress & Exposition*. Virginia: American Society of Civil Engineers, 2001, s. 1-18. DOI: 10.1061/40558(2001)15.

[7] ASTM C1383-15. *Standard Test Method for Measuring the P-Wave Speed and the Thickness of Concrete Plates Using the Impact-Echo Method*. 1. West Conshohocken: ASTM International, 2015.

[8] EPASTO, G., E. PROVERBIO a V. VENTURI. Evaluation of fire-damaged concrete using impact-echo method. *Materials and Structures* [online]. 2010, **43**(1-2), 235-245 [cit. 2016-12-19]. DOI: 10.1617/s11527-009-9484-0. ISSN 13595997. Dostupné z: http://www.springerlink.com/index/10.1617/s11527-009-9484-0

[9] ZHAO, Jie, Jian-jun ZHENG, Gai-fei PENG a Klaas VAN BREUGEL. A meso-level investigation into the explosive spalling mechanism of high-performance concrete under fire exposure. *Cement and Concrete Research*. 2014, **65**, 64-75. DOI:
10.1016/j.cemconres.2014.07.010. ISSN 00088846. Dostupné také z:
http://linkinghub.elsevier.com/retrieve/pii/S0008884614001501

EAN 2018
56th conference on experimental stress analysis

Plasma Modification of Microfiber Reinforcement to Improve Adhesion with Cement Matrix Containing Recycled Concrete

ĎUREJE Jakub[1,a], PROŠEK Zdeněk[1,2,a], TREJBAL Jan[1,c] and TESÁREK Pavel[1,d]

[1]CTU in Prague, Faculty of Civil Engineering, Thakurova 7, 166 29 Praha 6, Czech Republic

[2]CTU in Prague, UCEEB, Trinecka 1024, 273 43 Bustehrad, Czech Republic

[a]jakub.dureje@fsv.cvut.cz, [b]zdenek.prosek@fsv.cvut.cz, [c]jan.trejbal@fsv.cvut.cz, [d]tesarek@fsv.cvut.cz

Keywords: plasma fiber modification, fiber surface modification, oxygen plasma, polypropylene fibers, PP fibers.

Abstract. The article describes oxygen plasma surface modification of polypropylene (PP) microfibers in order to increase coherence with the cement matrix. Five different times of 30, 60, 120, 240 and 480 seconds of plasma modification were used. Changes on fiber surface were detected by SEM analysis, packed cell wettability and weight loss during modification. Selected modification times were chosen to produce test samples on which the modulus of elasticity was continuously measured, and then bending and compression tests were performed. The values were compared with the values for the reference samples.

Introduction

The major problem with cement composite materials is a fragility and low tensile or bend strength. [1] Therefore, the cement matrix should be suitably reinforced. There is possibility to reinforcing by scattered fibers. This method of reinforcement has been used for a long time. For example, straw has been used in the building industry for 3500 years ago to reinforce unfired clay bricks. The article deals with synthetic fibers that have been used since the early 1990s. [2] One of the most important properties of cement composites is cohesiveness of the fiber with the matrix. Cohesion is a property depends on an interfacial transition zone.

In order to increase (or reduce) the coherence of the fibers with the matrix, plasma modification of the fiber surface can be used. Plasma is produced by supplying sufficient energy to the gas. Therefore, plasma is referred to as the fourth state of matter. Plasma contains charged particles - positive ions and negative electrons and neutral particles. [4] For plasma fiber modification process is using low-temperature plasma because of low used material temperature resistance. Plasma acts on the surface of the fibers both mechanically and chemically. With the increasing time of plasma modification process are changes on the surfaces more significant. [5,6]

In previous works, the fibers have been modified to improve their coherence with various binders. Carbon fibers has been successfully modified for their application to the epoxy binders and polyetherimide (PEI) composites. [7,8] Polypropylene fibers have been modified for better applicability of colors and adhesives on their surface. [9] Plasma modification has successfully increased the resistance of sisal fibers in alkaline environments. [10]

Materials and Samples

Fibrofor polypropylene microfibers was chosen for plasma plasma oxygen modification. It is a common commercially reinforcement used in the building industry most often for reinforcing concrete as a secondary reinforcement. The mechanical properties of the fibers are shown in Table 1.

Tab.1 Fiber properties

Fibers	Company	Type	Length [mm]	Diameter [μm]	Tensile strength [MPa]	Modulus of elasticity [GPa]	Density [g/cm3]
PP	Fibrofor	High Grade	19	80	400	4,9	0,91

The fibers were plasma modified in the Tesla VT 214 for 30, 60, 120, 240 or 480 seconds. The plasma modification process was in the chamber of the Tesla VT 214 from which the air was first evacuated to a pressure of 22 ± 1 Pa and subsequently filled with oxygen at a pressure of 60 ± 1 Pa and then the plasma modification process was initiated for a set period of time. The low-temperature plasma was produced by radio frequency source with 100 W input power. The temperature of the substrate in the chamber gradually increased during the plasma modification, reaching the temperature at about 50 °C for the longest time. Based on the results of the measurements, fibers with the 60s and 480s were selected for reinforcement of the test samples. After modification, the fibers were stored in sealed bags to prevent air exchange and some fibers were stored in open bags for 5 days in a laboratory environment at 22 ± 1 ° C and a relative humidity of 50 ± 2%. (Table 2).

Tab. 2 Fibers modification

Designation	Fibers	Modification time [s]	Plasma gas	Storage
PP 60 s	PP	60	Oxygen	Sealed bags
PP 480 s	PP	480	Oxygen	Sealed bags
PP 480 s old	PP	480	Oxygen	Open bags

There were made four sets of samples, each set contains six testing samples. The dimensions of the test samples were 40 x 40 x 160 mm. Portland cement CEM 42,5R (Radotin), micro-milled concrete recyclate and plasma modified fibers were used to produce this samples. The composition of all test samples was the same, differing only by the modifications of the fiber surfaces (Table 3).

Tab.3 Composition of the samples

Designation	Cement [g]	Recyclate [g]	Water [g]	W/(C + R)	Fibers [%] volume	Fibers [g]
PP ref	400	100	205	0,41	2	5,13
PP 60 s	400	100	205	0,41	2	5,13
PP 480 s	400	100	205	0,41	2	5,13
PP 480 s old	400	100	205	0,41	2	5,13

Experimental Methods and Results

All fibers were weighed before and after their plasma modification. The fibers were weighed by Kern with accuracy 0.1 mg. The highest weight loss was 0.18% of their original weight at the modification time of 480 seconds. From the measured data it can be concluded that the loss of mechanical properties of the fibers (reduction of the fiber diameter) is almost negligible.

The fiber surface was examined using the Merlin ZEISS electron microscope, which allows the observed object to be magnified up to 2,000,000 times. Mechanical changes by plasma modification on fiber surfaces compared to reference fibers were observed by electron microscopy at time 120, 240 and 480 second (Fig. 1).

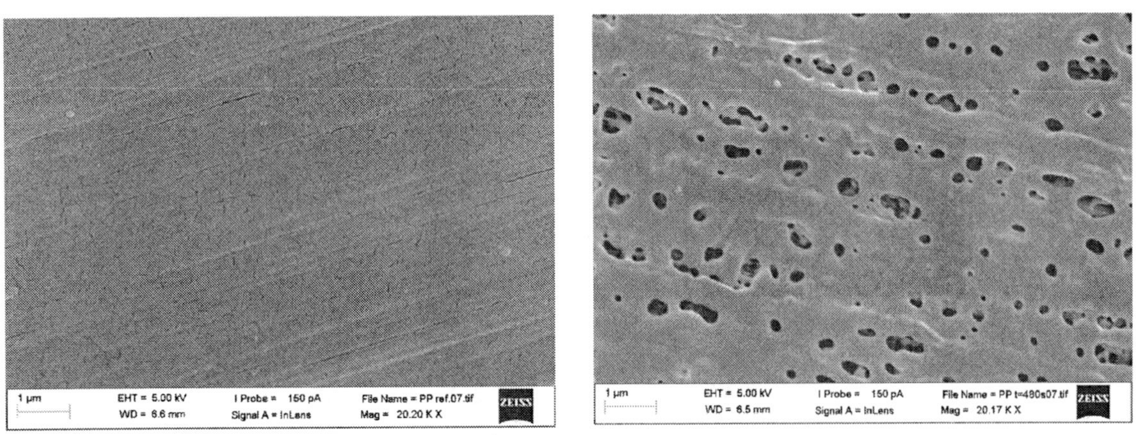

Fig.1 Comparsion of fibre surface, reference – left, plasma modified surface at 480 seconds - right

Fiber wettability was measured by an indirect method similar to the Packed-cell method. The fibers were placed into small container with a perforated bottom and lid. Subsequently, on the fibres were placed lead weights, consequently the fibers were compressed. The whole this container was weighed by Kern with accuracy 0.1 mg and then dipped under water up to the top edge for 30 seconds. Subsequently, the container was taken out of the water and after 120 seconds weighed. The container and lead were also weighed separately. From the measured data was calculated the percentage weight of the water in the container compared to the weight of the fibers (Figure 2).

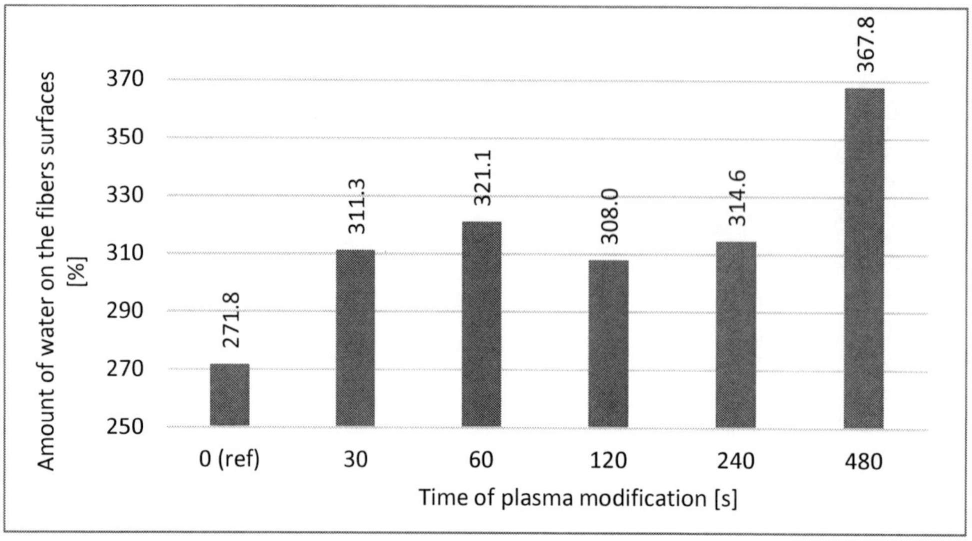

Fig.2 The weight of water between the fibers

The values of the compressive and the flexural strength of the samples were determined at the 24 days. Samples were tested in three-point bending test. They were loaded with a controlled displacement in a Heckert model FP100 hydraulic press at a rate of 0.4 mm per minute. The remaining half of the samples were tested in pressure test at a rate of 0.8 mm per second.
Flexural strength in plasma-modified fiber samples increased by 20-25% (Figure 3). In spite of the fact that the chemical effect of the plasma modification is not stable over time because of air humidity, we can conclude from the experimental results that after five days under normal conditions the chemical effect of the plasma modification almost did not degrade. Compressive strengths were nearly 75 ± 2 MPa for all sample sets.

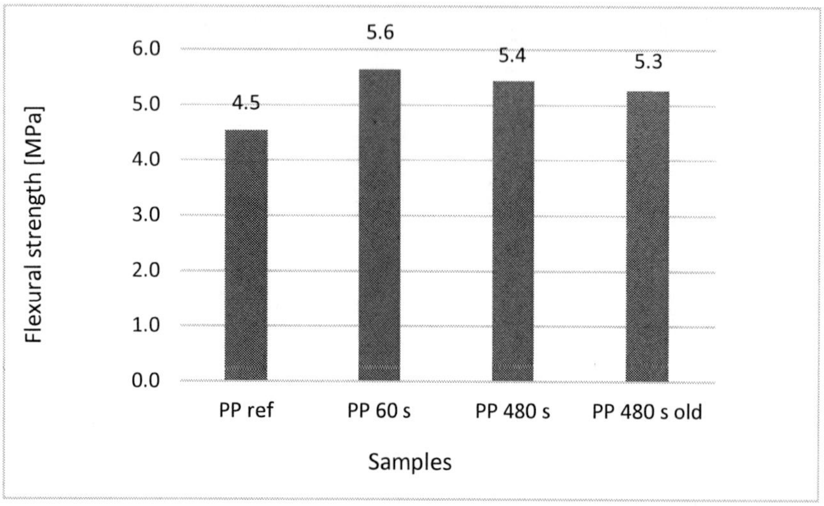

Fig.3 Comparison of Flexural strength

Conclusions

The plasma modification may suitably modify the surface of the polypropylene (PP) fibers. Increased bending strength of cement composite material occurred at shorter times of modification and, over longer periods of time, the strength did not increase, despite the fact that mechanical changes were observed on fiber surfaces by electron microscopy over longer periods of time. As an optimal modification with a given plasma modifying device setting is oxygen plasma modification for 60 seconds.

Acknowledgement

This paper was financially supported by the GA ČR research a project under the number 17-06771S and by Czech Technical University in Prague under No. SGS project SGS16/201/OHK1/3T/11.

References

[1] KERŠNER, Zbyněk. *Křehkost a lomová mechanika cementových kompozitů*. Brno, 2005. Habilitační práce. VUT v Brně.

[2] BHARATHI, D. The history of fibre reinforcement[online]. 2015 [cit. 2017-10-31]. Dostupné z: https://www.linkedin.com/pulse/history-fibre-reinforcement-bharathi-d

[3] BENTUR, Arnon. a Sidney. MINDESS. Fibre reinforced cementitious composites. 2nd ed. New York: Taylor, 2007. ISBN 978-020-3088-722.

[4] ROSE, Margaret. Definition Plasma [online]. 2017 [cit. 2017-10-31]. Dostupné z: http://whatis.techtarget.com/definition/plasma

[5] SABREEN, Scott R. Cold Gas Plasma Surface Modification – Optimize Plastics Bonding Adhesion [online]. 2012 [cit. 2017-10-31]. Dostupné z: http://www.adhesionbonding.com/2012/04/30/cold-gas-plasma-surface-modification-optimize-plastics-bonding-adhesion/

[6] CHRYSOSTOMOU, Demetrius. Plasma surface modification for the biomedical industry. California, 2009. Prezentace.

[7] MOOSBURGER-WILL, Judith, Matthias BAUER, Fabian SCHUBERT, et al., Methyltrimethoxysilane plasma polymerization coating of carbon fiber surfaces. Surface and Coatings Technology [online]. (2017), 223-230 [cit. 2018-01-06]. DOI: https://doi.org/10.1016/j.surfcoat.2017.01.017. Dostupné z: http://www.sciencedirect.com/science/article/pii/S0257897217300178?via%3Dihub

[8] EUNG-SEOK, Lee, Lee CHOONG-HYUN, Chun YOON-SOO, Han CHANG-JI a Lim DAE-SOON, Effect of hydrogen plasma-mediated surface modification of carbon fibers on the mechanical properties of carbon-fiber-reinforced polyetherimide composites. Composites Part B [online]. 2016 [cit. 2018-01-06]. DOI: https://doi.org/10.1016/j.compositesb.2016.10.088. Dostupné z: https://www.sciencedirect.com/science/article/pii/S1359836816312227

[9] PULÍČEK, Roman, 2016. Optimalizace procesu hydrofilizace polypropylenového textilu pomocí DBD atmosférické plazmy. Liberec. Diplomová práce. Technická univerzita v Liberci.

[10] BARRA, B.N., S.F. SANTOS, P.V.A BERGO, C. ALVES, K. GHAVAMI a H. SAVASTANO, Residual sisal fibers treated by methane cold plasma discharge for potential application in cement based material [online]. 2015 [cit. 2018-01-06]. Dostupné z: https://www.sciencedirect.com/science/article/pii/S0926669015302764#!

EAN 2018
56th conference on experimental stress analysis

Validation of Multiaxial Stress-based Analysis on Specimens from ČSN 41 1523

FOJTÍK František[1,a], PAPUGA Jan[2,b] and FUSEK Martin[1,c]

[1] VŠB – Technical University of Ostrava, Faculty of Mechanical Engineering, 17. listopadu 15, 708 00 Ostrava, Czech Republic

[2] FME Czech Technical University in Prague, Technická 4, 166 07 Prague 6, Czech Republic

[a]frantisek.fojtik@vsb.cz, [b]jan.papuga@fs.cvut.cz, [c]martin.fusek@vsb.cz

Keywords: multiaxial fatigue, high-cycle fatigue

Abstract. The paper describes the study on fatigue strength prediction criteria within the scope of a common high-cycle fatigue strength analysis including the effect of multiaxial loading. The study is based on own experiments performed on tubular specimens from ČSN 41 1523 structural steel in a combination of various load channels: push-pull, torsion, pressurizing, bending. The prediction methods are assessed in various fatigue lifetimes ranging from 100 000 cycles to 750 000 cycles. No substantial deviation of the methods behaviour in various lifetimes is observed, likely due to the stresses being low enough not to induce substantial plasticity effects. The PCRN method and the Liu-Zenner methods result in best predictions. Results of the recently proposed MMP method allowing the users to evaluate quickly multiaxial load states remain promising and only slightly worse than these two best methods.

Introduction

The paper extends the topic presented by authors in [1]. They reported on extensive experimental fatigue program on hollow specimens manufactured from ČSN 41 1523 structural steel, and they introduced a new MMP calculation method. This extension of the Manson-McKnight (MMK) method preserves its simplicity of the calculation, for which any simple spreadsheet program as e.g. MS Excel is sufficient. On the other hand, thanks to the substantial modification of the formula, MMP resulted in much better fatigue prediction compared with MMK. It was stated that the MMP method can reach quality of results comparable with much more complex multiaxial fatigue criteria.

The fatigue strength analysis provided in [1] was prepared at 750,000 cycles for all 24 S-N curves involved in the experimental campaign. The extension in the current paper concerns evaluating the same set in section cuts in the range from 100,000 to 750,000 cycles. The goal is to analyse, whether the quality of prediction by various evaluated calculation methods is in any way modified, if lower lifetimes are treated.

Experiments

The specimens tested in the experimental campaign are described in [1]. Three different types of hollow specimens were used with the major and minor diameters in the critical cross-section either D=20 mm and d=18 mm or D=11 mm and d=8 mm. They were manufactured from ČSN 41 1523 structural steel, properties of which are provided in Table 1.

Tab.1 Static material parameters and chemical composition of the studied material

Designation	Ultimate tensile strength [MPa]	Tensile yield stress [MPa]	Elongation at fracture [%]	Reduction of area at fracture [%]	True fracture strength in torsion [MPa]
ČSN 41 1523	560	400	31.1	74.0	516.6
Chemical composition:					
C [%]	Mn [%]	Si [%]	P [%]	S [%]	Cu [%]
0.18	1.38	0.4	0.018	0.006	0.05

The setup of experiments was extensively described in [1]. Here the schematic drawings in Fig. 1 show the setup of individual experiments. 24 different load cases were tested in the stress-controlled regime. Each final S-N curve was obtained by the linear (Basquin model) or non-linear (Kohout-Věchet model, see [4]) regression from at least 5 finished experiments. Various experimental machines had to be used to result in described load combinations (see also the setup of tests in Fig. 2 and Fig. 3) – the reconstructed Schenck PWXN machine, the INOVA ZUZ servo-hydraulic pulsator equipped with combined hydraulic cylinder for push-pull and torsion loading and the INSTRON 8802 biaxial servo-hydraulic pulsator.

Due to unusual load combinations, it is not immediately obvious in some cases, which surface of the hollow specimens is the more loaded one. Because of that, potential hot-spots on both surfaces had to be analysed during the fatigue prediction phase.

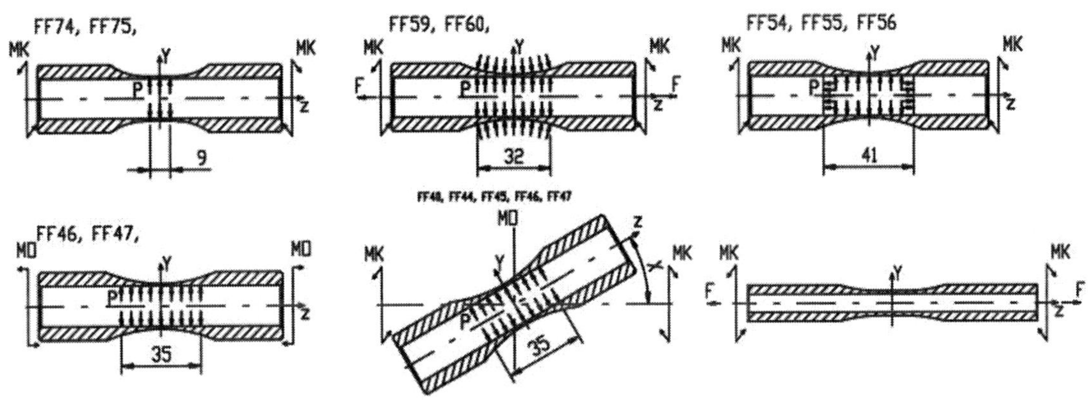

Fig.1 Overview of various setups of experiments. M0 ~ bending moment, MK ~ torque, P ~ pressure and F ~ push-pull

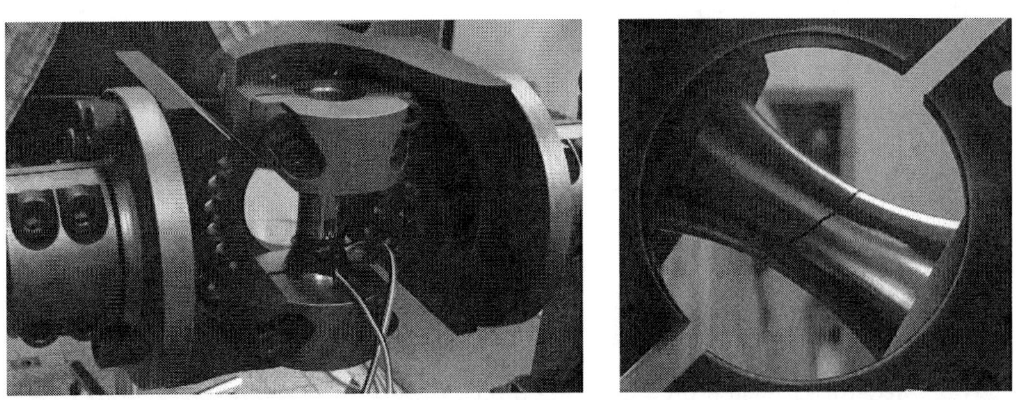

Fig.2 Setup of the experiments for load cases of torsion and bending on the torque controlled Schenck PWXN machine (bending load case FF040 left, combined load case FF044 right).

Fig.3 The load case of bending and pressurizing, with the pressure chamber and the pressure sensor visible – FF046 load case (left). The setup of the test case with the pressure chamber inducing inner and outer pressure – FF059 load case (right)

Calculations

Within the prediction evaluation, 27 calculation methods were analysed in total. Most of them are described in [2]. The stress states on individual load channels relevant to the durability at a given number of cycles N_x were calculated from the regression curves derived from the experimental data. The fatigue strength analysis was performed, i.e. the damage parameter D_P in the form of an equivalent axial stress amplitude was computed for various load combinations presented here and relevant to the evaluated number of cycles N_x. This equivalent stress amplitude was compared with the fully reversed axial fatigue strength p_{-1} (push-pull loading) or b_{-1} (bending) at the same N_x.

The analysis of the prediction quality was not evaluated on raw experimental data, but on fatigue strengths derived from the regression curves. The analyses were performed only on fatigue strengths that can be obtained within the interpolation regions of the individual S-N curves. The paper [1] referred to the fatigue strengths relevant to 750 000 cycles. Three more lifetimes are evaluated here: 100 000, 200 000 and 500 000 cycles.

The paper [1] comments the way of selecting the appropriate axial load mode for various load cases. At last, the bending fatigue strength was used only for cases loaded explicitly also in the bending load mode. The relative error between computed damage parameter D_P (equivalent stress amplitude) and e.g. the fully reversed push-pull fatigue strength p_{-1} defines the fatigue index error ΔFI:

$$\Delta FI = \left(\frac{D_P - p_{-1}}{p_{-1}}\right) \cdot 100\% \qquad (1)$$

The paper evaluates prediction quality based on this parameter. To enable the analysis for 24 different load cases at four different lifetimes, it must be processed statistically – here the mean value, standard deviation and the sum of its squares are evaluated.

Discussion of results

From results obtained for all tested prediction methods, only some are presented statistically in Table 2, Table 3 and in Fig. 4. Except of the MMP method (see [1]) and the updated PCRN method [3] derived from PCR [2], all prediction methods are described in [2]. The selection to highlight these particular methods was driven by the intention to show results of methods, which are used in available fatigue solvers (methods by Dang Van, McDiarmid, Findley, Sines and Matake), which present substantially better results than these methods (Liu-Zenner, MMP, PCR, PCRN), or which are often used in comparisons of prediction methods (Crossland and Papadopoulos).

Table 2 presents the summary in the sum of ΔFI squares, which describes well the remoteness of the general trend from the correct one. To distinguish the general shift of all results to either conservative or non-conservative prediction from the scatter around this mean value, Table 2 contains also the values of the mean relative error for individual lifetimes and for all of them. Table 3 then summarizes the information about the standard deviation of the same parameter, thus effectively showing the induced scatter of the prediction quality.

The most important question in the overall evaluation was, whether the individual methods will show a substantial difference at different lifetimes, i.e. if their results are stable within the high-cycle fatigue domain tested here. Table 2 and Table 3 show that the mean values of the fatigue index error change a bit. For most of the methods, the difference in the mean value is within 2%. The biggest differences can be observed for the McDiarmid, Findley and Matake methods. Compared with the overall scatter of the results, such shift in the mean value is not excessive.

Tab.2 Results of the ΔFI statistics for selected eleven methods at different N_x – sum of squares and mean values

Computational method	Sum of ΔFI squares	Mean value of ΔFI for N_x				
		All	100 000	200 000	500 000	750 000
Crossland	75%	-0.9%	-1.8%	-1.0%	-0.5%	-0.4%
Dang Van	84%	6.1%	4.8%	6.0%	6.6%	6.7%
Findley	351%	13.3%	11.1%	12.8%	14.2%	14.8%
Liu-Zenner	26%	1.8%	2.6%	2.2%	1.5%	1.1%
Matake	363%	12.4%	10.3%	12.0%	13.4%	13.9%
McDiarmid, MD var.	115%	4.1%	6.8%	5.1%	3.0%	1.9%
MMP	41%	2.5%	3.4%	3.1%	2.1%	1.4%
Papadopoulos	87%	0.2%	-0.7%	0.1%	0.7%	0.7%
PCR	43%	3.3%	3.2%	3.3%	3.4%	3.3%
PCRN	32%	3.9%	4.7%	4.1%	3.5%	3.3%
Sines	346%	10.1%	10.3%	11.0%	9.8%	9.2%

The change in the standard deviation is the most pronounced in the evaluation of the Findley and Matake methods, where it decreases with the decreasing lifetime. Because the mean value decreases in the same way, it can be stated that both these methods are more suitable for use at the lifetime 100 000 cycles than at higher lifetimes. Their results stay too scattered and too conservative anyway (see the histogram for the Findley method in Fig. 4).

If the overall prediction quality of individual methods is compared in all evaluated lifetimes, the methods by Liu-Zenner, PCRN, PCR and MMP provide better results than other methods evaluated here. If the standard deviation of the PCRN method is compared with the Dang Van method, which is the most important solution used in industrial practice, it can be noted that the standard deviation of the PCRN is 60% of the Dang Van solution. This means that the prediction results are less scattered, and they can be accepted with a better certitude (see again the comparison of histograms in Fig. 4).

The newer PCRN method [3] results in better prediction if compared with the original PCR [2]. It leads to a lower scatter of results, as can be documented by the lowest value of the overall standard deviation of ΔFI. Anyhow, its mean value is more distant from perfect prediction ($\Delta FI = 0$) than the mean value of the Liu-Zenner method (This larger deviation is at least in the conservative direction). Because of that, the sum of ΔFI squares of the Liu-Zenner method is the lowest one of all methods.

Tab.3 Results of the ΔFI statistics for selected eleven methods at different N_x – standard deviations

Computational method	Standard deviation of ΔFI for Nx				
	All	100000	200000	500000	750000
Crossland	8.9%	10.5%	9.0%	8.3%	8.3%
Dang Van	7.3%	8.2%	7.1%	7.0%	7.3%
Findley	14.1%	13.7%	13.5%	14.6%	15.2%
Liu-Zenner	5.0%	4.9%	4.5%	5.1%	5.6%
Matake	15.3%	14.6%	14.5%	15.9%	16.8%
McDiarmid, MD var.	10.3%	10.6%	10.0%	10.2%	10.4%
MMP	6.1%	6.8%	6.1%	5.8%	5.9%
Papadopoulos	9.7%	11.2%	9.7%	9.2%	9.2%
PCR	6.0%	6.6%	6.0%	5.8%	5.8%
PCRN	4.4%	4.3%	4.2%	4.6%	4.8%
Sines	16.4%	16.0%	16.3%	16.9%	17.4%

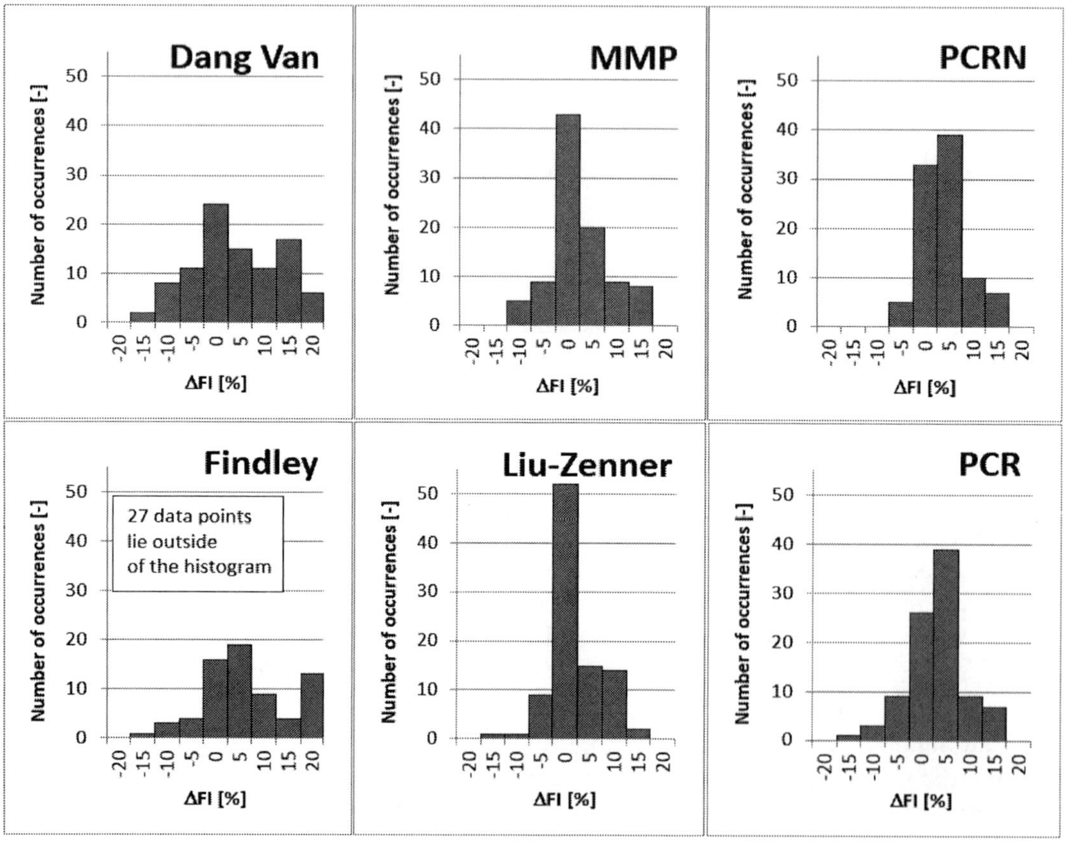

Fig.4 Examples of histograms of the fatigue index error over all experiments and all evaluated section cuts (100,000; 200,000; 500,000 and 750,000 cycles)

If the new MMP method [1] is compared with these two methods, it quite well resembles prediction quality of the original PCR. It has to be noted, that the simplicity of its computation is such, that the MMP can be solved within the MS Excel spreadsheet, while PCR(N) or Liu-Zenner methods require a dedicated software program.

Conclusions

The analysis presented in this paper concerned 27 different calculation methods of various types, which were tested on a test set of 24 S-N curves of different load cases. The experimental results described previously [1] were obtained on hollow specimens manufactured from ČSN 41 1523 structural steel.

Results of 11 selected calculation methods are provided here for comparison. It can be concluded that Liu-Zenner and PCRN methods lead to results with the lowest scatter, their standard deviation being close to one half of the standard deviation relevant to the industrially most often used Dang Van method. Within the range between 100 000 and 750 000 cycles, the prediction quality of individual methods stays relatively stable, and the observed trends (e.g. the change of the mean fatigue index error) is insignificant if compared to the total scatter of prediction results. The relatively narrow scope of the test set does not allow the researchers to generalize more detailed conclusions.

Acknowledgement

This work has been supported by the Ministry of Education, Youth and Sports of the Czech Republic within the framework of the National Programme for Sustainability (NPU II) project IT4 Innovations Excellence in Science – LQ1602, and by Specific Research project SP2018/63. This support is gratefully acknowledged. The authors also thank the Grant Agency of the Czech Technical University in Prague for its financial support for the research presented here within the framework of the grant No. SGS17/175/OH K2/3T/12.

References

[1] J. Papuga, F. Fojtík, Multiaxial fatigue strength of common structural steel and the response of some estimation methods. International Journal of Fatigue 104 (2017) 27-42.

[2] J. Papuga, A survey on evaluating the fatigue limit under multiaxial loading. International Journal of Fatigue 33 (2011) 153-165.

[3] J. Papuga, R. Halama, Mean stress effect in multiaxial fatigue criteria. Submitted to Archive of Applied Mechanics (2018).

[4] J. Kohout, S. Věchet, A new function for fatigue curves characterization and its multiple merits. International Journal of Fatigue 23 (2001) 175-183.

EAN 2018
56th conference on experimental stress analysis

Optimization of Circular Saw Blade Stress Fields Using Numerical Methods of Mechanics

FRANKOVSKÝ Peter[1,a], KICKO Michal[1,b], BOCKO Jozef[1,c], DELYOVÁ Ingrid[1,d] and KOSTAK Ján[1,e]

[1]Technical University of Košice, Faculty of Mechanical Engineering, Letná 9, 042 00 Košice, Slovak Republic

[a]peter.frankovsky@tuke.sk, [b]michal.kicko@tuke.sk, [c]josef.bocko@tuke.sk,[d]ingrid.delyova@tuke.cz, [e]jan.kostka@tuke.sk

Keywords: Circular saw blade, anti-vibration grooves, stress analysis, finite element method.

Abstract. The submitted article deals with the design of the cutting saw blade in terms of the location of the anti-vibration grooves on the surface of the disk. The position of the anti-vibration grooves affects vibrations, noise and stress distribution in the disk. Our concern is to investigate the best distribution of anti-vibration grooves to reduce the maximum stress in the critical parts of the cutting blade.

The article presents the numerical stress analysis of the saw blade without and with anti-vibration grooves for their suitable deployment in terms of reducing the stress around the teeth during their cutting process. A suitable layout of the anti-vibration grooves has a significant effect on the reduction of stress around the teeth. The individual steps of numerical analysis are detailly described in the article.

Introduction

Cutting blades whose geometry is provided with different grooves to reduce noise, vibration and stress distribution is at present an increasingly frequent phenomenon, especially in the area of profi-tools. This was one of the main impulses for making this contribution. As it is generally believed that such grooves have in particular, acoustic optimization effects, so they are primarily used to reduce noise. This contribution is also evidenced by their further optimization property, namely the reduction of stress fields caused by centrifugal forces but mainly the reduction of stresses in the area of cutting teeth.

Anti – vibration grooves

To reduce vibrations caused by centrifugal forces and unbalanced weight distribution, the cutting blades are provided with narrow grooves located in the region between the bearing and the teeth of the disc. Many manufacturers of circular saw blades are compensated with this problem by making holes, usually located under expansion grooves. These openings usually begin and end with rounding, with the laser cut in the shape of the letter "S" (Fig. 1). Laser cutting is essential for dimensional accuracy, as even small geometry inaccuracies can cause great vibration. Thus these holes help disperse vibrations. In some cases, vibration holes are filled with a soft material such as brass or copper, which helps absorb vibrations. Anti-vibration saw holes are designed to prevent unwanted effects and to perform smooth cuts without the use of stabilizers. Anti-vibration grooves also act as noise reduction elements.

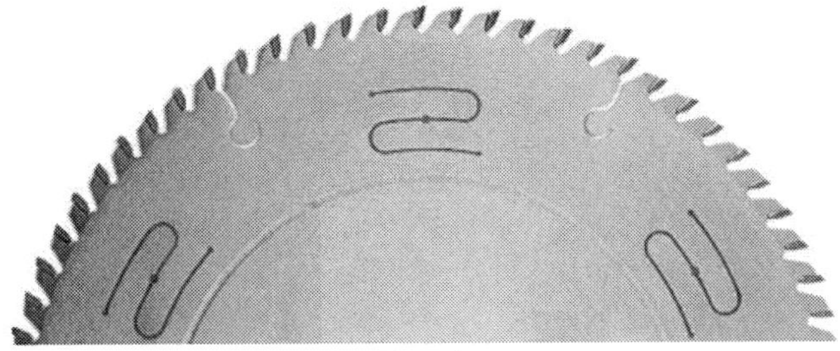

Fig. 1 Anti-vibration grooves

Design of the shape and dimensions of the circular saw blade

Fig. 2a shows a conventional wood cutting disc with a hole for bearing 20 mm in diameter and an outer diameter of 160 mm. By creating slides in the areas below the cutting teeth that will create adjacent grooves effect, it will be assumed that the stress drops primarily in the area of the cutting parts.

The proposed shape of the disc, which is shown in FIG. 2b will be made using laser cutting technology, with standardized steel 75Cr1 (DIN 1.2003), (STN 19418). In Fig. 2b two types of optimization elements are shown. The first is the five expansion grooves around the circuit of the disk that serve as a function of protecting the teeth (segments) against unwanted shape change or the tearing of the teeth when the workpiece is pierced during cutting. The other elements are anti-vibration grooves that fill the role of adjacent grooves effect to optimally distribute the principal stresses during cutting. If this design is certified numerically, it will be used to manufacture it from photosensitive material of PS-1A. Thus the produced specimen will then be subjected to experimental analysis by reflective photoelasticity using the LF / Z-2 reflection polariscop.

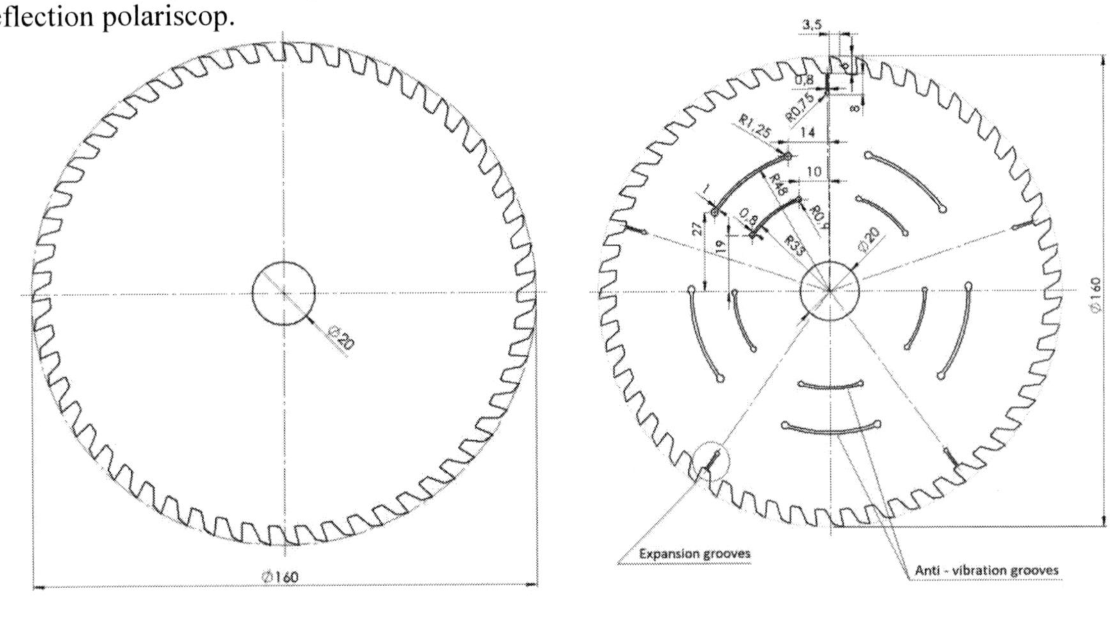

a) b)

Fig. 2 The shape and dimensions of the saw blades: a) conventional; b) with anti-vibration grooves

Numerical stress analysis

The shape and dimensions of the proposed anti-vibration grooves were analyzed and evaluated numerically in the Abaqus program. The first step was to simulate a conventional disk without optimizing grooves, and then calculate the stress fields in the optimized design of the cutting saw blade (Figure 2b). For the efficient calculation, both samples were created as shell elements [1-6].

The boundary conditions were chosen as follows: a reference point (RP) was created at the center of the bearing hole, to which the entire internal circumferential edge of the disk was attached (kinematic coupling) to simulate the hinge. In this RP, all degrees of freedom were taken apart excluding the rotation around the Z axis, thus simulating the possibility of movement during cutting. Subsequently, in RP was defined a very low unit of angular rotation for simulating the abovementioned movement of the disc.

The last step in defining boundary conditions was to prevent the movement in one of the teeth, namely by fixing the tip to simulate an engagement of tooth. In this way, the same trio of teeth for both conventional and optimized discs was used to evaluate and compare the results more objectively.

The results of the strength analysis, the fields of reduced stresses obtained with the HMH strength theory, are shown in Fig. 3. The relation for the calculation of the reduced stress according to the theory of strength HMH has the form

$$\sigma_{red}^{HMH} = \sqrt{\sigma_1^2 + \sigma_2^2 + \sigma_3^2 - (\sigma_1\sigma_2 + \sigma_2\sigma_3 + \sigma_1\sigma_3)}. \quad (1)$$

For better representation, the upper stress limit was set to 100 MPa in all cases.

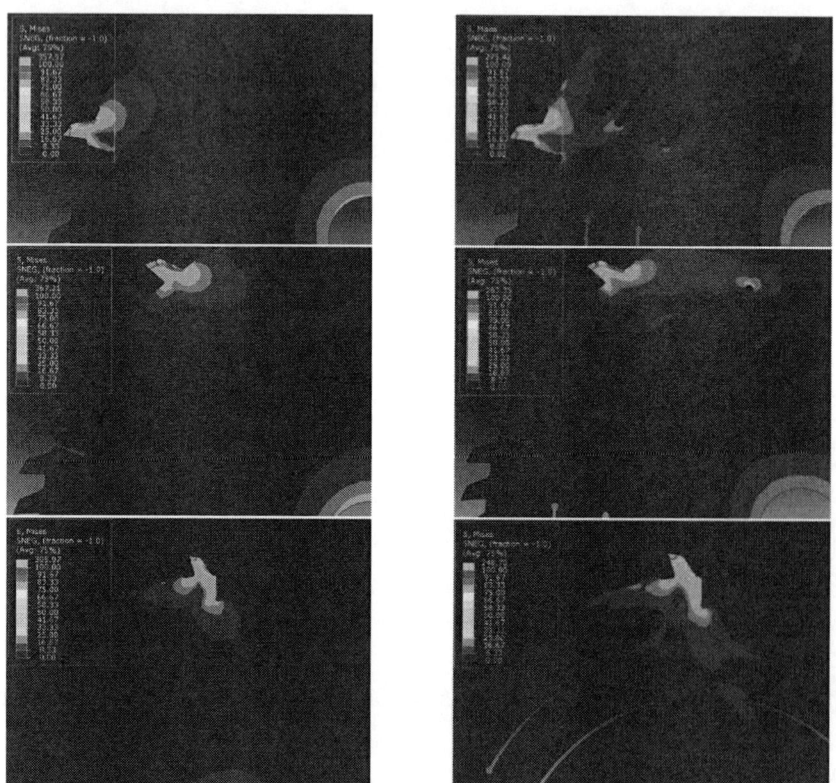

a) b)

Fig. 3 Comparison of stress fields for: a) Conventional disc; b) an optimized disc

The strength analysis results for all three of the above cases confirmed a significant reduction of maximum stress due to the application of optimization grooves within the conventional disk. In the first tooth we recorded a decrease of the maximum stress by 81 MPa, in the region of the second considered tooth by 104 MPa and in the last case by 57 MPa.

Modal analysis

By analyzing of eigenshapes, it has been shown that not only optimal distribution of stresses occurs but also the reduction of the noise of the cutting disc, that can be noticed by comparing the results according to the simulations shown in Fig. 4 for a conventional disc and Fig. 5 for optimized one.

Fig. 4 Mode shapes of conventional disc

Fig. 5 Mode shapes of disc with anti – vibration grooves

By comparing the results of the modal analysis in Tab. 1, it is obvious that the oscilation frequency magnitudes of the disc with the anti-vibration grooves are at a relatively equal RPM significantly smaller than in the disc with the constant cross-section.

Tab. 1 Oscilation frequencies: a) conventional disc; b) optimized disc

a)

Frame					
Index	**Description**				
0	Increment	0: Base State			
1	Mode	1: Value =	6.48817E+06 Freq =	405.40	(cycles/time)
2	Mode	2: Value =	6.49017E+06 Freq =	405.46	(cycles/time)
3	Mode	3: Value =	8.71841E+06 Freq =	469.94	(cycles/time)
4	Mode	4: Value =	1.38924E+07 Freq =	593.21	(cycles/time)
5	Mode	5: Value =	1.38941E+07 Freq =	593.25	(cycles/time)
6	Mode	6: Value =	6.01212E+07 Freq =	1234.1	(cycles/time)
7	Mode	7: Value =	6.01251E+07 Freq =	1234.1	(cycles/time)
8	Mode	8: Value =	1.73387E+08 Freq =	2095.7	(cycles/time)
9	Mode	9: Value =	1.73423E+08 Freq =	2095.9	(cycles/time)
10	Mode	10: Value =	2.49853E+08 Freq =	2515.7	(cycles/time)
11	Mode	11: Value =	3.09237E+08 Freq =	2798.8	(cycles/time)
12	Mode	12: Value =	3.62943E+08 Freq =	3032.1	(cycles/time)
13	Mode	13: Value =	3.63036E+08 Freq =	3032.5	(cycles/time)
14	Mode	14: Value =	3.81131E+08 Freq =	3107.1	(cycles/time)
15	Mode	15: Value =	3.88107E+08 Freq =	3135.4	(cycles/time)
16	Mode	16: Value =	5.94207E+08 Freq =	3879.6	(cycles/time)
17	Mode	17: Value =	5.94328E+08 Freq =	3880.0	(cycles/time)
18	Mode	18: Value =	7.21178E+08 Freq =	4274.1	(cycles/time)
19	Mode	19: Value =	7.21299E+08 Freq =	4274.4	(cycles/time)
20	Mode	20: Value =	1.12251E+09 Freq =	5332.3	(cycles/time)

b)

Frame					
Index	**Description**				
0	Increment	0: Base State			
1	Mode	1: Value =	5.76379E+06 Freq =	382.10	(cycles/time)
2	Mode	2: Value =	5.76630E+06 Freq =	382.18	(cycles/time)
3	Mode	3: Value =	7.65097E+06 Freq =	440.23	(cycles/time)
4	Mode	4: Value =	1.28611E+07 Freq =	570.77	(cycles/time)
5	Mode	5: Value =	1.28623E+07 Freq =	570.79	(cycles/time)
6	Mode	6: Value =	5.64858E+07 Freq =	1196.2	(cycles/time)
7	Mode	7: Value =	5.64927E+07 Freq =	1196.2	(cycles/time)
8	Mode	8: Value =	1.62179E+08 Freq =	2026.8	(cycles/time)
9	Mode	9: Value =	1.62212E+08 Freq =	2027.0	(cycles/time)
10	Mode	10: Value =	1.97337E+08 Freq =	2235.8	(cycles/time)
11	Mode	11: Value =	2.25614E+08 Freq =	2390.6	(cycles/time)
12	Mode	12: Value =	2.43671E+08 Freq =	2484.4	(cycles/time)
13	Mode	13: Value =	2.43798E+08 Freq =	2485.1	(cycles/time)
14	Mode	14: Value =	3.68532E+08 Freq =	3055.3	(cycles/time)
15	Mode	15: Value =	3.72468E+08 Freq =	3071.6	(cycles/time)
16	Mode	16: Value =	4.13024E+08 Freq =	3234.5	(cycles/time)
17	Mode	17: Value =	4.13152E+08 Freq =	3235.0	(cycles/time)
18	Mode	18: Value =	6.97304E+08 Freq =	4202.7	(cycles/time)
19	Mode	19: Value =	6.97429E+08 Freq =	4203.1	(cycles/time)
20	Mode	20: Value =	9.50907E+08 Freq =	4907.8	(cycles/time)

Thus, we can assert that by creating optimization grooves, stiffness of disc was decreased but on the other hand it has allowed dilation in several directions, thereby reducing the oscillation frequency and thus reducing vibrations and noise.

Conclusion

By quantifying and comparing the fields of the reduced stresses in the strength calculations and the oscillation frequencies in the modal analysis, a suitable layout and shape of the optimization grooves has been demonstrated which significantly affected not only the acoustic but also the stress values when comparing the conventional disc with the optimized [7-10].

Acknowledgment

This work was supported by grant projects VEGA No. 1/0290/18 and VEGA No. 1/0731/16.

References

[1] J. Bocko, Š. Segľa, Numerické metódy mechaniky tuhého a poddajného telesa. Košice: TU, 2016.

[2] F. Trebuňa, F., Šimčák, Príručka experimentálnej mechaniky. Košice: TU, 2002.

[3] O. Záhorec, S. Caban, Dynamika. Košice: TU, 2002.

[4] F. Trebuňa, F. Šimčák, Odolnosť prvkov mechanických sústav. Košice: TU, 2004.

[5] Abaqus 6.14 Documentation (Help) – Analysis – Materials – Elastic mechanical properties – Linear elasticity – Linear elastic behavior.

[6] Materiálový list ocele 75Cr1 (DIN 1.2003).

[7] P. Kováčiková, J. Vavro, A. Dubec, Numerical modal analysis for vibration-damping properties of ductile cast. In MATEC Web of Conferences, Vol. 157, p. 02019), 2018.

[8] Z. Paška, F. Fojtík, P. Ferfecki, Full Field Evaluation of the Stress Tensor Components in 2D Photoelasticity Via Computer Software. Strojnícky casopis–Journal of Mechanical Engineering, 66(2), 63-66, 2016.

[9] E. Tertel, P. Kuryło, J. Cyganiuk, J. The distribution of stress in a sandwich conical shell panel under combined load. Procedia Engineering, 96, 467-476, 2014.

[10] T. C. McCarthy, A. N. Annaidh, M. D. Gilchrist, On the sharpness of straight edge blades in cutting soft solids: Part II–Analysis of blade geometry. Engineering Fracture Mechanics, 77(3), 437-451, 2010.

EAN 2018
56th conference on experimental stress analysis

Evaluation of Multiaxial Fatigue Criteria on LCF Data of SS316L

FUSEK Martin[1,a], LIČKOVÁ Dagmar[2,b], HALAMA Radim[2,c], PORUBA Zdeněk[1,d] and FOJTÍK František[2,e]

[1] Department of Applied Mechanics, Faculty of Mechanical Engineering & National Supercomputing Centre IT4Innovations, VŠB - TU Ostrava, 17. listopadu 2172/15, 70833 Ostrava, Czech Republic

[2] Department of Applied Mechanics, Faculty of Mechanical Engineering, VŠB - TU Ostrava, 17. listopadu 2172/15, 70833 Ostrava, Czech Republic

[a]martin.fusek@vsb.cz, [b]dagmar.lickova@vsb.cz, [c]radim.halama@vsb.cz, [d]zdenek.poruba@vsb.cz, [e]frantisek.fojtik@vsb.cz

Keywords: fatigue, multiaxial loading, critical plane approach

Abstract. The paper (study) deals with the evaluation of 10 low cycle fatigue criteria (critical plane) for stainless steel SS316L under 7 different loading paths. Best results were obtained based on the Socie at all criterion. The criterion of Jiang´s provided slightly more conservative prediction. The advantage of Jiang´s criterion is its possibility to be used for general case of multiaxial loading including variable amplitude loading without use of rain flow method.

Introduction

Most machine parts are loaded in a combined mode. In the field of low cycle fatigue, stress-strain behavior plays a significant role, which is also influenced by the level of non-proportionality of loading. The effect of additional hardening due to non-proportional loading for stainless steels is well known. Within the project of the Grant Agency of the Czech Republic [1], the advanced concept of prediction of fatigue failure under general multiaxial loading was implemented according to Jiang [2]. Other aims of the project were to apply parallel algorithms to increase the speed of fatigue calculations, to evaluate available approaches for the elastoplastic correction at the root of the notch and to develop models of cyclic plasticity for selected metallic materials.

In this paper, we focused on the evaluation of selected low cycle fatigue criteria in the case of constant amplitude of strain. Ten selected criteria proposed for multiaxial fatigue according to various authors were evaluated. As an experimental set, the own data obtained for the 316L stainless steel under seven different strain paths served for evaluation.

Evaluation study description

Known algorithms have been implemented to determine the amplitude of normal and shear stress / strain. The algorithm of the circumscribed circle identification was used in presented calculations. In all cases, it is the criterion of the critical plane. Following lifecycle algorithms are presented:

- Jiang´s criterion [2],
- Socie et all [3],
- Fatemi-Socie [6],

- Brown-Miller [7],
- Kandil-Brown-Miller [8],
- Wang-Brown [9].

The Jiang's criterion includes the concept of the critical plane, the influence of the plastic deformation energy and the principle of the material memory from the mathematical modeling of cyclic plasticity. The combination of the energy approach with the material memory concept has the following form

$$dD = \langle \sigma_{mr} - \sigma_0 \rangle^m \left(1 + \frac{\sigma}{\sigma_f} \right) dY , \quad (1)$$

$$dY = a\sigma \, d\varepsilon^p + \frac{1-a}{2} \tau \, d\gamma^p . \quad (2)$$

Socie et all is the criterion of stress-strain models where the control parameter is deformation and the secondary parameter is stress. The effect of the mean normal stress $\sigma_{n,m}$ is included as follows

$$\gamma_{amax} + \varepsilon_{an} + \frac{\sigma_{n,mean}}{E} = \frac{\tau_f'}{G} (2N_f)^{b_\gamma} + \gamma_f' (2N_f)^{c_\gamma} . \quad (3)$$

Fatemi-Socie is a stress-strain criterion as well.

$$\frac{\Delta\gamma}{2} \left(1 + k \frac{\sigma_{n,max}}{R_e} \right) = f(N_f) . \quad (4)$$

Brown-Miller is a deformation criterion. The deformation is the damage critical parameter. The critical plane is defined as the plane where the shear deformation reaches the maximum value of

$$\frac{\Delta\gamma_{max}}{2} = \frac{\tau_f'}{G} (2N_f)^{b_\gamma} + \gamma_f' (2N_f)^{c_\gamma} . \quad (5)$$

Unlike the previous criterion, the effect of both normal and shear deformation is assumed in case of Kandil-Brown-Miller criterion.

$$\frac{\Delta\bar{\gamma}}{2} = \left(\frac{\Delta\gamma_{max}^\alpha}{2} + S\Delta\varepsilon_n^\alpha \right)^{\frac{1}{\alpha}} = f(N_f) . \quad (6)$$

Wang-Brown is a modification of the Kandil-Brown-Miller criterion, including the influence of mean stress, $\alpha = 1$

$$\frac{\Delta\bar{\gamma}}{2} = \frac{\Delta\gamma_{max}}{2} + S\Delta\varepsilon_n = f(N_f) . \quad (7)$$

a, m - material constant,
b_γ - fatigue shear strength exponent,

c_γ – fatigue ductility exponent,

dD - infinitesimal increment od fatigue damage,
D - fatigue damage,
E – Young´s modulus,

G - shear modulus,

k – material constant,
N_f - number of cycles to failure,

R_e - yield strength,
S – material constant,
Y - plastic strain energy on a material plane,
γ_{amax} – maximum plastic shear strain range,

$\acute{\gamma}_f$ - fatigue shear ductility coefficient,

γ^p - plastic shear strain corresponding to the shear stress on a material plane,
$\Delta\gamma$ – maximum shear strain range,
$\Delta\bar{\gamma}$- ekvivalent of maximum shear strain range,

ε_{an} – elastic-plastic notch tip strain,

ε^p - plastic strain corresponding to the normal stress on a material plane,
$\Delta\varepsilon_n$ - normal strain range on the plane of maximum shear strain,

σ - normal stress on a material plane,
σ_f - true fracture stress,
σ_{mr} - memory stress,
$\sigma_{n,max}$ – maximum normal stress on the plain of maximum shear strain

$\sigma_{n,mean}$ – mean normal stress on the plane of maximum shear strain,,

σ_0 - endurance limit,
τ - shear stress on a material plane,
$\acute{\tau}_f$ - torsional fatigue strength coefficient,

<> - MacCauley bracket.

Uniaxial fatigue parameters of the SS316L steel were determined by both conventional and 3D method. Six fatigue constants is usually determined by the conventional method based on the linear approximation of experimentally obtained data (three independent linear regressions). The main advantage of the new3D method is that it ensures the compatibility of six fatigue constants. The method is based on the linear approximation in space. The complete description of the 3D method is in [4].

Details about multiaxial experiments are listed in article [5]. The following loading spectre were considered:

- uniaxial (case A),
- torsional (B),
- proportional (C),
- square (D),
- rhombic (E),
- two blocks (F),
- circular (G),
- elliptical (H).

The elliptical path shape corresponds to the 45° out of phase fatigue test. Case (A) was used only to determine material parameters by the 3D method.

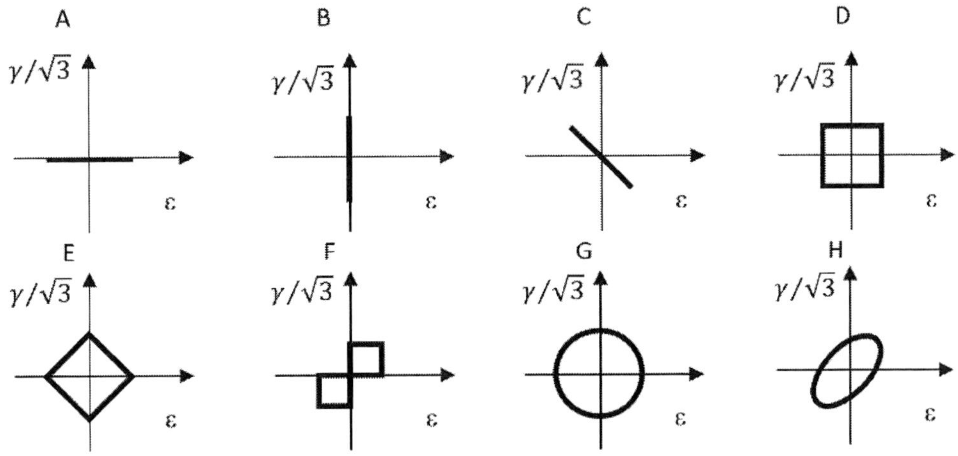

Fig.1 Strain path shapes in particular cases

The best results are obtained from theories according to Jiang [2] and Socie at all [3]. In order to evaluate the agreement between experimental measurements and results obtained from the predictions using individual criteria, charts were created in logarithmic coordinates, where values obtained from the experimental measurements N_{exp} are plotted on the horizontal axis and values obtained using individual criteria N_{pred} are plotted on the vertical axis. In the ideal case, points obtained should lie on a straight line that passes through the origin and is inclined at an angle of 45 degrees. Additionally, the Logarithmic Lifetime Ratio (LLR) criterion was selected to evaluate the agreement of individual criteria with experimental data, defined by the following expression

$$LLR = \log\left(\frac{N_{exp}}{N_{pred}}\right). \tag{8}$$

In the case of an ideal state - when the value obtained from the experiment and the value obtained using the criterion are the same, the value of LLR = 0. If values differ from one another, two cases may occur. If LLR is <0, the criterion predicts non-conservative results. On the other hand, if the value of LLR> 0, the criterion predicts conservative results. The value of the criterion (without considering the sign) shows then the quantitative match between results obtained by experiments and calculations.

Results presenting the comparison of the number of cycles predicted by criteria mentioned above with experimentally determined number of cycles are presented in Fig. 2.

Conclusions

Best results were obtained by the Socie at all criterion (see Figure 1). Jiang´s criterion gives slightly more conservative prediction. Despite this fact, the prediction capability of the Jiang´s criterion can be considered as very good too, compared to other eight criteria. Its advantage is the possibility of the application on the general multiaxial loading including the variable amplitude loading without the use the rainflow method.

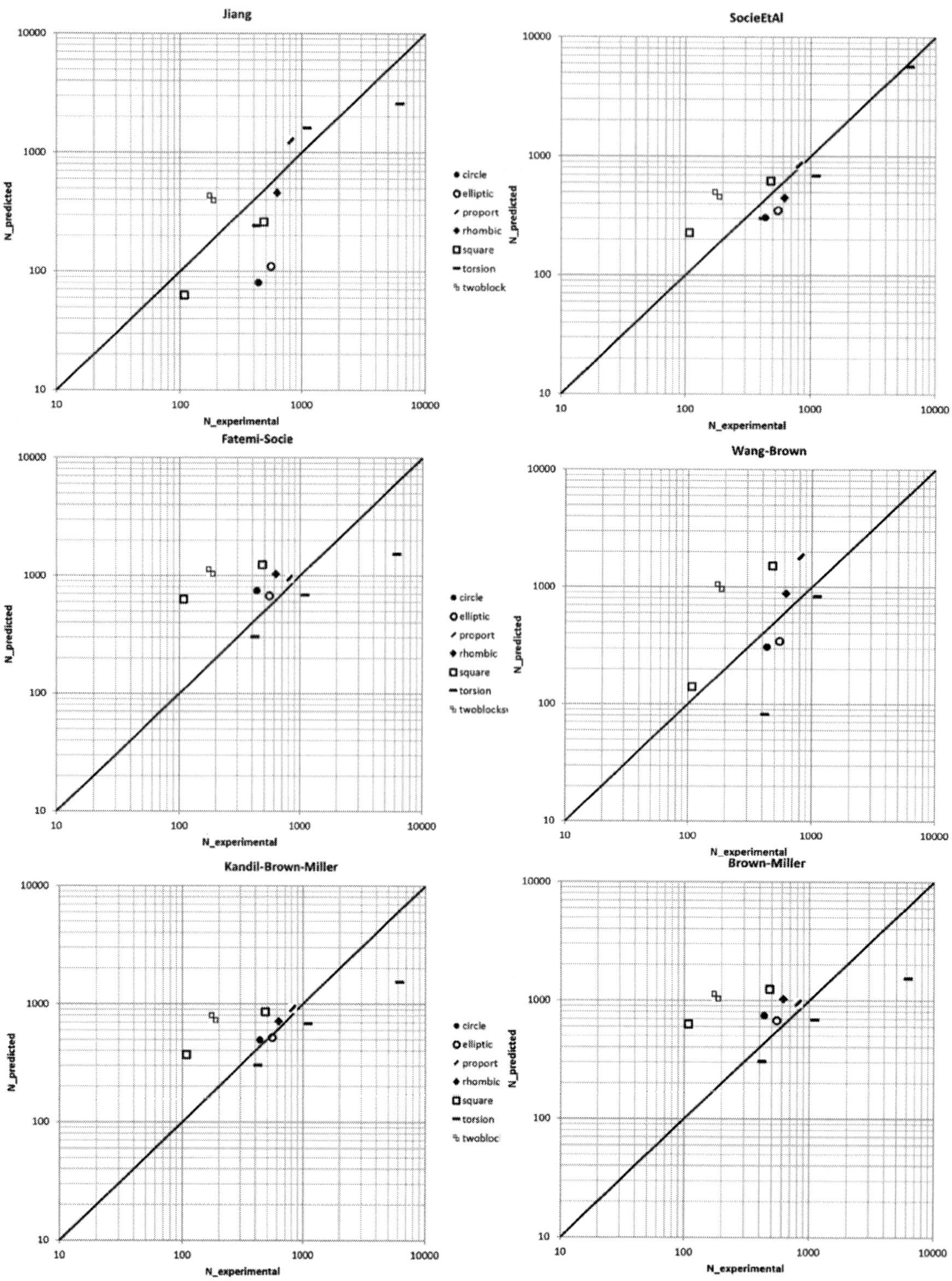

Fig.2 Comparison of predictions of multiaxial criteria with experiment

Acknowledgment

This work was supported by The Ministry of Education, Youth and Sports from the National Programme of Sustainability (NPU II) project „IT4Innovations excellence in science - LQ1602", the project of specific research under No. SP2017/136 and by the Grant Agency of the Czech Republic (GACR), project No. 15-18274S and by project "Experimental and computational modeling in application for selected tasks of technical practice" project No. SP2018/63.

References

[1] PAPUGA, J., HALAMA, R., FUSEK, M., ROJÍČEK, J., FOJTÍK, F., HORÁK, D., PECHA, M., TOMČALA, J., ČERMÁK, M., HAPLA, V., SOJKA, R., KRUŽÍK, J. Efficient Lifetime Estimation Techniques for General Multiaxial Loading. In: Proceedings of International Conference of Numerical Analysis and Applied Mathematics 2016, ICNAAM 2016; Rodos Palace Hotel Rhodes; Greece; 19 September 2015 through 25 September 2016. AIP Conference Proceedings, Volume 1863.

[2] JIANG, Y. A fatigue criterion for general multiaxial loading. *Fatigue & Fracture of Engineering Materials & Structures*, 2000, Vol. 23, 19-32.

[3] INCE, A. Development of Computational Multiaxial Fatigue Modelling For Notched Components. Waterloo, Ontario, Canada, 2012

[4] LIČKOVÁ D., HALAMA R. Identification of Fatigue Constants by means of 3D Method. Strojnícky časopis – Journal of Mechanical Engineering, 2016, Vol.66, No. 2, pp.107-116.

[5] HALAMA R., LIČKOVÁ D., MORALES J.J.R. Cyclic Hardening Behavior of Stainless Steel 316L - Experiments and Modeling. In: Proceedings of 54th International Scientific Conference EXPERIMENTAL STRESS ANALYSIS 2016, Srní, 30 May – 2 June 2016, Czech Republic.

[6] YOU, B.R., LEE, S.B. A Critical Review on Multiaxial Fatigue Assesments of Metals. Int.J.Fatigue. 1996, vol. 18, issue. 4, s. 235-244.

[7] REIS, L., LI, B., FREITAS, M. Crack initiation and growth path under multiaxial fatigue loading in structural steels: a complete restatement of the entire american law as developed by all reported cases : kept to date by cumulative annual pocket parts. International Journal of Fatigue. 2009, vol. 31, 11-12, s. 1660-1668.

[8] PAN, W., HUNG, CH., CHEN, L. Fatigue life estimation under multiaxial loadings. International Journal of Fatigue. 1999, vol. 21, issue 1, s. 3-10.

[9] SUN, G.Q., SHANG, D.G., BAO, M. Multiaxial fatigue damage parameter and life prediction under low cycle loading for GH4169 alloy and other structural materials. *International Journal of Fatigue*. 2010, vol. 32, issue 7, s. 1108-1115.

Experimental Analysis of Various Temperature Loadings Influence on Cement Composite

HOLČAPEK Ondřej[1, a], REITERMAN Pavel[1,b] and KONVALINKA Petr[1,c]

[1]Czech Technical University in Prague, Faculty of Civil Engineering, Experimental Centre, Thákurova 7, 166 29, Prague 6, Czech Republic

[a]ondrej.holcapek@fsv.cvut.cz, [b]pavel.reiterman@fsv.cvut.cz, [c]petr.konvalinka@fsv.cvut.cz

Keywords: fracture characteristics, mechanical properties, composite, aluminous cement, elevated temperatures.

Abstract. This contribution deals with the experimental investigation of mechanical, fracture and basic properties of refractory cement composite after the action of various temperatures loadings. Investigated refractory cement composite on aluminous cement based with high dosage of metakaolin was reinforced by basalt fibres. The temperature loading was especially focused on temperature between 700°C and 1000°C (grading after 100°C) and between 100°C and 700°C (grading after 200°C). The bulk density, dynamic modulus of elasticity, compressive strength, flexural strength and fracture energy were investigated after the action of mentioned temperatures. The action of temperature between 300°C and 500°C means the highest decrease of mechanical properties. The percentage decrease of flexural strength and fracture energy was higher than compressive strength. The decrease of investigated properties after the action of temperature higher than 700°C was gradual and not so rapid. The residual mechanical properties after the action of 1000°C achieved sufficient values for cement composite with natural aggregate.

Introduction

The issue of general cement composites′ reaction on the effect of elevated temperature has been investigated in various research works. The gradual chemical decomposition of CSH gels characterizes composites based on ordinary Portland cement. The portlandite $Ca(OH)_2$ is decomposited to quick-lime and CO_2 at 400°C [1]. The calcium aluminate cements characterize different phase composition and also the reaction and stability for elevated temperatures environments. The decomposition of crystalline gibbsite AH_3 starts at 300°C, while the formation of $C_{12}A_7$ starts at 500°C [2]. The growing temperature means conversion to calcium aluminate (CA) and calcium di-aluminate (CA_2) [3].

Description of performed experiments

Used refractory composite. Investigated refractory composite was on hydraulic base. Aluminous cement Secar®71 produced by Lafarge company in dosage of 675 kg/m^3 was supplemented by matakaolin Mefisto L05 in dosage of 225 kg/m^3. Natural crushed basalt aggregate 0/4 mm and 2/5 mm was used in total amount of 1100 kg/m^3. Totally 280 kg/m^3 of basalt fibers (combination of two lengths - 12.7 mm and 6.35 mm) were used as reinforcement. The water to binder ration achieved 0.25 level because of using of super-plasticizer SIKA.

Tab.5 Composition of investigated refractory composite

Aluminous cement	Metakaolin	Basalt aggregate		Basalt fibres	Water	Plasticizer
Secar®71	MefistoL05	0/4 mm	2/5 mm		potable	SVC 1035
kg/m^3						
675	225	880	220	29	225 kg/m^3	22.75

Temperature treatment and loading. Steel forms with fresh mixture were covered by foil for 3 days. Unformed specimens (40 × 40 × 160 mm) were stored in humidity conditions (approx. 90% of relative humidity and 20°C) till the age of 28 days. The drying process (100°C for three days) took place at the age of 28 days. Dried specimens (reference level) were exposed to thermal loading of different levels (300°C, 500°C, 700°C, 800°C, 900°C and 1000°C) for 3 hours (the temperature rise up in gradient 10°C/minute) in automatic electric furnace. After 3 hours the source of heat was automatically turn off and the furnace cooled down till laboratory conditions.

Investigated parameters. Measurement of weight took place at the end of thermal loading cycle, approximately at 60°C (to avoid the absorption of air humidity). The height of specimens was measured at laboratory conditions before drying and temperature loading. The remaining two dimensions were given by the size of used mould. From these values the average bulk density (ρ_m) was calculated. At this time the measurement of ultrasonic speed by PunditLab+ device took place. The test arrangement of specimen, producer and receiver of ultrasonic signal was one-dimensional and direct. The dynamic modulus of elasticity (E_{cum}) was calculated according to Czech standard [4]. The three-points bending test with clear distance of supports of 100 mm was performed by universal loading machine MTS100. Bending test was controlled by increase of deformation with speed 0.2 mm/min. The average flexural strength (f_{tm}) was calculated from maximal achieved force using theory of elasticity, as recommended in Czech Standard [5]. Fracture energy (G_{fm}) was calculated according to RILEM recommendation described by eq. (1) based on data record from bending test, where a and b (mm) are dimensions of cross-section, while the integral expression means the consumed work (area under the load-deflection diagram).

$$G_f = \frac{\int_0^{\delta_{max}} F(\delta)d\delta}{a \cdot b} \quad (1)$$

Finally the compressive test (f_{cm}) was performed on two fragments left from bending test. The loading machine EU 40 was used and the increase of deformation 0.2 mm/min controlled the test. The maximal testing area was defined by steel plate with dimensions 40 × 40 mm according to [5].

Results and discussion

All results of bellowed described parameters are summarized in Table 1. From each parameter the average calculated values are shown and the percentage influence of temperature loading influence is expressed. The comparable level (100%) corresponds to the action of 100°C (drying).

The highest decrease of bulk density takes occur between 300°C and 500°C (5.1%) and between 100°C and 300°C (3.1%). The decrease of bulk density between laboratory conditions and drying at 100°C was not the aim of interest (this decrease is connected with loss of free and part of physically bounded water). The recorded bulk density drop between 100°C and 1000°C corresponds to 89.8%.

The first thermal loading after drying (300°C) means decrease of dynamic modulus of elasticity by 32.6% of reference value. A decrease of approximately 77% characterizes the action of 700°C. It should be noted the values of dynamic modulus of elasticity depend on bulk density. This parameter is suitable for description of temperature influence trend.

The action of 300°C means only 19% decrease of compressive strength, while the highest decrease expressed by 58.1% of the original value takes occur between 300°C and 500°C. The decrease between 700°C and 900°C is expressed by approx. 10%, while the action of 1000°C characterizes decrease of 12.3% of the value of 900°C. The relatively small decrease of compressive strength after the action of 300°C is describe for ordinary concrete in [6].

The dependence of flexural strength and fracture energy on the effect of various high temperatures is similar. Unlike compressive strength, the highest decrease of flexural strength and especially fracture energy takes place after the action of 300°C. The fracture energy is expressed by 45.2%, while flexural strength means 57.7% of the original value. The increasing of temperature characterizes the gradual decrease of mentioned parameters (see Fig. 2 and Fig. 4).

Tab.6 Summarization of investigated parameters (values and percentage expression)

Parameters		100°C	300°C	500°C	700°C	800°C	900°C	1000°C
Bulk density - *(ρ_m)*	[kg/m³]	2265	2195	2080	2065	2055	2040	2035
	[%]	*100*	*96.9*	*91.8*	*91.2*	*90.7*	*90.1*	*89.8*
Dynamic modulus of elasticity - *(E_{cum})*	[GPa]	42.7	28.8	17.6	10.0	9.6	8.4	8.5
	[%]	*100*	*67.4*	*41.2*	*23.4*	*22.5*	*19.7*	*19.9*
Flexural strength - *(f_{tm})*	[MPa]	17.5	10.1	7.4	6.9	6.0	5.5	5.0
	[%]	*100*	*57.7*	*42.3*	*39.4*	*39.3*	*31.4*	*28.6*
Compressive strength - *(f_{cm})*	[MPa]	104.5	85.1	60.7	56.6	52.5	42.6	29.8
	[%]	*100*	*81.4*	*58.1*	*51.2*	*50.2*	*40.8*	*28.5*
Fracture energy - *(G_{fm})*	[J/m²]	644.5	291.3	257.5	203.6	182.7	197.8	153.4
	[%]	*100*	*45.2*	*40.0*	*31.6*	*28.3*	*30.7*	*23.8*

Graphs on Fig. 1, Fig. 2, Fig. 3 and Fig. 4 provide the graphical evaluation of investigated parameters and bring the information about gradual decrease of mentioned properties.

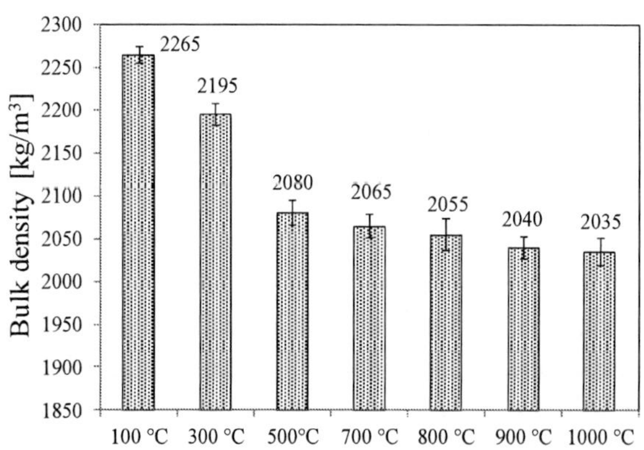

Fig.1 Evaluation of bulk density

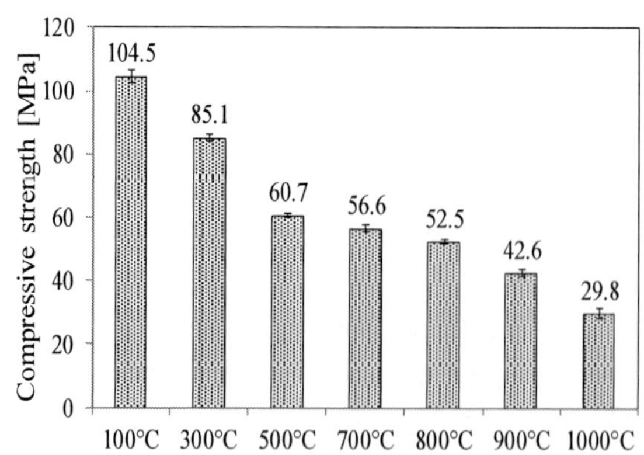

Fig.2 Evaluation of compressive strength

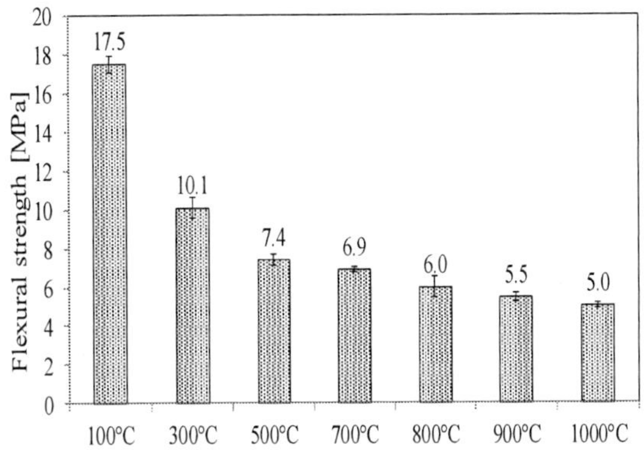

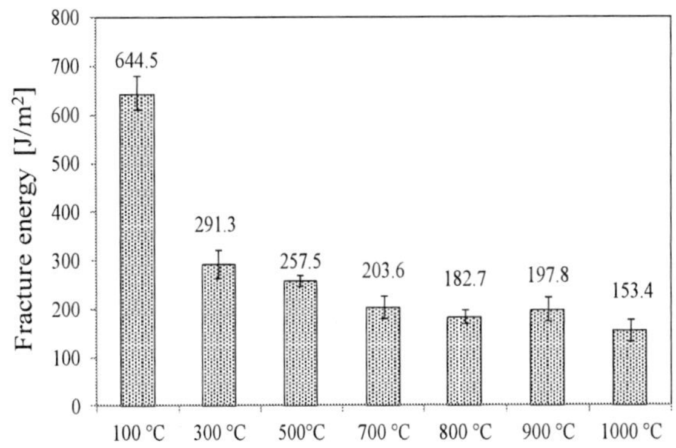

Fig.3 Evaluation of flexural strength

Fig.4 Evaluation of fracture energy

Conclusions

Described parameters were tested after the action of elevated temperature of various levels. The highest decrease of basic and mechanical parameters takes occur between 100°C and 500°C. This research was specifically focused on the temperature range between 700°C and 1000°C. The decrease of bulk density is connected with evaporation of free, physically and chemically bounded water. All investigated parameters gradually decrease. The highest decrease was observed in the case of dynamic modulus of elasticity, which caused by decrease of bulk density. For practical purposes the flexural and compressive strength mean the main parameter. Based on performed experimental program we can conclude that mentioned parameters are on sufficient level even after the action of 1000°C.

Acknowledgement

This research was financially supported by CTU project No. SGS16/199/OHK/3T11 – "Multicomponent binder systems for composite building materials".

References

[1] Thermal stability of C-S-H phases and applicability of Richardson and Groves′ and Richardson C-(A)-S-H-(I) models of synthetic C-S-H, E.T. Rodriguez, K. Garbev, D. Merz, L. Black, I.G. Richardson, Cement and Concrete Research, vol. 93, pp. 45-56, 2017.

[2] High temperature material properties of calcium aluminate cement concrete, W. Khaliq, H.A. Khan, Construction and Building Materials, vol. 94, pp. 475-487, 2015.

[3] High temperature durability of fiber reinforced high alumina cement composite, E. Vejmelková, D. Koňáková, L. Scheinherrová, M. Doležalová, M. Keppert, R. Černý, Construction and Building Materials, vol. 162, pp. 881-891, 2018.

[4] Czech Standard CSN 73 1371 Non-destructive testing of concrete – Method of ultrasonic pulse testing of concrete.

[5] Czech Standard CSN EN 196-1 Methods of testing cement – Part 1: Determination of strength, 2005.

[6] Properties of Concrete at Elevated Temperatures, V. Kodur, Civil Engineering, pp. 1-15, 2014.

EAN 2018
56th conference on experimental stress analysis

Effect of Activation Recycled Concrete using Different Admixtures on the Strength and Modulus of Elasticity of the Resulting Cement Composite

HRBEK Vladimír[1,a], ZOBAL Ondřej[1,2,b], PROŠEK Zdeněk[1,2,c], ĎUREJE Jakub[1,d] and TESÁREK Pavel[1,e]

[1]CTU in Prague, Faculty of Civil Engineering, Department of Mechanics, Thákurova 7, 166 29 Prague 6, Czech Republic

[2]University Centre for Energy Efficient buildings, Czech Technical University in Prague, Třinecká 1024, 273 43 Buštěhrad, Czech Republic

[a]vladimir.hrbek@fsv.cvut.cz, [b]ondrej.zobal@fsv.cvut.cz, [c]zdenek.prosek@fsv.cvut.cz, [d]jakub.dureje@fsv.cvut.cz, [e]pavel.tesarek@fsv.cvut.cz

Keywords: recycled concrete, cement, lime, fly ash, slag, mechanical activation, modulus of elasticity, compressive strength, tensile strength in bending.

Abstract. The demanding process of recycling building waste, including concrete, is a very topical topic. One of the possibilities of processing of waste concrete and subsequent use is its milling, i.e. mechanical activation. Concrete powder can be thermally activated, but it is an energy-intensive process and is produced CO_2. This paper deals with the improvement of mechanical activation with help of various admixtures and their influence on the material and strength characteristics.

Introduction

The mechanical activation of recycled concrete involves a very fine fraction of the waste concrete (< 1 mm). Coarse fractions are commonly used as aggregate or filler in concrete, wider use of very fine fraction (powder) has not yet been found. The cement paste forms a larger part of the fine fraction when recycling concrete and causes greater water absorption. However, other crushing and milling of powder can eliminate this problem [1]. The old cement paste contains unhydrated grains. The milling to a finer fraction can help activate these grains, so we talk about mechanical activation. The this way treated material can then be used as a substitute for cement in concete, but maximal 15%, with higher substitution there is a significant decrease in compressive strength and an increase in water absorption [2,3]. A fine fraction of recycled concrete can be successfully used to reinforce the underlying layers of railway constructions [4].

Concrete powder can be thermally activated, but it is an energy-intensive process and is produced CO_2. It has been proven that a new "clinkering" at temperatures of 500 to 800 ° C has a positive effect on the fine fraction of recycled concrete, because it is possible to replace the cement with this dehydrated powder while maintaining the strength of the concrete. As other results show, when combined dehydrated concrete with fly ash or slag can be strength of concrete further increased [5-9].

This document shows a combination of mechanical activation of the powder from recycled concrete and various additives with goal to improve material and strength characteristics of cement composite.

Materials and test specimens for the experiments

The basic reference mixture is made of cement paste from cement with designation CEM I 42.5 R The second reference mixture is composed of 50 % cement and 50 % recycled concrete from the railway sleepers. Lime hydrate, slag and fly ash were used as additives for further cement substitution. The chemical name of lime hydrate is calcium hydroxide and is prepared by firing pure gypsum, subsequently it is grinded and hydrated. Calcium hydrate has strong alkaline properties and is used in the building industry for the production of dry plasters and mortar mixtures [10]. The slag forms a waste material in the process melting of iron ore in blast furnaces, but due to its hydraulic properties it is widely used in the building industry for the production of mixed cements [11]. Fly ash is a waste material from coal combustion in thermal power plants, but just as the slag is used for its hydraulic properties and the ability to reduce the hydration temperatures in the building industry in cement and concrete. For the experiment, standard fly ash was used according to ČSN EN 450-1 from locality Mělník [12,13].

Altogether, 11 different mixtures were prepared - 2 references mixtures with designation CEM and CEMREC-5050 and 9 mixtures with other additions. The designation of the other mixtures depends on the type and quantity of the additive used - the lime hydrate Ca, the fly ash FA and the slag S. The substitution was 5, 10 and 15 % weight of cement. Water and binder ratio was same for all mixtures and his value was 0.35. Table 1 summarizes the designation and composition of the mixtures. All specimens have the shape of beams with dimensions of 40 × 40 × 160 mm and 6 pieces of each mixture have been made. Individual beams were taken from molds after 24 hours and stored in a water bath at 20 °C for 28 days.

Tab.1 Designation and composition of the mixtures.

Designation of the mixture	Composition of the mixture [%]					w/b [-]
	Cement	Rec. concrete	Lime	Fly ash	Slag	
CEM	100	0	0	0	0	0.35
CEMREC-5050	50	50	0	0	0	0.35
CEMREC-5050-Ca05	45	50	5	0	0	0.35
CEMREC-5050-Ca10	40	50	10	0	0	0.35
CEMREC-5050-Ca15	35	50	15	0	0	0.35
CEMREC-5050-FA05	45	50	0	5	0	0.35
CEMREC-5050-FA10	40	50	0	10	0	0.35
CEMREC-5050-FA15	35	50	0	15	0	0.35
CEMREC-5050-S05	45	50	0	0	5	0.35
CEMREC-5050-S10	40	50	0	0	10	0.35
CEMREC-5050-S15	35	50	0	0	15	0.35

Methodology and evaluation of experiments

For all fresh mixtures, the consistency and workability was determined using a flow test on Hägermann's table [14]. On all test samples, the dynamic and shear modulus of elasticity was evaluated using a non-destructive resonance method [15]. On half of specimens, the compressive and tensile strength after 28 days was tested using destructive experiments according ČSN EN 196-1 and verified procedures [16,17].

Consistency. The flow test on the Hägermann's table verified the consistency and workability of the individual fresh mixture. The Table 2 shows average volume density and evaluation of consistency and workability.

Tab.2 Average values of the volume density and consistency and workability

Designation of the mixture	ρ_{avg} [kg/m^3]	Spillage - avg [mm]
CEM	2088.4 ± 11.2	218.5
CEMREC-5050	1977.7 ± 7.0	191.5
CEMREC-5050-Ca05	1958.0 ± 4.0	176.5
CEMREC-5050-Ca10	1906.0 ± 17.1	168.5
CEMREC-5050-Ca15	1876.8 ± 4.5	138.5
CEMREC-5050-FA05	1926.7 ± 4.4	209.5
CEMREC-5050-FA10	1907.6 ± 3.0	199.5
CEMREC-5050-FA15	1866.8 ± 6.6	212.5
CEMREC-5050-S05	1965.2 ± 5.2	218.5
CEMREC-5050-S10	1959.6 ± 5.6	239.0
CEMREC-5050-S15	1956.0 ± 2.0	236.0

Average spillage of reference mixture CEM was 218.5 cm. The substitution of 50 % weight cement with recycled concrete reduced spillage of 12.4 %. The individual additives had different effects on consistency and workability. As expected, lime hydrate degraded the workability significantly, when using 5 % by 19.2 % and using 15 % even by 36.6 %. The mixture with fly ash varied to 10 % less, and when slag added, the spillage increased by almost 10 %.

Dynamic and shear modulus of elasticity. All 66 test specimens were dried, measured and weighed after taking out from the water bath, so that the required parameters were obtained - dimensions and volume density - for later evaluation of the modules. The influence of additives at substitution max 15 % to the volume density is minimal, as well as the 50% substitution of cement by the recycled concrete itself (difference maximal 11 %).

The transverse frequency for each test body was measured. The measurement itself was non-destructive and was made after 7, 14, 21 and 28 days to observe the impact of the additives. Tables 3 and 4 show the average values of the dynamic modulus of elasticity E_{dyn} and the shear modulus of elasticity G and their standard deviations (for most mixtures it is 1 to 2 %, for some mixtures with lime hydrate and fly ash it is up to 5 %). Graphs on Fig. 1 and 2 depict changes on modules values within 28 days.

Tab.3 Average values of the dynamic modulus of elasticity and standard deviation

Designation of the mixture	E_{dyn} [GPa]							
	$E_{dyn,7}$	$\sigma_{x,Edyn,7}$	$E_{dyn,14}$	$\sigma_{x,Edyn,14}$	$E_{dyn,21}$	$\sigma_{x,Edyn,21}$	$E_{dyn,28}$	$\sigma_{x,Edyn,28}$
CEM	22.719	0.188	24.765	0.186	25.714	0.188	26.335	0.219
CEMREC-5050	17.265	0.063	18.678	0.056	19.149	0.045	19.450	0.057
CEMREC-5050-Ca05	15.699	0.083	17.148	0.086	17.638	0.089	17.891	0.116
CEMREC-5050-Ca10	13.421	0.094	15.301	0.728	15.118	0.107	15.347	0.104
CEMREC-5050-Ca15	12.183	0.204	13.365	0.242	13.748	0.245	13.946	0.213
CEMREC-5050-FA05	15.105	0.058	15.980	0.842	17.092	0.049	17.415	0.029
CEMREC-5050-FA10	14.339	0.082	15.847	0.056	16.389	0.052	16.776	0.057
CEMREC-5050-FA15	13.055	0.087	14.510	0.090	15.087	0.098	15.570	0.089
CEMREC-5050-S05	14.855	0.013	16.915	0.019	17.528	0.003	17.884	0.023
CEMREC-5050-S10	13.338	0.059	15.283	0.072	15.918	0.055	16.323	0.076
CEMREC-5050-S15	12.377	0.029	14.164	0.049	14.780	0.037	15.181	0.038

Tab.4 Average values of the shear modulus of elasticity and standard deviation

Designation of the mixture	G [GPa]							
	G_7	$\sigma_{x,G,7}$	G_{14}	$\sigma_{x,G,14}$	G_{21}	$\sigma_{x,G,21}$	G_{28}	$\sigma_{x,G,28}$
CEM	8.997	0.087	9.824	0.053	10.021	0.115	10.646	0.381
CEMREC-5050	6.939	0.023	7.534	0.042	7.679	0.063	7.744	0.072
CEMREC-5050-Ca05	6.364	0.101	7.039	0.049	7.189	0.088	7.221	0.095
CEMREC-5050-Ca10	5.455	0.086	6.266	0.367	6.131	0.104	6.102	0.070
CEMREC-5050-Ca15	4.765	0.236	5.335	0.054	5.331	0.024	5.507	0.059
CEMREC-5050-FA05	6.077	0.021	6.571	0.310	6.820	0.071	6.952	0.058
CEMREC-5050-FA10	5.863	0.047	6.440	0.075	6.808	0.303	6.673	0.092
CEMREC-5050-FA15	5.331	0.033	5.964	0.096	6.054	0.103	6.256	0.049
CEMREC-5050-S05	5.820	0.094	6.656	0.037	6.808	0.049	7.037	0.071
CEMREC-5050-S10	5.249	0.119	5.945	0.078	6.133	0.039	6.385	0.030
CEMREC-5050-S15	4.746	0.049	5.587	0.075	5.691	0.060	5.888	0.119

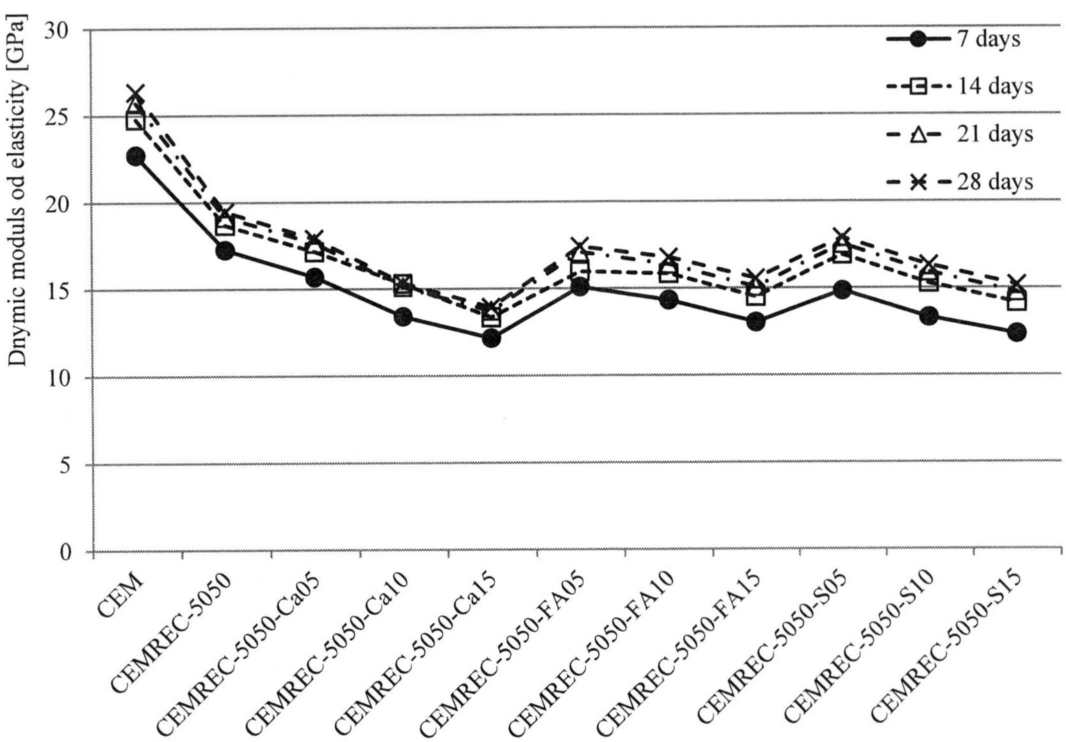

Fig. 1 Development of dynamic modulus of elasticity of individual mixtures during 28 days

Both the evaluation of the dynamic module and the shear modulus showed a great influence of 50% substitution of cement with recycled concrete. The decrease in the dynamic modulus was 22.8 %, in the shear modulus was almost the same, 23.3 %. The largest decrease in module values occurred in mixtures with 15 % lime hydrate and slag, ie mixture CEMREC-5050-Ca15 and CEMREC-5050-S15. All values increase over time. The dynamic module increased for reference mixture CEM by 15.9 % and by 12.7 % for CEMREC-5050 after 28 days. The shear module increased for reference mixture CEM by 11.8 % and by 11.6 % for CEMREC-5050 after 28 days.

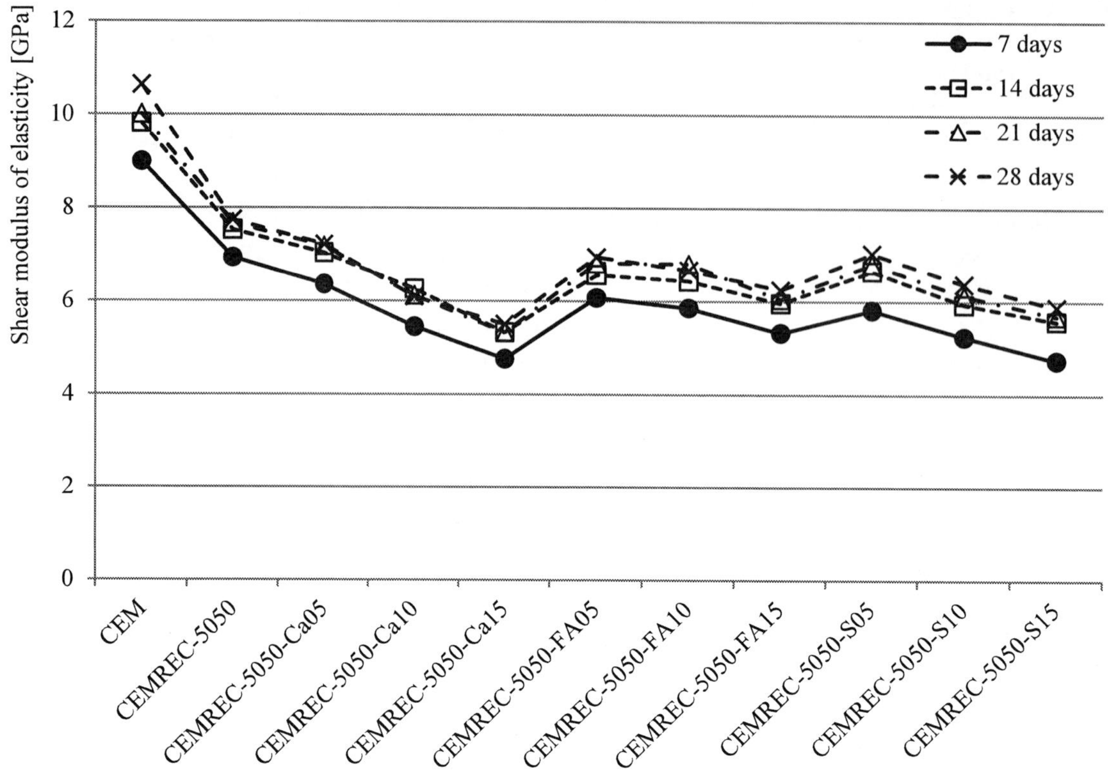

Fig. 2 Development of shear modulus of elasticity of individual mixtures during 28 days

Compressive and tensile strength. Half of test beams with dimensions 40 × 40 × 160 mm, ie 33 pieces, were used for destructive measurement of tensile strength after 28 days. The other half will be tested after 90 days. On the broken test samples after the tensile test, the compressive strength was tested (ie 6 bodies for each mixture). The measurement of tensile strength was proveded as a three point test. Distance of supports was 100 mm and test was calculated by help of the maximum force. The area under compressive load had size 160 mm^2 and has been demarcated by the loading device.

Tab.5 Average values of the compressive and tensile strength and their standard deviation after 28 days

Designation of the mixture	Compressive and tensile strength [MPa]			
	$\sigma_{c,28}$	$\sigma_{x,\sigma c,28}$	$\sigma_{t,28}$	$\sigma_{x,\sigma t,28}$
CEM	115.228	6.659	7.131	0.905
CEMREC-5050	50.256	1.455	6.169	0.470
CEMREC-5050-Ca05	44.065	2.527	5.926	0.308
CEMREC-5050-Ca10	36.298	1.041	5.671	0.409
CEMREC-5050-Ca15	30.755	1.361	4.161	1.620
CEMREC-5050-FA05	44.150	2.425	6.287	0.717
CEMREC-5050-FA10	38.712	3.140	6.188	0.344
CEMREC-5050-FA15	34.201	1.645	5.615	0.484
CEMREC-5050-S05	44.907	1.484	7.474	0.269
CEMREC-5050-S10	36.785	1.371	6.272	0.460
CEMREC-5050-S15	32.532	1.497	6.329	0.369

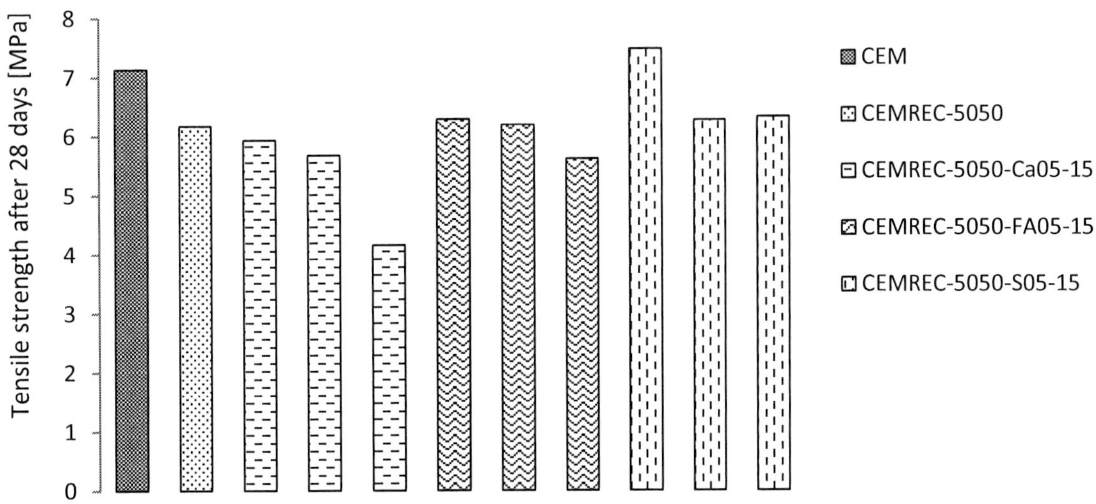

Fig. 3 Average values of tensile strength after 28 days

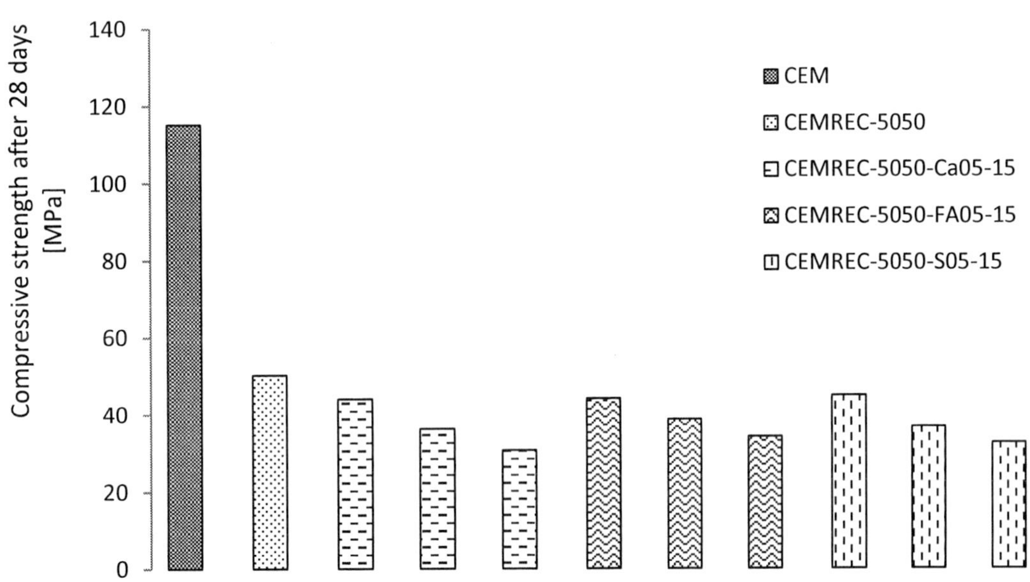

Fig. 4 Average values of compressive strength after 28 days

Table 5 summarizes the average values of compressive and tensile strengths after 28 days and their standard deviation. Fig. 3 and 4 show graphically the development of the strengths of the individual mixtures. The compressive strength of the reference mixture CEM is 115.2 MPa, by the mixture CEMREC is 50.3 MPa. The compressive strength values for mixtures with additives are between 30.8 and 44.9 MPa. A similar trend of development of strength can be observed for all mixtures with additives.

The tensile strength of the reference mixture CEM is 7.1 MPa, by the CEMREC mixture has value 6.2 MPa. The development of tensile strength values for mixtures with additives has a different trend over compressive strength. The mixture CEMREC-5050-S05 has a value of tensile strength even higher than the reference mixture (7.5 MPa).

Conclusions

Achieved results indicate that milled recycled concrete powder can be used for special applications in cement and concrete. The problem can occur with the consistency. The various grain curves of the individual components of the composite have certainly influence on consistency. The influence of particle size on the properties of the fresh and hardened mixture is also solved [18]. It is necessary to find the optimum between good workability and sufficient strength and material characteristics. When 50 % of cement was replaced with recycled concrete, the compressive strength decreased by more than 50 % but still reached values of about 40 to 50 MPa after 28 days. The tensile strength had not a dramatic decrease and the some additives had the other way around a positive effect. The increase in compressive strength over time is assumed for mixtures with fly ash and slag. [19,20].

Acknowledgement

This paper has been supported by the the Grant Agency of the Czech Republic GA ČR, project No. 17-06771S.

References

[1] C.-S. Poon et al., Effect of microstructure of ITZ on compressive strength of concrete prepared with recycled aggregates. The Hong Kong Polytechnic University, Hong Kong 2002.

[2] Y. J. Kim et al., Utilization of waste concrete powder as a substitution material for cement. Construction and Building Materials 30 (2012) 500–504.

[3] Q. Liu et al., Investigation of using hybrid recycled powder from demolished concrete solids and clay bricks as a pozzolanic supplement for cement. Construction and Building Materials 73 (2014), 754–763.

[4] M. Lidmila et al., Utilization of Recycled Binder for Improvement of Subsoil under Railway Sleepers. 51nd International Conference on Experimental Stress Analysis 2014.

[5] Z. Shui Z et al., Cementitious characteristics of hydrated cement paste subjected to various dehydration temperatures. Construction and Building Materials 23 (2009), 531–537.

[6] M. V. A. Florea et al., Activation of liberated concrete fines and their application in mortars. Construction and Building Materials 50 (2014), 1–12.

[7] F. Tomosawa et al., Towards completely recyclable concrete. Integrated design and environmental issues in concrete technology. E & FN Spon, London, UK (1996), 263–272

[8] J. R. Costes et al., Total recycling of concrete, 2007, http://omogine.blogspot.com/.

[9] E. Anastasiou et al., Utilization of fine recycled aggregates in concrete with fly ash and steel slag. Construction and Building Materials 50 (2014), 154–161.

[10] Information on https://www.cemix.cz/produkty/vapenny-hydrat#soubory

[11] ČSN EN 197-1 Cement – Part 1: Composition, specifications and conformity criteria for common cements, Prague, 2011

[12] W. Berg, Handbuch Flugasche im Beton, Dusseldorf Verlag Bau+Technik GmbH, 2008

[13] ČSN EN 450-1 Fly ash for concrete – Part 1: Definition, specifications and conformity criteria, Prague, 2013

[14] ČSN EN 459-2 Building lime – Part 2: Test methods, Prague, 2011

[15] Plachý, T.; Tesárek, P.; Ťoupek, R.; Polák, M. Nondestructive Determination of Young's Modulus of Gypsum Specimens Using Impulse Excitation Method In: Experimental Stress Analysis 2010. Olomouc: Univerzita Palackého, 2010. pp. 339-344. ISBN 978-80-244-2533-7.

[16] ČSN EN 196-1 Methods of testing cement – Part 1: Determination of strength, Prague, 2005.

[17] O. Holčapek, P. Reiterman, P. Konvalinka, Fracture characteristics of refractory composites containing metakaolin and ceramic fibers, Advances in Mechanical Engineering 7 (3) (2015) 1-13.

[18] J. Topič; Z. Prošek, Properties and microstructure of cement paste including recycled concrete powder, Acta Polytechnica. 2017, 57(1), 49-57. ISSN 1210-2709.

[19] O. Zobal, V. Šmilauer, L. Kopecký a kol., Analýza betonu z přehrady Orlík po 50 letech, Beton TKS 14, 2014

[20] A. Neville, Properties of concrete, 2009

Micromechanical Characterization and Modelling of Recycled Concrete

HRBEK Vladimír[1,a], TESÁREK Pavel[1,b], PROŠEK Zdeněk[1,c], NEŽERKA Václav[1,d] and HLŮŽEK Radim[1,e]

[1] Faculty of Civil Engineering, Czech Technical University in Prague, Thákurova 7, 166 29 Prague, Czech Republic

[a]vaclav.nezerka@fsv.cvut.cz, [b]tesarek@fsv.cvut.cz, [c]zdenek.prosek@fsv.cvut.cz, [d]vladimir.hrbek@fsv.cvut.cz, [e]radim.hluzek@fsv.cvut.cz

Keywords: Recycled concrete, Micromechanics, Homogenization, Microstructure, Nanoindentation, Microscopy.

Abstract. The purpose of this contribution is to outline a basic framework of identification of phases, determination of elastic properties, and micromechanical modeling of cementitious composites based on recycled concrete. It was found that micromechanical modelling can help to understand the role of individual phases, establish relationships between input materials and effective properties of the hardened composites, and assess the impact of individual phases.

Introduction

During the past century, concrete became the most widely used construction material [1]. The production of Portland cement is energetically very demanding [2] and largely contributes to the global CO_2 emissions [3, 4]. For that reason, combined with economic benefits, the partial substitution of cement with recycled reactivated concrete is very appealing. The effect of such recycling would be twofold as the use of the disposed of material would also help to reduce waste production. In the Czech Republic alone, over 23 million tons of concrete were disposed of in 2012, while European Union produced over 2.5 billion tons of concrete waste.

To maximize the incorporation of the recycled material into newly produced mixes, one has to understand the mechanisms and consequences. If, e.g., some minor component of the disposed of material appears to deteriorate mechanical strength or durability of the newly produced concrete, it can be eliminated or modified.

To investigate the behavior of the reactivated particles at the microscale, grid indentation and scanning electron microscopy with energy-dispersive X-ray spectroscopy (SEM, EDX) were employed. Also, we formulated a simple micromechanical model based on Mori-Tanaka scheme, validated by macroscopic measurements.

The individual components of the recycled concrete can deteriorate, but also enhance the properties of the newly produced composites in two ways. First, fine powder can act as an inert micro-filler that contributes to the reduction of porosity and supporting agent for clinker hydration. Second, the non-hydrated phases can further contribute to the matrix strengthening. These mechanisms were revealed recently by Chen et al. [6] or Kim et al. [7].

Materials and Experimental Methods

An epoxy resin bonded samples of finely ground old concrete from disposed of railway sleepers. Such an approach allowed polishing, microscopy investigation, and consequent nanoindentation.

The Scanning Electron Microscopy (SEM) was accomplished using a MIRA II LMU (Tescan corp., Brno) microscope. The chemical composition of each phase characterized by a distinct level of grayscale (Fig. 1). The micrographs were taken using EDX (Bruker Corp. Berlin). The SEM EDX analysis revealed five main phases present in the recycled material. These are low- and high-density C-S-H gels, calcium hydroxide, non-hydrated clinker, and aggregate fragments (Fig. 2).

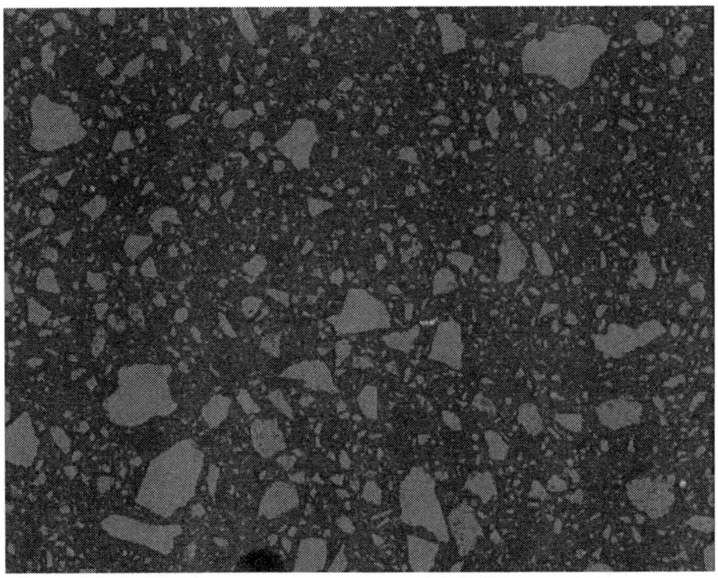

Fig. 1 Mapping of phases in SEM EDX images of micronized recycled concrete

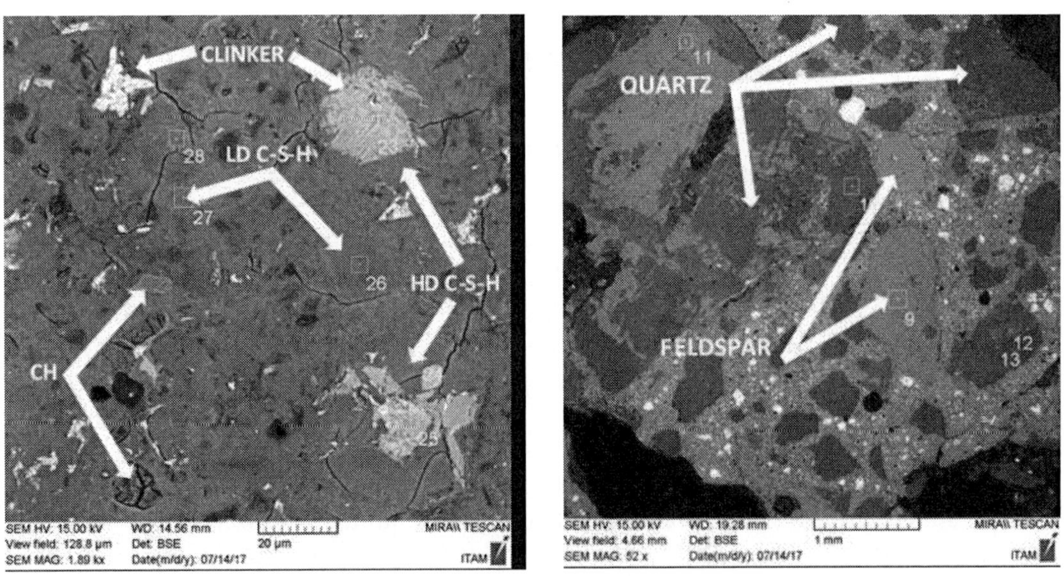

Fig. 2 Micrographs at distinct scales and identification of individual phases

Quasi-static nanoindentation was addressed to accomplish mapping of elastic stiffness of the pastes within selected areas. To that purpose, we used nanoindentor Ti 7500 (Hysitron Inc.). The measurement was performed at two scales (Level I and II), corresponding to further micromechanical modeling. At Level I the nanoindentation grid consisted of 25 by 25 indents with the spacing of 2 μm. Such a setting allowed to investigated the phases constituting the matrix at Level II. At Level II, a displacement-controlled quasi-static grid indentation was performed in the 21 by 21 raster with the spacing of indents equal to 5 μm.

After deconvolution of indentation distribution functions the stiffness of individual phases was established as follows: low-density C-S-H 19.17 ± 3.54 GPa, high-density C-S-H 39.75 ± 5.98 GPa, calcium hydroxide 58.59 ± 6.34 GPa, aggregate 79.78 ± 3.11 GPa, clinker 118.35 ± 16.85 GPa. The map of distribution of Young's moduli is presented in Fig. 3. The results are consistent with previously published research, e.g., [7-9].

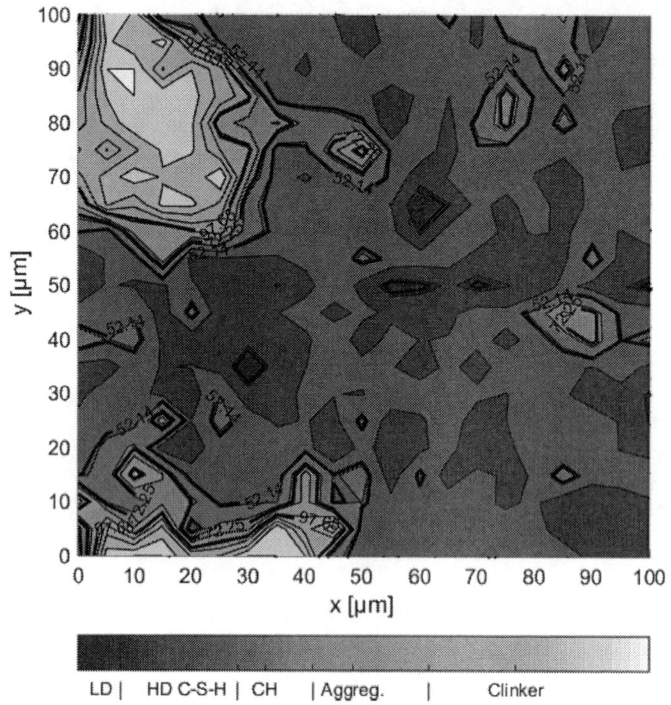

Fig. 3 Contour maps of Young's moduli distribution over the indented area at Level II

Micromechanical Modeling

Micromechanical modeling will be employed to confirm the assumptions about the impact of weak transition zones formed around inert phases and assess the impact of finely ground recycled material. The model will be based on Mori-Tanaka scheme with coated inclusions and combined with von Mises failure criterion to estimete the compressive strength at macro scale. Such an approach was recently used for micromechanical characterization of cement pastes with marble powder [10]. Unlike in that study, the model will operate at two scales as in the case of nanoindentation.

A preliminary study was carried out on pastes containing micronized recycled concrete (Level I). The study revealed that if the model does not take into account improvement of ITZ or the matrix itself by the presence of unhydrated reactivated clinker, it underestimates the elastic stiffness, especially in the pastes rich in recycled concrete powder (see Fig. 4). Further research is needed to establish the effect of recycled concrete on macroscopic properties of cementitious composites containing recycled concrete.

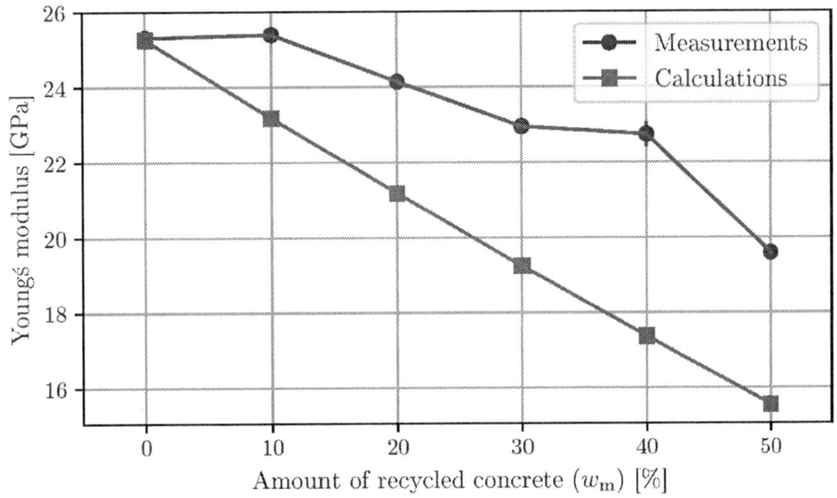

Fig. 4 Preliminary outcomes of micromechanical modeling of cementitious pastes containing recycled concrete powder; the model underestimates the elastic stiffness, indicating that the unhydrated phases present in the recycled concrete can be activated and improve the mechanical properties of cementitious composites by means of ITZ or matrix strengthening

Conclusion

Measurements at micro scale were carried out to assess the elastic properties of individual phases in cementitious composites containing recycled concrete. The material was finely ground to allow secondary hydration of unhydrated clinker. Micromechanical model was suggested to revaeal the role of unhydrated phases and weak interfacial transition zones.

After the model validation, it can be used for designing mixes with a suitable volume/mass ratio of individual components. However, the issue cannot be addressed only from the point of view of micromechanics, but study on chemical processes in recycled and finely ground concrete will be needed..

Acknowledgment

This research was supported by the GA ČR research a project under the number 17-06771S. The authors thank the Center for Nanotechnology in Civil Engineering at the Faculty of Civil Engineering CTU in Prague, and LAVARIS Ltd for providing the micronized concrete.

References

[1] N. Gautam, V. Krishna, A. Srivastava, Sustainability in the concrete construction, International Journal of Environmental Research and Development 4 (2014) 81-90.

[2] R. Siddique, Performance characteristics of high-volume Class F fly ash concrete, Cement and Concrete Research 34 (2004) 487-493.

[3] H. Y. Aruntac, M. Gürü, M. Dayi, I. Tekin, Utilization of waste marble dust as an additive in cement production, Materials {\&} Design 31 (2010) 4039-4042.

[4] V. Nagarajan, S. Devi, S. Manohari, M. Santha, Experimental study on partial replacement of cement with coconut shell ash in concrete, International Journal of Science and Research 3 (2014) 651-661.

[5] M. Chen et al., Potential of recycled fine aggregates powder as filler in asphalt mixture, Construction and Building Materials 25 (2011) 3909-3914.

[6] Y. J. Kim et al., Utilization of waste concrete powder as a substitution material for cement, Construction and Building Materials 30 (2012) 500-504.

[7] O. Bernard, F.-J. Ulm, E. Lemarchand, A multiscale micromechanics-hydration model for the early-age elastic properties of cement-based materials, Cement and Concrete Research 33 (2003) 1293-1309.

[8] W. Zhu, J. J. Hughes, Nenad Bicanic, Ch. J. Pearce, Nanoindentation mapping of mechanical properties of cement paste and natural rocks, Materials Characterization 58 (2007) 1189-1198.

[9] L. Sorelli, G. Constantinides, F.-J. Ulm, F. Toutlemonde, The nano-mechanical signature of Ultra High Performance Concrete by statistical nanoindentation techniques, Cement and Concrete Research 38 (2008) 1447-1456.

[10] V. Nežerka et al., Micromechanical characterization and modeling of cement pastes containing waste marble powder, *submitted to* Journal of Cleaner Production.

EAN 2018
56th conference on experimental stress analysis

Fatigue Loading of Polymer Composite Lap Joints with Washers Inserted in the Lay-up

HRON Robin[1, a], KADLEC Martin[1, b] and BĚLSKÝ Petr[1,c]

[1]VZLU – Czech Aerospace Research Centre, Beranových 130, Prague, Czech Republic

[a]hron@vzlu.cz, [b]kadlec@vzlu.cz, [c]belsky@vzlu.cz

Keywords: composite, joint, washer, fatigue

Abstract. Carbon fibre-reinforced composites are still more often incorporated into aircraft structures, primary structures included. This also increases the need to join individual elements of the aircraft construction into complex components. Single lap adhesive bonded joints with one fastener were evaluated. A new approach of reinforcing the bolted hole using an inserted aluminium washer in the composite lay-up was used. Two types of washers were used – smooth washer and 3D printed spiked washer. A cross-section analysis showed some minor resin rich areas and voids on the washer contact with the composite. Fatigue crack fronts in the adhesive during loading were marked based on single transducer ultrasonic probe. The C-scan revealed also crack initiation from the washer border. The washer reinforced series had lesser final crack acceleration and thus showed improved fatigue behaviour in comparison with the reference specimen. Spiked 3D print washer improved the fatigue life by 28 %. The failure mechanism was also explained. Bonded joints currently do not meet airworthiness requirements and it is the motivation for the innovative research of the composite joints.

Introduction

Carbon fibre-reinforced polymer (CFRP) laminates are used in aerospace due to their effective strength to weight ratio. The major classification of composite joints in aircraft structures are the bonded and bolted joints. Regarding the bolted joints, drilling holes for fasteners cause fibres interruption and thereby weakens the construction. In addition, holes can act as stress concentrators and composite bearing strength is also quite low which is connected with the compression strength that is lower than in tension. On the other hand, bonded joints has lower weight and very good sealing to the structure. They also distribute the load uniformly over the bonded area [1, 2]. However, bonded joints are not effective in transferring high loads because the composite material itself has lower resistant to interlaminar shear stresses and also bonded joints are very sensitive to environmental conditions.

Composite aircraft structures are currently primarily bonded by riveting or bolting. Adhesive bonding of thin-walled composite structures is a very promising method that could replace the fasteners. Adhesively secondary bonding would considerably contribute to the weight and cost reduction. Therefore, it is the motivation for the current research of the bonding joints. For secondary structures, adhesive bonding is a common practice but until today the certification rules that are applicable for primary bonded structures prevent the use of bolt-free bonded joints for primary structures, as a result of earlier experiences, where the interpretation of the rules led to in-service premature failure incidents on adhesively bonded joints [3]. This is because no non-destructive method can evaluate quality of the joint in terms of strength.

Using fasteners as crack stopping features is one way to achieve the certification [4, 5]. This work investigates possible innovations for the bolted connection using inserted washers to reinforce the bolted area. Other ways of the reinforcement can be titanium-lamella reinforced CFRP bolted joints [6] or metallic inserts with variable thickness used for multiple bolt configuration [7]. These authors stated that the reinforced samples improved the both the strength and total displacement at final failure. Titanium insert hole repairs were also investigated in Ref. [8]. The inserts improved the bearing and pull-through strength and the primary failure did not involve the insert.

This article deals with fatigue life enhancement of composite joints using reinforced holes by metal washers inserted into the lay-up.

Materials and Methods

Single lap adhesive bonded joints with one fastener were used. Two adherents of 20 and 13 plies were joint for each specimen. Length of each adherent was 205 mm, thickness of the non-reinforced 20 ply adherent was 5.7 mm with the lay-up of [(-45/+45)/(+45/-45)/(90/0)/(+45/-45)/(-45/+45)/(0/90)/(-45/+45)/(90/0)/(+45/-45)/(0/90)]s and thickness of the washer reinforced 13 ply adherent was 3.7 mm with the lay-up of [(+45/-45)/(-45/+45)/(0/90)/(90/0)/(+45/-45)/(-45/+45)/(0/90)/(-45/+45)/(+45/-45)/**(90/0)/(0/90)**/(-45/+45)/(+45/-45)]s. The washer was inserted in the two (90/0) cut-out layers highlighted by the bold text. Overlap length was 60 mm, and width of the specimen was 35 mm. The joints were bonded by film adhesive (REDUX FM300) and fastened by a fastener (ABS1707 B C2 V1 A – 5 mm in diameter).

The reference sample and two types of innovative technologies for joining of CFRP laminates were evaluated. Two types of washers inserted into the carbon fibre fabric laminate during manufacturing were used for the innovative specimens - smooth washer and spiked 3D print washer. The washers were made of aluminium alloy. Surface of the smooth washer was free from any protrusion, roughness or any other inequalities (Fig. 1). The thickness was not constant with lower thickness far from the hole. The cross-section in Fig. 1 showed significant resin rich areas on both surfaces of the smooth washer and also a significant resin pocket on the washer border. Spiked 3D print washer had on the surface hooks, which should ensure better cohesion with laminate (Fig. 2). Unlike the smooth washer, the contact area with the composite was without resin rich areas. However, there were some voids around certain hooks.

Fig.1 Cross-section of a composite adherent with inserted aluminium smooth washer

Fig.2 Cross-section of a composite adherent with inserted aluminum 3D spiked washer

Loading machine Instron 55R1185 with 100 kN load cell was used for quasi-static testing. Crosshead displacement rate was set to 1 mm/min. The extensometers were used for displacement measurement of static tests.

The fatigue tests were performed on INOVA ZUZ 100 kN and Hydropuls Schenck 250 kN with maximum load of 8 600 N and coefficient of asymmetry $R = 0.1$. In selected intervals C-scan using a phased array probe was performed (Olympus – Omniscan). The crack front was monitored using a single probe A-scan in regular intervals.

Results

The static testing load-displacement curves are shown in Fig. 3. There is significant load drop after the peak at approx. 25 kN where the adhesive failed. This load drop was not depended on the reinforcement as the adhesive was the same for all the series. After the drop to, the load was carried only by the fastener. Spiked washer reinforced series had higher bearing load (10 kN) than reference series by approximately 35%. This was caused by the higher cross-section surface as the thickness around the reinforcement was increased by the washer.

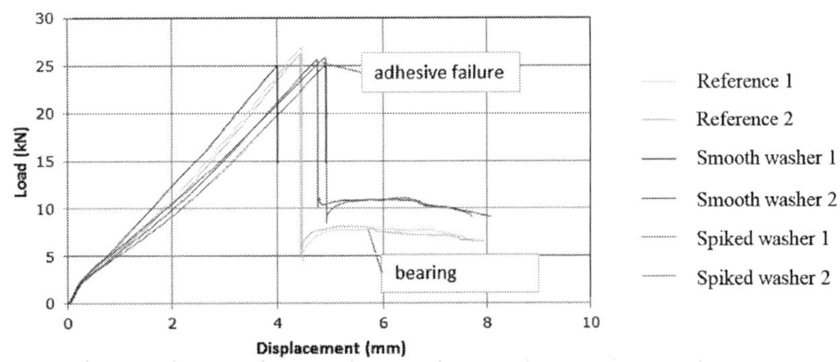

Fig.3 Results of quasi-static testing of a single lap joint with adhesive bonding and a fastener

The results of fatigue crack growth of a single lap joint with adhesive bonding and a fastener are shown in Fig. 4. Fatigue crack fronts during loading were marked based on single transducer ultrasonic probe. Typically the crack propagated first from the bottom edge and then from the upper. Near the lower edge corner, there was an insert to initiate the crack. However, there was no proof that this insert really initiated the crack because the crack front was typically straight and the insert seemed to be also bonded to the laminate. The C-scan revealed also crack initiation from the washer border. This could be caused by the two layers that were cut-out before inserting the washer. The border around the washer than behaved as a stress concentrator.

The crack growth rate was evaluated based on the delaminated area that was divided by the specimen width. The delamination curves are shown in Fig. 4 where both lower and upper cracks are included. There is higher crack growth rate for the lower edge in the beginning that could be caused either by the insert or by the higher peel forces in the area more far from the fastener. Then, there was stable crack growth until 350 000 cycles where the both lower and upper crack accelerated. The washer reinforced series had lesser final crack acceleration and thus showed improved fatigue behaviour in comparison with the reference specimen. There was 5 % fatigue life improvement for smooth washer specimens and 28 % fatigue life improvement for spiked 3D print washer joints when compared to reference specimens.

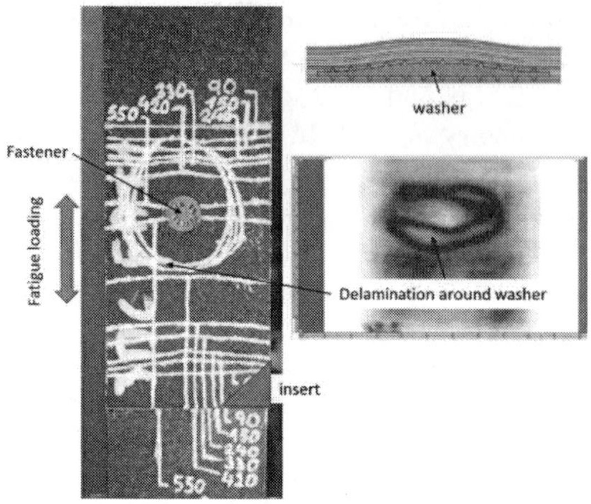

a) Fatigue crack fronts during loading and a delamination initiation from the washer.

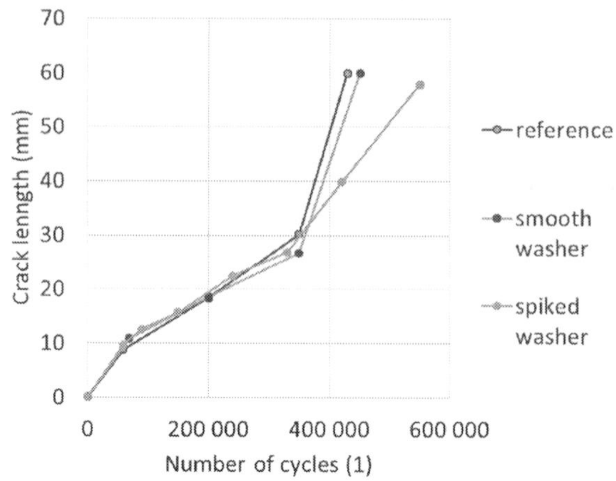

b) Fatigue crack growth curves for the reference and two reinforced joints.

Fig.4 Results of fatigue crack growth of a single lap joint with adhesive bonding and a fastener

A macro photography of fracture surfaces revealed fibre and adhesive dominated fracture which corresponded with crack growth and shearing direction. The mechanism of angled cracks is explained in Fig. 5. The cracks initiated at both free edges and then they were driven diagonally toward one of the bonded surface due to the shear forces that caused 45° major stress direction. The cracks propagated along the interfaces until they reached the bolt. The crack propagation in both planes caused the splitting of the fibre dominated (black) and adhesive dominated (green) fracture surface.

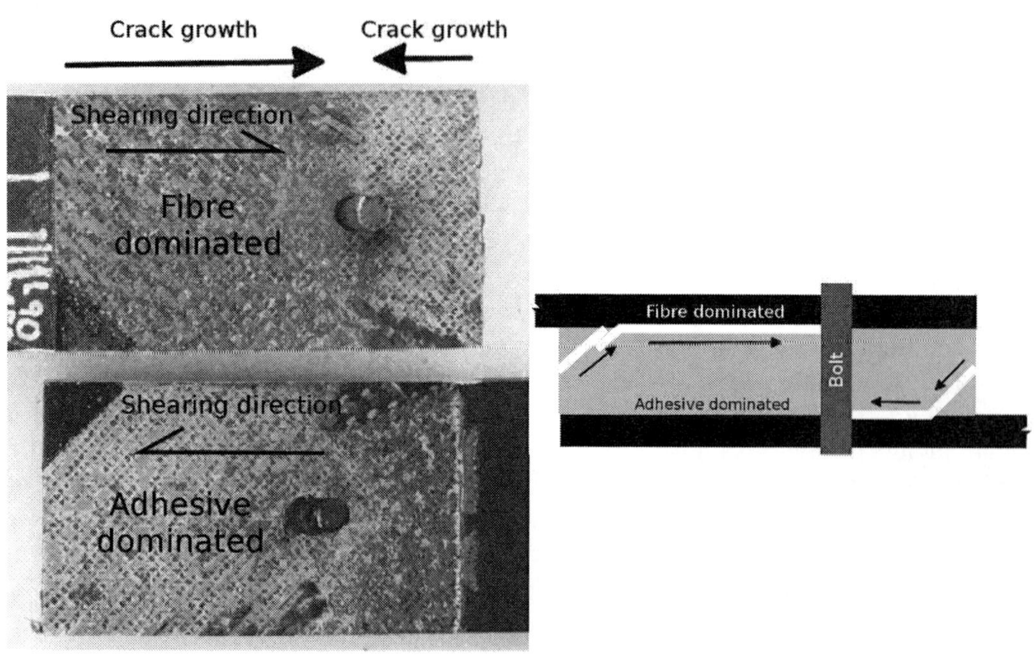

Fig 5 Fatigue crack fracture surface with explained angled crack mechanism leading to fibre and matrix dominated fracture

Conclusion

To conclude, the results of fatigue loading of single lap shear joints showed that the reinforcement influenced only the final stage of crack growth where it hindered the crack and increased the joint lifetime by 30%. The fatigue crack growth mechanism leading to fibre and matrix dominated fracture surface was also explained. The crack stopping features have great potential to certificate adhesive joints for primary aerospace structures.

Acknowledgement

This work was funded by the institutional support for the long-term conceptual development of a research organization provided by the Ministry of Industry and Trade of the Czech Republic.

References

[1] Sp. Pantelakis, K.I. Tserpes, Boltless joining of composite aerostructures: From secondary to primary parts: International Conf. On Airworthiness & Fatigue – 7th ICSAELS Series Conf. 25-27 March, 2013 Beijing, China.

[2] R. Jones, A.J. Kinloch, W. Hu, Cyclic-fatigue crack growth in composite and adhesively-bonded structures: The FAA slow crack growth approach to certification and the problem of similitude, International Journal of Fatigue 88 (2016) 10-18, http://dx.doi.org/10.1016/j.ijfatigue.2016.03.008.

[3] Banea MD, da Silva LFM. Adhesively bonded joints in composite materials: an overview. *Proc. IMechE Part L: J. Materials: Design and Applications, 2009*; **223**: 1-18.

[4] K.I. Tserpes, G. Peikert, I.S. Floros, Crack stopping in composite adhesively bonded joints through corrugation, Theoretical and Applied Fracture Mechanics 83 (2016) 152-157, http://dx.doi.org/10.1016/j.tafmec.2015.10.003.

[5] T. Kruse, T. Körwien, R. Růžek, Fatigue behaviour and damage tolerance design of composite bonded joints for aerospace application, ECCM 2016 - Proceeding of the 17th European Conference on Composite Materials2016; Code 126913, ISBN: 978-300053387-7

[6] G. Kolks, K.I. Tserpes, Efficient progressive damage modeling of hybrid composite/titanium bolted joints, Composites Part A: Applied Science and Manufacturing 56 (2014) 51-63, http://dx.doi.org/10.1016/j.compositesa.2013.09.011.

[7] V. Kradinov, E. Madenci, D.R. Ambur, Bolted lap joints of laminates with varying thickness and metallic inserts, Compos Struct 68 (2005) 75–85. doi:10.1016/J.COMPSTRUCT.2004.03.00.

[8] D. Falconieri, F. Franco, The effect of titanium insert repairs on the static strength of CFRP coupons and joints, Composite Structures 134 (2015), 799-810.

EAN 2018
56th conference on experimental stress analysis

Measurement of Vibration Responses of a Rotating Disc Using DIC Method

HUŇADY Róbert[1, a], PAVELKA Peter[1,b] and KAĽAVSKÝ Adam[1,c]

[1]Technical University of Košice, Letná 9, Košice, Slovakia

[a]robert.hunady@tuke.sk, [b]peter.pavelka@tuke.sk, [c]adam.kalavsky@tuke.sk

Keywords: vibration analysis, modal analysis, digital image correlation, rotating disc

Abstract. The paper deals with measurement of operational deflection shapes and modal shapes of a rotating disc using high-speed 3D DIC system. Since, the system is able to capture the motion of an object that is composed of a rigid body motion (rotation and translation) and flexible body motion (deflections caused by inertia forces or excitation forces), there is necessary to separate these two motion components to obtain the responses corresponding only to vibration. The method how to do it is explained in the theoretical part of the paper. The use of this method is presented in the practical part that describes three experiments: run-up analysis, ODS analysis, and operational modal analysis of a rotating disc.

Introduction

Rotating disk-like structures have a relevant interest in many engineering fields, where it is of paramount importance to determine their natural frequencies in order to avoid resonance problems that could cause an undesirable behavior or a critical failure. Experimental vibration analysis as well as modal analysis of a rotating object is a demanding task that requires the use of special measurement approaches such as Continuous-Scan Laser Doppler Vibrometry, Pulsed ESPI or high speed Digital Image Correlation [1]. In all these cases it is appropriate to exclusion rigid body motion. It can be realized by several ways, for example, by using a stroboscopic light source, frequency of which is synchronized with rotating frequency. The object observed by cameras then appears to be in steady state [2]. When the responses are measured by laser scanning vibrometer, the use of device called an optical derotator is necessary [3]. The derotator tracks the motion of the rotating object, resulting in a steady position of the laser beam. Another possibility is a numerical approach based on post-processing of measurement data when the rotation matrix and translation vector are calculated for each time step. There are several methods that make it possible [4-9]. The method used in this study uses singular value decomposition and least-squares estimation [9] to calculate rigid transformation between two positions of a rotating object. After the separation of rigid body components from the measured displacement fields, the remaining data correspond only to vibration responses of the disc. As will be shown below, it leads to more pronounced response spectrums in which the components of higher harmonics frequencies or higher modes are easily identifiable.

Elimination of rigid body movements

A rigid body motion from a position P into another position Q can be characterized by a translation vector $\mathbf{t}$ and a rotation matrix $\mathbf{R}$. Let $P=\{\mathbf{p}_1, \mathbf{p}_2,...,\mathbf{p}_n\}$ and $Q=\{\mathbf{q}_1, \mathbf{q}_2,...,\mathbf{q}_n\}$

to be the sets of radius vectors describing the position of n corresponding points of the same body in three dimensional space. A rigid transformation that optimally aligns the two sets can be found by least-square method

$$(\mathbf{R},\mathbf{t}) = \arg\min \sum_{i=1}^{n} w_i \left\|(\mathbf{R}\mathbf{p}_i + \mathbf{t}) - \mathbf{q}_i\right\|^2, \tag{1}$$

where $w_i > 0$ are weights for each point pair. Detailed procedure of the derivation of $\mathbf{R}$ and $\mathbf{t}$ is given in [9]. We provides only summary of the steps to computing the optimal translation vector and rotation matrix that minimize Eq. 1.

1. Compute the weighted centroids of both point sets:

$$\overline{\mathbf{p}} = \frac{\sum_{i=1}^{n} w_i \mathbf{p}_i}{\sum_{i=1}^{n} w_i}, \quad \overline{\mathbf{q}} = \frac{\sum_{i=1}^{n} w_i \mathbf{q}_i}{\sum_{i=1}^{n} w_i}. \tag{2}$$

2. Compute the centered vectors:

$$\mathbf{a}_i := \mathbf{p}_i - \overline{\mathbf{p}}, \quad \mathbf{b}_i := \mathbf{q}_i - \overline{\mathbf{q}}, \quad i = 1, 2, ..., n. \tag{3}$$

3. Compute the $d \times d$ covariance matrix:

$$\mathbf{C} = \mathbf{A}\mathbf{W}\mathbf{B}^T, \tag{4}$$

where $\mathbf{A}$ and $\mathbf{B}$ are $d \times n$ matrices that have $\mathbf{a}_i$ and $\mathbf{b}_i$ as their columns, respectively, and $\mathbf{W} = \mathrm{diag}(w_1, w_2, ..., w_n)$.

4. Compute the singular value decomposition $\mathbf{C} = \mathbf{U}\boldsymbol{\Sigma}\mathbf{V}^T$. The rotation we are looking for is then

$$\mathbf{R} = \mathbf{V}\begin{pmatrix} 1 & & & \\ & \ddots & & \\ & & 1 & \\ & & & \det(\mathbf{V}\mathbf{U}^T) \end{pmatrix}\mathbf{U}^T \tag{5}$$

5. Compute the optimal translation as:

$$\mathbf{t} = \overline{\mathbf{q}} - \mathbf{R}\overline{\mathbf{p}}. \tag{6}$$

To suppress rigid body movements, we consider vectors $\mathbf{p}_i$ to be corresponding to initial position of the disc, and vectors $\mathbf{q}_i$ to all subsequent positions. Since the highest stiffness of the disc is near the center of rotation we introduced

$$w_i = \frac{1}{\|\mathbf{q}_i\|}. \tag{7}$$

After the computation **R** and **t** for all time steps, the new point coordinates eliminating rigid body movements can be determined on the basis of inverse transformation

$$\mathbf{q}_i^* = \mathbf{R}^{-1}\mathbf{q}_i - \mathbf{t} - \mathbf{p}_i \,. \tag{8}$$

The last adjustment is to shift the center of the disk to origin point and rotate the axis of rotation to match the z-axis of the coordinate system

$$\mathbf{q}_i^* = \mathbf{S}\,\mathbf{q}_i^* - \mathbf{r}_0 \,, \tag{9}$$

where $\mathbf{S}$ is a rotation matrix defining the transformation between vector $[0\ 0\ 1]^T$ and average normal vector of the disc in its initial position, $\mathbf{r}_0$ is a position vector of the center of the disk in its initial position related to the origin point. The Fig. 1 shows the disc at the same time step before and after the elimination of rigid body movements.

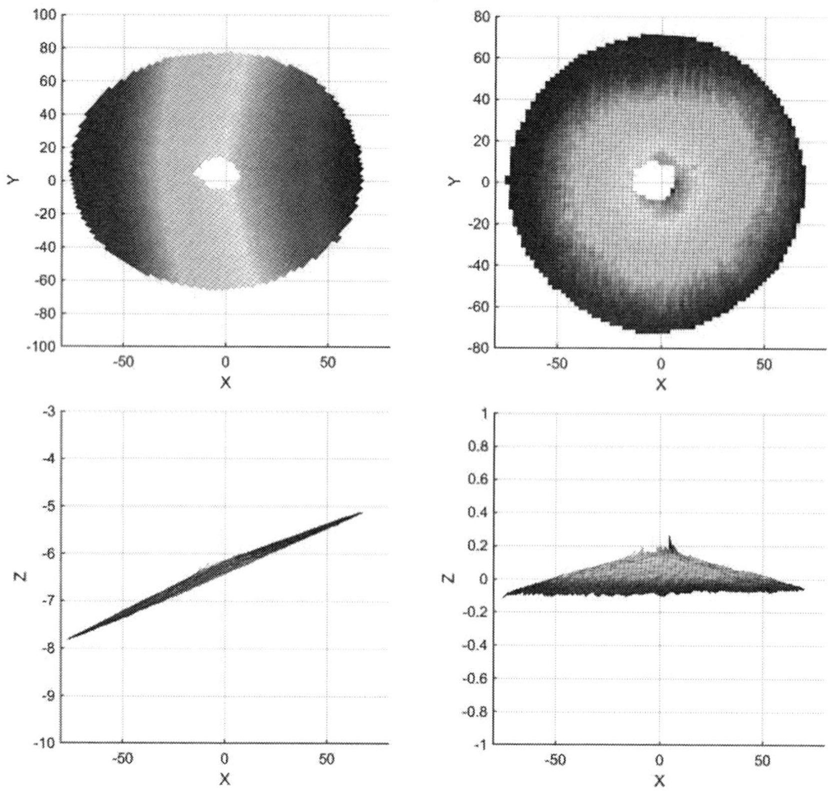

Fig. 1 Disc at the same time step before and after the elimination of rigid body movements

Experimental measurements

The object of measurement was a plane circular disc made of plastic material PS-1 commonly used in photoelasticimetry. The dimension of the disc and its material properties are shown in Fig. 2. The disc was attached to the servomotor's shaft by means of a screw connection. The servomotor was powered by DC voltage source. A thin preprinted vinyl foil with a stochastic speckle pattern was applied on the front disc's surface for the purpose of an image correlation process. The responses were measured by system Q-450 Dantec Dynamics with two high-speed cameras Phantom SpeedSense 9060. Due to high angular velocity of the disc, an extremely short shutter time 20 ns had to be used to capture sharp images. This required the use of two powerful light reflectors. The experimental setup and cameras' views can be seen in Fig. 3.

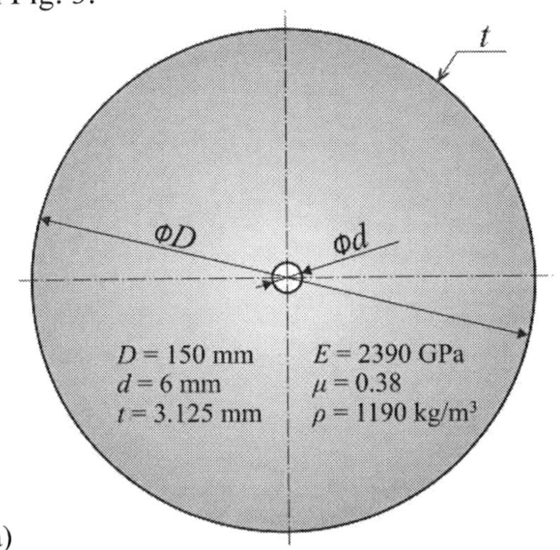

a)

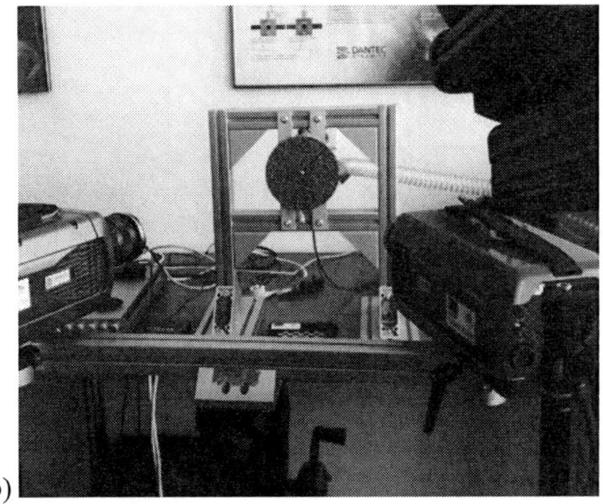

b)

Fig. 2 a) Dimensions and material properties of the disc, b) Experimental setup

ODS analysis

The aim of the operational deflection shape analysis was to determine operational shapes of vibration of the rotating disc that is excited by inertial forces caused by its imbalance. The measurement was performed at constant revolutions 4700 rpm. A sampling frequency of the cameras was approx. 4000 fps. An image correlation analysis was carried out in Istra4D software. The obtained displacement fields were then imported to Matlab for post-processing. DICMAN 3D [10] application was used for spectral analysis and to obtain operational deflection shapes.

Fig. 3 shows the combined frequency spectrums of the responses of the rotating disc measured in z-direction. Fig. 3a relates to displacements including rigid body movements, Fig. 3b belongs to displacements after their numerical elimination. From the comparison it is apparent that the higher harmonic and sub-harmonic components are easier identifiable in the frequency response spectrum in which the rigid body movements are suppressed.

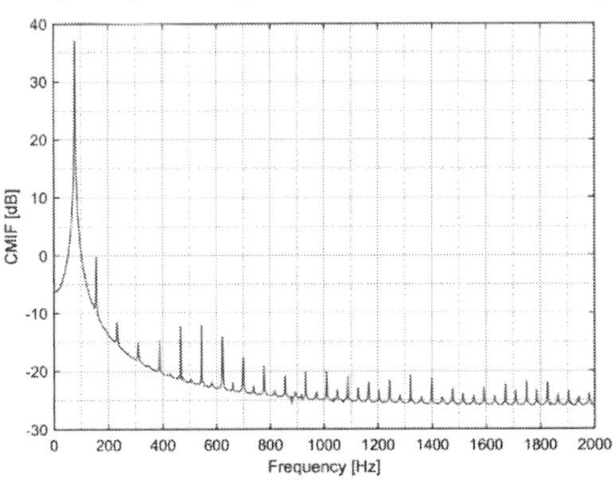

a)

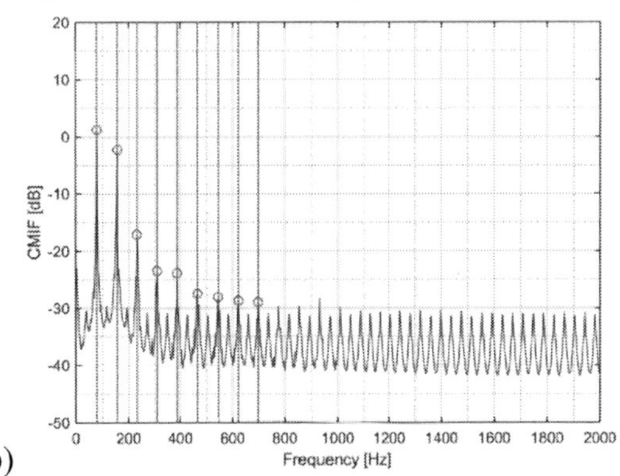

b)

Fig. 3 Combined frequency spectrums of the responses a) before, and b) after the elimination of rigid body movements

The operational deflection shapes for the first 8 harmonic frequencies are shown in Fig. 4.

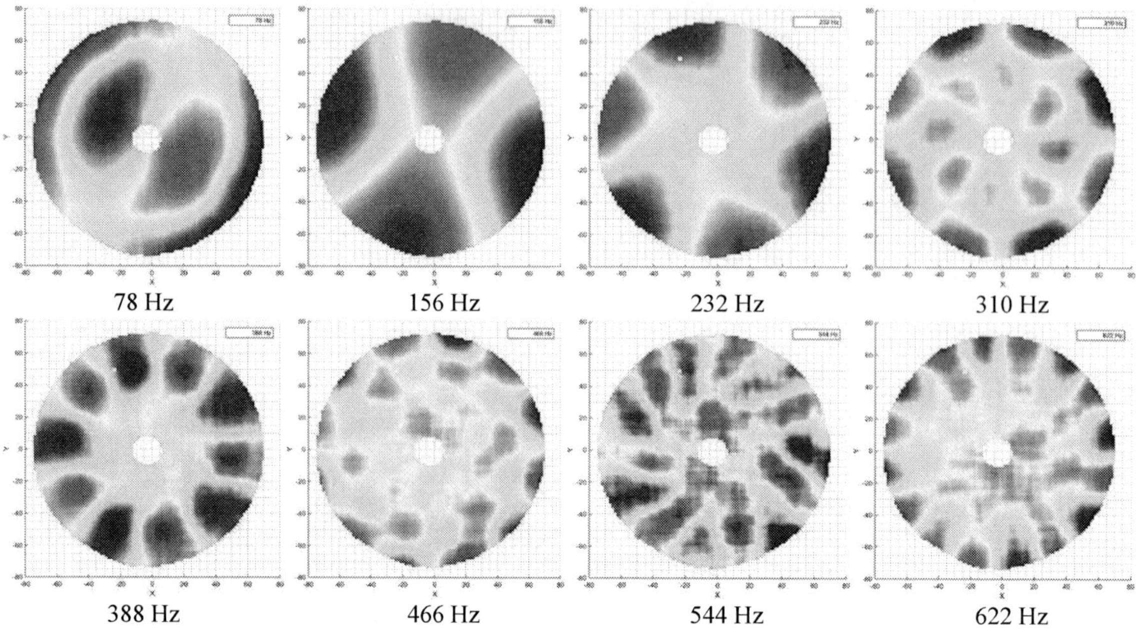

Fig. 4 Operational deflection shapes of the disc rotating at 4700 rpm

Run-Up analysis

Run-up analysis is commonly carried out to investigate dynamic behavior of rotating parts or to identify their failures. The aim of the experiment was to present the possibilities of using the presented method also in this area. The measurement was performed under the same conditions, except that the sampling frequency was 1500 fps and the maximum disc speed was about 4800 rpm. The result of the analysis is given by diagram shown in Fig. 5.

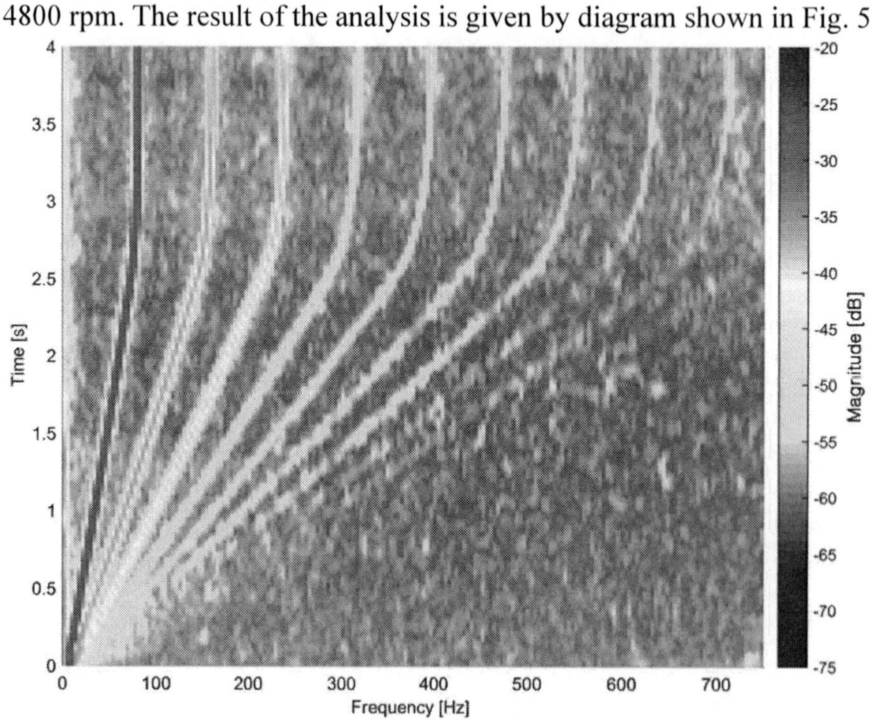

Fig. 5 Run-up diagram

The diagram shows the frequency spectrum of combined responses over time. The responses are given by displacements measured in z-direction. Their magnitudes are expressed by color field. The colored lines that can be seen in the diagram belong to fundamental frequency and its higher harmonics frequencies of the rotating disc.

Operational modal analysis

The third experiment was focused to determine modal parameters of the rotating disc. During the measurement, the disc was rotating at 4800 rpm and continuously excited from an acoustic source of sound. Fig. 6 shows the location of an excitation nozzle. A broadband excitation was ensured by the white noise signal. It should be noted that the excitation signal was not measured, i.e. modal parameters were estimated from only the response signals. The sampling frequency was set to 3000 fps.

Fig. 6 Excitation of the disc

The measurement was processed in Matlab and evaluated in DICMAN 3D. Fig. 7 shows the CMIF plot of the disc. In the plot, there are peaks corresponding to modes of the disc and peaks corresponding to vibration components caused by rotation. The so called Modal Phase Collinearity (MPC) criterion can be used to distinguish them each other. The criterion expresses the linear functional relationship between the real and the imaginary parts of the unscaled mode shape vector [11]. MPC can take values from 0 to 1. The normal modes usually get values close to 1. In the frequency range up to 1500 Hz, nine modes of the disc were identified. Their natural frequencies, MPC values, and mode shapes are shown in Fig. 8.

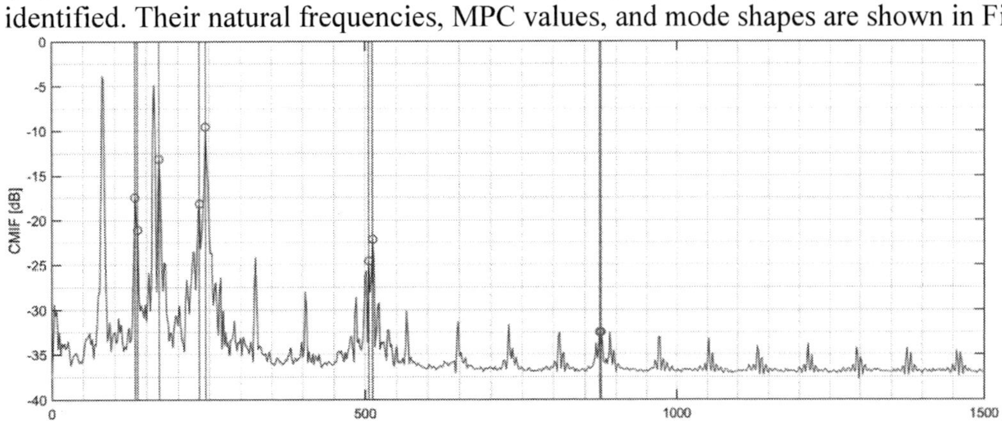

Fig. 7 CMIF plot of the rotating disc

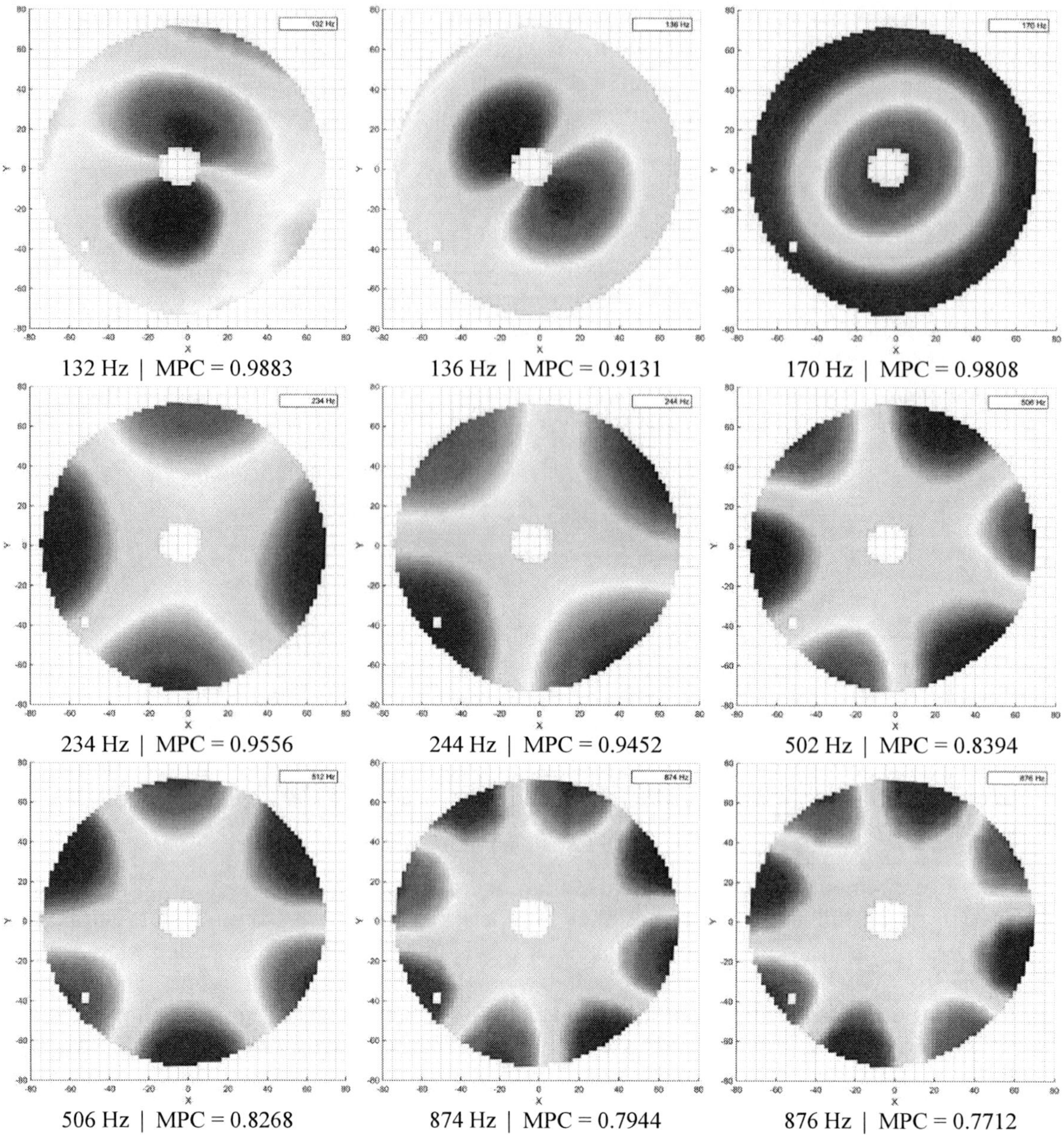

Fig. 8 Mode shapes of the disc rotating at 4800 rpm

Conclusions

The paper presented the procedure that allows high-speed digital image correlation to be effectively used in vibration analysis of rotating objects. Its base is a suppression of rigid body movements, such as rotation and translation, which are primarily contained in measured responses. The elimination of rigid body components is given by inverse transformation of coordinates from current to initial position of the body. For that purpose, a rotation matrix and a translation vector have to be determined. The method used in this study for that purpose uses singular value decomposition and least-squares estimation. The results of the experiments described in the paper show that the elimination leads to a higher accuracy of measurement and more pronounced frequency response spectrums.

Acknowledgment

This research was supported by project VEGA 1/0355/18 and VEGA 1/0731/16.

References

[1] F. Trebuňa, M. Pástor, R. Huňady, P. Frankovský, M. Hagara, Optical methods in mechanics (in Slovak), TUKE, Typopress, 2017.

[2] A.J. Molina-Viedma, L. Felipe-Sesé, E. López-Alba, F. Díaz, High frequency mode shapes characterisation using Digital Image Correlation and phase-based motion magnification, Mechanical Systems and Signal Processing, Volume 102, (2018), 245-261.

[3] P. Šároši, T. Harčarík, R. Huňady, Vibrational study of the spinning disc using LDV technique, Applied Mechanics and Materials 816 (2015), 469-473.

[4] A. Mukerjee and N. Mittal, Qualitative models of 3-D frame transformation motions, in Proc. IEEE Int. Conf. Robot. Autom., Nagoya, Japan, May 1995, 2279-2284.

[5] S. Umeyama, Least-squares estimation of transformation parameters between two point patterns, IEEE Trans. Pattern Anal. Mach. Intell., vol. 13, no. 4, (1991), 376-380,

[6] K. S. Arun, T. S. Huang, and S. D. Blostein, Least-squares tting of two 3-D point sets, IEEE Trans. Pattern Anal. Mach. Intell., vol. PAMI-9, no. 5, (1987), 698-700.

[7] B. K. P. Horn, H. M. Hilden, and S. Negahdaripour, Closed-form solution of absolute orientation using orthonormal matrices, J. Opt. Soc. Amer. A, Opt. Image Sci., vol. 5, no. 7, (1988), 1127-1135.

[8] Cashbaugh, C. Kitts: Automatic Calculation of a Transformation Matrix Between Two Frames, IEEE Access, Vol. 6, (2018), 9614-9622.

[9] O. Sorkine-Hornung, M. Rabinovich, Least-Squares Rigid Motion Using SVD, Technical notes, ETH Zurich, 2017.

[10] R. Huňady, M. Hagara, A new procedure of modal parameter estimation for high-speed digital image correlation, Mechanical Systems and Signal Processing. 93 (2017) 66-79.

[11] W. Leurs, F. Deblauwe, F. Lembregts, Modal parameter estimation based on complex mode indicator functions, Eleventh IMAC, Orlando, Florida, February 1993, 1035-1041.

EAN 2018
56th conference on experimental stress analysis

Plantar Pressure Distribution during Walking after Specific Soft Tissue Heel Injury – Case Study

JANDOVÁ Soňa

Technical University of Liberec, Faculty of Mechanical Engineering, Department of Applied Mechanics, Studentská 2, 461 17 Liberec, Czech Republic

sona.jandova@tul.cz

Keywords: dynamography; trajectory; centre of pressure; asymmetry

Abstract. Calcaneal injuries are injuries, affecting the all-aged population. The abnormalities in the gait pattern after calcaneal injuries are very common many months after the injury. These abnormalities can be evaluated by different biomechanical methods. Pedobarography detects differences in plantar pressure distribution between injured (INJ) and uninjured foot (UIN) while walking. The aim of our study was to compare plantar pressure distribution and trajectory of centre of pressure (CoP) in the stance phase during walking 6 and 12 months after the heel injury in a patient after the necrotic tissue removal. This study was a case interpretation. The participant was 9-years old young girl after a car accident. The measurement was performed 6 and 12 months after the injury and after a soft necrotic tissue removal. Measurement of plantar pressure distribution and pressure values was performed while walking on the pedobarography measurement platform (EMED ®-c50/R, Novel, Munich, Germany) with the frequency of 50 Hz. The measurement was done during the stance phase while walking. For each step, loading variables and temporal variables and also trajectory of CoP were observed. The mean of 5 steps for each variable was used to compare INJ and UIN. The loading variables normalized to body mass included peak pressure, representing the maximum pressure recorded under the foot, maximum vertical force and trajectory of CoP. Peak pressure values were higher on the UIN (6 months: 10.4% and 12 months: 8.2%). At maximum vertical force, the values on the UIN were lower (6 months: 3% and 12 months 2.5%). Peak pressure values under the heel were higher on the UIN in both measurements. Six and 12 months after the injury, the trajectory of CoP on the INJ was similar. It started in the midfoot, continued in the direction to the hindfoot and then tended over the metatarsal head to the toes area. Trajectory of CoP in UIN has changed when 6 months after the injury the trajectory was similar to the healthy subjects (starting in the hindfoot and continuing through the midfoot, metatarsal heads to the big toe area) but 12 months after the injury was the trajectory of CoP more similar to the trajectory at the INJ. Plantar pressure distribution asymmetry between the INJ and UIN occurred in both measurements (6 and 12 months follow up). The trajectory of CoP on the UIN in 6 months follow up measurement was similar to the healthy subjects in previous studies but it has changed during the next 6 months. Unfortunately the correct gait stereotype on the UIN has lost and patient tended to use the wrong gait stereotype according the INJ. According to our study we recommend to use orthopaedic insoles with various heel support to eliminate the differences between the plantar pressure distribution under the injured and uninjured foot and to ensure the right gait stereotype.

Introduction

Calcaneal injuries are complex injuries, typically affecting the middle-aged active population. The main causes include falls and landing from heights or car accidents. In most cases, soft tissues are harmed.

The results of calcaneal treatment are assessed in several ways. It is in most cases clinical examination, X-ray images or computed tomography [1]. The abnormalities in the gait pattern after calcaneal injuries are very common many months after the injury [2,3]. These abnormalities can be evaluated by different biomechanical methods. Such methods include pedobarography, which allows plantar pressure analysis at various time intervals after the injury [2-4]. With this method, differences in the plantar pressure distribution between injured and uninjured foot while walking can be detected. These abnormalities include usually decreased pressure under the injured foot, shortened stance phase and load shift from the calcaneus to the midfoot [4].

One aspect of physiological walking is symmetry in the action of the lower limbs and symmetry of the ground reaction forces [5]. High degree of loading symmetry between the left and right feet during walking in healthy subjects have been observed [6,7].

The purpose of this study was to compare the plantar pressure distribution in the stance phase during walking 6 and 12 months after the calcaneal injury in a patient after the necrotic tissue removal.

Methods

The aim of our study was to compare plantar pressure distribution and trajectory of Centre of Pressure (CoP) during walking 6 and 12 months after the injury. It was a case study.

Patient

The participant was 9-years old young girl with necrotic tissue on the calcaneus caused by a car accident (Fig. 1). The measurement was performed 6 and 12 months after the injury.

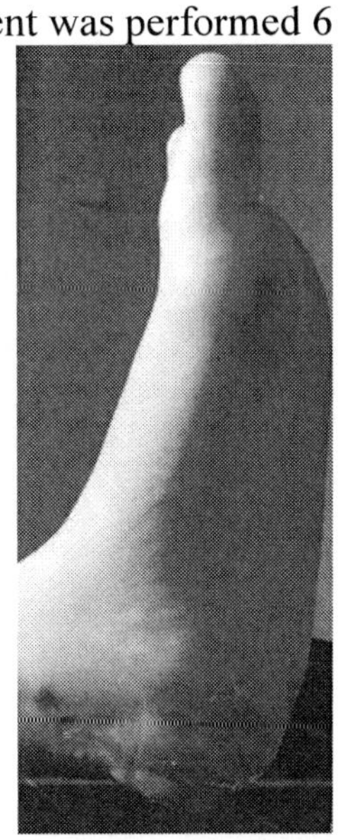

Fig. 1 Heel bone after necrotic tissue removal

Six and twelve months following operative treatment, the patient was contacted to participate in the measurement. It was performed during the period 7/2015 – 7/2016. Participant and also her parents were familiarized with the design of the measurement and parents signed an informed consent. Analysis of plantar pressure distribution while walking was then performed. All experiments were approved by the Institutional Review Board of Technical University of Liberec and in accordance with the Declaration of Helsinki.

Plantar Pressure Analysis Protocol

Measurement of the plantar pressure distribution and pressure values was performed while walking on the pedobarography measurement platform (EMED ®-c50/R, Novel, Munich, Germany) with the frequency of 50 Hz.

The measurement was preceded by a 5-minute training walk over a measurement platform. The measurement was then initiated, aimed at analysing the plantar pressure distribution during the stance phase while walking. The participant was asked to walk barefoot at a self-selected speed that she uses in everyday life over a platform. She took at least three steps before and after the platform to ensure recording of a step in full gait. The measurement was repeated until a total of five individual steps of injured (INJ) and five individual steps of uninjured (UIN) limbs were recorded.

For each step, loading variables and temporal variables and also trajectory of CoP were observed. The mean of 5 steps for each variable was used to compare INJ and UIN. The loading variables normalized to body mass included peak pressure, representing the maximum pressure recorded under the whole foot, maximum vertical force and also the trajectory of centre of pressure (CoP).

Results

Comparison of plantar pressure variables between the injured and uninjured limbs 6 and 12 months after the surgery is shown in Figure 2.

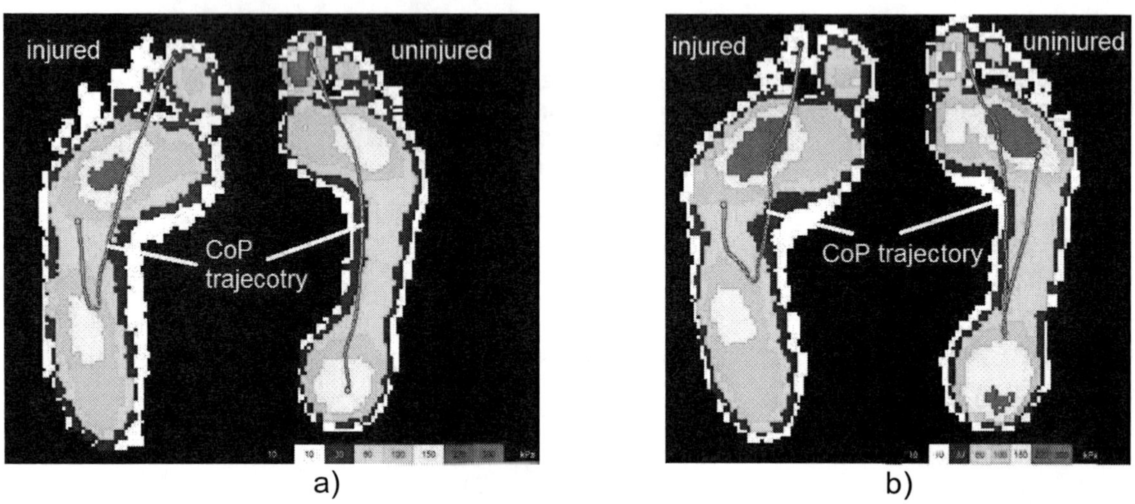

Fig. 2 Plantar pressure distribution and trajectory of CoP 6 (a) and 12 (b) months after the injury

Peak pressure values were higher for both measurements on the uninjured limb (6 months: 10.4% and 12 months: 8.2%). At maximum vertical force, the values on the injured limb were lower (6 months: 3% and 12 months 2.5%). Peak pressure values under the heel were significantly higher on the uninjured limb in both measurements.

CoP trajectory was different for both limbs in both measurements as well. Six months after the injury, the trajectory of CoP on the injured limb started in the midfoot, continues very shortly in the direction to the rearfoot and then tends over the middle metatarsal head to the toes area. Twelve months after the injury is the trajectory of CoP on the injured limb very similar to the trajectory 6 months after the injury. The interesting change of the trajectory of CoP in uninjured limb occured twelve months after the injury when 6 months after the injury the trajectory is as in healthy subjects (starting in the rearfoot and continuing through the midfoot, metatarsal heads to the big toe area) but 12 months after the injury is the trajectory of CoP more similar to the trajectory at the injured foot (starts under the metatarsal heads, continuous to the heel direction and then tends to the forefoot - big toe area). Unfortunately the subject lost the correct gait stereotype on the uninjured limb and tends to use the wrong gait stereotype according the injured foot between the 6 and 12 months measurement.

Discussion

The aim of our study was to compare plantar pressure distribution and trajectory of CoP during walking 6 and 12 months after the car accident and injury of the heel tissue.

One aspect of physiological walking is symmetry in the action of the lower limbs and symmetry of the ground reaction forces [5]. High degree of loading symmetry between the left and right feet during walking in healthy subjects have been observed [6,7]. In general, lateralization of peak pressure distribution and temporal parameters presented in our study for the injured and uninjured foot are similar with the values found in the literature for calcaneal fractures [4,6,8,9].

Six and 12 months after surgical treatment of tissue on the heel significant differences between the injured and uninjured limb in all of the observed variables (maximal pressure, vertical force, trajectory of CoP) were found. These side-to-side differences have been reported in the literature to remain for quite a long time after surgery when the calcaneal bone was injured [6,9]. Although in our study the tissue on the heel was necrotic and the replacement will be done not before 10 years, the symmetrical plantar pressure distribution can not be expected.

The time of the step on the rearfoot of the injured limb was shorter according to previous study carried out on the subjects after calcaneal fractures [10]. Thus, unlike the uninjured foot, the stance phase and heel contact of the injured foot was shortened [4].I

As for dynamic parameters, the values on the injured foot were lower than those for uninjured foot regardless of the follow up duration. The peak pressure measured under the whole foot during the stance phase was different between both limbs in the 6 months follow up measurement, in which the injured foot values were 10.4% lower. In the 12 months follow up measurement, the peak pressure values in the injured foot were lower by 8.2%.

This finding points to the fact that 6 and 12 months after injury, the patient - with respect to maximum pressure under the heel - tended to step more and more cautiously on the injured heel. This overall trend of a load shift towards lateral regions of the injured limb is supported by differences in other areas. This indicates a persisting deviation of the gait pattern long after injury [6] and in this case it can occur at least until the tissue replacement will be done.

On the injured foot, plantar pressure is shifting towards the forefoot. The displacement of pressure toward the midfoot can be explained by the limited range of motion in the subtalar and tibiotalar joints [11]. Post-traumatic changes after this type of injury may lead to a breakdown of the gait stereotype and result in a flat foot [9].

In terms of time in specific zones (heel, midfoot, forefoot and toes), the CoP trajectory was significantly different in the injured and the uninjured foot in 6 months and 12 months follow up measurements. CoP trajectory in the injured foot started in the area of the midfoot, continues slightly in the posterior direction and then tends over the middle metatarsal head to the toes area. It does not reach the heel zone. This fact is likely to be related to the overall CoP shift toward the forefoot and relieving the heel due to the absent tissue in the area under the heel bone when stepping. The trajectory of CoP in the uninjured foot 6 months after the injury started in the heel zone as in the healthy subjects and then is directed more toward the anterior part of the foot. In our study, there was a significant change in the trajectory of CoP in the uninjured foot in the 12 months follow up comparing to the 6 months follow up measurement. The trajectory tends to be like in the injured foot and the gait stereotype is changed. This fact is very undesirable from the prospective point of view when the tissue replacement will be done in the future. According to our study we recommend to use orthopaedic insoles with various heel support to eliminate the differences between the plantar pressure distribution under the injured and uninjured foot (Fig. 3) and to ensure the right gait stereotype.

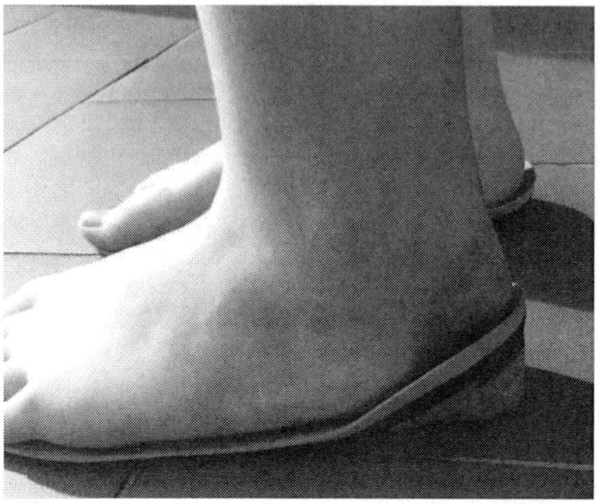

Fig. 3 Formthotics insoles with heel support

Limitations of this study include the fact, that due to the low frequency and oneness of this injury, it was not possible to conduct the measurements in other patients. Generalisability of these results should be done carefully. Another limiting factor in this study is that the participant walked at her preferred speed. Because the measurement was taken at a relatively early stage of recovery, it was not possible to influence the walking speed. The study was performed not in a group of healthy subjects, but according to previous studies the human gait in healthy adults is symmetrical [5].

However, despite our appreciation of the limitations of our investigation, we believe that the results of this study could be useful for the posttraumatic therapy and the results could be used in the practice of gait stereotype.

References

[1] Hirschmüller A, Konstantinidis L, Baur H, Müller S, Mehlhorn A, Kontermann J. Grosse U, Südkamp NP, Helwig P. Do changes in dynamic plantar pressure distribution, strength capacity and postural control after intra-articular calcaneal fracture correlate with clinical and radiological outcome? Injury 42(10):1135-1143, 2011.

[2] Jandova, S, Pazour, J. Limited versus extended lateral approach for osteosynthesis of calcaneal fractures – comparison of temporal and dynamic parameters of the gait cycle, Acta Chir Orthop Traumatol Cech. 85/1, 57-61, 2018.

[3] Schepers T, Van der Stoep A, Van der Avert H, Van Lieshout EM, Patka P. Plantar pressure analysis after percutaneous repair of displaced intra-articular calcaneal fractures. Foot Ankle Int 29(2):128-35, 2008.

[4] Hetsroni I, Ben-Sira D, Nyska M, Ayalon M. Plantar pressure anomalies after open reduction with internal fixation of high-grade calcaneal fractures. Foot Ankle Int 35(7):712-8, 2014.

[5] Hamill J, Bates BT, Knutzen KM. Ground reaction force symmetry during walking and running. Med Sci Sports Exerc 15:170, 1983.

[6] Rosenbaum D, Lübke B, Bauer G, Claes L. Long-term effects of hindfoot fractures evaluated by means of plantar pressure analyses. Clin Biomech10(7):345-351, 1995.

[7] VanZant RS, McPoil TG, Cornwall MW. Symmetry of plantar pressures and vertical forces in healthy subjects during walking. J Am Podiatr Med Assoc 91(7):337-42, 2001.

[8] Jansen H, Frey SP, Ziegler C, Meffert RH, Doht S. Results of dynamic pedobarography following surgically treated intra-articular calcaneal fractures. Arch Orthop Trauma Surg 133(2):259-265, 2013.

[9] Besch L, Radke B, Mueller M, Wredenhagen MD, Waroga D, Hilgert RE, Mathiak G, Oehlert K, Seekamp A. Dynamic and functional gait analysis of severely displaced intra-articular calcanus fractures treated with a hinged external fixator or internal stabilization. J Foot Ankle Surg 47(1):19-25, 2008.

[10]Rammelt S, Amlang M, Barthel S, Gavlik JM, Zwipp H. Percutaneous treatment of less severe intraarticular calcaneal fractures. Clin Orthop Relat Res 468:983-990, 2010.

[11]Allmacher DH, Galles KS, Marsch JL. Intra-articular calcaneal fractures treated nonoperatively and followed sequentially for 2 decades. J Orthop Trauma 20:464-469, 2006.

EAN 2018
56th conference on experimental stress analysis

Advanced Methods in Crash Safety Testing - Active Lateral Intrusion Simulation

JELÍNEK Jakub[1,2,a], RŮŽIČKA Milan[2,b] and KALINSKÝ Michal[1,c]

[1]TÜV SÜD Czech, Novodvorská 994/138, Praha 4, Czech Republic

[2]Czech Technical University in Prague, Faculty of Mechanical Engineering, Technická 4, 166 07 Praha 6, Czech Republic

[a]Jakub.Jelinek@tuv-sud.cz; Jakub.Jelinek@fs.cvut.cz, [b]Milan.Ruzicka@fs.cvut,cz, [c]Michal.Kalinsky@tuv-sud.cz

Keywords: side crash, ALIS, DYCOT, dummy, biomechanical loads, FEM

Abstract. This work is to give an overview of currently used methodology in the field of passive safety, ie. crash tests. There are two types of crash tests (mandatory, consumers) with many setup scenarios. As the OEMs aim to cover all legislative and get top score in consumer tests, there is a high pressure on safety. So far the physical crash testing has been mainly done on full vehicles.

Introduction

Main idea is to come up with an improved testing methodology of simulating crash tests. It is supposed to decrease development time, while increasing volume of work done and variability altogether with cost reduction. Such an approach has been allowed by a new catapult system DYCOT that was opened by TÜV SÜD Czech in Mladá Boleslav last year. It enables an advanced approach to performing side crash physical testing by using sled test and unique "Active Lateral Intrusion Simulation" (ALIS).

The basic principle will be described as well as first steps to the overall results. ALIS has to be designed and simulated first before physical validation and testing will take a part.

Such a combination of DYCOT and ALIS presents unique testing capabilities at least within the Europe region and should significantly improve restraint systems tuning and hence improve biomechanical loads in general.

Crash tests

The principle of the crash test is to understand mechanics, kinematics and dynamics of the impact itself. With first crash tests in the second half of the last century, they have assessed structural behaviour of the vehicle only with output parameters such as deceleration and intrusion. Later on the effect of the crash on human body has been put in front and so the first dummies have been developed. Nowadays there are many types of dummies. Each of them represents different "human body" and allows different biomechanical parameters to be measured and it is intended for different crash event.

Mandatory and consumer crash tests

Mandatory tests have been established by governments to ensure minimal safety and all vehicles have to meet their criteria. Each country/region has its own tests (e.g. EU, USA,

China,...), nevertheless they differ only in several cases. The base is pretty much the same everywhere and so is the occupant protection. Mandatory test have two results – pass or fail. Overview of the crash tests around the world is shown in Figure 1.

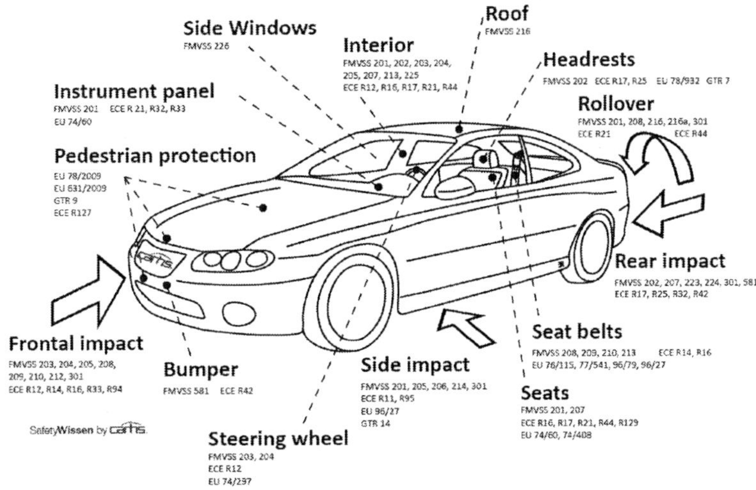

Fig.1 Overview of crash tests [1]

Consumer tests have been additionally established in order to compare cars among OEMs and so push them to the improved safety through better score. They are usually stricter than mandatory tests. The results scale is different to mandatory tests and it uses stars, when 5-star is a top score. The score is based on car safety performance during all types of crash tests, presence of other safety features (e.g. seatbelt reminder) and others.

The main objective of the crash tests is to assess car primary structure performance in terms of injury risk. Main criteria are biomechanical loads that apply to human during crash. The whole complexity of the system is assessed via parameters (biomechanical loads) measured on several parts of body. These parts are as follows (standard for ECE legislation and EuroNCAP):

- Head
- Face
- Neck
- Chest
- Abdomen + Pelvis
- Spine
- Lower extremities

Restraint systems

Restraint system is a passive system in car that does not require any action from driver or passenger to get it work. There are two main sub-systems:

- Airbags
- Seatbelts

Airbag is a vehicle safety device that is based on rapidly quick inflation and deflation. It consists of flexible bag from fabric, inflation module, sensor and soft cushion. When the vehicle crashes with sufficient severity, sensor will trigger inflation module and the cushion is inflated. This inflated cushion decreases head acceleration and generally softens the head impact pulse and absorb some head kinetic energy. It is important as it does not allow the head to come into the contact with steering wheel and dashboard (front crash) and other structural parts, including trim (side crash).

Seatbelt is a safety feature that "controls" movement of the pelvis and torso. There are many types of the seatbelts (2-point, 3-points, 6-points, etc.) but the common one is 3-points lap belt. It has got two additional features. Load-limiter triggers when the maximum load in seatbelt is reached, it slightly releases a seatbelt and engages again. One could call this principle as ABS for seatbelts. Pre-tensioner is triggered just like the airbag by a sensor, but this sensor derives the severity of the crash event. It measures pressure drop when closing to another object and ensures that in the early stage of the crash the body will be tightened to the seat via pre-tensed seatbelt. This reduces further injury risk due to excessive body motion.

There are some others types as well but there are not that widely used compared to airbags and seatbelts. All restraint systems are very important for minimizing the injury risk during crash and so they will be included in further simulations and testing.

New approach

- **Current status**

At the moment airbags setting has been found as very important, but difficult. Simulations have shown certain level of accuracy, but for correct validation many expensive crash tests have to be carried out. In the whole world, the absolute majority of tests have been done with full vehicles. The procedure is sketched in Figure 2.

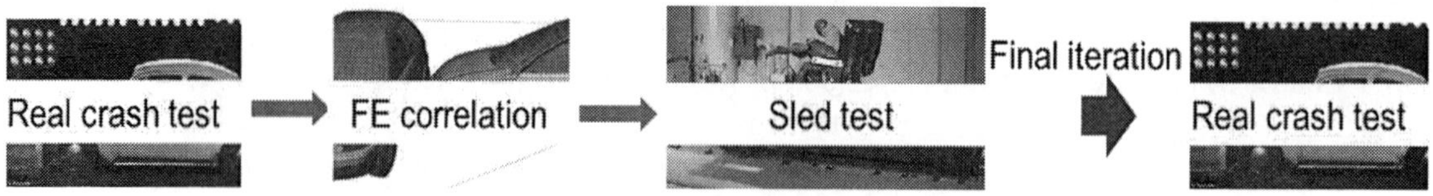

Fig.2 Current development approach

That means expensive and time-consuming. With increasing testing requirements, people have tried to find an alternative approach that would offer a sufficient level of predictability and accuracy with time and money reduction. Physical testing of sub-systems has been on market for a while, but it has not been considered to be used for side crash that much.

There has been a request on global market to develop such a testing device and method to incorporate all mentioned above. Such a device should be able to test sub-systems and also simulate side acceleration pulse. Usually catapult that is able to develop precise negative pulse that imitates side crash acceleration has been used. As it has been known, sled tests have been covering mostly front and rear crash scenarios. Usage of sled test for side impact is rather rare and it is not the main task of it.

There are several test labs around world that attempt to simulate side crash via sled tests. It consists of sled platform and additional system for lateral loading – pushing from side. This method is rather new. Currently one hydraulic cylinder is being used to imitate door intrusion mechanism.

DYCOT

TÜV SÜD Czech has recently invested a large sum to test lab equipped with sled system (catapult) – DYnamic COmponent Testing (DYCOT). The capabilities of such test device have been significantly increased by adding Advanced Lateral Impact System (ALIS) into serie, right next to the sled platform. It uses up to 6 hydraulic cylinders in order to correctly simulate the door intrusion kinematics during the side crash. It enables one to use only small part of the car together with dummies and restraint systems and carry out simulation of the side crash with focus on restraint system and biomechanical loads.

The system may seem as a "train of trolleys". The driven sled trolley is mounted to the main hydraulic system that generates the main acceleration pulse. ALIS is mounted on the separate trolley, attached to the sled one. The whole structure is shown on the Figure 3, where main components are identified. The lateral system consists of additional hydraulic system, several hydraulic cylinders, linear guiding system and "impact break-in structure".

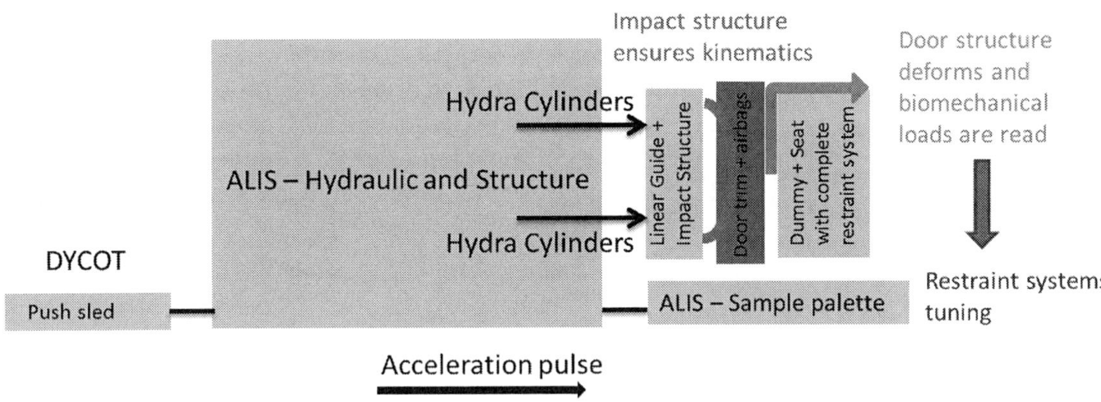

Fig.3 DYCOT + ALIS concept

As this equipment is supposed to cover and be useable for most of the typical cars, it has the ability to move all cylinders within a working space. This forms a boundary box that has been overlaid by typical car sketches as shown on Figure 4. Note that the space covers area between front driver and rear passenger.

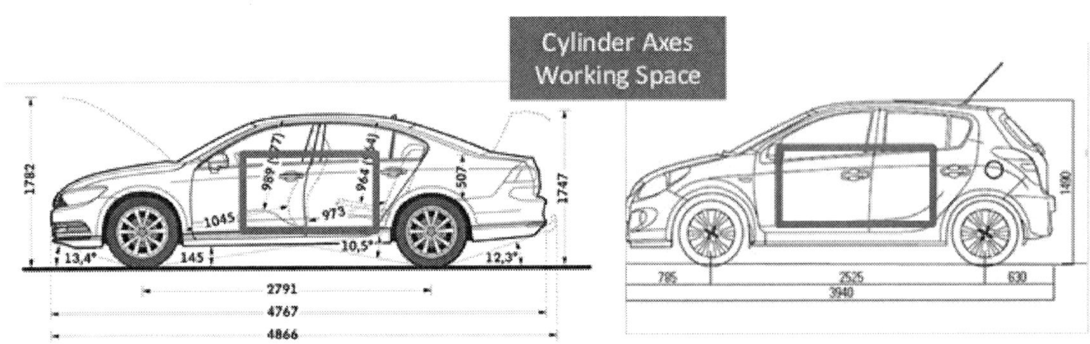

Fig.4 Typical car vs ALIS and MDB (sketch) [2,3]

This is very important as according to the latest ECE and EuroNCAP standards, the rear passenger will be evaluated as well. So far all test methods and systems have been focused only on the single passenger, mainly driver. ALIS presents a unique opportunity to capture also interaction between driver and front passenger and also between rear seat passengers. It has not been assessed so far and offers a great way to evaluate a lot of new data, biomechanical loads and mechanisms within one test run or simulation.

Principle of ALIS

The basic principle is to accelerate the sled with ALIS attached according to the real side crash pulse while the cylinders push forward through linear guide into the "impact structure". It should then cause exactly the same intrusion and kinematics of the door system, ie. biomechanical loads (=replication of the physical test). The source data will be extracted from full vehicle crash test and sled test Finite Element (FE) simulations.

The main idea is to perform simulations before the physical testing loop to ensure the correct kinematics and structural behaviour reflect physical test. The simulation would determine parameters such as amount and position of used cylinders; timing, shape and magnitude of the cylinder pulses. These will be then used for the physical test of reduced model. The method is unique due to its limitless options of simulations. It will save time, money and help engineers with restraint systems tuning. Currently the process of tuning side, curtain and head airbags is extremely time-consuming and expensive (painful). The approach is based on using only part of a car and it is a combination of physical and virtual methods. It is clear that every vehicle will require unique set of input parameters as well as impact structure.

Methodology

The whole process starts with prototype and is shown in Figure 5. When the first one is built and crashed, biomechanical loads are recorded. Following that correlation of FE simulation is the next step. Output is to be biomechanical loads, intrusion and kinematics of important structural parts such as doors, A- and B-pillars.

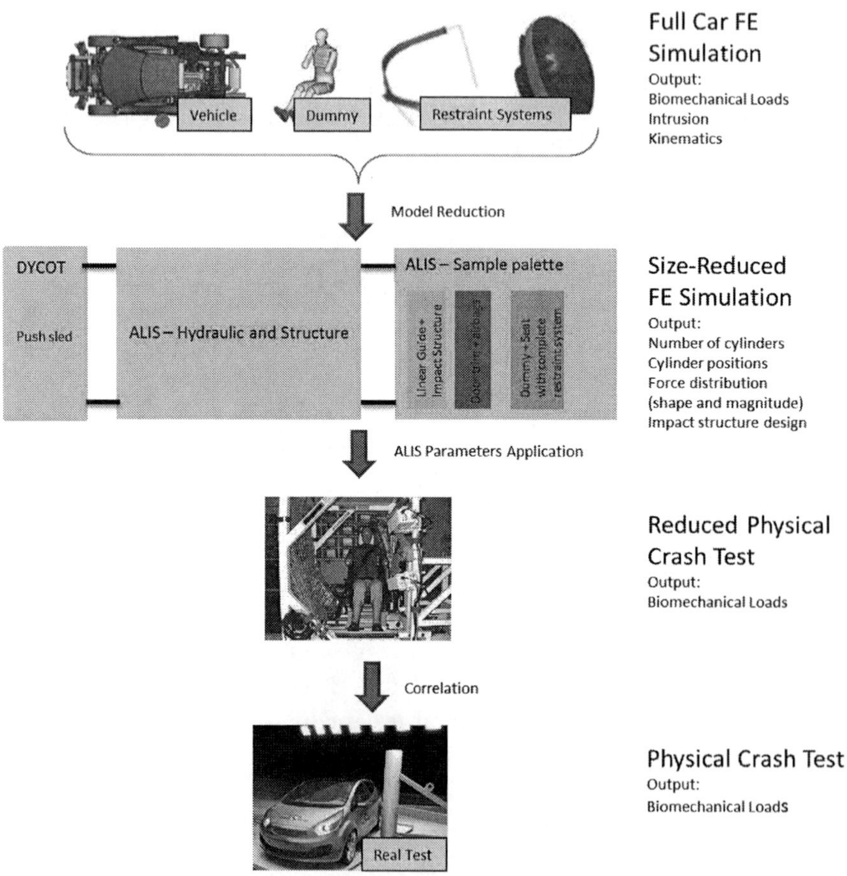

Fig.5 Real crash to ALIS reduction procedure [4,5]

Size reduction of FE model comes next. The most important outcome of this phase is determination of the ALIS settings. This includes number of cylinders used, their timing and also design of the impact structure. Amount of input parameters is countless.

Stages and partial objectives

Complete process covers many intermediate steps that need to be met in predefined sequence. The complex problem has been split into several major steps as follows:

- Take over initial complete side crash simulation of virtual car – pole strike / barrier (from OEM)
- Model size reduction and evaluation of biomechanical loads
- Creation of ALIS virtual model and setup (kinematics, interactions,...)
- Initial determination of physical setup input parameters of ALIS (@biomechanical loads)
- Correlation of physical and virtual test
- Final assessment of the setup for ALIS and its definition
- Successful physical test

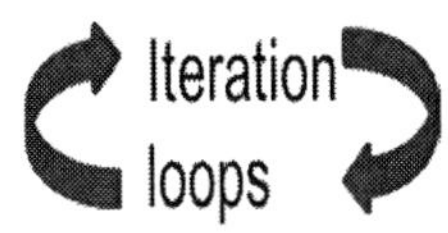

There are several available approaches, but main two have been selected also known as "Design of Experiment" (DoE) and Step-by-Step iteration with subsequent physical correlation.

Design of experiment is advanced mathematical approach, where there are ranges of input parameters. Based on measured responses, the mathematical n-dimensional response surface is created. Such a method has been used especially for sensitivity and robustness studies and would reduce number of runs and predicts multiple results based on input parameter combination as shown in Figure 6.

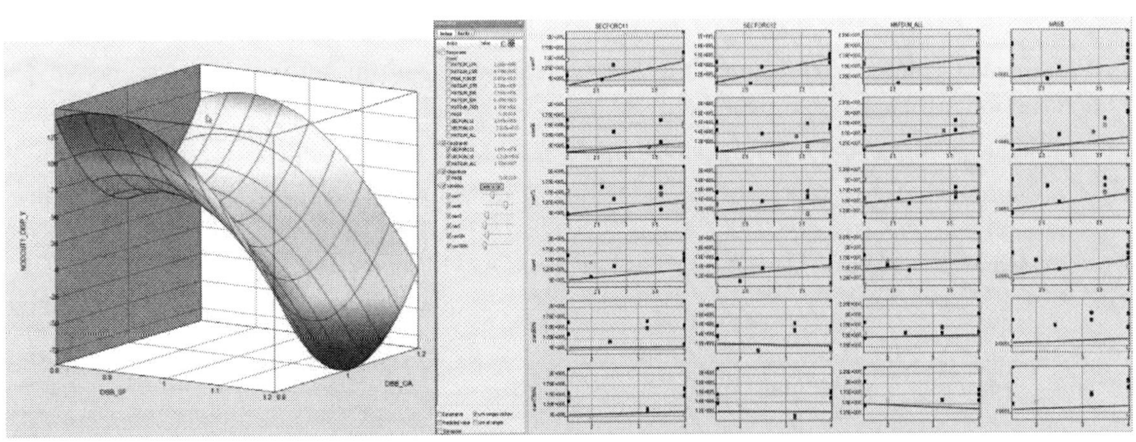

Fig.6 "DoE" Response surface (left) and sensitivity study (right)

Step-by-step iteration presents traditional approach where every change of parameter is being investigated and therefore know-how base is built. The complete process, including iteration loop, is shown in Figure 7.

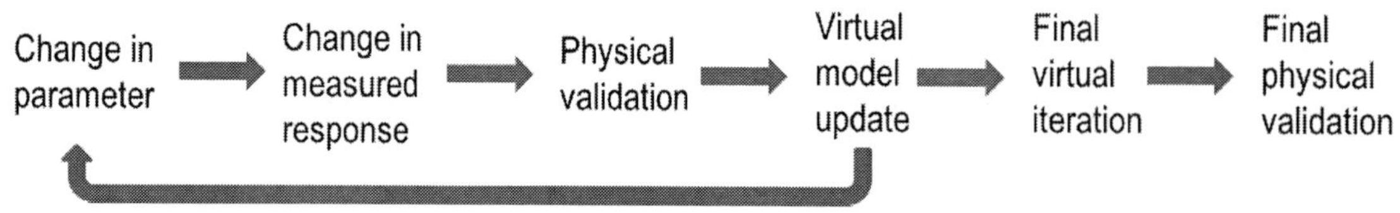

Fig.7 Step-by-Step iteration scheme

Ultimate idea of both methods is to be able virtually reflect and simulate real physical behaviour of the system. Due to many unknown variables this task represents very complex problem that needs to be understood and patiently tuned.

Conclusions

The main objective is to develop a virtual method that would allow reducing full crash into sled crash via ALIS, defining complete ALIS setup and give highly accurate results, while reducing costs.

This new approach to such a problem size is a unique in at least the whole Europe and has never been applied before in such scale. It combines physical and virtual testing and should lead to sufficient and efficient support in terms of restraint system settings. It should bring reduced costs, decreased time of the product development cycle, increased predictability in early stage of the project and quick and cheap tuning of restraint systems.

References

[1] CARHS magazine: Safety Companion 2016, 2016, pages 4-39

[2] CARWOW, VW Passat sizes and dimensions guide, figure online: https://www.carwow.co.uk/blog/2015-volkswagen-passat-dimensions-interior-and-exterior-sizes

[3] Hyundai forum; Online figure: http://www.team-bhp.com/forum/indian-car-scene/31563-hyundai-launch-b-segment-car-i20-soon-8.html;

[4] Eichberger A., Wörgötter F., Haberkorn G., Fellner B., 2016. A new side impact sled test rig at Magna Steyr for validation of side impact development, ATZ online; figure online: www.atzonline.com/Artikel/3/7889/A-NEW-SIDE-IMPACT-SLED-TEST-RIG-AT-MAGNA-STEYR-FOR-VALIDATION-OF-VIRTUAL-SIDE-IMPACT-DEVELOPMENT.html

[5] EuroNCAP, Side Pole, 2015; figure online: www.euroncap.com/en/vehicle-safety/the-ratings-explained/adult-occupant-protection/side-pole/

EAN 2018
56th conference on experimental stress analysis

Influence of Non-alcoholic Beverages on Micromechanical Properties of Human Enamel

JÍRA Aleš[1,a], TOMANEC Martin[1,b] and HÁJKOVÁ Petra[1,c]

[1]Czech Technical University in Prague, Faculty of Civil Engineering, Department of Mechanics, Thákurova 7, 166 29 Prague, Czech Republic

[a]jira@fsv.cvut.cz, [b]martin.tomanec@fsv.cvut.cz, [c]petra.hajkova@fsv.cvut.cz

Keywords: enamel, degradation of tooth, non-alcoholic beverage, reduced modulus, hardness, nanoindentation.

Abstract. Human enamel is a part of tooth which protects dentin and pulp against excessive mechanical stress during mastication and other outer stimuli (chemical, thermal etc.). If the enamel is damaged, as a result of change in mineral content and consequent decrease of mechanical resistance, cavities can initiate easier and it can lead to extraction of the whole tooth. Aim of the study is to show negative impact of beverages with a high content of sugar and a low pH value on micromechanical properties and thickness of the human enamel. The study points out that even short-term contamination (10 s) by the selected beverage causes decrease of surface hardness (H_{it}) and reduced modulus (E_r) of the enamel. The contamination for 5 min by coke caused decrease of the enamel thickness ~ 400 μm and decrease of the hardness ~ 22 %.

Introduction

Enamel is the hardest tissue in a human body. The hardness of enamel is a result of its composition – 96 % inorganic mineral in the form of hydroxyapatit and 4 % water and organic matter [1]. The enamel's primary purpose is to protect dentin and pulp against exposure to the outer stimuli. Among the most common stimuli are mechanical stress during mastication and exposure to the chemical processes during consumption of foods and beverages. Damage to the brittle enamel may then cause easier initiation of cavities and subsequent damage to the whole tooth and need of extraction.

Generally known fact is that excessive consumption of food and beverages with a high content of sugar has negative impact on oral health. In recent years, authors [2, 3] dealt with influence of coke on dental erosion and hardness of the enamel because the coke is favourite drink of adults and kids, despite the fact that contains much sugar and has a low pH value. The authors [3, 4] warn that contamination by the coke causes demineralization and softening of the enamel, although saliva slows down this action. People, who want to live healthy, can be threatened by damage of the enamel caused by consumption of drinks too. Consequences of sport drinks, fruit juices [4, 5], black and herbal tea [6] are often discussed.

This work is dedicated to evaluation of the influence of different common beverages that humans consume on the values of micromechanical properties and thickness of the enamel. Some of the analysed beverages are coke, orange juice, tea with honey or water. Due to the low volume of analysed material, we chose the nanoindentation method to determine the values of micromechanical properties. The analysed properties are hardness (H_{it}) and reduced modulus (E_r).

Methods

Micromechanical tests were carried out on a series of molars extracted due to paradontal reasons. The test specimens were carefully chosen so as to contain minimal visible damage to the enamel layer. The chosen molars were embedded in epoxy resin and subsequently cut into specimens 1 cm thick. The specimens were then grounded and polished in order to reach the occlusal surface. The complex morphology of the occlusal surface does not enable for direct indentation on the outer rim of enamel, but the tooth rather needs to be polished into a flat area. The specimen preparation is based on previous experiences in nanoindentation of teeth and materials of dental implants [7, 8].

The indentation was split into three parts. The first measurement was made on a clean specimen. The second measurement was made after contaminating the specimen for 10 seconds. The third measurement was made after 5 minutes of contamination. All specimens were individually contaminated by 20 μl of the liquid. For the purpose of this study, the most basic liquids (tap water, coca-cola, orange juice and tea+honey) were chosen. The values of pH (Tab. 1) were determined by pH meter (Hanna Instruments, USA). The contents of sugar (Tab. 1) were determined by nutrition labels, the only exception was a sugar content in the tea+honey which had to be calculated from the ratio tea/honey and the sugar content in the honey itself.

Tab. 2 The values of pH and sugar contents of the selected beverages

Beverage	pH	Sugar content [g/100 ml]
coke	2.74	11.20
orange Juice	3.97	10.00
tea+honey	4.58	8.90
H_2O	7.74	0.00

The measurements were performed on the CSM Instruments Nano Hardness Tester (CSM Instruments SA, Switzerland) equipped with a cube corner indentation tip. We used the mode of directed force with maximum force of 15 mN and loading speed of 90 mN/min. Evaluation of results was made in accordance with the standard Oliver&Pharr method [9].

Results

Individual measurements were made in matrices of 5x5 indents. Subsequently, the changes in values of the micromechanical properties (the hardness, the reduced modulus) caused by chemical impact of the different commonly consumed beverages were evaluated. Impact of the beverages on degradation of specimen surface and enamel thickness was compared as well. The degradation of the surface was analyzed by an optical microscopy (1000x magnification) and an atomic force microscope (AFM). Fig. 1 shows obvious difference in roughness of the polished specimens.

without contamination 10 s 5 min

01 coke

02 orange juice

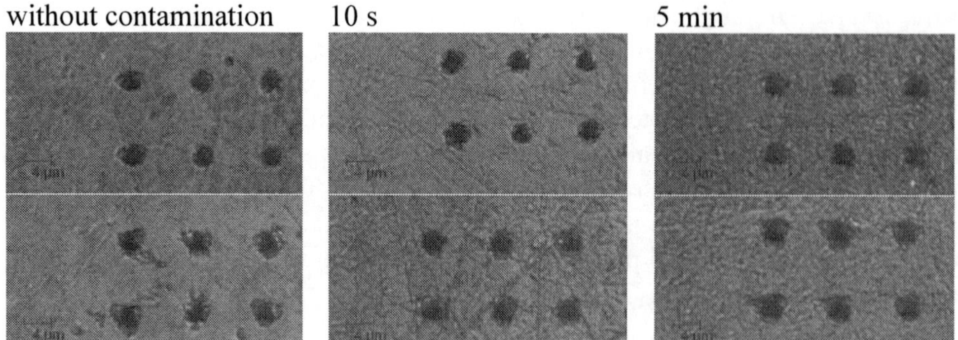

03
H_2O

04
tea+honey

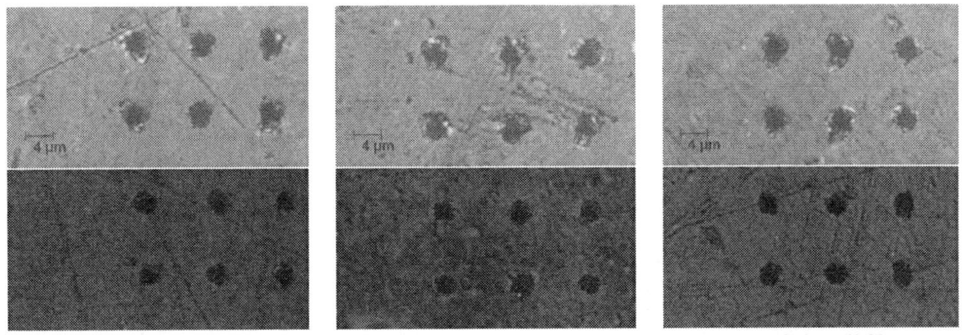

Fig.1 Degradation of the specimen surfaces caused by chemical impact of the different commonly consumed beverages. The degradation is compared for the specimen surfaces without contamination and contaminated for 10 s or 5 min

The determination of decrease in thickness of the enamel which is part of the specimen embedded in epoxy resin was a difficult task. Therefore, the nanoindentation was used for approximate evaluation. A matrix of indents with depths ~ 1240, 1040, 810, 730, 680, 550 and 400 nm was performed by different maximum loading forces – 25, 15, 10, 8, 6, 4 and 2 mN. Based on observability of the indents after 5 min of the contamination was possible to estimate decrease in the enamel thickness. Fig. 2 shows that the indent with depth of 400 nm (2 mN) is almost invisible which was caused by etching. The greatest decrease in thickness ~ 300–500 nm and the most degraded surface was observed after contamination by the coke and the juice for 5 min. On the contrary, after contamination by the water no degradation of the surface was observed. The change in smoothness of the specimen surface caused by short-term contamination (10 s) was the most obvious for the juice (Fig. 1).

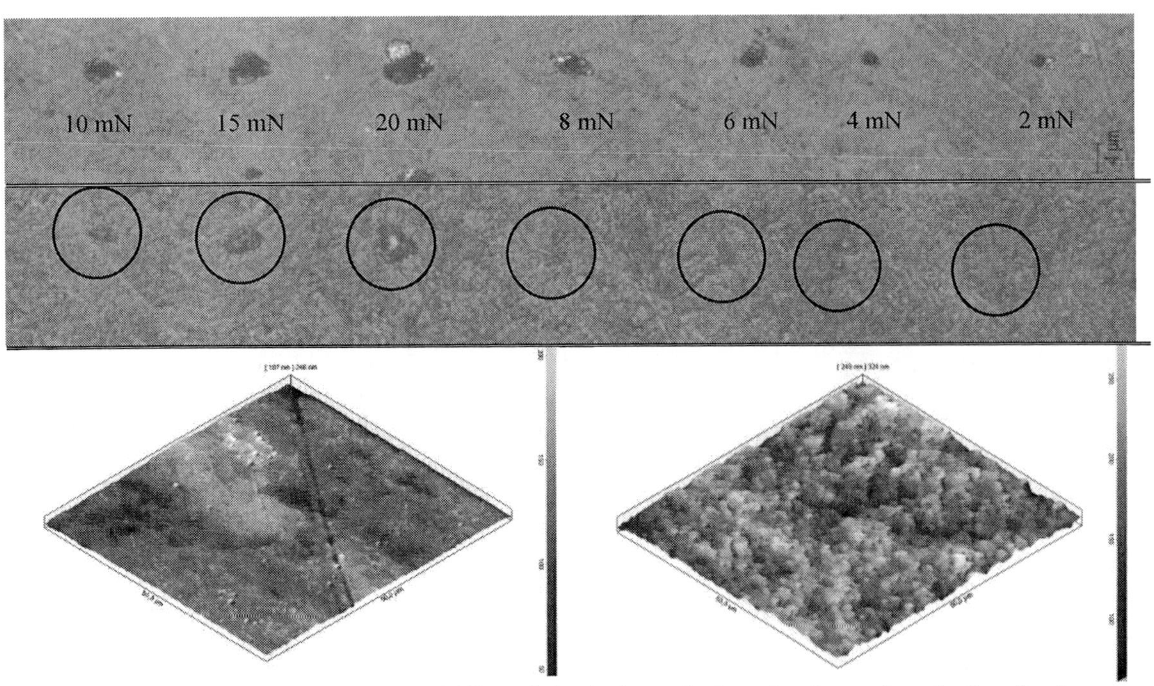

Fig.2 Comparison of observability of the indents before (at the top) and after 5 min contamination by the coke (in the middle). The figures were performed by optical microscope (1000x magnification). Based on the observability it was possible to estimate decrease in the enamel thickness ~ 300–500 nm. Pictures performed by AFM (at the bottom) show difference in surface roughness before (on the left) and after (on the right) contamination

The measurements of the micromechanical parameters, such as the hardness (H_{it}), the reduced modulus (E_r) and a contact depth (h_c), were statistically determined and overall results are showed in Tab. 2. The obvious changes were observed for every parameter but the difference in hardness (Fig. 3) was dominant. The hardness of enamel decreased ~ 10–20 % for specimens 01 and 02 (the coke and the juice) after 5 min of contamination. Short-term contamination (10 s) caused decrease ~ 3% for the same specimens.

Tab. 2 Reduced modulus (E_r), hardness (H_{it}) and contact depth (h_c) without and with contamination for 5 min

		01 – cola		02 – orange juice		03 – H_2O		04 – tea+honey	
		0 s	5 min	0 s	5 min	0 s	5 min	0 s	5 min
H_{it} [MPa]	Mean	3378	2645	2907	2603	2918	2926	3508	3379
	Std Dev	182	167	283	185	146	252	170	133
[%]	Decrease	21.70		10.46		–			3.68
E_r [Gpa]	Mean	96.13	91.36	89.25	88.29	89.56	87.94	101.75	94.82
	Std Dev	6.6	7.6	7.7	4.5	5.3	4.9	8.1	3.8
[%]	Decrease	4.97		1.08		–		6.81	
h_c [nm]	Mean	1030	1169	1115	1179	1110	1107	1010	1031
	Std Dev	29	37	55	43	29	50	25	21

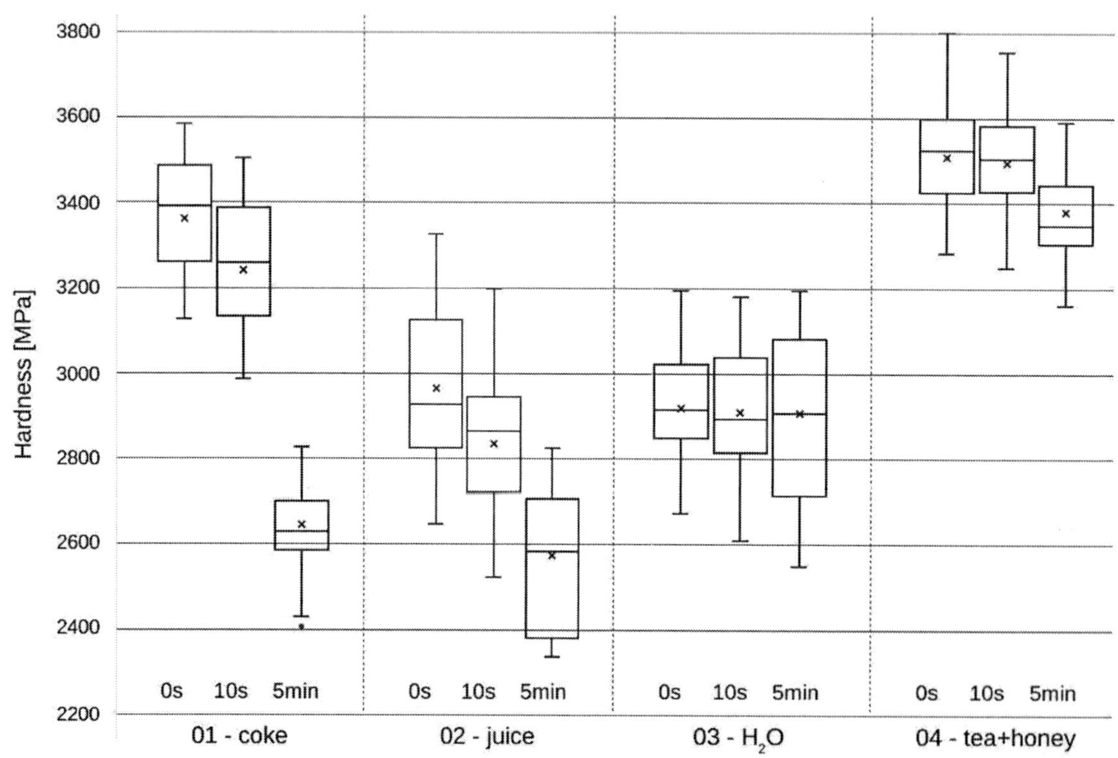

Fig.3 The statistically determined values of the hardness (H_{it}) before and after contamination (10 s, 5 min) by the coke, the juice, the H_2O and the tea+honey

Conclusions

The measurements proved that contamination by the commonly consumed beverages (the coke, the orange juice) can negatively affect the human enamel. The contamination by H_2O was observed as well for verification of the measurements. As expected, the water had no effect on the micromechanical parameters and the degradation of specimen surface. The analysis showed that drinks with low values of pH can cause significant decrease in enamel thickness and degrade surface of the enamel. As a result, quality of teeth and resistance to cavities can be lowered. The question is how much the degradation depends on the high sugar content of the observed beverages.

Acknowledgement
The financial support by the Faculty of Civil Engineering, Czech Technical University in Prague (SGS project No. SGS17/168/OHK1/3T/11) is gratefully acknowledged.

References

[1] J. K. Avery, D. J. Chiego, Essentials of oral histology and embryology: A clinical approach, third ed., Elsevier, St. Louis, 1992.

[2] H. Devlin, M. A. Bassiouny, D. Boston, Hardness of enamel exposed to Coca-cola and artificial saliva, Journal of Oral Rehabilitation 33 (2006) 26-30.

[3] G. Maupomé et al., In vitro quantitative assessment of enamel microhardness after exposure to eroding immersion in a cola drink, Caries Research 32 (1998) 148-153.

[4] S. Wongkhantee et al., Effect of acidic food and drinks on surface hardness of enamel, dentin, and tooth-coloured filling materialsm, Journal of Dentistry 34 (2006) 214-220.

[5] A. Lussi, T. Jäggi, S. Schärel, The influence of different factors on in vitro enamel erosion, Caries Research 27 (1993) 387-393.

[6] P. A. Brunton, A. Hussain, The erosive effect of herbal tea on dental enamel, Journal of Dentistry 29 (2001) 517-520.

[7] P. Hájková, A. Jíra, Micromechanical analysis of complex structures by nanoindentation, Key Engineering Materials 731 (2017) 60-65.

[8] L. Řehounek, A. Jíra, Experimental and numerical analyses of 3D-printed titanium trabecular dental implants, Acta Polytechnica 57 (2017) 218-228.

[9] W. C. Oliver, G. M. Pharr, An improved technique for determining hardness and elastic modulus using load and displacement sensing indentation experiments, Journal of Material Research 7 (1992) 1564-1583.

Vibration Isolation of Electric Golf Cart Seat

JÍROVÁ Radka[1, a], SVOBODA Roman[2, b] and PEŠÍK Lubomír[3, c]

[1] Faculty of Mechanical Engineering, Technical University of Liberec, Czech Republic

[2] BAG DEVELOPMENT s. r. o., Mladá Boleslav, Czech Republic

[3] Faculty of Mechanical Engineering, Technical University of Liberec, Czech Republic

[a]radka.jirova@tul.cz, [b]roman.svoboda@afd.cz, [c]lubomir.pesik@tul.cz

Keywords: Vibration Isolation, Golf Cart Seat, Minimization of Vibration, Suspension Seat.

Abstract. The paper is focused on minimization of vibration transmitted from electric golf cart wheels to seat and seated person. This vibration appears during golf cart driving on uneven terrain of the golf court. This problem can be solved according to [2] using the elastic and damping link between cart frame and seat tightly connected to battery of electric power. For evaluation of designed solution is used mathematical calculation describing golf cart as system of masses connected with elastic and damping links. The result of mathematical description of designed solution is reference step response of golf cart seat.

Introduction

Suspension seats are used in a wide range of vehicles, especially for the purpose of protecting person's health exposed to whole-body vibration during travel, and also for comfortability increase. Driving on uneven terrain of golf court leads to high-intensity vibration of wheels which is transmitted to frame, seat and seated person. Conventional battery powered electric golf carts do not have suspension seats, for the purpose of transmitted vibration reduction to seat and seated person the elastic and damping link between chassis and frame of golf cart is being used. These links reduce transmission of vibration only partially, so the drive in golf cart can get uncomfortable.

Effects of Vibration on Comfort

Evaluation of effect of vibration on comfort of seated person during the golf cart drive was made in accordance with [1]. Result of evaluation of effect of vibration is frequency-weighted r.m.s. acceleration of measured acceleration as a function of time during the golf cart drive and comparison this acceleration value with table in [1]. Frequency-weighted r.m.s. acceleration for the one-third octave band is shown in Fig. 1.

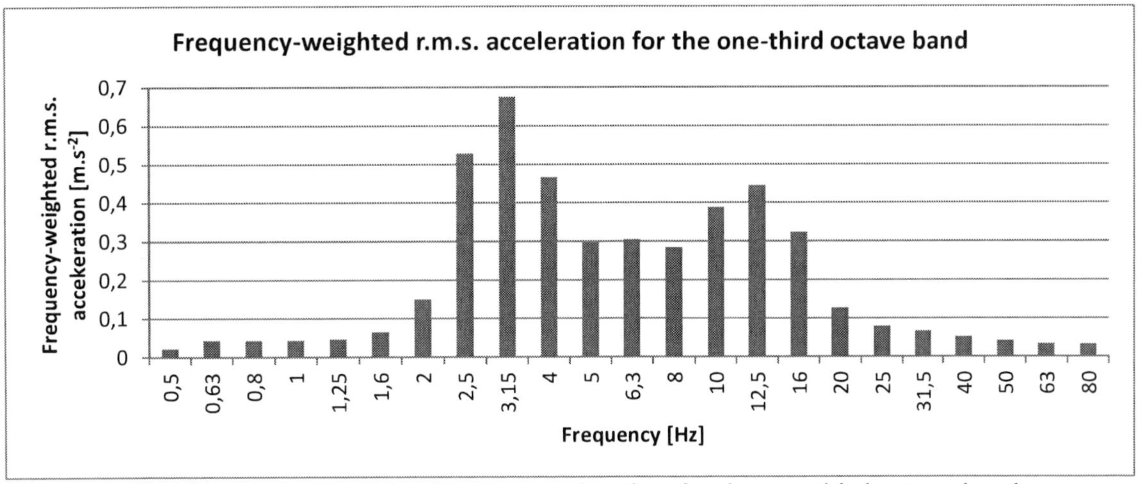

Fig.1 Frequency-weighted r.m.s. acceleration for the one-third octave band

Comparison of frequency-weighted r.m.s. acceleration with table in [1] shows that effect of vibration on comfort during golf cart drive is in interval of uncomfortable and very uncomfortable.

Minimization of Transmitted Vibration

Minimization of vibration, transmitted from golf cart wheels to seat and seated person, is based on principle described in patent application [2]. Transmitted vibration can be reduced according to [2] using the elastic and damping link between cart frame and seat tightly connected to battery of electric power. Battery of electric power tightly connected to seat increases inertial effect of mass, which reduces seat vibration. The principle is shown in Fig. 2.

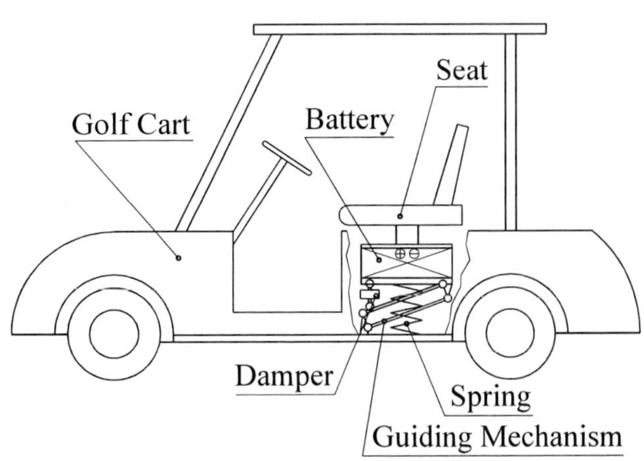

Fig.2 Principle of vibration isolation of seat

Calculation of Transmitted Vibration

The system in Fig. 2 can be described in simplified way by differential equations:

$$m_R.\ddot{x}_R+b_R.(\dot{x}_R-\dot{u})+b_{SB}.(\dot{x}_R-\dot{x}_{SB})+k_R.(x_R-u)+k_{SB}.(x_R-x_{SB})=0, \qquad (1)$$

$$m_{SB}.\ddot{x}_{SB}+b_{SB}.(\dot{x}_{SB}-\dot{x}_R)+k_{SB}.(x_{SB}-x_R)=0, \qquad (2)$$

where m_R is mass of frame, m_{SB} mass of seat, battery and seated persons, b_R damping coefficient of chassis, b_{SB} damping coefficient between seat with battery and frame, k_R stiffness of chassis, k_{SB} stiffness between seat with battery and frame, u deflection of chassis, x_R deflection of frame and x_{SB} deflection of seat with battery.

In matrix form can be the system in simplified way described by equation:

$$\vec{q}=(\mathbf{K}+i.\omega.\mathbf{B}-\omega^2.\mathbf{M})^{-1}.\vec{f}, \qquad (3)$$

where $\vec{q}$ is vector of deflection, **K** matrix of stiffness, **B** matrix of damping, **M** matrix of mass, $\vec{f}$ vector of force substituting deflection u, ω angular frequency and i imaginary unit.

In case of using elastic and damping link between golf cart frame with tightly connected battery and seat, the differential equations are:

$$m_R.\ddot{x}_R+b_R.(\dot{x}_R-\dot{u})+b_S.(\dot{x}_R-\dot{x}_S)+k_R.(x_R-u)+k_S.(x_R-x_S)=0, \qquad (4)$$

$$m_S.\ddot{x}_S+b_S.(\dot{x}_S-\dot{x}_R)+k_S.(x_S-x_R)=0, \qquad (5)$$

where m_S is mass of seat and seated persons, b_S damping coefficient between seat and frame, k_S stiffness between seat and frame and x_S deflection of seat.

Differential equations Eq. 1, Eq. 2, Eq. 4 and Eq. 5 are described in Fig. 3 on the left, equation Eq. 3 is described in Fig. 3 on the right.

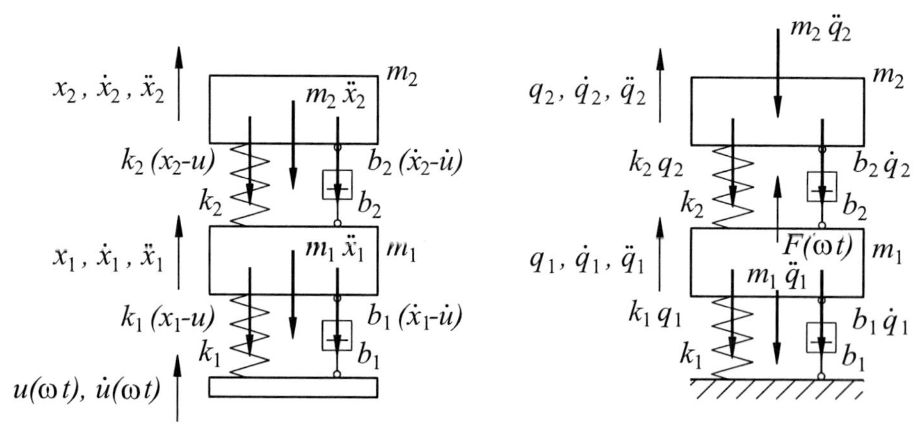

Fig.3 Description of system with two DOF

The result of solved equations Eq. 1 – Eq. 5 is amplitude-frequency response shown in Fig. 4 and reference step response of systems (Fig. 5 and Fig. 6).

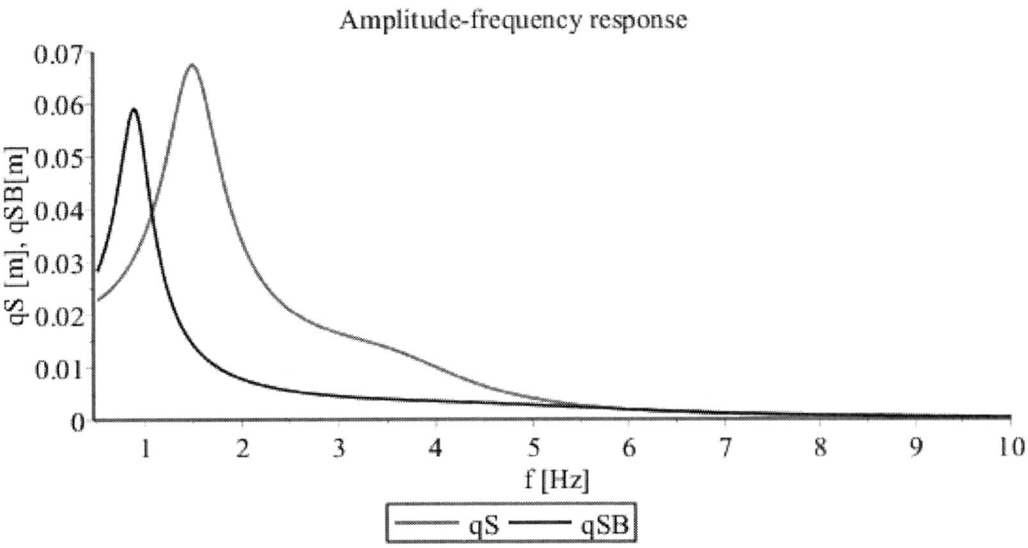

Fig.4 Amplitude-frequency response

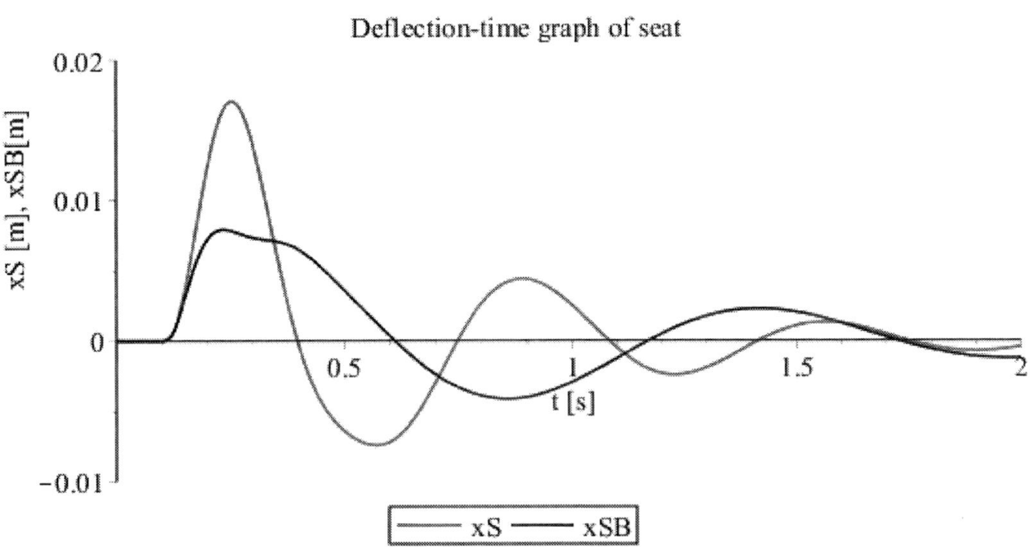

Fig.5 Step response of systems – deflection-time graph

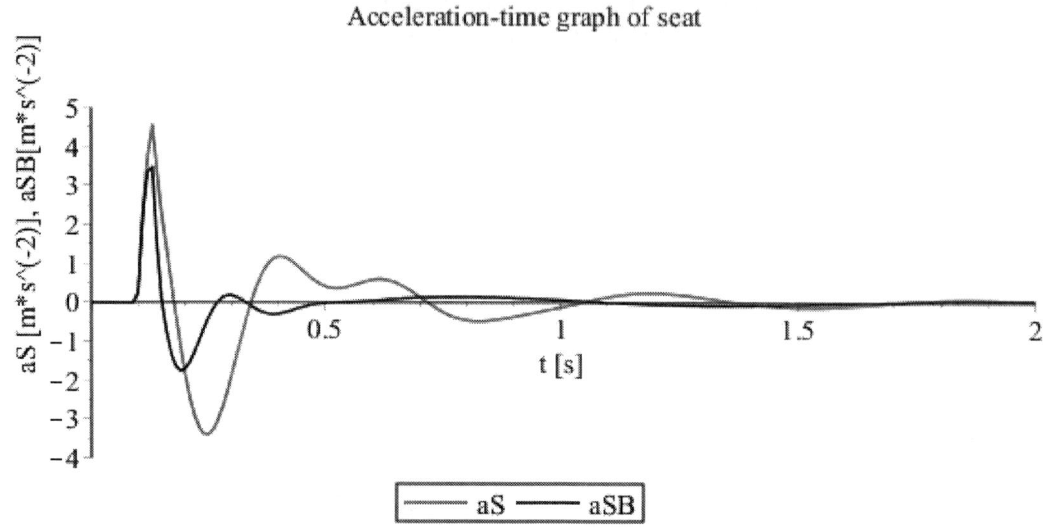

Fig.6 Step response of systems – acceleration-time graph

Driving golf cart on uneven terrain of golf court leads to non-periodic and non-harmonic vibration. In simplified way this vibration can be represented by separated step functions. The deflection u in differential equations Eq. 1, Eq. 2, Eq. 4 and Eq. 5 can be substituted by reference step signal. Decisive for evaluation of effect increasing seat inertia is reference step response of systems, amplitude-frequency response shown in Fig. 4 should be taken as additional.

For evaluation of effect of vibration on comfort in accordance with [1] is decisive acceleration-time graph of seat as reference step response (Fig. 6).

Evaluation of Transmitted Vibration

The system of electric golf cart without suspension seat can be described in simplified way by differential equation:

$$m_R.\ddot{x}_{RS}+b_R.(\dot{x}_{RS}-\dot{u})+k_R.(x_{RS}-u)=0, \qquad (6)$$

where x_{RS} is deflection of frame with seat.

In Fig. 7 the differential equation Eq. 6 is described.

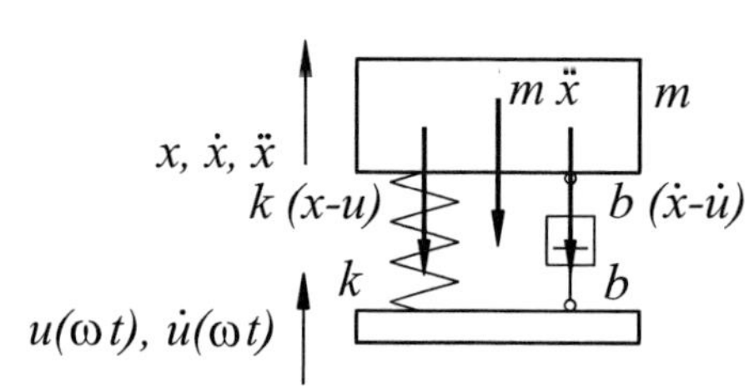

Fig.7 Description of system with one DOF

The result of solved differential equations Eq. 1, Eq. 2 and Eq. 6 is reference step response of systems. Acceleration-time graph of suspension seat and seat tightly connected to golf cart frame is shown in Fig. 8.

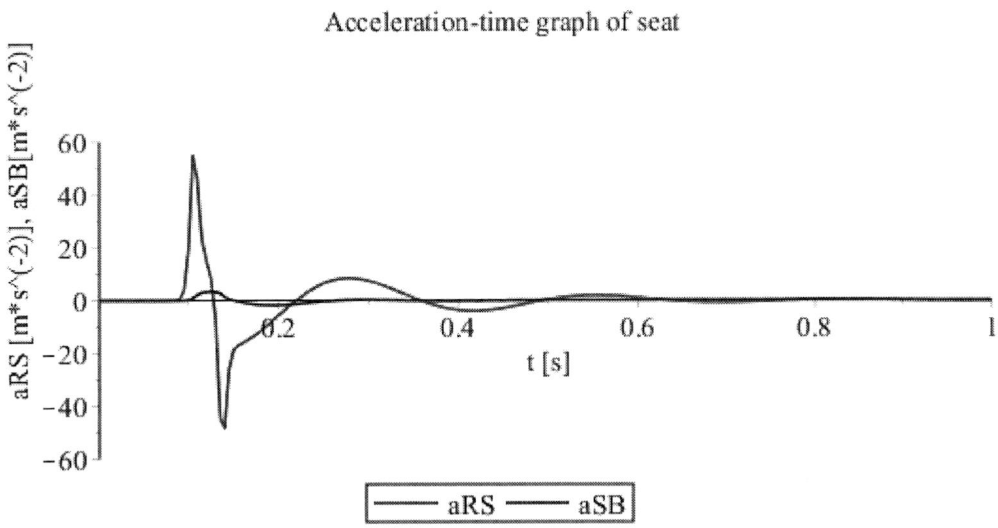

Fig.8 Step response of systems – acceleration-time graph

Conclusion

The paper deals with current issue of vibration isolation of golf cart seat. The solution uses invention idea of increase inertia of tightly connected mass to seat. This allows significantly reduce vibration transmitted from golf cart wheels to driver and passenger.

Acknowledgements

This publication was written at the Technical University of Liberec as part of the project "Innovation of products and equipment in engineering practice" with the support of the Specific University Research Grant, as provided by the Ministry of Education, Youth and Sports of the Czech Republic in the year 2018.

References

[1] ČSN ISO 2631-1. *Vibrace a rázy – Hodnocení expozice člověka celkovým vibracím – Část 1: Všeobecné požadavky*. Praha: Český normalizační institut, 1999.

[2] BAG DEVELOPMENT s.r.o. Golfový vozík s bateriovým elektrickým pohonem a sedákem. R. Svoboda, R. Jírová a L. Pešík. Česká Republika. Přihláška vynálezu PV 2018-112. 8.3.2018.

[3] Pešík, L., Skarolek, A.: Tuning of Vibration Isolation Differential Pneumatic System by Means of Trottle Valve. In *Transactions of the Universities of Košice: Research reports from the Universities of Košice*. 2nd edition. Košice: Technická univerzita v Košiciach, 2011. s. 191-196. ISSN 1335-2334.

[4] Pešík, L., Skarolek, A.: Heat Transfer Effects on Vibration Isolation Differential Pneumatic System. In *Transactions of the Universities of Košice: Research reports from the Universities of Košice*. 3rd edition. Košice :Technická univerzita v Košiciach, 2012. s. 109-114. ISSN 1335-2334.

[5] Dresig, H.; Holyweißig, F.: *Maschinendynamik*. 8. Auflage.SpringerVerlag. Berlin, 2008.

EAN 2018
56th conference on experimental stress analysis

Delamination Growth under Mixed Mode I - Mode II Fatigue Loading in Polymer Composite

KADLEC Martin[1,a] and ŠEDEK Jakub[1,b]

[1]VZLU – Czech Aerospace Research Centre, Beranových 130, 199 05 Prague – Letňany, Czech Republic

[a]kadlec@vzlu.cz, [b]sedek@vzlu.cz

Keywords: Polymer composite, delamination, mixed-mode, fatigue, aerospace.

Abstract. The characterization of material parameters on mixed mode bending (MMB) specimens was needed for modelling of crack growth in the composite air inlet of new L-39 NG jet trainer. The paper describes methods and results of the delamination growth in a carbon fibre-reinforced polymer composite under mixed mode I (opening) – mode II (shear) quasi-static and fatigue loading. This article summarizes the results of 15 quasi-static and 3 fatigue specimens tested in 3 different mode mixtures – 25%, 50% and 80%. The fracture toughness was increasing with the increasing mode mixture. The scatter of values was increasing with the mode II preference. The curves of crack growth rate vs. strain energy release rate were obtained and fitted by a regression model considering the mode mixtures. The dependence of crack growth law parameters according to Paris equation on the mode mixtures was non-linear. Thus, unique material characteristics needed for modelling were obtained.

Introduction

Fatigue damage accumulation and crack growth in metallic and composite materials are critical for the structural integrity evaluation. In recent years, composite materials have been widely used for structural components, especially in the field of aerospace engineering.

The delamination between plies of the laminate is one of the most important damage mechanisms. The air inlet has curved shape and so the mixed mode I (opening) and mode II (shear) loading can be expected when the delamination occur. The mode III contribution in delamination growth is typically quite small for composite structures due to the constraints of adjacent plies [1] and therefore it was not considered [1].

Several analytical models have been proposed and experimentally evaluated [1-3]. The results showed general non-linear relation of parameters used for the crack growth description in the form of Paris law. The FE-modelling employing cohesive zone models is also appropriate. It was concluded that only with implementing a cyclic damage variable in the cohesive interface element the experimentally observed crack growth and stiffness degradation can be captured properly [4].

The aim of this work is to determine the input data for the modelling of damage evolution in the composite air inlets of the newly emerging aircraft L-39NG. The data will be used to evaluate acoustic fatigue properties of the inlet using finite element (FE) analyses.

Methods

Material. Carbon fibre-reinforced composite Hexply AGP 193PW/8552S RC40 with the 16 layers lay-up of $[45/0]_{4S}$ was manufactured and MMB specimens were extracted with dimensions of 185 x 25 x 3.3 mm. During manufacturing, a non-adhesive plastic foil sheet was embedded into the plate near the edge during manufacturing to the neutral axis to initiate the delamination growth between the 8th and 9th layer. A pair of aluminium alloy blocks with dimensions of 25 mm x 20 mm x 10 mm was bonded using R 21 MP-2 instant cyanoacrylate adhesive to the end of each specimen on the side with non-adhesive insert. These blocks are used to apply the load to the specimens. Then a white paint marker was coated on the specimen edges to reveal the crack front during loading. Both edges of the specimen were marked using black lines with the spacing of 25 mm.

A series of 10 specimens was used for quasi-static testing and 5 specimens were used for fatigue testing. The specimens were divided to cover pure Mode I and three different mode mixtures – G_{II}/G = 25%, 55% and 85% (shear component of G / total strain energy release rate G).

Test setup. Tests were performed on non-aged specimens at room temperature, ambient atmospheric pressure and ambient relative humidity. A special fixture designed according to ASTM D6671M-06 [5] standard was employed (Fig. 1). Three test rig configurations were used to achieve different loading modes. Double cantilever beam (DCB) method was used to induce pure mode I. A pair of video cameras with the resolution of 2048 x 1536 pixels was focused onto the edges to cover the area between supports.

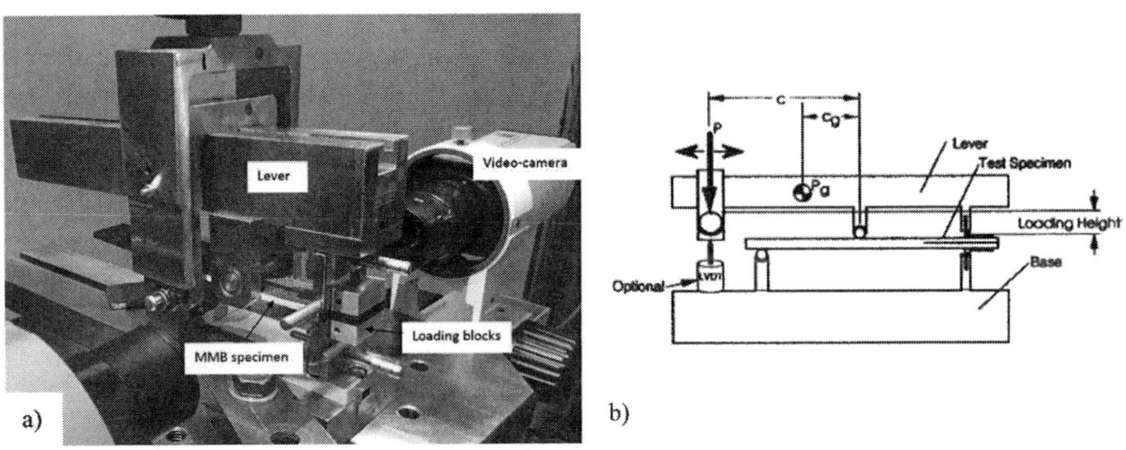

Fig.1 a) Mixed-mode bending test set-up and b) schematic of the mixed mode bending [5]

The ASTM D6671M-06 standard was used for evaluation of the quasi-static test and the maximum strain energy release rate G in the cycle was determined. The load was performed with a constant speed of 1 mm/min until the crack propagated by at least 25 mm. During loading the crack propagation was monitored using the cameras at 12.5 frames per second.

The values of G_I and G_{II} corresponding to the crack length a were calculated from Eq. 1 and Eq. 2, where parameters of fixture L, c, c_g were set.

$$G_I = \frac{12[P(3c-L) + P_g(3c_g - L)]^2}{16b^2h^3L^2E_{1f}}(a + \chi h)^2 \tag{1}$$

$$G_{\mathrm{II}} = \frac{9[P(c + L) + P_g(c_g + L)]^2}{16b^2h^3L^2E_{1f}}(a + 0.42\chi h)^2 \tag{2}$$

Variables E_{1f}, χ and Γ were calculated according to Eq. 3 - 7, where E_{11} = 68 000 MPa, E_{22} = 68 000 MPa and G_{13} = 4 000 MPa. The parameter m_{cal} in Eq. 6 was obtained from a metal specimen calibration. P denotes the force, b is the specimen width and h is the specimen thickness.

$$E_{1f} = \frac{8(a_o + \chi h)^3(3c - L)^2 + [6(a_o + 0.42\chi h)^3 + 4L^3](c + L)^2}{16L^2bh^3\left(\frac{1}{m} - C_{sys}\right)} \tag{3}$$

$$\chi = \sqrt{\frac{E_{11}}{11G_{13}}\left\{3 - 2\left(\frac{\Gamma}{1+\Gamma}\right)^2\right\}}, \quad \Gamma = 1.18\frac{\sqrt{E_{11}E_{22}}}{G_{13}} \tag{4, 5}$$

$$C_{sys} = \frac{1}{m_{cal}} - C_{cal}, \quad C_{cal} = \frac{2L(c + L)^2}{E_{cal}b_{cal}t^3} \tag{6, 7}$$

The relations between the components G_{I} and G_{II} and the total value of G are as follows (Eq. 8, and Eq. 9.):

$$G = G_{\mathrm{I}} + G_{\mathrm{II}} \quad \frac{G_{\mathrm{II}}}{G} = \frac{G_{\mathrm{II}}}{G_{\mathrm{I}} + G_{\mathrm{II}}} \tag{8, 9}$$

For fatigue, the test was driven by displacement amplitude δ_{max} = 6 mm and δ_{min} = 0.6 mm with frequency of 5 Hz. Currently there is no standard for fatigue delamination during mixed-mode loading and therefore the quasi-static based ASTM D6671M-06 standard was used for the evaluation. The maximum fracture toughness in the cycle was determined according to Eq. 1- 9. The crack propagation during the following loading was monitored using video cameras.

The crack propagation length for the individual specimens used for the compliance calibration was determined from the snapshots of the video recording. The length was measured from the picture using an in-house software with a virtual gauge. The author invented VZLU modified compliance calibration (VMCC) that was used for data reduction and fitted better to the experimental results than classic MCC method. The constants, A_1 and k from Eq. 10, were determined for individual specimens using a least-square fit applied to the plot of the optically observed delamination lengths (a/h) squared versus the cube root of the corresponding compliance C. For each specimen, the constants were used specifically for one specimen. The relationship between the compliance and the delamination length for the VMCC solution is:

$$(a/h)^2 = A_1 \cdot C^{\frac{1}{3}} + k, \tag{10}$$

where the compliance C is δ_{max}/P_{max} and h is the specimen thickness. The fitted constants A1 and k were used for the crack length a determination for each cycle using Eq. 10, based on the continuous P_{max} recorded from the test machine. For this purpose, the displacement δ_{max} was constant during loading.

The specimen was cycled until unstable crack growth occurred and the crack reached the middle support making the crack front not visible. Polynomial method was used to calculate the crack growth rate $\Delta a/\Delta N$. The method calculated $\Delta a/\Delta N$ by fitting a second-order polynomial to each set of 29 successive data points. The polynomial was used to calculate the value at the midpoint of the 29-point dataset.

The values of $(G_I)_{max}$ and $(G_{II})_{max}$ corresponding to the crack growth rate were calculated from Eq. 2 – Eq. 10.

FEM modelling. Performed static tests were evaluated using finite element approach utilizing cohesive zone model (CZM) implemented in ABAQUS 2017 software. In the analysis, the implicit solver ABAQUS/standard was used. Shell FE model consisted from 4 node thin shell elements with reduced integration scheme (S4R ABAQUS element type). Shell element included 3 through-thickness integration points. Characteristic FEM mesh size was 2 mm.

Model was symmetrically supported like in the three-point bending test. Loading was realized through kinematic coupling of loading points (Fig. 2). Nonlinearity due to displacement is taken into account. The loading was prescribed in displacement in the direction perpendicular to the specimen´s flat plane in the initial position. Two parts of the model were connected in the mid-plane by cohesive interaction. A nonlinear constitutive law represented the response of the material at the interface between plies. The linear-elastic softening behaviour was used in the analysis.

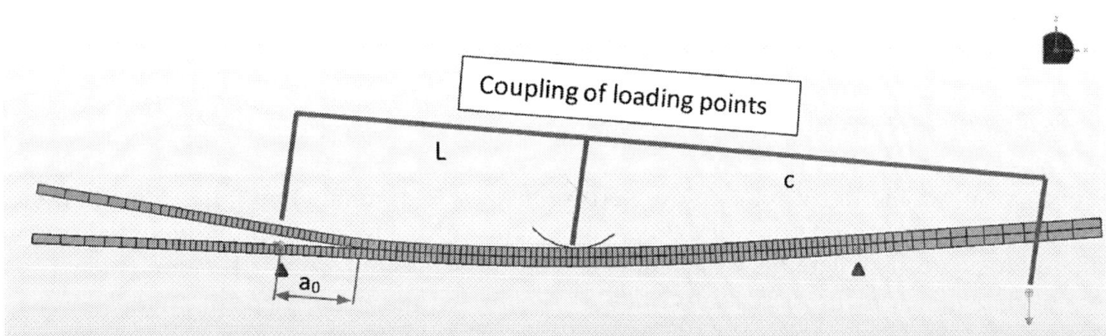

Fig.2 Finite element mesh boundary conditions and loading concept

In order to define the interfacial behaviour with a linear softening cohesive law, the initial stiffness K_i, critical values of driving force (delamination fracture toughness) G_{iC} in modes I, II and III and interlaminar strengths τ_i^0 in directions n, s and t are required. The values of listed parameters were used as follows: $K_i = 1.10^6$ N/mm^3, $\tau_n^0 = 10$ N/mm^2, $\tau_s^0 = 20$ N/mm^2, $\tau_t^0 = 20$ N/mm^2, $G_{IC} = 650$ J/m^2, $G_{IIC} = 2300$ J/m^2, $G_{IIIC} = 2300$ J/m^2.

Quadratic traction relationship is used for combination of cohesive strengths and tractions in the form of Eq. 11. If the relation equals 1, the softening in connection begins.

$$\left(\frac{\langle \tau_n \rangle}{\tau_n^0}\right)^2 + \left(\frac{\langle \tau_s \rangle}{\tau_s^0}\right)^2 + \left(\frac{\langle \tau_t \rangle}{\tau_t^0}\right)^2 \tag{11}$$

Results

For quasi-static testing, the individual values of the fracture toughness showed exponential increase of fracture toughness (Fig. 3). The scatter of values was increasing with the G_{II} mode prevailing.

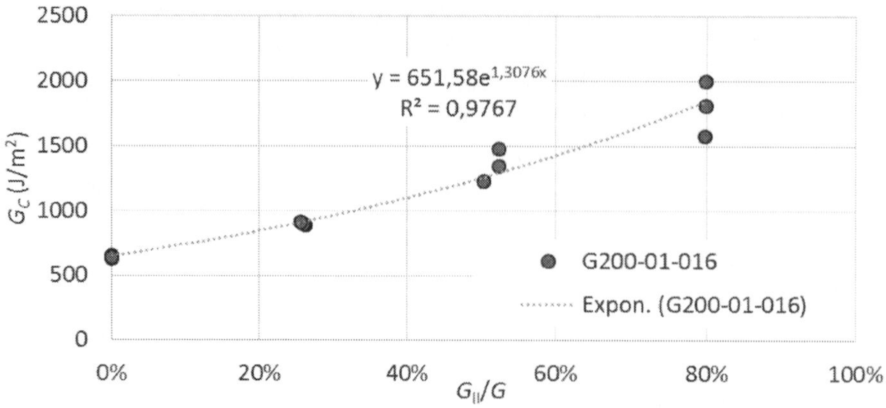

Fig.3 Total mixed-mode fracture toughness; experimental data and fitted exponential curve

For fatigue testing, the plots of the crack growth rate $v = \Delta a/\Delta N$ vs. $(G_I)_{max}$ were generated for all the fatigue specimens (Fig. 4). The curves are wavy which is caused by the carbon fabric structure that locally affects the crack growth rate [6]. It is clear that as mode II contribution increased (or mode I contribution decreased) the data moved consistently to the left (i.e. crack growth rate increasing with increasing mode II component of loading). This suggests that the mode II component has a significant influence on the crack growth rate.

These curves were used for generating the coefficients of the crack growth rate described by Eq. 12 in the form similar to Paris law for individual specimens.

$$\frac{\Delta a}{\Delta N} = A(G_i)_{\max}{}^{r} \quad (12)$$

Fig. 5a shows the linear relation of coefficient log A with mode mix ratio. The coefficient A is increasing with higher mixture. Fig. 5b presents Paris law exponent polynomial relation with mode mix ratio where the exponent is decreasing with higher mode II component. Tab. 1 presents the values of fitted constants of the Paris law for individual mode mixtures.

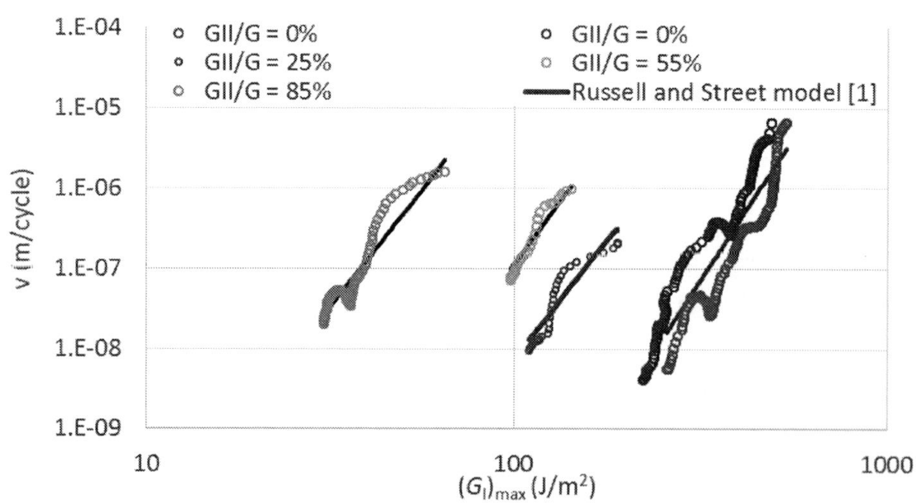

Fig.4 Crack grow rate data; experimental data and Russell and Street model fit [1]

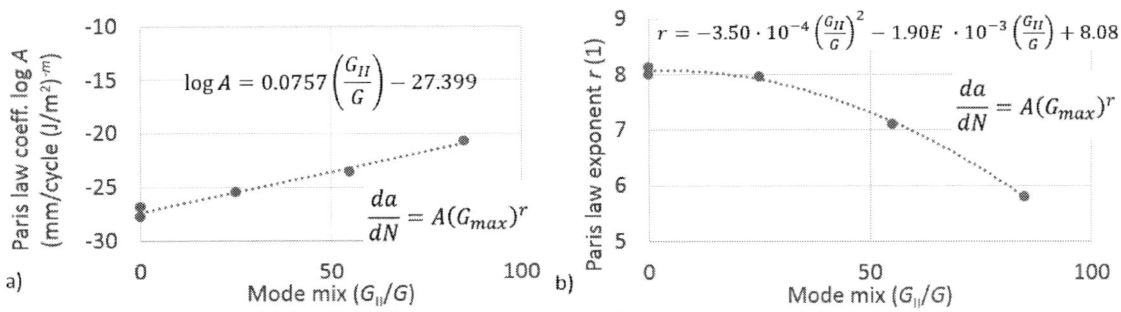

Fig.5 Paris law exponent relation with mode mix ratio

Tab. 1 Coefficients for Paris law at different mode-mixtures

Sample ID	Actual G_{II}/G	Paris law log A	Paris law r
009	0 %	-26.8677	8.002
006	0 %	-27.8142	8.134
014	24 - 28 %	-25.4931	7.957
016	53 - 55 %	-23.5969	7.112
010	83 - 85 %	-20.7375	5.796

Finally, all data were filtered and fitted by the Russell and Street analytical model (Fig. 4) obtained from the article [1], nevertheless other more recent models can be found [2, 3]. The model with 4 constants (A_I, A_{II}, r_I, r_{II}) is presented by Eq. (13). G_i denotes to maximum strain energy release rate in the cycle and ΔG_i to the range of G in the cycle. G_{ic} is fracture toughness.

$$\frac{da}{dN} = \left(\frac{G_I}{G_I + G_{II}} A_I + \frac{G_{II}}{G_I + G_{II}} A_{II}\right) \left(\frac{\Delta G_I}{G_{Ic}} + \frac{\Delta G_{II}}{G_{IIc}}\right)^{\left(\frac{G_I}{G_I+G_{II}} r_I + \frac{G_{II}}{G_I+G_{II}} r_{II}\right)} \tag{13}$$

The mode mixture is slightly changing during the crack growth as the fixture setup is fixed. Actual mode mixture range where the crack growth rate and the coefficients were evaluated is stated in Tab. 2.

Tab. 2 Coefficients for fatigue crack growth at different mode-mixtures determined by Eq. 12 fitting based on total energy release rates components of ΔG.

	A_I	A_{II}	r_I	r_{II}
Russell and Street model [1]	2.813E-05	4.976E-03	7.209	4.839

A finite element numerical simulation of quasi-static tests utilizing cohesive zone modelling was also evaluated. The analysis was performed for three values of G_{II}/G: 26, 54, and 80%. The results are presented in Fig. 6 together with only one representative experimental result for each G_{II}/G ratio (dashed lines).

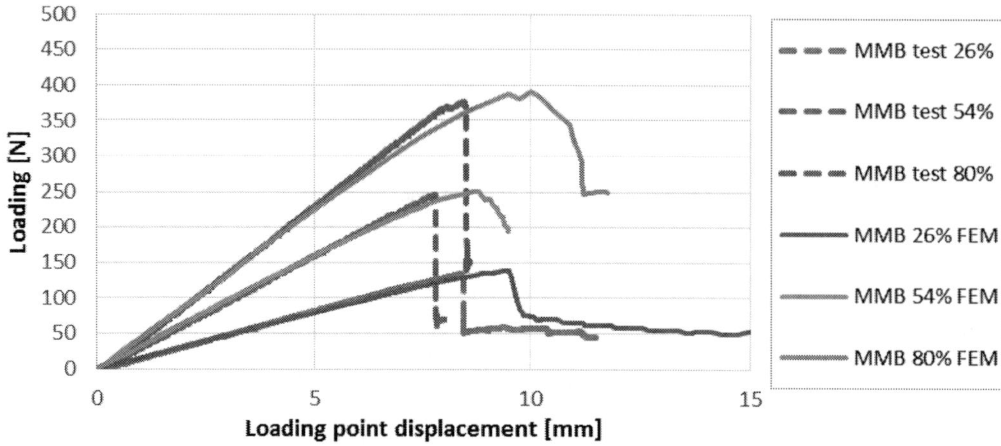

Fig.6 Load-displacement curves; FE analysis and representative experimental data

The initial stiffness is in good correlation with experimental data. Before the maximum load-carrying capacity the softening is present. The effect is more severe with G_{II}/G ratio increase. The maximum level of loading corresponds with the maximum test load, but displacement is greater in the FE analysis. The loss of load-carrying capacity is sharp in the case of experiment, but FE analysis shows gradual load decrease. It is also more severe with increase of G_{II}/G ratio. This response can be caused by gradual development of cohesive zone before crack advance; especially in mode II.

Conclusions

A method for fatigue crack growth evaluation under mixed mode I/II was established. The curves of crack growth rate vs. strain energy release rate were obtained and fitted by the analytical regression model considering the mode mixtures. In the FE-analysis, the cohesive zone modelling technique was used to represent the cohesive behaviour of interlaminar interface. The comparison of experimental data and computed load-displacement curves shows initial corresponding development, but slow crack advance in comparison with the experimental rapid crack growth. The experiment provides unique composite material data that can be sparsely found in the literature.

Acknowledgment

The authors would like to thank the Technology Agency of the Czech Republic for supporting this research with project no. TE02000032.

References

[1] N. Blanco, E.K. Gamstedt, L.E. Asp, J. Costa, Mixed-mode delamination growth in carbon–fibre composite laminates under cyclic loading, Int. J. Solids Struct. 41 (2004) 4219–4235.

[2] L. Zhao, et al., A novel interpretation of fatigue delamination growth behavior in CFRP multidirectional laminates, Compos. Sci. Tech. 133 (2016) 79-88.

[3] Y. Liu, C. Zhang, A critical plane-based model for mixed-mode delamination growth rate prediction under fatigue cyclic loadings, Compos. Part B. 139 (2018), 185-194.

[4] P. Naghipour, M. Bartsch, H. Voggenreiter, Simulation and experimental validation of mixed mode delamination in multidirectional CF/PEEK laminates under fatigue loading, Int. J. Solids Struct. 48 (2011), 1070-1081.

[5] ASTM Standard D6671M-06, Mixed Mode I-Mode II Interlaminar Fracture Toughness of Unidirectional Fiber Reinforced Polymer Matrix Composites, ASTM International, West Conshohocken, PA, 2006.

[6] M. Kadlec, L. Nováková, I. Mlch, L. Guadagno, Fatigue delamination of a carbon fabric/epoxy laminate with carbon nanotubes, Compos. Sci. Tech. 131 (2016) 32-39.

EAN 2018
56th conference on experimental stress analysis

Determining the Young's Modulus of a Specimen Using Experimental Modal Analysis

KAĽAVSKÝ Adam[1,a], HUŇADY Róbert[1,b] and PAVELKA Peter[1,c]

[1]Department of Applied Mechanics and Mechanical Engineering, Faculty of Mechanical Engineering, Letná 9, 042 00 Košice, Slovakia

[a]adam.kalavsky@tuke.sk, [b]robert.hunady@tuke.sk, [c]peter.pavelka@tuke.sk

Keywords: young modulus, modal analysis, CMIF

Abstract. The paper deals with determining the Young's modulus of a steel specimen that is calculated on the basis of natural frequencies determined by experimental modal analysis. The paper described two different approaches to do it. The first approach is based on analytical calculation of fundamental bending frequencies of a beam with a rectangular cross section. The paper provides results for two cases: free-free beam and fixed-free beam. The second approach is based on FEM analysis combined with parameter optimization method. This approach is applied on a free rectangular plate the analytical solution of which is more complicated.

Introduction

We know a lot of engineering problems where the exact knowledge of material parameters such as Young's modulus or Poisson ratio is essential for their solution or for an accuracy of numerical models. The material parameters are commonly determined by standard material tests. Alternatively, they can be determined by experimental modal analysis where the measured natural frequencies are subsequently used for an analytical solution. In the case of isotropic materials, it is preferable to perform an analysis of bending beam vibration because it is very easy to obtain accurate analytical solution for different boundary conditions [1, 2]. In general, it is recommended to analyze a free-free beam because the real stiffness of constraints is quite difficult to take into account in the calculation. Moreover, we can very well model a free suspension. In the case of orthotropic materials, the solution of plate vibration is essential [4]. Unfortunately, there exists no closed form solution for the case of a rectangular Kirchhoff plate with free boundaries, but several approximate methods have been proposed and applied. Warburton [4] used characteristic beam vibration functions in Rayleigh's method [5] to obtain a useful and simple approximate expression for the natural frequencies of vibration of thin isotropic plates. His work was extended by Hearmon [6] and applied to special orthotropic plates and again by Dickinson [7] to include the effect of uniform in-plane loads. Kim and Dickinson [8] provided an improved approximate expression, where they use Rayleigh's method in connection with the minimum potential energy theorem. Iguchi [9] gave solutions for an isotropic rectangular plate, however, he limits his determination to squared plates only. Leissa [10] presented comprehensive and accurate analytical results for the free vibration of rectangular plates. He applied the Ritz method [11] and compared the results with the method of Warburton [4]. However, his work is limited to isotropic plates. Leissa's work was extended by Deobald and Gibson [12] who applied the Rayleigh Ritz method to orthotropic plates as well. Gorman [13, 14] solved the differential equation for isotropic as well as orthotropic plates using a superposition method, which allows one to full fill the boundary conditions with desired accuracy. Wang and Lin [15] presented a systematic analysis for solving boundary value problems in structural mechanics, where a weighted residual form of the differential equations is used with sinusoidal weighting functions. Recently, this approach has been extended for calculating the eigenfrequencies and eigenmodes of an orthotropic plate with completely free boundaries using an exact series solution by Hurlebaus et al. [16]. As the quasi-analytical methods can be used the concept of sinusoidal equivalent length, the one term Rayleigh method, the three term Rayleigh method, the Rayleigh-Ritz method, the superposition method, and the exact series solution [17].

Procedures for obtaining Young's modulus

A. Beam vibration

The equation of motion for the bending vibration of a beam with constant cross-section and material properties is given by [16]

$$c^2 \frac{\partial^4 w(x,t)}{\partial x^4} + \frac{\partial^2 w(x,t)}{\partial t^2} = 0 \quad \rightarrow \quad c^2 \frac{\partial^4 w(x,t)}{\partial x^4} - \omega^2 w(x,t) = 0\,, \tag{1}$$

where $w(x,t)$ is the transverse vibration of the beam at position x at time t, ω is the natural frequency of the beam, and c is given by

$$c = \sqrt{\frac{EI}{\rho A}}\,, \tag{2}$$

in which I is the cross-sectional area moment of inertia, ρ is the mass per unit volume, and A is the cross-sectional area. The natural frequencies of the beam are from

$$\omega_n = (\beta_n \ell)^2 \sqrt{\frac{EI}{\rho A \ell^4}}\,, \tag{3}$$

where the subscript n is the bending mode number and β_n is the constant depending on boundary conditions (see Tab. 1). Young's modulus E of the beam material can be calculated from Eq. (3) as follows

$$E = \frac{\omega_n^{\,2} \rho A \ell^4}{(\beta_n \ell)^4 I}\,. \tag{4}$$

Tab. 1 Boundary condition values for the first four bending modes of the beam

Support condition	$(\beta_1 \ell)$	$(\beta_2 \ell)$	$(\beta_3 \ell)$	$(\beta_4 \ell)$
Free - Free	4.7300	7.8532	10.9956	14.1371
Clamped - Free	1.8751	4.6941	7.8539	10.9956

The natural frequencies of the beam were determined by experimental modal analysis. The object of measurement was the beam with rectangular cross section made of steel. The dimensions of the beam and calculated geometric characteristics and material properties are listed in Tab. 2. The mass of the beam was determined by weighing.

Tab. 2 Dimensions and calculated geometric characteristics and material properties of the beam

Dimensions		Geometric characteristics
$\ell = 0.401$ m $b = 0.040$ m $h = 0.008$ m	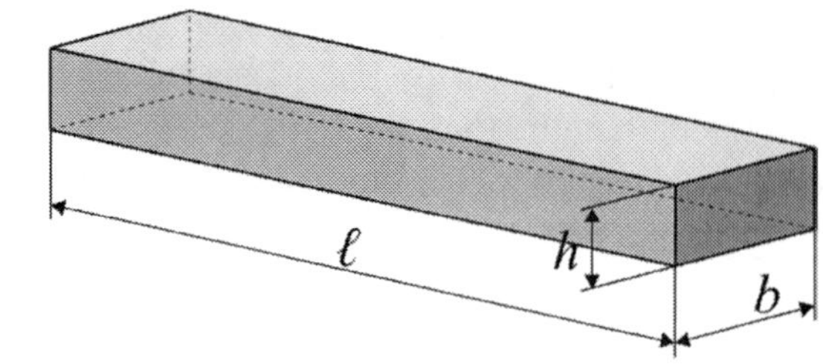	$A = bh = 3.2 \cdot 10^{-4}\ \mathrm{m}^2$ $I = \dfrac{bh^3}{12} = 1.7067 \cdot 10^{-9}\ \mathrm{m}^4$
		Material properties
		$m = 1.004\,\mathrm{kg}$ $\rho = \dfrac{m}{A\ell} = 7832\,\mathrm{kg \cdot m^{-3}}$

Two types of support conditions of the beam were investigated: free-free and clamped-free (Fig. 1). The free-free supporting was realized by two elastic ropes. The experimental modal analysis was performed using the measurement system Pulse LAN XI. The responses were measured at one point by laser Doppler vibrometer Polytec PDV-100. The beam was excited at 18 points by impact hammer Bruel&Kjaer 8206 with aluminum tip.

Fig. 1 Measurement setup for free-free and clamped-free supporting of the beam

Rational Fraction Polynomial-Z method was used to estimate modal parameters. In both cases, the first 5 modes were found, including 4 bending modes and 1 torsion mode. Their natural frequencies and mode shapes are shown in Tab. 3.

Tab. 3 Natural frequencies and mode shapes of the beam

Free - free beam				
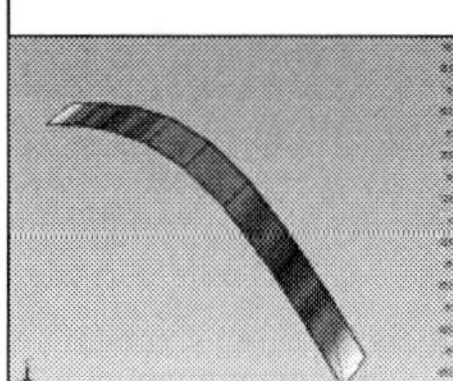	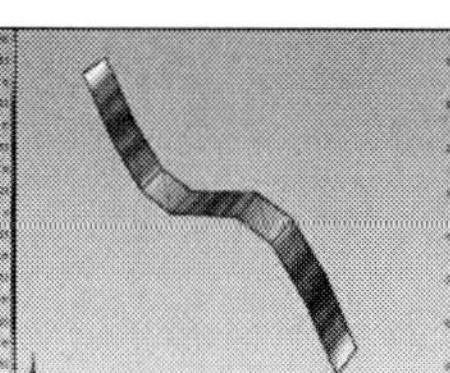	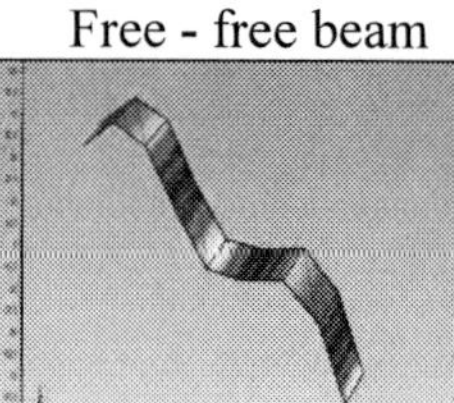	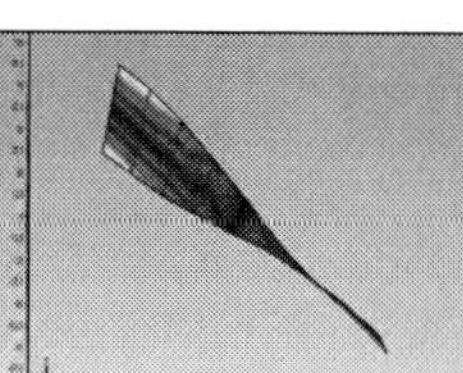	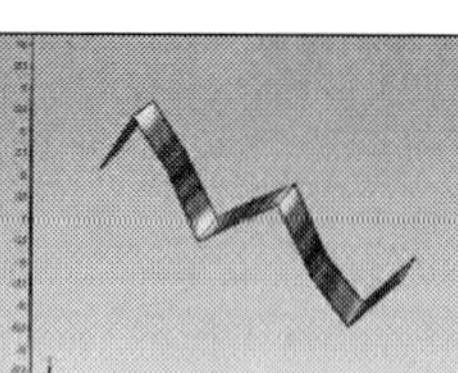
1: 265,44 Hz	2: 731,27 Hz	3: 1429,97 Hz	4: 1496,95 Hz	5: 2355,81 Hz
Clamped - free beam				
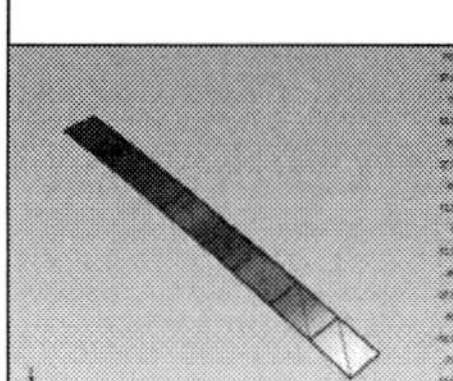	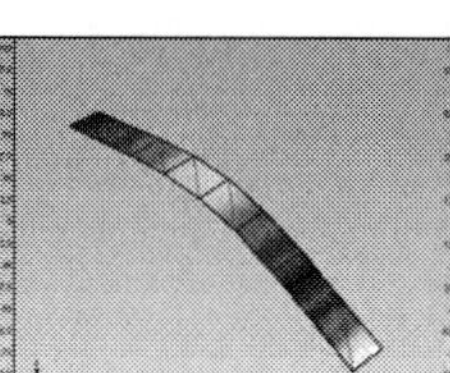		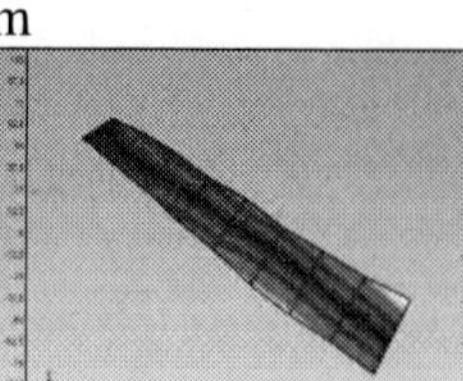	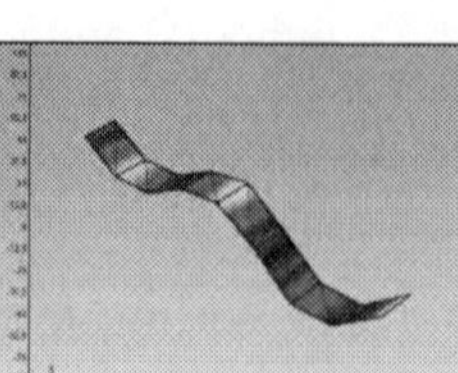
1: 40,528 Hz	2: 252,50 Hz	3: 699,08 Hz	4: 716,36 Hz	5: 1333,02 Hz

After substituting the values from Tab. 2 and the natural frequencies of bending modes into Eq. 4, Young's modulus of the beam material was calculated. Its values are listed in Tab. 4.

Tab. 4 Young's modulus of the beam

Support condition	1. mode	2. mode	3. mode	4. mode	Mean value
Free - Free	211.81 GPa	210.76 GPa	209.69 GPa	208.28 GPa	210.14 GPa
Clamped - Free	199.19 GPa	196.84 GPa	192.54 GPa	182.21 GPa	192.70 GPa

Since the analytical solution is expressed for the ideal boundary conditions and does not take into account the real stiffness of the support, Young's modulus values determined for the free supported beam are more in line with the expected value than the values determined for the clamped beam. From this point of view, it is more preferable to determine material constants on the basis of natural frequencies of a free supported specimen. The conditions of this type of support can be achieved relatively easily by hanging the specimen with elastic ropes or by laying it on a foam pad. A necessary condition is that the frequency of the last rigid body mode must be at least ten times less than the frequency of the first flexible mode [18].

B. Isotropic plate vibration

The second procedure is based on finite element analysis that is combined with parametric optimization. The object of interest was a rectangular plate of constant thickness made of steel. The plate was considered as free on all edges. The free support was realized by putting the plate horizontally on three stretched elastic ropes. The responses were measured at one point by laser Doppler vibrometer Polytec PDV-100. The plate was excited at 54 points by impact hammer Bruel&Kjaer 8206 with aluminum tip. The measurement setup and the basic dimensions of the plate can be seen in Fig. 2.

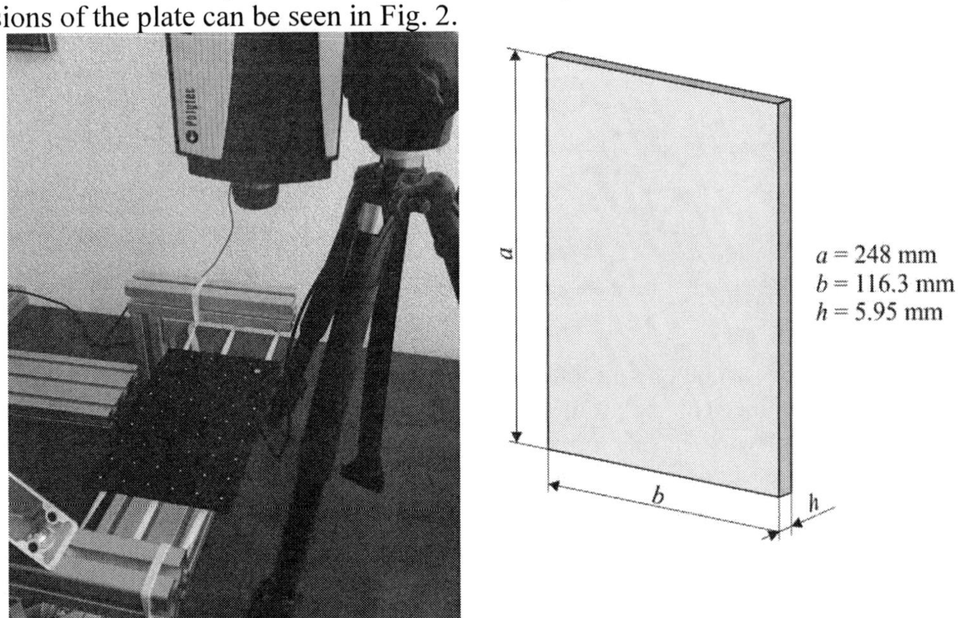

Fig. 2 Measurement setup and the basic dimensions of the plate

Modal parameters of the plate were determined by Rational Fraction Polynomial-Z method. Fig. 3 shows the plot of Complex Mode Indicator Function peaks of which correspond to the natural frequencies. The first peak represents rigid body mode. The natural frequencies and mode shapes of flexible modes are shown in Tab. 5.

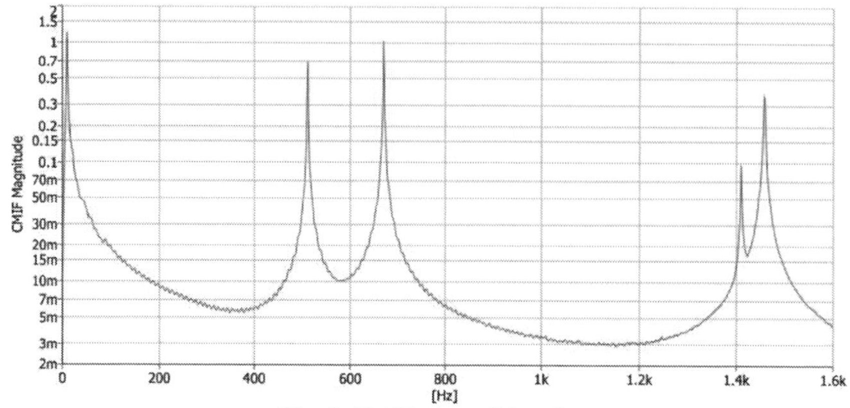

Fig. 3 CMIF plot of the plate

Tab. 5 Natural frequencies and mode shapes of the plate

	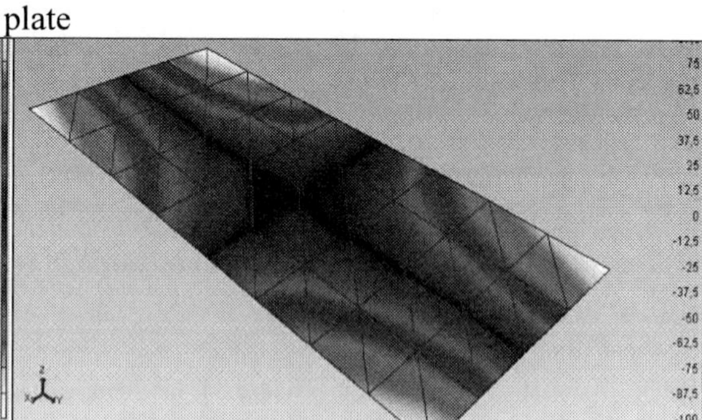
1: 511,58 Hz	2: 671,15 Hz
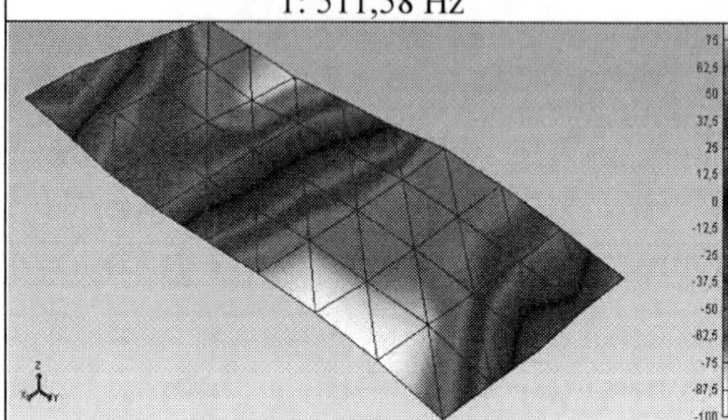	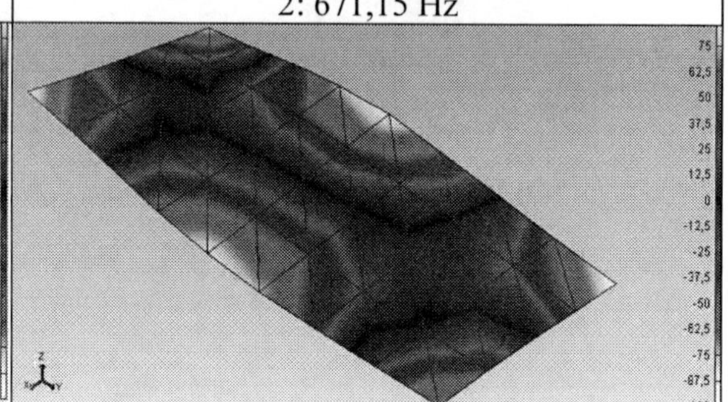
3: 1409,09 Hz	4: 1457,48 Hz

As mentioned above, Young's modulus value was estimated numerically by parametric optimization. The numerical solution was carried out in NX Nastran software. The optimization process was based on eigenvalue solution provided by SOL103 Real Eigenvalues.

As the first, 2D CAD model of the plate was created. Shell elements CQUAD8 of approximated width 4 mm were used to mesh the model. Subsequently, the material properties were associated to FE model. The material density of the plate was determined by calculation as follows

$$\rho = \frac{m}{abh} = \frac{1.324}{1.7161 \cdot 10^{-4}} = 7715 \text{ kg} \cdot \text{m}^{-3},$$

where m is the mass determined by weighing. The initial Young's modulus value was set to 150 GPa. Poisson's ratio was 0.29. There were defined no boundary conditions.

Geometry optimization module was used in the optimization process. Young's modulus was defined as a design variable that can acquire values from 100 GPa to 300 GPa. The objective function was the natural frequency of chosen mode, so the optimization process was carried out 4 times. Fig. 4 shows the course of the optimization where target function was the frequency of the first mode 511.58 Hz.

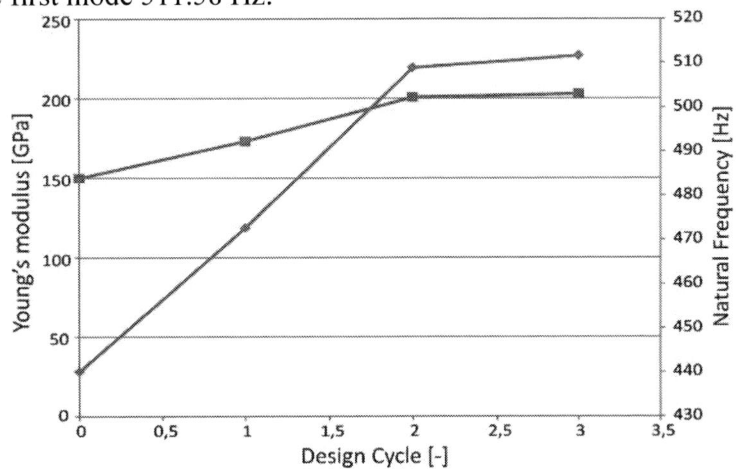

Fig. 4 Target and design variable depending on design cycle

The results of the optimization process for each target function are listed in Tab. 6. The values of Young's modulus are in the range from 200.81 to 205.63 GPa. Its mean value is 203.45 GPa. Tab. 7 shows the absolute percentage error of the natural frequency values. The highest mean absolute percentage deviation (MAPD) 0.65% was reached in the third optimization process. However, such deviations are entirely acceptable. The first four mode shapes obtained by FEM analysis are shown in Fig. 5.

Tab. 6 Results of the optimization process

Target value of objective function	1. mode	2. mode	3. mode	4. mode	E [GPa]
1. mode: 511,58 Hz	**511.59 Hz**	666.37 Hz	1415.73 Hz	1450.53 Hz	202.70
2. mode: 671,15 Hz	515.28 Hz	**671.16 Hz**	1425.90 Hz	1460.96 Hz	205.63
3. mode: 1409,09 Hz	509.20 Hz	663.25 Hz	**1409.11 Hz**	1443.75 Hz	200.81
4. mode: 1457,48 Hz	514.04 Hz	669.55 Hz	1422.49 Hz	**1457.45 Hz**	204.64
				Mean value:	203.45

Tab. 7 Percentage error and mean absolute percentage deviation of the natural frequency values

Target value of objective function	1. mode	2. mode	3. mode	4. mode	MAPD
1. mode: 511,58 Hz	0.00 %	0.71 %	0.47 %	0.48 %	0.41 %
2. mode: 671,15 Hz	0.72 %	0.00 %	1.19 %	0.24 %	0.54 %
3. mode: 1409,09 Hz	0.46 %	1.19 %	0.00 %	0.95 %	0.65 %
4. mode: 1457,48 Hz	0.48 %	0.24 %	0.95 %	0.00 %	0.41 %

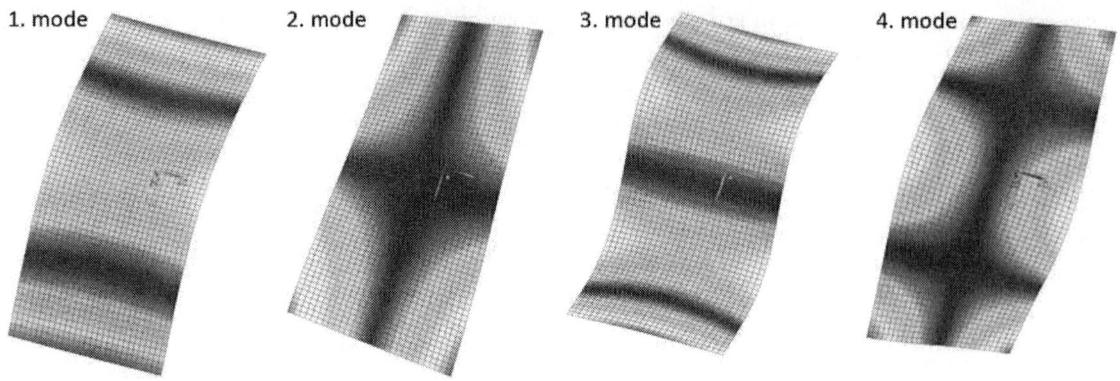

Fig. 5 The first four mode shapes of the plate obtained by FEM analysis

Conclusions

The paper presented two methods to determine Young's modulus of an isotropic material. The first of them is a combination of experimental modal analysis and analytical solution, in which the known values of natural frequencies are substituted into the well-known equation for calculation of bending mode frequencies of a beam. It has been shown that the free support of the beam is more advantageous for determining Young's modulus value. The second method uses the parametric optimization that is applied on FE model of a plate and the eigenfrequency of a given mode is used as an objective function to calculate Young's modulus. The obtained results confirm that this approach is sufficiently precise and effective.

Acknowledgement

This research was supported by project VEGA 1/0355/18.

References

[1] D. J. Inman, Engineering Vibration, Second edition, 2001, (Prentice Hall, Inc. Upper Saddle River, NJ), 446-470 (2001)

[2] J. Bocko, Š. Segľa, R. Huňady, Vibration of mechanical systems (in Slovak), TUKE, Typopress, Košice, (2016).

[3] S. Chakraverty, Vibration of Plates, CRC Press, (2008).

[4] G.B. Warburton, The vibration of rectangular plates. In: Proceedings of the Institution of Mechanical Engineers 168, 371–384 (1954).

[5] J.W.S. Rayleigh, The Theory of Sound (Dover, New York, 1945).

[6] R.F.S. Hearmon, Journal of Applied Mechanics 26, 537–540 (1959).

[7] S.M. Dickinson, Journal of Sound and Vibration 61(1), 1–8 (1978).

[8] C.S. Kim and S.M. Dickinson, Journal of Sound and Vibration 103(1), 142–149 (1985).

[9] S. Iguchi, Ingenieur-Archiv 21(5/6), 303–322 (1953).

[10] A.W. Leissa, Journal of Sound and Vibration 31(3), 257–293 (1973).

[11] W. Ritz, Journal für die reine und angewandte Mathematik 135, 1–61 (1909).

[12] L.R. Deobald and R.F. Gibson, Journal of Sound and Vibration 124(2), 269–283 (1988).

[13] D. J. Gorman, Journal of Sound and Vibration 165(3), 409–420 (1993).

[14] D.J. Gorman, Vibration Analysis of Plates by the Superposition Method (World Scientific Publishing, Singapore, 1999).

[15] J.T.S. Wang and C.C. Lin, Journal of Applied Mechanics 66, 380–387 (1999).

[16] S. Hurlebaus, L. Gaul, and J.T.S. Wang, Journal of Sound and Vibration 244(5), 747–759 (2001).

[17] S. Hurlebaus, Eigenfrequencies of orthotropic plates, ZAMM · Z. Angew. Math. Mech. 87, No. 10, 762 – 772 (2007).

[18] D.J. Ewins, Modal Testing: Theory, Practice and Application, 2nd edition, Research Studies Press, (2000).

EAN 2018
56th conference on experimental stress analysis

Experimental Analysis of Modern Screw Fasteners

KANAVAL Jan[1,a], CÉZOVÁ Eliška[1,b], STARÝ Frantšek[1,c], DYNYBYL Vojtěch[1,d] and KLÍMA Vít [1,e]

[1]Czech Technical University in Prague, Faculty of Mechanical Engineering, Technická 4, 166 07 Prague 6, Czech Republic

[a]jan.kanaval@fs.cvut.cz, [b]eliska.cezova@fs.cvut.cz, [c]frantisek.stary@fs.cvut.cz, [d]vojtech.dynybyl@fs.cvut.cz, [e]vit.klima@fs.cvut.cz

Keywords: prestressed bolted connection; advanced coatings and lubricants; friction and force conditions; modular designed testing stand, Standard VDI 2230, KissSoft program.

Abstract. This contribution describes analytical evaluation and experimental verification of the friction and force conditions in the prestressed bolted connections. In the case of using standard fasteners (bolts and nuts) there were evaluated mainly influence of advanced coatings using and lubricants into the threaded surfaces of bolted joints. To achieve the set goals modular design testing stand was designed. Achieved analytical findings and experimental verification will allow to refine the technological processes of assembly of bolted joints, which will have a major impact on the strength, durability and reliability of the screw connections (fasteners). Discovered knowledge is not currently available for design engineers during the designing of bolted joints. Design engineers typically use only recommendation of retailers and manufacturers when designing of bolted joints. In industrial practice, it is now very often required to calculate the screws according to the standard VDI 2230. [2] Standard VDI 2230 specifies a procedure for the strength calculation of prestressed screw connections. The calculation is usually done on a personal computer using the KissSoft program.

Introduction

The process of tightening threaded fastener assemblies, especially for critical bolted joints, involves controlling both input torque and angle of turn to achieve the desired result of proper preload of the bolted assembly. Understanding the role of friction in both the underhead and threaded contact zones is the key to defining the relationship between torque, angle, and tension. There can be as many as 200 or more factors that affect the tension created in a bolt when tightening torque is applied. Fortunately, torque-angle signature curves can be obtained for most bolted joints. By combining the torque-angle curves with a few simple calculations and a basic understanding of the engineering mechanics of threaded fasteners, you can obtain the practical information needed to evaluate the characteristics of individual fastener tightening processes. The torque-angle curves can also provide the necessary information to properly qualify the capability of tightening tools to properly tighten a given fastener. These findings were described by Shoberg. [1]

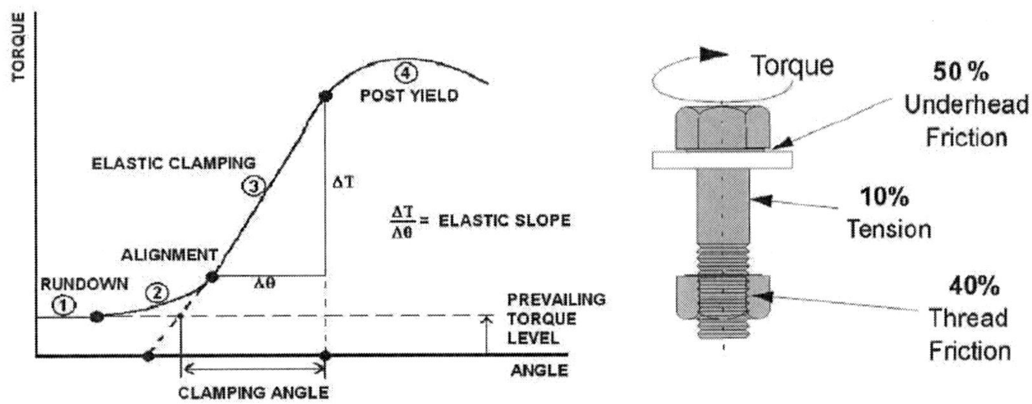

Fig. 1 Measured Torque-angle Curve. Four Zones of the Tightening Process. [1]

Prestressed Bolted Joints Theory

There are shown a prestressed bolted connection loaded with a centric axial force and diagram of prestressed bolted connection force - deformation in Fig. 2.

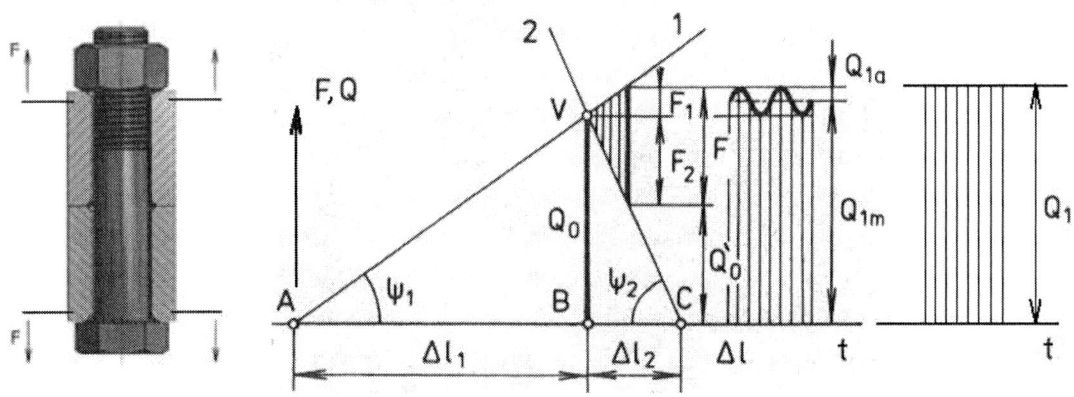

Fig. 2 Prestressed Bolted Connection Loaded with a Centric Axial Force (on the left side); Diagram of Prestressed Bolted Connection Force - Deformation (on the right side). [5]

The static strength calculation of the prestressed screw connection is performed according to the following equations (Eq. 1 through Eq. 17):

$$\gamma = arctg \frac{P}{\pi \cdot d_2} \tag{1}$$

$$f' \approx \frac{f}{\cos \beta} \tag{2}$$

$$M_{KK} = M_K + M_{TM} = Q_o \cdot \frac{d_2}{2} \cdot tg(\gamma + \varphi') + Q_o \cdot \rho_{TM} \cdot f_{TM} \tag{3}$$

$$Q_o = \frac{M_{KK}}{\frac{d_2}{2} \cdot tg\left(\gamma + \varphi'\right) + \rho_{TM} \cdot f_{TM}} \tag{4}$$

$$c = \frac{E \cdot S}{l} \tag{5}$$

$$\frac{1}{c_s} = \frac{1}{c_{sd}} + \frac{1}{c_{sz}} \tag{6}$$

$$\frac{1}{c_1} = \frac{1}{c_s} + \frac{2}{c_{lDa}} + \frac{1}{c_M} \tag{7}$$

$$c_2 = c_{lDb} \tag{8}$$

$$F_1 = \frac{c_1}{c_1 + c_2} \cdot F \tag{9}$$

$$F_2 = \frac{c_2}{c_1 + c_2} \cdot F \tag{10}$$

$$Q_1 = Q_o + F_1 \tag{11}$$

$$\sigma_{Q1} = \frac{Q_1}{S_3} = \frac{4 \cdot Q_1}{\pi \cdot d_3^2} \tag{12}$$

$$\tau_K = \frac{M_K}{W_K} = \frac{16 \cdot M_K}{\pi \cdot d_3^3} = \frac{16 \cdot Q_o \cdot \frac{d_2}{2} \cdot tg(\gamma + \varphi')}{\pi \cdot d_3^3} \tag{13}$$

$$\sigma_{red} = \sqrt{\sigma_{Q1}^2 + 3 \cdot \tau_K^2} \tag{14}$$

$$k_s = \frac{R_{p0,2}}{\sigma_{red}} \geq k_{s\min} = 1,2 \tag{15}$$

$$p_Z = \frac{Q_1}{\pi \cdot d_2 \cdot H_1 \cdot z} \geq p_{Zdov} \tag{16}$$

$$p_M = \frac{4 \cdot Q_1}{\pi \cdot (D^2 - d^2)} \geq p_{Mdov} \tag{17}$$

γ - thread pitch angle [°]; β - half angle of the thread profile [°]; φ' - friction angle in thread [°]; P - pitch [mm]; d_2 - flank diameter of the bolt thread [mm]; d_3 - core diameter of the bolt thread [mm]; ρ_{TM} - friction radius under the nut [mm]; f - coefficient of friction [1]; f' - coefficient of friction in thread [1]; f_{TM} - coefficient of friction under the nut [1]; M_{KK} - tightening torque [Nmm]; M_K - torque [Nmm]; M_{TM} - friction torque under the nut [Nmm]; Q_o - mounting - pretension force [N]; c - constants of stiffness [Nmm^{-1}]; E - Young's modulus of individual parts [Nmm^{-2}]; S - cross section of individual parts [mm^2]; l - length of individual parts [mm]; F - external axial force [N]; F_1, F_2 - components of external axial force [N]; Q_1 - maximum internal force of the bolt [N]; σ_{Q1} - tensile stress [Nmm^{-2}]; τ_K - share stress [Nmm^{-2}]; σ_{red} - equivalent stress [Nmm^{-2}]; $R_{p0,2}$ - yield point [Nmm^{-2}]; k_S - safety against yield point [1]; p_Z - contact pressure in the thread [MPa], p_M - contact pressure under the nut [MPa], D, d - outer and inner diameters of the replacement tube [mm].

There is defined the concept of controlled tightening of the screw connection in the technical support of the company Bossard. [3] This method of tightening the screw connection respects the variable value of the friction coefficient in the thread and the reduction of prestress of the screw connection after assembly. Actual tensile bolt prestress after controlled tightening ensures correct functioning of screw joints during operation. The tensile prestress value of the bolt must be greater than the minimum value in terms of the correct operation of the screw connection and at the same time tensile prestress value of the bolt must by less than the maximum permissible value in relation to the achieve the yield strength of the bolt (see Fig. 3).

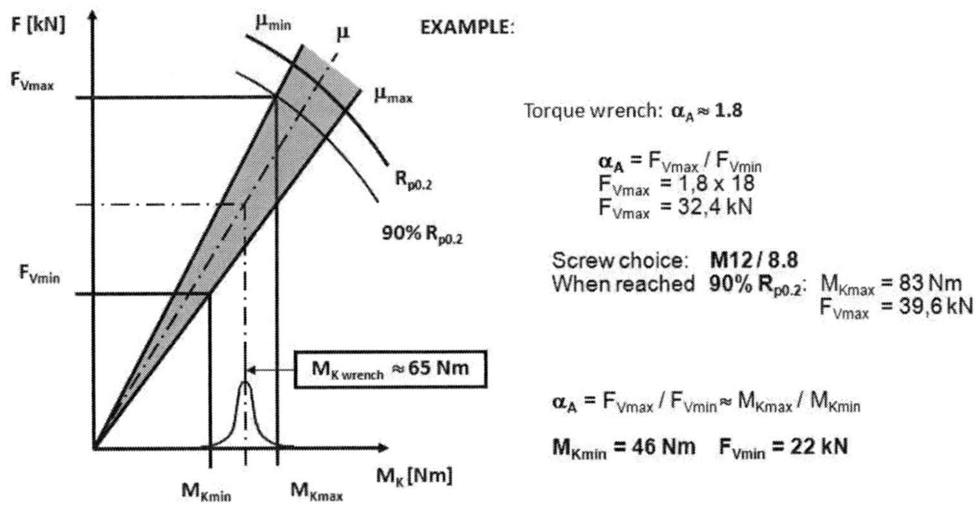

Fig. 3 Controlled Tightening Process of the Prestress Screw Joints. The Calculation Procedure of the Bolted Joints and these Diagrams are Presented in the [1,2,3]

Due to the contact of multiple surfaces with real surface roughness, the tension axial prestress is reduced after the screw connection is assembled. This phenomenon is caused by plastic deformation of individual contact surfaces. Reduction of the axial tensile stress is dependent on the number of contact surfaces of the screw joint and on their surface roughness (see Fig. 4).

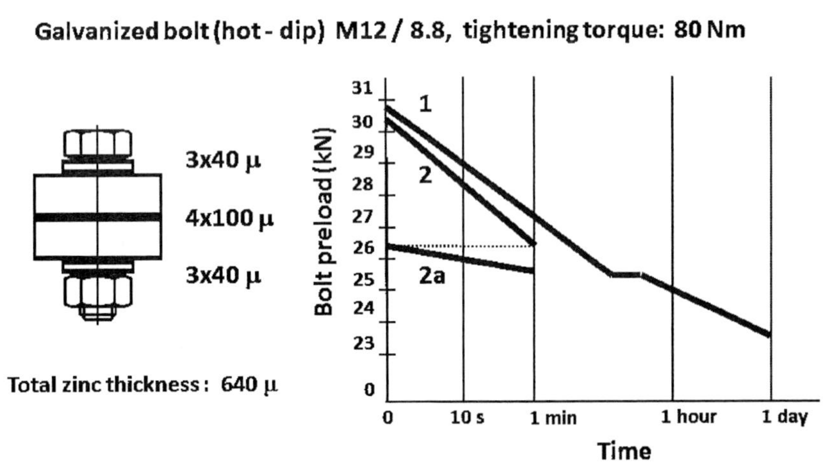

Fig. 4 Tension Axial Prestress Reduction of the Screw Joint after Assembly in Time. [2,3]

Experimental Analysis

The simple testing equipment for evaluation of threaded contact friction between the bolt and the nut was designed and manufactured at Department Designing and Machine Components of the Czech Technical University in Prague and simple experiments there were realized too. The results of these experiments are shown in Table 1.

Fig. 5 Simple Testing Equipment for Evaluation of Threaded Contact Friction, (on the left side), HBM Torque Sensor T20WN, 20 Nm, (on the right side). These photographs were taken by the corresponding author (2015, 2016)

These simple realized experiments verified the chosen measurement methodology. The aim of the experiments was to determine the magnitude of the friction coefficient in the thread depending on the surface (coating) of the screw and the nut. The experience gained in this way was used to design testing equipment for complex screw joint testing (see Fig. 6) The test equipment is currently being implemented by the prepared structural design.

Tab. 1 Table Presenting Measured and Calculated Values for Bolt Size M12. This table was prepared by the corresponding author (2016)

Table Presenting Measured and Calculated Values for Bolt Size M12
(Bolt: stainless steel A2-80, Nut: stainless steel A4-80, coating - Delta Seal Black)

Hinge mass:		2,86kg				
Nut friction moment M_p:		0Nmm				
Weight mass (including hinge) m [kg]	Tensile force Q_o [N]	Wrench torque moment M_{kk} [Nmm]	Wrench torque moment without nut friction moment M_k [Nmm]	Calculated thread friction angle φ' [°]	Calculated thread friction coefficient f'[1]	Calculated friction coefficient f [1]
0	0	0	0	0	0	0
161,51	1584,41	1000	1000	9,50769	0,16748	0,14504
220,46	2162,71	1750	1750	11,32404	0,20026	0,17343
240,14	2355,77	2500	2500	13,8561	0,24666	0,21362
245,51	2408,45	2500	2500	13,62506	0,24239	0,20991
	Statistic evaluation		Sample average μ:	12,08	0,214	0,186
			Sample standard deviation σ:	2,06	0,04	0,03

New Testing Equipment for Complex Analysis of Screw Fasteners

On this testing equipment can be detected:
1) The value of the friction coefficient in the thread.
2) The value of the friction coefficient under screw head (under nut).
3) The tension axial prestress reducing after the screw connection is assembled.
4) To evaluate the characteristics of fastener tightening processes (Torque-angle curves).
5) Verifying the strength of the screw connection.

Technical parameters of this testing equipment:

Maximum testing torque moment	185 Nm
Maximum testing tension axial force	100 kN
Maximum testing rotation speed	97 min^{-1}
Maximum testing stroke	300 mm

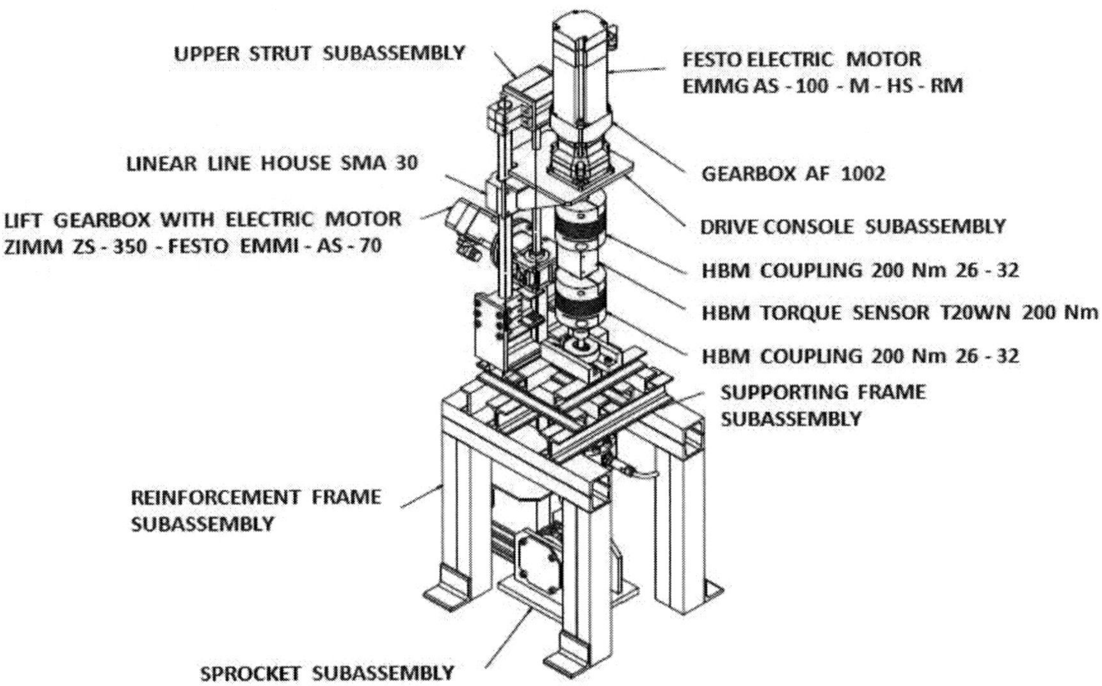

Fig. 6 Testing Equipment for Complex Analysis of Screw Fasteners. This image was processed by the corresponding author (2017)

Experimental Determination of the Friction Coefficient on the Tribometer

Tribo tests were performed to experimentally investigate the magnitude of the frictional coefficient of the selected material pair (see Table 2, Fig. 7 and Fig. 8).

Tab.2 Parameters of Tribometer Test (material pair - stainless steel A2-80/A4-80, with application of dry lubricant coating - "Gleitmo"). This table was prepared by the corresponding author (2016*)*

Tribo parameters

Tribometer module / Version 4.4.N	**Sample**	**Static partner**
Acquisition	- Coating : -	- Substrate : Steel
- Radius : 9,50 [mm]	- Substrate : -	- Cleaning : Acetone
- Lin. Speed : 6,03 [cm/s]	- Cleaning : Acetone	- Dimension : 6,00 [mm]
- Normal load : 10,00 [N]		- Geometry : Ball
- Stop condit. : 700,0 [lap]		
- Effective Stop : Laps		
- Acquisition rate : 10,0 [hz]		

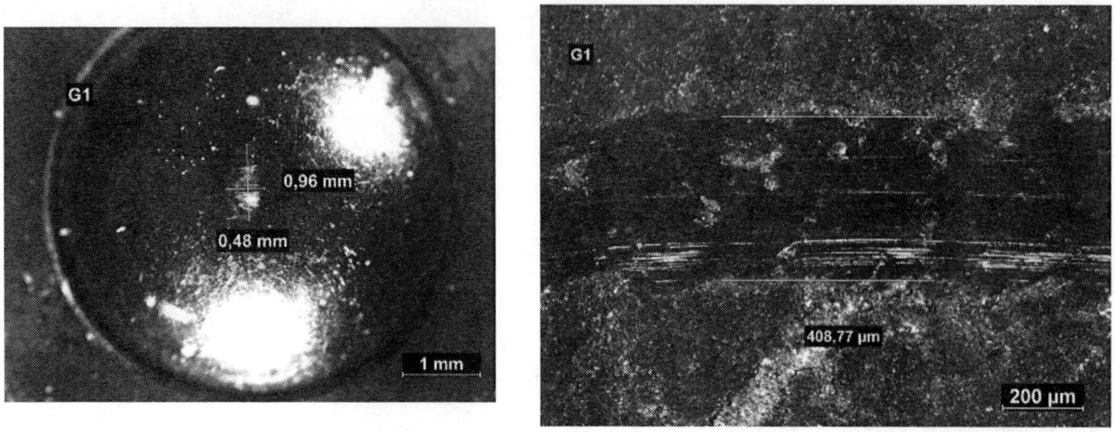

Fig. 7 Abrasive Wear of Selected Material Pair - Test Ball (A2-80) and Tested Surface (A4-80), with Application of dry lubricant coating - "Gleitmo". This image was processed by the corresponding author (2017)

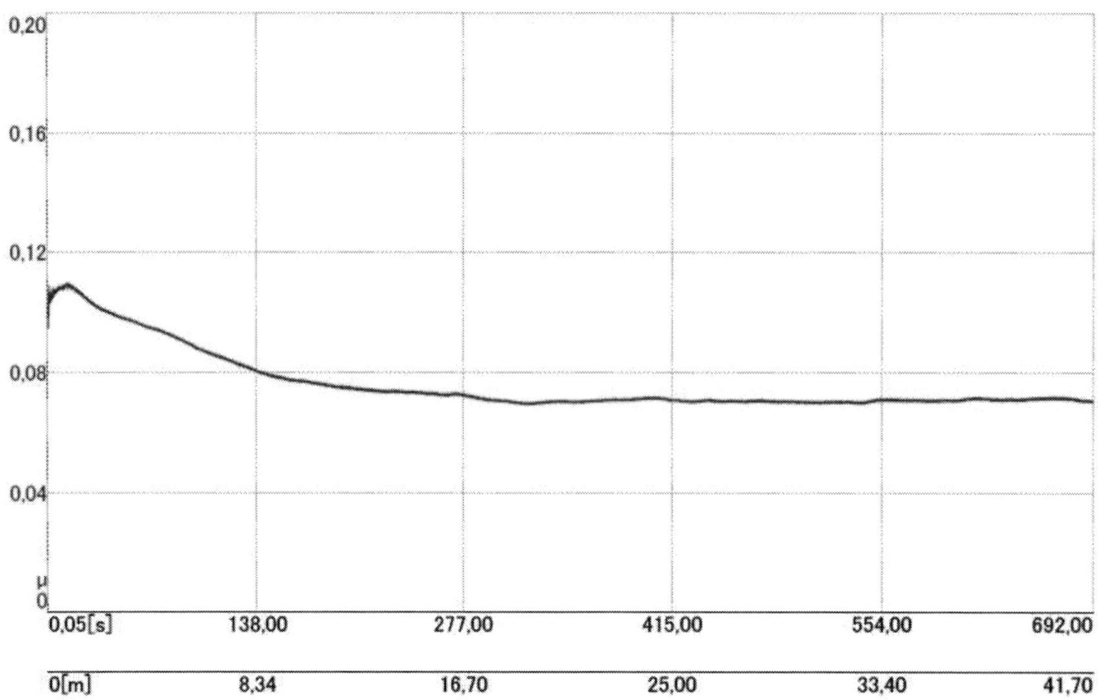

Fig. 8 Course of the Experimentally Determined Values of the Friction Coefficient of Selected Material Pair - test ball (A2-80) and tested surface (A4-80), with application of dry lubricant coating - "Gleitmo". This image was processed by the corresponding author (2017)

Evaluation of Plastic Deformations of the Threaded Surfaces

Plastic deformations of the threaded surfaces of bolt and nut after tightening and loading were evaluated by using metallographic cuts too (see Fig. 9).

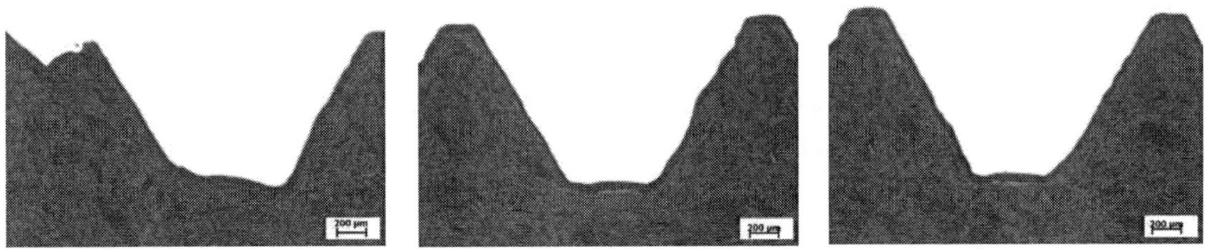

Fig. 9 Metallographic Cut of Threaded Surfaces of the M16 Nut (A4-80, with application of dry lubricant coating - "Gleitmo"). From the left side - 1. loaded thread, 2. loaded thread and 3. loaded thread. This photographs was processed by the corresponding author (2017)

Conclusions

Aim of the experiments was to determine the magnitude of the friction coefficient in the thread depending on the surface (coating) of the screw and the nut. Measured data are presented in the Table 1 (only one example). Experimentally determined value of the friction coefficient in the thread (0,214) corresponds to the value reported in the technical literature.

The simple realized experiments verified the chosen measurement methodology. The experience gained in this way was used to design testing equipment for complex screw joint testing. The test equipment is currently being implemented by the prepared structural design.

Tribo tests were performed to experimentally investigate the magnitude of the frictional coefficient of the selected material pairs and plastic deformations of the threaded surfaces of bolt and nut after tightening and loading were evaluated by using metallographic cuts too.

Acknowledgment

This study was supported by Student grants competition of the Czech Technical University in Prague, Faculty of Engineering, no. SGS16/145/OHK2/2T/12

References

[1] R. Shoberg, Engineering fundamentals of threaded fastener design and analysis, MI 48335, USA: RS Technologies, a Division of PCB Load & Torque, Inc., 2010.

[2] Guideline VDI 2230, Systematic calculation of high duty bolted joints, Part 1, Joints with one cylindrical bolt, 2003.

[3] Bossard CZ s.r.o., Technical support, Information on https://www.bossard.com/cs/aplikacni-inzenyrstvi/technické-zdroje.aspx, Brno, 2018-04-18.

[4] Fabory CZ Holding s.r.o., Technical support, Information on https://www.fabory.com/cs, 2018-04-18.

[5] Vl. Švec, Parts and mechanisms of machines, Joints and connecting parts, Prague: CTU, 2008.

EAN 2018
56th conference on experimental stress analysis

Possibility of Application of the Simulation Based Assessment Method in Modelling of Structures

KESL Petr[1,a] and PLÁNIČKA František[1,b]

[1]University of West Bohemia in Pilsen, Faculty of Applied Sciences, Department of Mechanics, Univerzitní 8, 301 00 Plzeň, Czech Republic

[a]pkesl@kme.zcu.cz, [b]planicka@kme.zcu.cz

Keywords. Probabilistic SBRA Method, Monte Carlo simulation, serviceability of structure, strain, stress, modeling, steel beam, experimental results, results obtained by SBRA Method.

Abstract. The article deals with possibility of application of the simulation based reliability assessment SBRA Method in modeling of structures in the field of judgement of their serviceability. A steel supported beam of rectangular cross sectional area was taken into account. Strains of the model were determined using electrical resistance straingange method were determined. Because of the same material of the model and real beam, strains and stresses are the same in both cases.
At the first step a model of the beam was created and necessary corresponding station modulus was determined for real beam using modeling rules.
It means that inverse procedure was used in that case. Stresses of the model and corresponding beam were determined using SBRA when fallowing random variable quantities as section modulus of the beams, Young´s modulus of elasticity E and evading force were taken into account.

Introduction

The article deals with an ability to assess the reliability of structural element on its model by means of the simulation based reliability assessment SBRA Method. The aim is to present a possibility of judgement of probability of failure of real component on its model. Steel beam (Fig. 2) was taken into account as a model of real steel beam. Stresses of steel beam and its model were determined using SBRA Method too. Obtained results were compared.

The simulation Based Reliability Assessment Method is a probabilistic method using the Monte Carlo simulation [1, 3]. Probability of failures of model and real beam were determined using Anthill software [3].

Modelling of engineering problems can be a way to solve them. This method is generally based on the conditions

$$(\pi_i)_S = (\pi_i)_M, \quad i = 1, 2, ..., m \tag{1}$$

where π_i are so called dimensionless parameters for structure (subscript S) and model (subscript *M*). Probability of failure is guided by so called safety function.

$$P_{f(i)} = R_{(i)} - S_{(i)} \quad (2)$$

Where **S** is the load effect And **R** is the structural resistance. The probability of failure P_f of a body can be then expressed as a ratio between the number N_f of results that do not fulfil the defined before by safety function Eq.3 and the total number of results of safety function N_t [1], see Fig. 1. Then

$$P_f = N_f / N_t \quad (3)$$

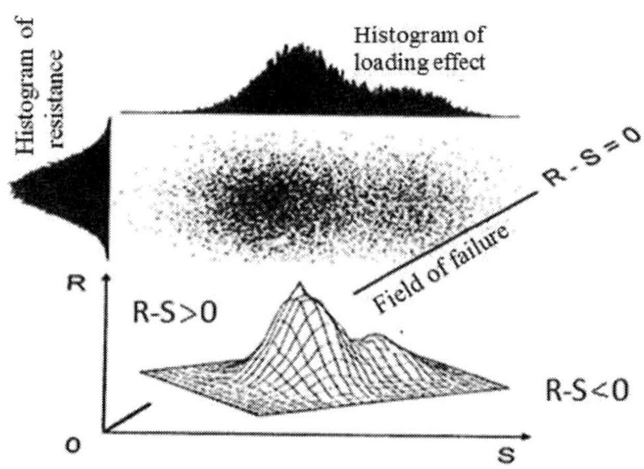

Fig. 1 Probability of failure

Experimental results

The used steel model in mm is stated in Fig.2. Loading forces were F=100, 200, 300, 400, 450 N. Strains were measured using electrical-resistance gages. Fig. 1. Recorded values of strains and corresponding stresses are in the Tab. 1. To determine corresponding state in real steel beam length of l=2500 mm rules of modelling were used and loading forces F_B and section

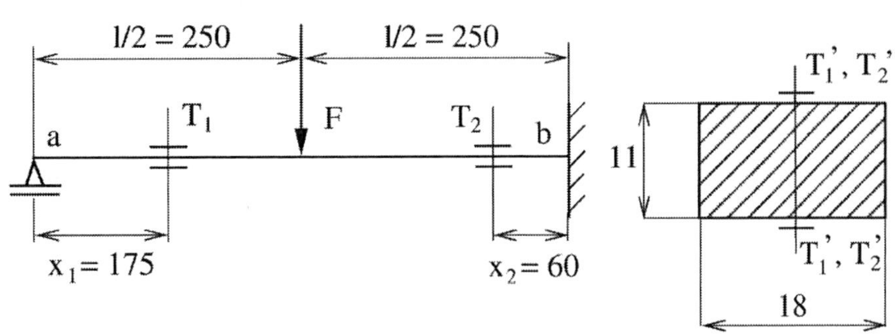

Fig. 2 Model of the beam

modulus W_B were determined. Material of model and of real beam was the same therefore stresses in the model and in the real beam are the same, steel S235 with Young´s modulus of elasticity E =2.1x10^5 MPa was used, see the Tab. 1.

Tab.1 Results of the model and of the beam

	F_m[N]	300	450
Model experiment	$2\ \varepsilon_{1m}$ x10^{-3}	0.225	0.33
	σ_{1m} [Mpa]	47.25	69.30
	$2\ \varepsilon_{2m}$x10^{-3}	-0.165	-0.26
	σ_{2m} [Mpa]	-34.65	-54.6
Beam	F_B[kN]	7.5	11.25

Results of model and SBRA simulation is obvious in Tab. 2.

Tab.2 Results of model and of SBRA simulation

F_m[N]	300			450		
F_B[kN]	7.5			11.25		
	MODEL exp.	SBRA-m	SBRA-B	MODEL exp.	SBRA-m	SBRA-B
$2\varepsilon_{1m}$ x10^{-3}	0.225	-	-	0.33	-	-
$2\varepsilon_{2m}$ x10^{-3}	- 0.165	-	-	-0.26.	-	-
σ_1 [MPa]	47.25	45.190	45.202	69.30	67.79	67.80
σ_2 [MPa]	-34.7	-34.29	-34.19	-54.60	-54.48	-54.32
$\sigma_{1Det.}$[MPa]	45.2			67.79		

m - **Model,** B – **Beam,** Det. – **beam determined**

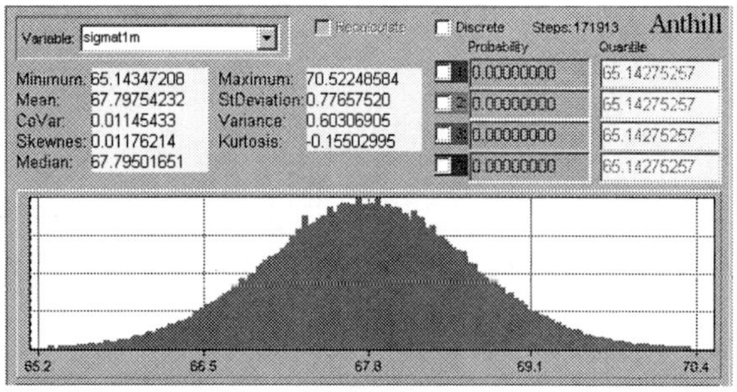

Fig.3 The resulting stress σ_{1m} [MPa]

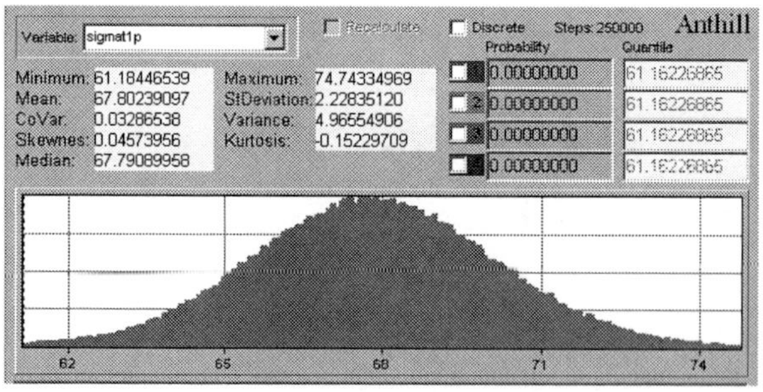

Fig.4 The resulting stress σ_{1B} [MPa]

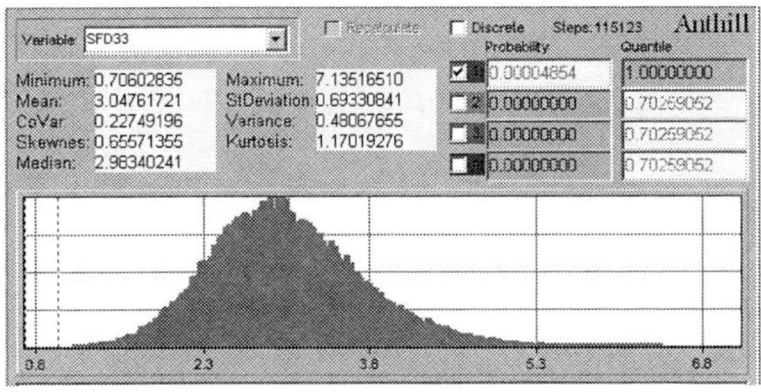

Fig.5 Probability of failure Pf(i), SF(i)

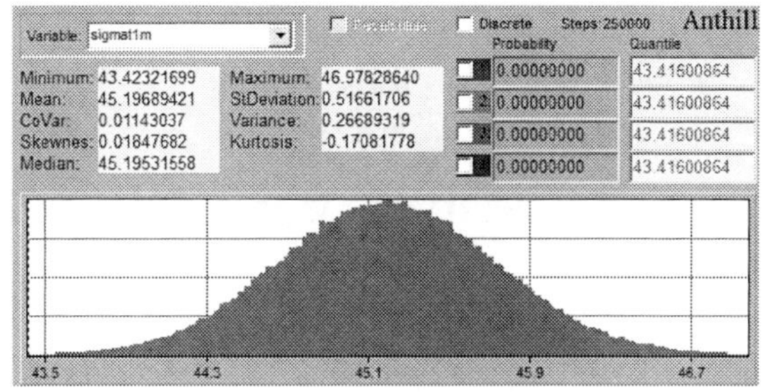

Fig.6 The resulting stress σ_{1m} [MPa]

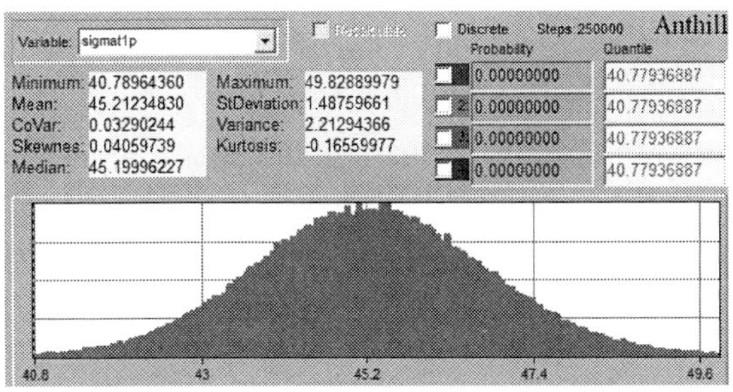

Fig.7 The resulting stress σ_{1B} [MPa]

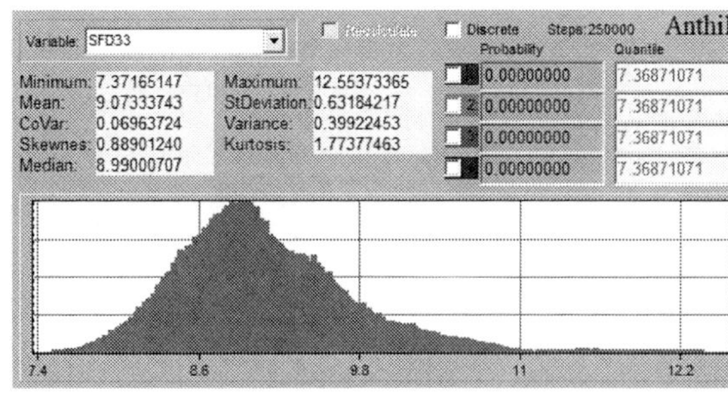

Fig.8 Probability of failure Pf(i), SF(i)

In Fig.3. and Fig.4. there are for illustration presented histograms of the resultant stresses and section modulus and in Fig. 5. is presented histogram of probability of failure all for the loading of F_m= 300 N and figures 6. ,7. and 8. for the load F_m= 450 N.

Probability of failures of the beam determined using SBRA method were $\mathbf{P}_{f(i)B} = 4{,}8 \cdot 10^{-5}$ and for the model $\mathbf{P}_{f(i)m} = 4{,}103 \cdot 10^{-5}$.

Obtained results show very good correspondence. It gives a sure possibility for determination of failure probability of structure to determine it using corresponding model.

Acknowledgements

The research work was supported by the Project SGS-2016-038.

References

[1] Plánička, F., Šolc, M., Marek, P.: From Partial Factors Design to the SBRA Probabilitistic Method, Acta Mechanica Slovaca Vol.15, No 2, 2011, ISSN 1335-2393, pp. 22-27.

[2] Plánička, F., Krystek, J.: Dimensional Analysis and Modelling in Mechanics of Materials, Proceedings of 28th Danubia-Adria Symposium on Advances in Experimental Mechanics 2011, Siófok, Hungary, ISBN 978-963-9058-32-3, pp. 229-230.

[3] www.sbra-anthill.com/lit/publications.

EAN 2018
56th conference on
experimental stress analysis

Mechanical Behaviour of Foam Used in Joint Protectors

KOTTNER Radek[1, a], KAŇÁKOVÁ Sandra[2,b] , ŠOLTÉS Lukáš[1,c], BOŃKOWSKI Tomasz[1,d] and KRYSTEK Jan[1,e]

[1]NTIS - New Technologies for the Information Society, Faculty of Applied Sciences, University of West Bohemia, Technická 8, 301 00 Pilsen, Czech Republic

[2] Department of Mechanics, Faculty of Applied Sciences, University of West Bohemia, Technická 8, 301 00 Pilsen, Czech Republic

[a]kottner@kme.zcu.cz, [b]kanaksan@students.zcu.cz, [c]soltesl@ntis.zcu.cz, [d]tomasz@ntis.zcu.cz, [e]krystek@ntis.zcu.cz

Keywords: Compression test, foam, impact test, joint protector, tensile test.

Abstract. This work demonstrates experiments carried out on a foam, which is used in joint protectors. Three types of experiments were carried out: a compression test, a tensile test and an impact test. The compression and the tensile tests were each realized for two different strain rates. The impact test was carried out for three impact velocities, for which ten repetitions were realized. Tested samples were cut from the shoulder protector by CNC milling machine. Obtained data were evaluated and subsequently can be used as a target for material model optimization.

Introduction

The best possible means of protection are not only matter of professional sportsmen and military services, but also motorcyclists. Motorcyclists are frequently exposed to dangerous high-speed situations, so it is natural that they demand effective protective equipment. However, creating a reliable protective equipment can be time consuming and expensive concern. Therefore, it is convenient to be able to simulate a mechanical response of the used material [1]. This ability subsequently allows easier improvement of the protector design.

This work pursue the previous research [2] and demonstrate the behaviour of an energy absorbing foam in the course of different loading and validate different types of means used for the force determination. Tested foam is widely used in motorcyclist protective equipment, e.g. shoulder and spine protectors. Used samples were cut from the central part of the shoulder protectors.

The aim was to obtain the force response during the different type of the foam loading, namely for compression, tension, and impact tests. Emphasis was put on the impact test, because this type of test matches presumed loading of the protector. Even though, the tension, especially for higher value deformation, is not usual type of loading of this material. Unfortunately, the tension stress/deformation curves are required for some material models. Therefore, it was necessary to realize this test. An example of the material model created from a series of stress/strain curves is shown in [3].

Obtained material data can be used either for the comparison with another type of material or its material model identification. Using the model of the foam together with the model of the leather [4] will allow the creation and evaluation of different designs of motorcyclist garments including the protectors. Furthermore, these models will be coupled with the model of a motorcyclist body and together will be used for simulations of motorcycle accidents [5].

Experiments

Three types of experiments were carried out: a compression test, a tensile test and an impact test. The samples used for each experiment were cut from the shoulder protector by a CNC milling machine. All the experiments were realized in the temperature equal to 23 ± 1 °C and the humidity equal to 5 ±6 %.

Compression test. The experiment was carried out in the ZWICK/ROELL Z050 universal testing machine. Two different deformation velocities were selected to demonstrate the strain rate dependency, which is typical for energy absorbing materials. It was 0.00909 s^{-1} for the slow test and 0.90909 s^{-1} for the fast test. Measured samples were in a truncated cone shape. The geometry of the samples is specified in Table 1. Samples were compressed until their deformation reached the value equal to 0.8, then they were left 60 s to relax, and subsequently they were relieved. Force response was detected during the process.

Tab.1 Compression test samples parameters
($\dot{\varepsilon}$ is the strain rate, r_1 the upper base radius, r_2 the lower base radius, l_0 is the sample height)

$\dot{\varepsilon}$ [s^{-1}]	0.00909	0.90909
r_1 [mm]	14.0	14.5
r_2 [mm]	15.5	15.0
l_0 [mm]	11	11

The results of the compression test are displayed in Fig. 1. It is necessary to convert the force response to the nominal stress in order to compare detected values for each velocity. The strain rate dependency is evident from Fig. 1. The maximal nominal stress reaches the value equal to 2.422 MPa for the slow test and the value equal to 8.170 MPa for the fast test. Ability of the foam to relax is also apparent from Fig. 1, where is a considerable decrease of the nominal stress value in the case of the maximal deformation.

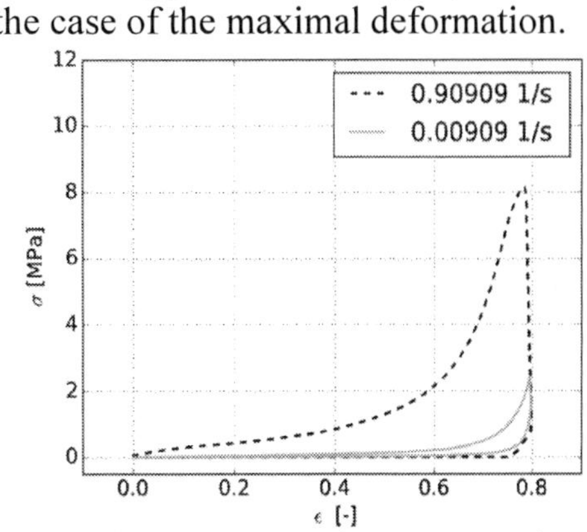

Fig.1 Compression tests result

Tensile test. This experiment was designed to correspond with the compression test, therefore, it was carried out for the same strain rates, 0.00909 s^{-1} for the slow test and 0.90909 s^{-1} for the fast test. The slow test was performed in the ZWICK/ROELL Z050 universal testing machine and the fast test in the INSTRON 8850 testing machine. The sample had the rectangular base of 11 x 20 mm and its length was 180 mm. The length measured by an extensometer was 80 mm for the fast test and 40 mm for the slow test.

The results are displayed in Fig. 2. The negative stress values are notable for the slow test, which was caused by the inability of the foam to relax fast enough. The maximal nominal stress is 0.312 MPa and the minimal nominal stress is -0.022 MPa. The peak at the beginning of the fast stress/deformation curve was caused by the INSTRON jaw inertia. Unfortunately, a sample failure occurred at the end of the sample loading, therefore there are no relaxation and unloading data. There is a marginal decrease of nominal stress, apparent in Fig. 2. Due to the character of the stress/strain curve, the maximal nominal stress was reached before achieving the deformation of 0.8. Its value is 0.812 MPa.

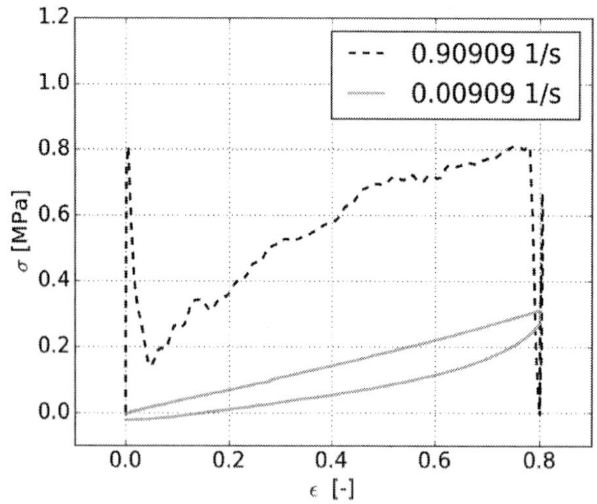

Fig.2 Tension test results

Impact test. This experiment was performed in a drop tower designed by the authors. The impact test was carried out for three impact velocities, namely 1 m·s^{-1}, 2 m·s^{-1}, and 3 m·s^{-1}, for which 10 repetitions were realized. The impactor with a flat steel head weighted 2.454 kg. The impactor displacement, the force response and the impactor acceleration were measured during the experiments. The reaction force was determined in three ways, directly by the force cell KISTLER 9712B5000 and indirectly as the multiple weight of the impactor and its acceleration. The acceleration was measured by the accelerometer KISTLER 8702B500M1T in one case and it was calculated as the second derivative of the impactor displacement in the another one. The displacements were measured using the Micro-Epsilon optoNCDT 2300-50 laser. Sampling frequency was 26 kHz.

The impactor releasing height was determined as:

$$h = \frac{v^2}{2g} + l_0, \tag{1}$$

where v is the impact velocity, g is the gravitational acceleration equal to 9. 81 m·s^{-2} and l_0 is the sample height. Its values are stated in Table 2.

Tab.2 Impact test parameters
(v is the impact velocity, b_1 and b_2 are the rectangular base length, l_0 is the sample height, h is the impactor releasing height, m is the sample mass)

v [m·s^{-1}]	1	2	3
b_1 [mm]	21	20	21
b_2 [mm]	20	19	20
l_0 [mm]	11	11	11
h [mm]	62	218	470
m [g]	1.30	1.18	1.27

Fig. 3 shows the maximum deformation for 10 repetitions of each impact velocity. In the legend, the index symbol denotes the impact velocity value in m·s^{-1}. It is obvious that the maximal deformation values increase with the increasing velocity.

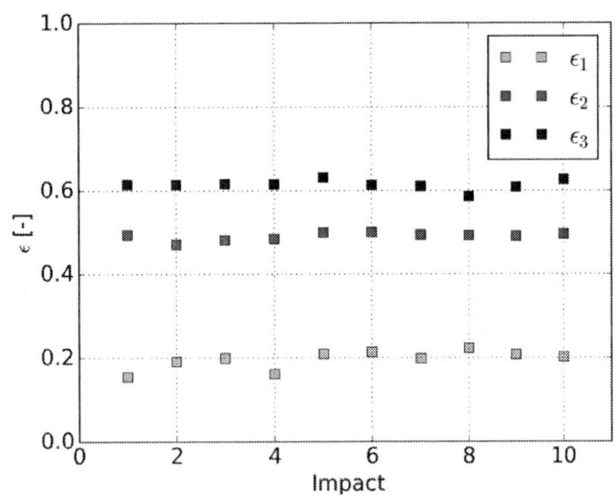

Fig.3 Maximum deformations during impact test

Fig. 4 shows maximum forces during the impact test for the three impact velocities. It also shows the comparison of the three ways of the force response determination. In the legend, the index symbol denotes the device used for the force response determination, index f belongs to the force cell, a to the accelerometer and d to the laser (the displacement second derivation). The number in the legend determines the value of the impact velocity in m·s^{-1}. It is obvious that the used accelerometer is not suitable tool for the reaction force determination. However, disunity of the measured values could be caused by the accelerometer fastening, which contained from the magnet. It is possible that there was an unpredictable movement during the impact. On the other hand, the force cell and the laser appear to be eligible for this task. The force response, determined with the second derivation of the displacement, is slightly higher than from the force cell. This phenomenon does not have to be necessarily incorrect. Thus the determined force response can be considered safer. From the increasing trend of the force response values, the most notable for impact velocity of 3 m·s^{-1}, it is apparent that 60 s relaxation period was not sufficient. Force response values for the first impacts are stated in Table 3.

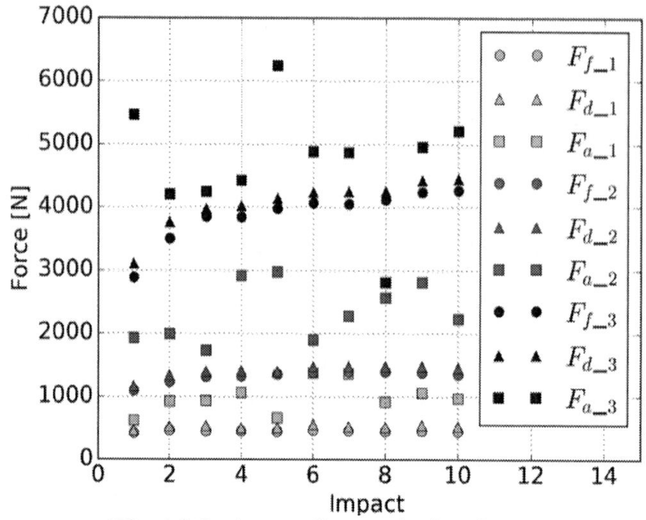

Fig.4 Maximum forces during impact test

Tab.3 First impact test results
(v is the impact velocity, F_f is the force response measured by force cell, F_d is the force response determined by second displacement derivation and F_a is the force response calculated by measured acceleration)

v [m·s^{-1}]	1	2	3
F_f [N]	430	1099	2895
F_d [N]	468	1168	3110
F_a [N]	622	1927	5464

Conclusions

The accelerometer was not proved as a suitable tool for the reaction force determination. However, the force determined as a multiple of the impactor mass and the second derivation of the displacement, appears to match directly to the measured force response.

The energy dissipation, which significantly rises with the increasing strain rate value, proves the foam suitability as a protector. Although the increasing trend of the maximal force response values was registered, the foam still protects even when the impact is repeated. This is the distinctive benefit over the plastic protector, which can get destroyed after the first impact.

The obtained data serves as the target for the identification of the material model. Since the higher temperature can negatively affect the energy dissipation ability and the higher temperature exposure is a real possibility for motorcyclists, the experiments will be also carried out under different temperature conditions in the future.

Acknowledgement

The work was supported by the research project LTC17001 "Exploitation of virtual human model for reducing injury risk of PTW riders" as the national link to the COST Action TU1407 "Scientific and technical innovations for safer Powered Two Wheelers (PTW)" and the internal grant project SGS-2016-059 "Computer modelling and monitoring of human body used for medicine". The authors would like to thank PSí Hubík for providing foam samples.

References

[1] The P. Talaia, D. Moreno, D., M. Hajžman, L. Hynčík, A 3D model of a human for powered two-wheeler vehicles, 23rd International Conference on Noise and Vibration Engineering (2008) 2229-2238.

[2] L. Šoltés, T. Bońkowski, R. Kottner, L. Hynčík, Drop test of foams used in motorbike protectors, 55th Conference on Experimental Stress Analysis (2017).

[3] G. Milne, C. Deck, N. Bourdet, Q. Alline, A. Gallego, R.P. Carreira, R. Willinger, Assessment of bicyclist head injury risk under tangential impact conditions, IRCOBI Conference (2013).

[4] T. Bońkowski, L. Šoltés, L. Hynčík, R. Kottner, P. Kochová, Leather for motorcyclist garments: Multi-test based material model fitting in terms of Ogden parameters, Applied and Computational Mechanics 11 (2017) 129-136.

[5] L. Hynčík, T. Bońkowski, R. Kottner, Virtual assessment of motorcycle helmet contribution to decreasing injury risk in impact, 37th FISITA World Automotive Congress (2018).

EAN 2018
56th conference on experimental stress analysis

Stiffness Reduction of Glass/Epoxy Laminate Under Fatigue Loading

KRÁL Michal[1, a] and CABRNOCH Bohuslav[1, b]

[1]Czech Aerospace Research Centre, Department of Composite Technologies, Beranových 130,199 05 Praha - Letňany, Czech Republic

[a]kral@vzlu.cz, [b]cabrnoch@vzlu.cz

Keywords: fiber reinforced plastic, cycling loading, fatigue damage, stiffness reduction, test specimen

Abstract. The importance of proper dimensioning of machines parts and assemblies is well-known. The assessment of fatigue strength and determination of lifetime is a necessary part of the design process of most components including composite structures. The main objective of this paper is to describe the behavior of Glass/Epoxy laminate under tensile fatigue loading. The emphasis was placed on observation of stiffness reduction during lifetime. Three stacking sequences were tested to get a satisfactory overview of laminate behavior. In one case unexpected stiffness increasing was observed and further investigated in order to determine the effect on real structures. It was necessary to propose methodology for residual modulus estimation before experimental measurement. The description of used methodology based on standards ASTM D3039 and ASTM D3518 is also the objective of this paper.

Introduction

Despite a long term research of fatigue of composite materials a satisfactory methodology which can describe this phenomenon including outer signs of damage has not been found yet. Stiffness and strength reduction can be included between outer signs of damage affecting the structure very distinctly. In general, the reduction of mechanical properties causes that design reserve factors are not valid during a whole lifetime. Stiffness also affects other properties of the structure such as resistance against buckling or natural frequencies.

The motivation for the research of methodology for measurement of laminates stiffness reduction under cycling loading stems from the requirements of customers from many industrial sectors. In the case of structures which must have sufficient stiffness during their lifetime it is necessary to estimate residual stiffness after defined number of cycles. As an example of typical composite structures with defined requirements on stiffness properties which are subjected to cyclic loading can be mentioned e.g. blades of axial fan, composite springs of loom and blades of wind turbines.

Basic review of current state of research

Research and modeling of fatigue in composite materials has been the objective of research for more than 40 years. In comparison with metal materials, fatigue damage in composites is more complicated and worse describable. Fatigue damage in laminates is cumulated in a whole volume of material and is accompanied by more damage mechanisms. The consequence of the development of these damage mechanisms is the reduction of mechanical properties.

Growth of damage mechanisms and therefore also stiffness reduction is dependent on many factors. These factors can be divided to three main groups. The first group can be called material factors and includes e.g. stacking sequence or type of reinforcement (unidirectional, fabric, NCF). The second group can be called the parameters of loading and includes e.g. frequency of loading, type of cycling loading (alternating, tensile only, compressive only), randomness of loading and character of loading (force controlled, deformation controlled loading). The third category can be called external influences and includes temperature, humidity etc.

The division of models for fatigue life prediction is always a question of personal opinion, but generally prediction models can be divided into several basic categories according to used approach of modeling. Division proposed by well-known experts in material science can be found e.g. in [1], [2] and [3]. Models based on residual stiffness are usually classified as phenomenological models of residual material properties. It must be mentioned that stiffness of material in degradation models is very often expressed through Young's modulus. These models are capable to estimate damage parameter D like in the following model (1) proposed by Liu and Lessard [4]. Damage parameter D is usually expressed using equation (2).

$$\frac{dD}{dN}=\frac{A.(\sigma_{max})^c}{B.D^{B-1}}. \tag{1}$$

$$D=\frac{E}{E_0}. \tag{2}$$

Symbol D is damage parameter, σ_{max} is peak stress, A, B and C are model coefficients, E is residual modulus and E_0 is virgin modulus of material. In general, damage parameter can take values from 0 to 1. As already mentioned, stiffness reduction under sufficient level can be defined as limit state. In this case, the knowledge of residual stiffness can be directly used for dimensioning.

Frequently discussed question is the measurement of residual stiffness. During unidirectional tensile experiments is possible to measure residual Young's modulus because the stress field in test specimen is uniform and can be expressed using one value. In the case of test specimens with complex shape or loading conditions, it is necessary to measure residual stiffness because the value of Young's modulus can be different in each point of the structure. This paper is focused on the measurement of residual stiffness during unidirectional tensile experiments and therefore related methods will be discussed. The measurement of residual modulus during loading depends on the character of loading. Considering deformation controlled loading, the force needed for defined deformation achieving is measured. This force decreases due to modulus reduction. The situation is more complicated in the case of force controlled loading. This type of loading is also less used by research teams and proposed methodology is described in following text.

A typical shape of residual stiffness curve is shown in Fig. 1. It can be seen, that stiffness reduction during lifetime is divided into three stages and forms a typical S shaped curve [5]. Each stage is characteristic by development of certain failure modes. The size of individual stages is dependent on more factors but the most important is the level of cyclic stress. On low levels, the third stage may not arise. On high levels the first stage passes to the third stage very fast and the second stage is very short.

The main goal of a number of authors is a modeling of stiffness reduction using finite element method. This approach is based on implementation of degradation models to FE code. A long-term research of this method is performed e.g. by Degrieck and Paepegem, see [6] and [7]. FE modeling is limited by the complexity of fatigue damage in composites and

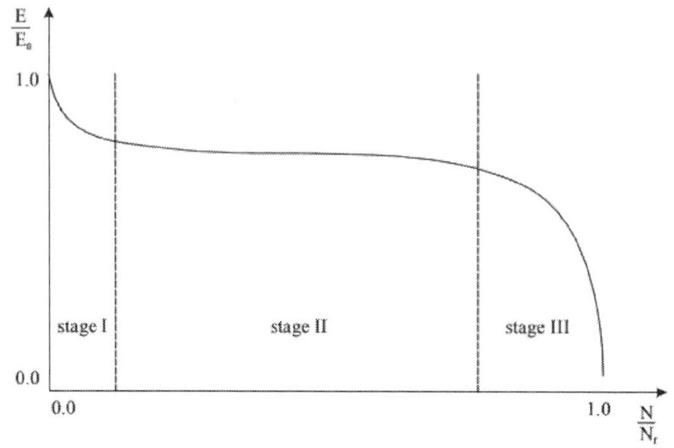

Fig.1 A typical shape of residual stiffness curve [5]

a very bad possibility of generalization. Identified model implemented in FE code will always be useable only for certain conditions which are usually similar to the conditions applied during measuring of data used for identification of implemented model.

Description of proposed methodology for residual stiffness measurement

As already mentioned, the motivation for this research stems from the requirements to structures which must have sufficient stiffness during their lifetime. Between these structures designed at the Composite Technologies Department in Czech Aerospace Research Centre it is possible to include e.g. blades of axial fan. It is generally known that the properties of composite structure can be programmed by the stacking sequence. In the case of non-autoclave technologies, fabrics are generally used as reinforcement. The orientation is usually 0/90°, ±45° and their combinations. Therefore, during experiments three stacking sequences $[(0/90)_8]$, $[(\pm45)_8]$ and [(0/90, ±45, 0/90, ±45) s] were tested in order to get a sufficient review of material behavior and effect of the layers with individual orientations.

Using stacking sequence $[(0/90)_8]$ was possible to determine Young's modulus and using stacking sequence $[(\pm45)_8]$ was measured shear modulus. Test specimens were subjected to shape optimization to prevent from tab failures. Finally, the shape of dog bone was chosen as optimal, see Fig. 2. Glass / Epoxy composite was selected as a material of tabs. The reason is relatively small stiffness and resulting low stress concentration at the end of tabs.

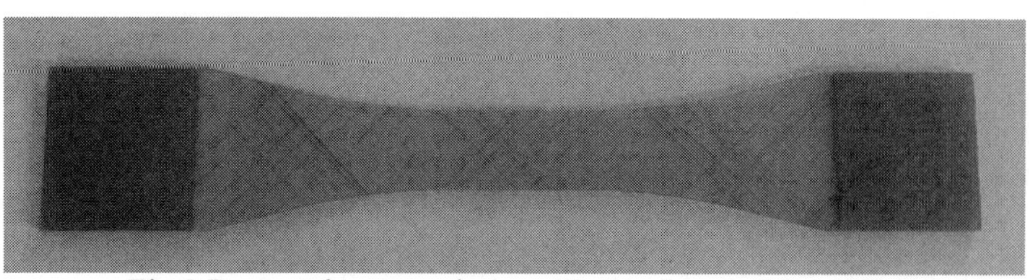

Fig.2 Proposed test specimen with the shape of dog bone

The loading was realized as unidirectional tension with frequency of 10 Hz. With respect to structures, usually designed at the Composite Technologies Department, was necessary to use force controlled loading. Cyclic loading was performed on servo-hydraulic machine MTS with hydraulic grips modified for flat specimens.

Measurement method for individual stacking sequences is in detail described in following text. Residual modulus was measured several times during cyclic loading up to 5.10^5 cycles. The load levels were chosen in order to achieve only the first and the second stage of stiffness reduction curve. With respect to the shape of this curve, the points of measurement were distributed unevenly with more points at the beginning of the curve. Residual modulus was measured with respect to stacking sequence using extensometer or biaxial extensometer. Modulus reduction was expected approximately 15% and it was necessary to measure residual values as exact as possible. At each point the modulus was measured three times to get greater statistical sample and estimate residual modulus in current point with low error. On each level three test specimens were tested. Using described procedure, residual modulus in each point was estimated on the basis of nine independent values.

Modulus reduction of stacking sequence [(0/90)$_8$]

Cyclic loading of specimens with stacking sequence [(0/90)$_8$] was realized on three stress levels with peak stress equal to 169 MPa, 135 MPa and 101 MPa. The behavior of material was in line with expectation and stiffness reduction was more pronounced on higher stress levels, see Fig. 3. The modulus was measured on the basis of standard ASTM D3039. For strain measurement an axial extensometer was used. Data obtained during these experiments can be directly used for FE calculations, because with this stacking sequence can be on the basis of standard ASTM D3039 determined mechanical properties of single layers.

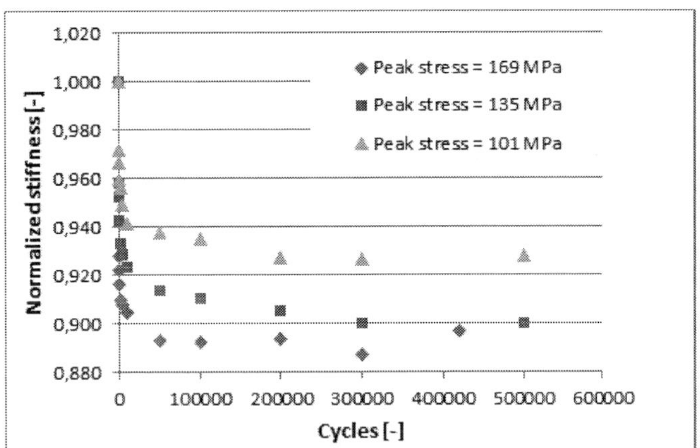

Fig.3 Stiffness reduction of test specimens with stacking sequence [(0/90)$_8$]

Modulus reduction of stacking sequence [(±45)$_8$]

Cyclic loading was expected also on three stress levels. Shear modulus was determined on the basis of standard ASTM D3518. Longitudinal and transversal moduli were measured using biaxial extensometer. During testing of the first stress level with peak stress equal to 40 MPa (relative to a cross-section of test specimen), the behavior was not in line with expectations. Shear modulus was increasing and after 5.10^5 cycles, residual modulus was 115% of the virgin value, see Fig. 4. The attention was focused on finding out the cause of this behavior.

In the case of stacking sequence with dominating layers oriented ±45°, matrix is more involved in transmission of forces. Possible change of mechanical properties of the matrix can therefore cause considerable change in mechanical properties of the whole composite. It is well known that composites get warm during cyclic loading.

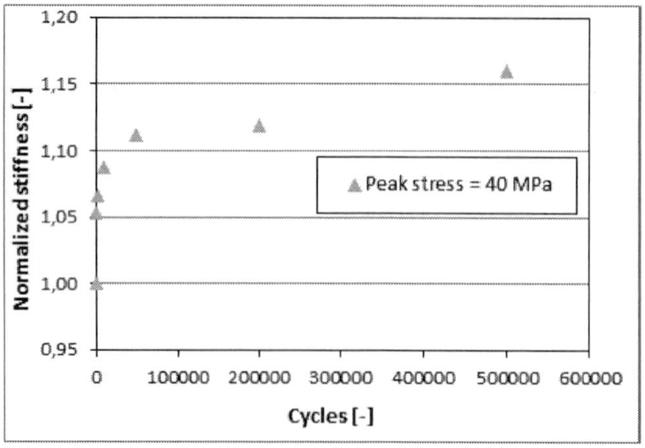

Fig.4 Stiffness change of test specimens with stacking sequence [(±45)8]

The possibility, if this warming can cause post-curing of the matrix, was verified using simple comparison of test results of test specimens cured at standard and significantly higher temperatures (matrix glass transition temperature was increased from 60°C to 120°C). The test results comparison shown no difference in test specimens behaviour and thus the effect of matrix post-curing has been ruled out. The situation is demonstrated in Fig. 5. The specimens post-cured at higher temperature exhibited higher modulus increase.

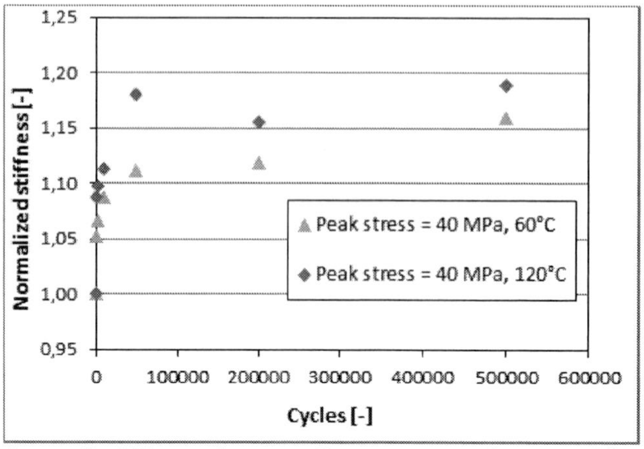

Fig.5 The comparison of stiffness change of two test specimens with stacking sequence [(±45)8] cured at different temperatures

Another evaluated reason of modulus increase was a creep strain. Creep strain in the case of stacking sequence ±45° can cause reorienting of fibers to lower angles than 45° and thus induced the change of test specimen stiffness in direction of loading. To verify a range of creep strain the observation of test specimen using optical microscope was performed. Prepared test specimen is shown in Fig. 6. Two sets of lines with defined distance of 25 mm and 50 mm were drawn on specimens. The distance between the lines was measured before and during cyclic loading at 2 000, 50 000 and 500 000 cycles.

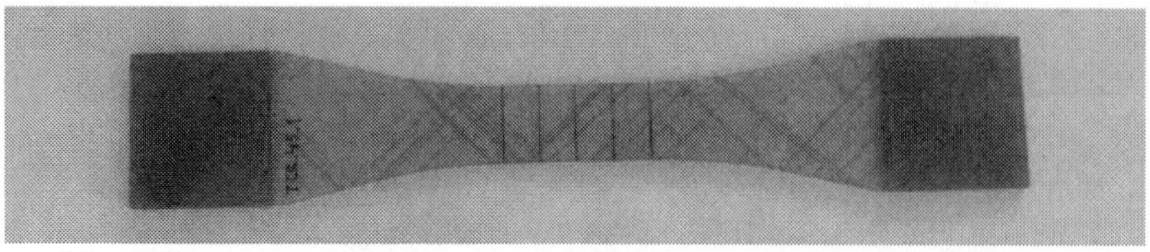

Fig.6 Test specimen proposed for creep strain measurement

Measured permanent strain was 2000 με on the base of 25 mm and 1000 με on the base of 50 mm. The peak strain of test specimens during cyclic loading is about 4000 με and therefore the permanent strain reaches 25 % and 50 % of this value. In Tab. 1 there is shown the comparison of distances measured before and after cyclic loading. Measured creep strain influences the behaviour of the tested material and causes stiffness increase.

Tab.1 The change of distance of two sets of lines caused by creep strain in specimens with stacking sequence [(±45)8]

Distance	Virgin value	2 000 cycles	50 000 cycles	500 000 cycles	Resultant increase
25 mm	24.79 mm	24.81 mm	24.84 mm	24.84 mm	0.05 mm
50 mm	49.53 mm	49.56 mm	49.61 mm	49.58 mm	0.05 mm

Modulus reduction of stacking sequence [(0/90, ±45, 0/90, ±45)s]

Cyclic loading was realized on three stress levels with peak stress equal to 127 MPa, 102 MPa and 76 MPa (relative to a cross-section of test specimen). The behavior of test specimens was in line with expectations. Stiffness was measured again on the basis of ASTM D3039, but in the case of this quasi-isotropic stacking sequence, the resultant modulus was measured. It is impossible to estimate the behavior of single layers on the basis of these results. It can be seen that the stiffness reduction is between 2% and 13% in dependence on stress level, see Fig. 7.

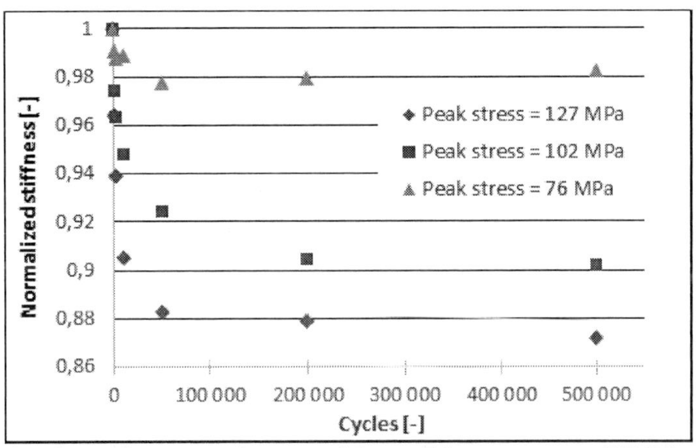

Fig.7 Stiffness reduction of test specimens with stacking sequence [(0/90, ±45, 0/90, ±45)s]

Conclusion

From presented results can be seen that behavior of composite materials under fatigue loading is quite complex issue. Results can be generalized only in the case of very similar boundary conditions. Stiffness reduction on stress levels which were tested in presented research is from 2% to 15%. Stress levels were chosen with respect to real engineering applications and all levels were under 60% of static strength. The question if stiffness reduction of 15% is for real structures danger is not easy to answer. In the case of structures which must be prevented from resonant stages and must have exactly defined natural frequencies, the change of stiffness within this range can be unacceptable.

It was expected to determine the reduction of shear modulus of tested material using stacking sequence [(±45)8]. However, the results shown that the change of shear modulus is not easy to describe. It must be mentioned that behavior of single layer with this orientation may not be the same when the layer is embedded in laminate and when the layer is unrestrained like in discussed experiment. In stacking sequence where both orientations are combined, the layers with orientation 0/90° will not allow layers with orientation ±45° to reach so significant creep strain. During cyclic loading the layers with orientation ±45° will be influenced by two main mechanisms – creep strain and fatigue damage. Creep strain can cause increase of stiffness and fatigue damage can cause reduction of stiffness. Resultant behavior will be defined by the ratio of these two mechanisms and the ratio will be different for unrestrained layer and for layer embedded in composite.

Experiment with quasi-isotropic stacking sequence [(0/90, ±45, 0/90, ±45)s] was performed in order to verify above mentioned hypothesis, that in stacking sequence, where both orientations are combined, the effect of stiffness increase of layers oriented ±45° is not so expressive. To evaluate the behavior of layers oriented ±45°, the experiment, where the loading will be deformation controlled will be proposed. Loading with this character is closer to real loading of layer in final laminate.

Acknowledgements

The research was performed with financial institutional support from government budget through the Ministry of Industrial and Trade.

References

[1] J. Degrieck, W. van Paepegem, Fatigue damage modeling of fiber-reinforced composite materials: Review. Applied Mechanics Reviews. 2001, 54 (4) 279-300. ISSN 00036900.

[2] J. Montesano, J. K. Behdinan, Z. Fawaz, Ch. Poon, Considerations for Progressive Damage in Fibre-Reinforced Composite Materials Subject to Fatigue. In: High Performance Computing Systems and Applications. Berlin, Heidelberg: Springer Berlin Heidelberg, 2010, p. 148-164. ISBN 978-3-642-12658-1.

[3] G.P. Sendeckyj, Life Prediction for Resin-Matrix Composite Materials. Fatigue of Composite Materials. Composite Materials Series, 1991, p. 431-483. ISBN 9780444705075.

[4] B. Liu, L.B. Lessard, Fatigue and damage-tolerance analysis of composite laminates: Stiffness loss, damage-modeling, and life prediction. Composites Science and Technology. 1994, 51(1) 43-51. ISSN 02663538.

[5] W. van Paepegem, A new coupled approach of residual stiffness and strength for fatigue of fibre-reinforced composites. International Journal of Fatigue. 24(7) 747-762. ISSN 01421123.

[6] W. van Paepegem, J. Degrieck, Experimental set-up for and numerical modelling of bending fatigue experiments on plain woven glass/epoxy composites. Composite Structures. 2001, 51(1) 1-8. ISSN 02638223.

[7] W. van Paepegem, J. Degrieck, Modelling damage and permanent strain in fibre-reinforced composites under in-plane fatigue loading. Composites Science and Technology. 2003, 63(5) 677-694. ISSN 02663538.

EAN 2018
56th conference on experimental stress analysis

Materials Selection for Efficient Cross-section of Components

KRATOCHVÍL Martin[1,a], MAZÍNOVÁ Ivana[1,b] and HRDLIČKA Filip[1,c]

[1]University of West Bohemia, Pilsen, Czech Republic

[a]kratochv@kks.zcu.cz, [b]mazini@kks.zcu.cz, [c]ficeriov@kks.zcu.cz

Keywords: material indices, shape factor, ANSYS, FEM

Abstract. Materials selection is an essential part of the product design process as materials influence most of the product's properties. Currently design engineers are able to choose from up to 160 000 engineering materials. We use the methodology of prof. Ashby and CES EduPack software to material selection.

Introduction

This paper introduces methods for co-selecting material and section shape. In many applications section shape is not a variable. But when components carry bending, torsion or compressive loads, both the area and shape of the cross-section are important. By shape we mean that the cross-section is formed to a tube, I-section or the like. Efficient shapes use the least material to achieve a given stiffness or strength. So material and shape are coupled, requiring a method of choosing them together.

However, shape factors are only developed for simple loading states. Derivation of shape factors for combined loads is one of the aims of this paper.

Selection of Material and Shape

Shape Factor. We have a many types of profiles as I-sections, rods, tubes, etc. which we can use for designing, but which one is the best? That depends on three things, type of load, shape of the profile and material. For this purpose we will neglect the material because we will compare only the profiles with the same material. Loads we can divided into 5 categories, tension, compression, torsion, bending and shear but we will work only with compression, torsion and bending because in tension and shear shape of the profile is not important. Also as in the previous case, we will compare only the profiles with the same loads.

Every profile have a volume defined by a cross section and length. Together with the density of the material we can get a weight of the profile. If we want compare two different shapes of the profiles, we need to do for the constant weight. As we mentioned above, the material is constant and also the cross section and length must be constant too.

From previous information we can see that the only variable which we will use, will be the shape of the cross section of the profiles which is define by dimensions. By comparison of two different shapes we can get the shape factor which is dimensionless number and this number expresses how much is the compared shape better (bigger than 1) or worse (lower than 1) than the reference one.

For every comparison is important to choose the reference shape with which we will compare the new one. As the best option is a solid square rod with cross section area A_0.

If we want to make a comparison of two different cross sections, then firstly we need to say if we are searching shape factor for elastic loading or shape factor for onset of plasticity (failure). For elastic loading we need to compare second moment of area, for failure we need to compare section modulus of area.

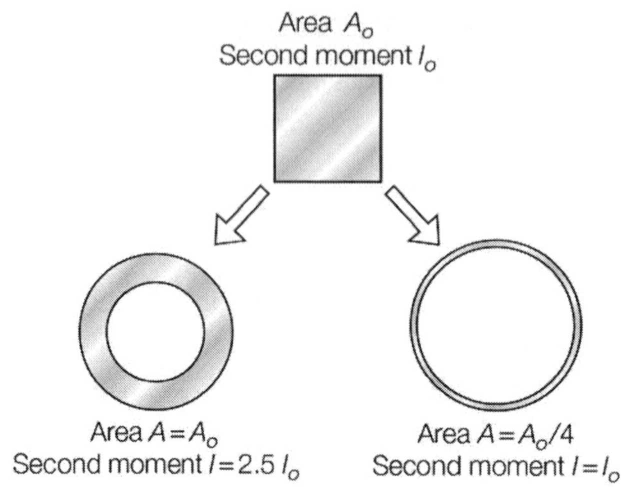

Fig. 32 Principle of the shape factor [1]

As we can see in the Fig. 1, when we compare the shape of the solid square rod with the shape of the tubes we can get different results. The tube in the left has the same cross section area as the square rod but it is 2.5 times stiffer. The tube in the right has the same stiffness as the square rod but it is 4 times lighter.

As we mentioned in previous chapter, we have two types of shape factors, the shape factor for elastic loading and shape factor of failure.

The shape factor for elastic bending (ϕ_B^e) is define as a ratio of bending stiffness of solid square rod (S_0) and compared profile (S) where:

$$S \propto \frac{E \cdot I}{L^3} \tag{1}$$

Here E is Young's modulus, I is the second moment of area and L is length.

$$\phi_B^e = \frac{S}{S_0} = \frac{\frac{E \cdot I}{L^3}}{\frac{E \cdot I_0}{L^3}} = \frac{I}{I_0} \tag{2}$$

$$I_0 = \frac{{b_0}^4}{12} = \frac{A^2}{12} \tag{3}$$

$$\phi_B^e = \frac{12 \cdot I}{A^2} \tag{4}$$

The factor is dimensionless number that mean that the value of the factor does not depend on scale, only on shape.

The shape factor of failure for bending (ϕ_B^f) we can get from equation for bending stress (σ_b). Bending stress is a ratio of the bending moment (M_B) and the section modulus (Z). If we compare the bending cross section modulus for solid square rod (Z_0) with the bending cross section modulus of any other shape (Z) we will gain the shape indicator for strength efficiency of the shaped beam ϕ_B^f.

$$\sigma_B = \frac{M_B}{Z} \leq \sigma_{B\,max} \tag{5}$$

$$z_0 = \frac{b_0 \cdot h_0}{6} \tag{6}$$

Because Z_0 is section modulus for square rod, then $b_0 = h_0$ and then:

$$z_0 = \frac{{b_0}^3}{6} = \frac{A^{\frac{3}{2}}}{6} \tag{7}$$

$$\phi_B^f = \frac{Z}{Z_0} = \frac{6 \cdot Z}{A^{\frac{3}{2}}} \tag{8}$$

Similar shape factors characterize stiffness and strength in torsion. For elastic twisting:

$$\phi_T^e = 7.14 \cdot \frac{K}{A^2} \tag{9}$$

And for failure in torsion:

$$\phi_T^f = 4{,}8 \cdot \frac{Q}{A^{\frac{3}{2}}} \tag{10}$$

Where K is torsional moment of area and Q is torsion cross section module.

Maximum values for shape factors depends on materials and we can gain theoretical maximum values from equation:

$$\phi_e = 2\sqrt{\frac{E}{\sigma_y}} \tag{11}$$

Where E is Young modulus and σ_y is Yield strength. Maximum theoretical values of shape factors for few basic material are listed in the Table 1.

Tab. 3 Limits for Shape factors [1]

Material	**Max ϕ_e**	**Max ϕ_f**
Steels	65	13
Aluminium alloys	44	10
GFRP and CFRP	39	9
Unreinforced polymers	12	5
Woods	8	3
Elastomers	< 6	-

Material Indices that Include Shape

This paper introduces methods for co-selecting material and section shape. We use the methodology of prof. Ashby and CES EduPack software to material selection [2]. The methodology of materials selection is based on material indices. This chapter explains the ground of the material indices and their interconnection with the shape factors. A wider explanation is in the book [1].

Materials selection is based on and driven by engineering design process of a given product. The requirements on the product are the inputs for this process. The requirements can be divided into three categories: function, constraints, goals and free variables. The function is defined by the purpose of the part, e.g. to support load. The constraints are conditions that must be met, e.g. maximum deflection or maximum dimensions. During the design process there are some goals to be achieved. We might want the product to be as light or as cheap as possible. Some parameters might be adjusted to maximize the fulfilment of goals. These are free variables.

In Fig. 2 engineering design-driven materials selection scheme is shown. Design requirements are translated into product specification suitable for materials selection. This can be done by deriving of material indices. The constraints set out limit values of certain properties. The goals define the material indices for which we seek extreme values. If a goal is not bound with a constraint the material index becomes a simple material property. Otherwise the index becomes a group of properties.

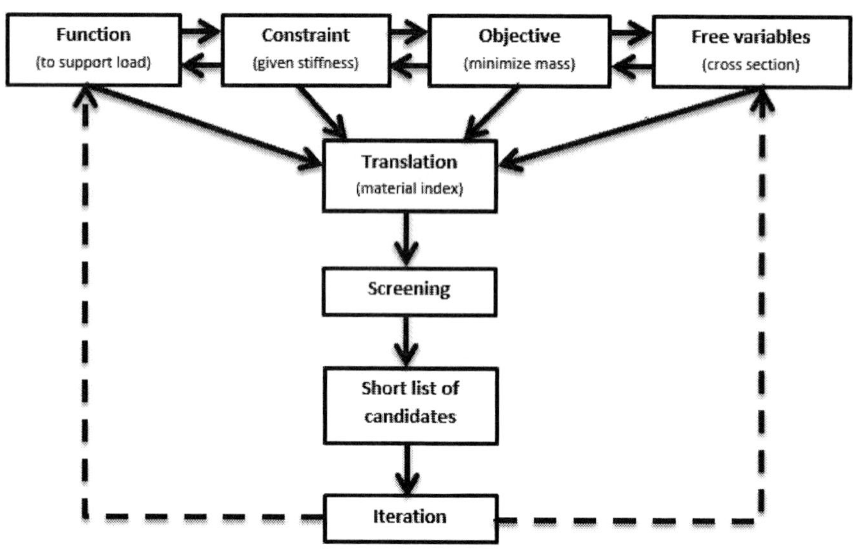

Fig. 33 Engineering design-driven materials selection

The performance of a part is given by three parameters: functional requirements (F), geometrical dimensions (G) and material properties (M). It can by expressed by the equation below:

$$P = f_1(F) \cdot f_2(G) \cdot f_3(M) \tag{12}$$

The performance for is maximized when we maximize $f_3(M)$ which is called material index. Each combination of function, goal and constrain leads to a material index which makes this method general an applicable for a wide range of problems.

Material index for a stiff, light (bending) beam. The most usual mode of loading of engineering structures is bending. All these structures carry bending moments, they are beams. The requirement here is for a beam of specified stiffness and minimum mass.

Function: Beam

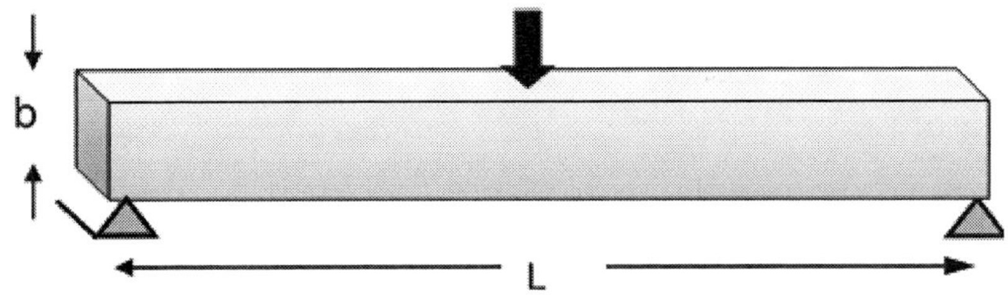

Constraints:
Length L is specified (geometric constraint)
Bending stiffness is specified as S* (functional constraint)

$$S = \frac{CEI}{L^3} = \frac{CEA^2}{12L^3} \geq S^* \tag{13}$$

$$I = \frac{b^4}{12} = \frac{A^2}{12} \tag{14}$$

Objective:
Minimize mass m of the beam

$$m = A\,L\,\rho \tag{15}$$

Performance metric m:

$$m = \left(\frac{12L^5S}{C}\right)^{\frac{1}{2}} \left(\frac{\rho}{E^{\frac{1}{2}}}\right) \tag{16}$$

Material index for light, stiff beam is:

$$\left(\frac{E^{\frac{1}{2}}}{\rho}\right) \tag{17}$$

This is material index that does not depend on shape.

Material index that include shape. But stiffness and strength-limited design depends on shape of cross-section of components. For stiff, light beam material and shape are coupled, requiring a method of choosing them together.

For the square beam of

$$I_0 = \frac{A^2}{12} \tag{18}$$

whereas for the shape beam

$$I = \frac{\phi_e \cdot A^2}{12} \tag{19}$$

The resulting equation for the metric m shows how the mass of the beam depends on the mechanical constraint it must meet (S) on the section shape (ϕ_e) and on the material of which its made (ρ/E1/2). The quantity (ρ /$(\phi_e E)^{1/2}$) can be thought as a "shaped-material index".

Material index that include shape for light, stiff (bending) beam is:

$$\left(\frac{\phi_e E^{\frac{1}{2}}}{\rho}\right) \tag{20}$$

Selecting material-shape combinations

For mechanical designers is the most often goal of designing to reach a minimum mass of construction. From this reason is useful to connect shape factor indicator with material indicator together. If we do this, we will get the indices as you can see in the Tab. 2.

Tab. 4 Indices with shape [1]

Component Shape, Loading, and Constraints	Stiffness-limited Design*	Strength-limited Design*
Tie (tensile member) Load, stiffness, and length specified, section-area free	$\frac{E}{\rho}$	$\frac{\sigma_f}{\rho}$
Beam (loaded in bending) Loaded externally or by self-weight, stiffness, strength and length specified, section area and shape free	$\frac{(\phi_B^e E)^{1/2}}{\rho}$	$\frac{(\phi_B^f \sigma_f)^{2/3}}{\rho}$
Torsion bar or tube Loaded externally, stiffness, strength, and length specified, section area and shape free	$\frac{(\phi_T^e E)^{1/2}}{\rho}$	$\frac{(\phi_T^f \sigma_f)^{2/3}}{\rho}$
Column (compression strut) Collapse load by buckling or plastic crushing, strength and length specified, section area and shape free	$\frac{(\phi_B^e E)^{1/2}}{\rho}$	$\frac{\sigma_f}{\rho}$

For example, materials for stiff, shaped beams of minimum weight:

- Fixed shape (φ_e fixed): choose materials with low $\frac{\rho}{E^{\frac{1}{2}}}$
- Shape φ_e a variable: choose materials with low $\frac{\rho}{(\phi_e \cdot E)^{\frac{1}{2}}}$

Tab. 5 Suitability of materials [1]

Material	$\rho\ [Mg/m^3]$	$E\ [GPa]$	$\phi_{e,max}$	$\rho/E^{\frac{1}{2}}$	$\rho/(\phi_{e,max} \cdot E)^{\frac{1}{2}}$
1020 Steel	7,85	205	65	0,55	0,068
6061 T4 Al	2,70	70	44	0,32	0,049
GFPR	1,75	28	39	0,35	0,053
Wood (oak)	0,9	13	8	0,25	0,088

For fixed shape (up to $\phi_e = 8$) is wood the best solution. For maximum shape ($\phi_e = \phi_{e,max}$) are Al-alloys the best solution. Steel recovers some performance through high $\phi_{e,max}$.

Shape factors created by numerical analysis

Theory of material indices developed by professor Ashby assumes developing of indices based on analytical solution [1]. It needs to know analytical (mathematical equations) model of a chosen physical phenomenon. But in many cases, only numerical solution is known or it is difficult to create analytic solution or arrange parameters during process of creation a material index or shape factor.

The method of creating shape factors using numerical solution is based on analysis of FEM solver output data. During FEM output analysis, changes of a solved object parameters against changes of calculated deformation or stress are compared. The aim of comparison is to find equation of a factor using regression, logarithm or elementary equations. This paper describes creation of shape factors of fixed beam with rectangle cross section.

First step of analysis is to define scope of analysis:

- Function: beam, one end fixed, other end free and loaded by moment (bending, torsion or its combination)
- Objective: minimize stress (von Mises)
- Constrains: linear elastic material, length
- Free variables: width, height, load

Tab. 6 Analysis parameters

Parameter	Symbol	Type	Bending factor	Torsion factor	Combined factor
Width	B	Shape parameter	Variable	Variable	Variable
Height	H	Shape parameter	Variable	Variable	Variable
Length	L	Constant	Constant	Constant	Constant
Material properties	E, υ	Constant	Constant	Constant	Constant
Bending moment	M_b	Load	Variable	0	Variable
Torsion moment	M_t	Load	0	Variable	Variable

First step is modelling a parametric geometry of the beam. In this case, Design Modeller application from ANSYS bundle was used. The geometry was linked to Static Structural module. Material of beam must be linear elastic, steel was selected. Application of loads and constrain is described on the picture (Fig. 3):

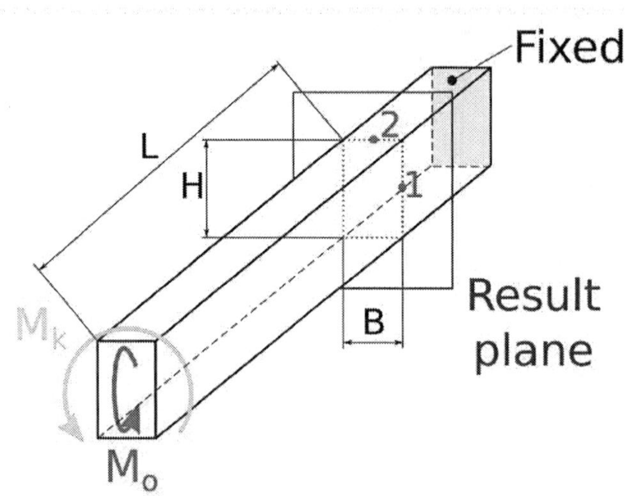

Fig. 34 Loads and constrains in FEM analysis

The analysis does not consider the effect of fixed constrain. Parametric model of the beam is a little bit longer than considered, but stress results were calculated at a result plane. The result plane is normal to beam axis and its distance from free end of the beam is the same as considered length L of a beam (Fig. 3).

It is also necessary to create design points, which are used to changes of simulation parameters. Design points must be created according change of the shape of cross section and table Tab. 4 above. Shape of rectangle cross section is defined by equation:

$$shape = \frac{H}{B} \tag{21}$$

For all cases in this paper:

$$H \geq B \tag{22}$$

It is necessary to eliminate influence of cross section size, so the cross section dimensions should pass the condition:

$$H_1 B_1 = H_2 B_2 = A_0 \tag{23}$$

Simulations of examples bellow were calculated using FEM system ANSYS Workbench 17. Result analysis of FEM output was done in Python 3 with math libraries developed under SciPy project.

Shape factor for pure bending. Creation of bending shape factor is relatively simple, because there is direct dependence of normal stress on the dimensions of cross section.

ANSYS output data was analyzed using equation:

$$\lambda_p = \ln \frac{\sigma_{i+1}}{\sigma_i} \cdot \left(\frac{p_{i+1}}{p_i}\right)^{-1} \tag{24}$$

where p is parameter which influence is analyzed (in this case B or H) and λ_p its estimated exponent. The result of analysis of FEM output are values:

$$\lambda_B = 1 \tag{25}$$

$$\lambda_H = -1 \tag{26}$$

Estimated shape factor should be something as:

$$\psi = B^{\lambda_B} H^{\lambda_H} = \frac{B}{H} \tag{27}$$

where ψ is estimated solution for bending factor. It is necessary to verify slope of equation of estimated factor ψ:

$$\psi = B^{\lambda_B} H^{\lambda_H} = \frac{B}{H} \tag{28}$$

where λ_s is slope correction exponent and its value is:

$$\lambda_s = 0.5 \tag{29}$$

Result of previous analysis is *shape* factor *for beam loaded by bending moment* (in physics mean, lower value is better design):

$$\zeta_1 = \psi^{\lambda_s} = \left(\frac{B}{H}\right)^{0.5} = \left(\frac{H}{B}\right)^{-0.5} \tag{30}$$

The shape factor of bended beam was created from numeric values, in this case better design has lower value of shape factor (lower stress). According to prof. Ashby's theory [1], it was transformed to get the highest value for the best design:

$$\phi_B^f = \zeta_1^{-1} \tag{31}$$

$$\phi_B^f = \sqrt{\frac{B}{H}} \tag{32}$$

where ζ sign is shape factor with physics mean (better design has lower factor values) and ϕ sign the prof. Ashby form of shape factor (better design has highest factor value). Prof. Ashby published the same form of shape factor for beam loaded by bending moment [1].

Shape factor for pure torsion. Torsion shape factor was created similarly as bending shape factor. After preliminary analysis, solution was estimated in form:

$$\chi = \frac{H}{B} + 1 \tag{33}$$

The result of second analysis is *shape factor for the beam loaded by torsion* (in physics mean, lower factor value is better design):

$$\zeta_2 = \left(\frac{H}{B} + 1\right)^{\frac{3}{7}} \tag{34}$$

Torsion shape factor modified to prof. Ashby form:

$$\phi_T^f = \zeta_2^{-1} = \left(\frac{H}{B} + 1\right)^{-\frac{3}{7}} \tag{35}$$

Prof. Ashby published this factor with different equation [1]:

$$\phi_{T\,Ashby}^f = 1.6 \cdot \sqrt{\frac{B}{H}} \cdot \frac{1}{1 + 0.6 \cdot \frac{B}{H}} \tag{36}$$

Comparison of values of prof. Ashby's version of shape factor and torsion shape factor based on FEM (booth in physical mean, Fig.4):

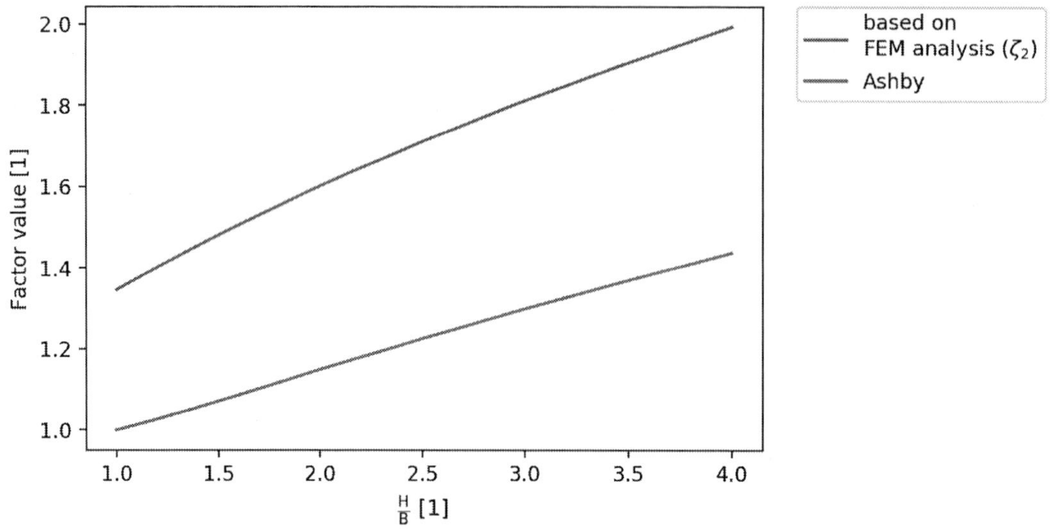

Fig. 35 Comparison of values of prof. Ashby and FEM factors

One of the properties of shape factors is presented below - ability to determine stress of a loaded object. Suppose that there is any beam loaded by torsion moment, maximal shear stress τ_1 is known. The goal is to create a new better design. Shear stress τ_2 of a new design should be determined using equation:

$$\tau_2 = \tau_1 \cdot \frac{\phi_{T2}^f}{\phi_{T1}^f} \tag{37}$$

Tab. 7 Stress prediction comparison of prof. Ashby and FEM factors

Design	B	H	factor type	factor value	Calculated stress (FEM)	Stress estimated by factor
Old	20 mm	80 mm	Ashby	1.437	192 kPa	-
			FEM	1.993		-
New	34.3 mm	46.6 mm	Ashby	1.050	139 kPa	140.3 kPa
			FEM	1.445		139.2 kPa

In this example (Tab. 5), booth shape factors has similar results. Unlike the previous example, there are two different equations, each of them provides different values. It is a question if this phenomenon is desirable due to the current state of the theory, where shape factors are compared against a given base geometry during creation [1]. Analysis of FEM data is based on another principle – comparison changes of stress result against given input parameters. However, the next example shows benefit of some relationship of FEM based shape factor value on value of calculated stress.

Combined bending-torsion shape factor. Creation of combined bending-torsion shape factor is more difficult than previous shape factors. The reason is stress distribution of rectangle cross section shape of the beam. There are two different types of stress in two areas. Let's see on the picture bellow (Fig. 5). There are two areas market by point 1 and 2. Highest stress could be in one of them or booth, depending on values of bending moment and torsion moment.

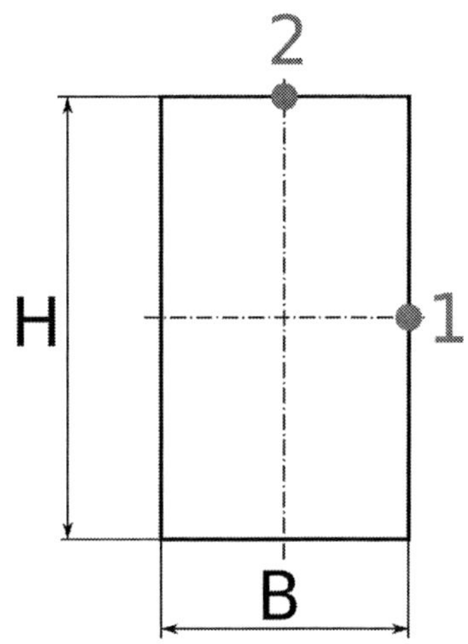

Fig. 36 Areas of higher stress

In the case of dominant bending moment (or pure bend), it is important to evaluate normal stress and shear stress in the point 2. For dominant torsion moment (or pure torsion) of the beam, there is shear stress in the point 1. In this case, von Mises stress in point 1 is higher than stress (von Mises) in point 2. The situation is explained on the picture (Fig. 6) bellow.

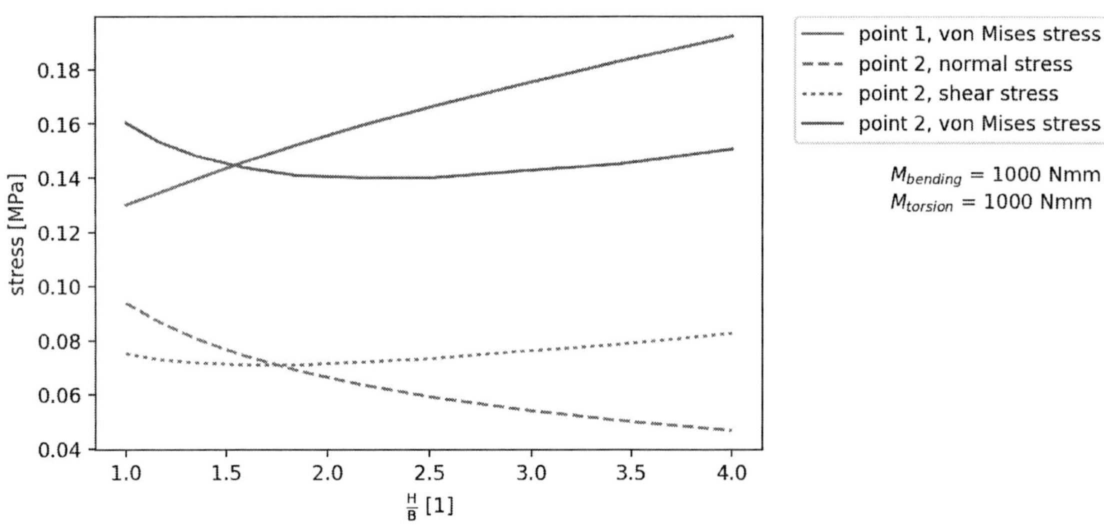

Fig. 37 Stress in point 1 and 2

The solution of this problem is a little bit complicated, but can be solved by von Mises yield criterion. Shear stress in point 1 is described by shape factor ζ_2. Shape factor ζ_1 describes normal stress in point 2. It remains to create shape factor for shear stress in point 2. From the picture (Fig. 6) above is evident, that shear stress in point 2 drops (when shape ratio of cross section grows). There is minimum of shear stress around 1.6 value of cross section shape ratio. Then shear stress increases (if shape ratio still grows). It seems an equation that describes this phenomenon could be:

$$\omega = \left(\frac{H}{B} - C\right)^2 \tag{38}$$

As mentioned, von Mises yield criterion will be used. To be used correctly, all (particular) shape factors (ζ_1, ζ_2, ζ_3) must be normalized. It means values of all shape factors must have the same relative ratio against stresses described by them. Factors ζ_1 and ζ_2 have relative ratio 10 (compared to von Mises stress), it means they are 10 times higher than stresses described them, if the beam is loaded by 1000 N.mm bending or torsion moment. Particular factors ζ_3 should be created with the same relative ratio requirement. After some analysis, final version of shape factors ζ_3 was created:

$$\zeta_3 = \frac{\frac{H}{B} + \left(\frac{H}{B} - 2\right)^2}{5 \cdot \left(\frac{H}{B} + 1\right)} + 1.1 \tag{39}$$

Normalization is explained in the picture (Fig. 7) below, all stresses were multiplied by 10:

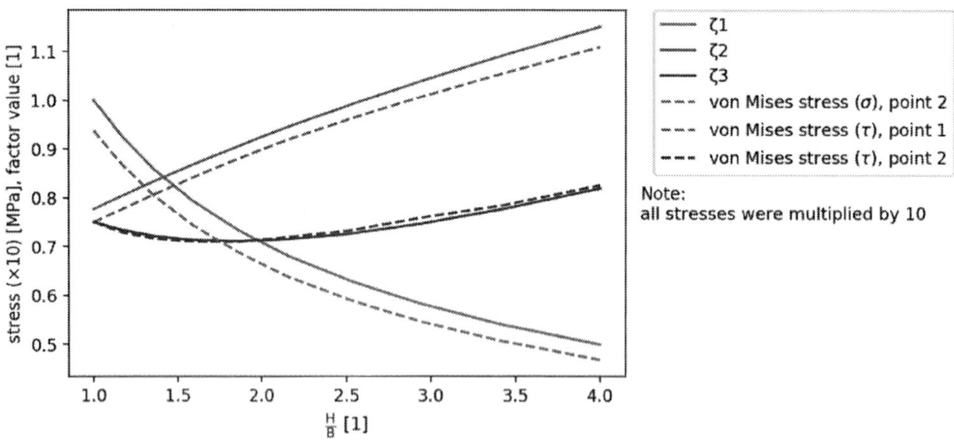

Fig. 38 Calculated stress compared to factors prediction

Relative size of bending moment and torsion moment described by coefficients:

$$k_b = \frac{M_b}{M_b + M_t} \tag{40}$$

$$k_t = \frac{M_t}{M_b + M_t} \tag{41}$$

Formula of combined bending-torsion shape factor based on von Mises yield criterion (equation which multiplies the set makes relative ratio 10 against von Mises stress):

$$\zeta_c = \max\left\{\sqrt{3 \cdot \left(\zeta_2 \frac{k_t}{\sqrt{3}}\right)^2}, \sqrt{(\zeta_1 \cdot k_b)^2 + 3 \cdot \left(\zeta_3 \cdot \frac{k_t}{\sqrt{3}}\right)^2}\right\} \cdot \sqrt{1+3} \tag{42}$$

Equation bellow is shape factor for beam loaded by combination of bending moment and torsion moment (in physics mean, lower value is better for design):

$$\zeta_c = 2 \cdot \max\left\{k_t \cdot \left(\frac{H}{B} + 1\right)^{\frac{3}{7}}, \sqrt{\left(k_b \cdot \left(\frac{H}{B}\right)^{-0.5}\right)^2 + \left(k_t \cdot \left(\frac{\frac{H}{B} + \left(\frac{H}{B} - 2\right)^2}{5 \cdot \left(\frac{H}{B} + 1\right)} + 1.1\right)\right)^2}\right\} \tag{43}$$

where M_b is bending moment and M_t torsion moment. Combined bending-torsion factor in the form of prof. Ashby form can be calculated using equation:

$$\phi_{BT}^{f} = \frac{1}{\zeta_c} \tag{44}$$

Equation of the combined shape factor is relatively complicated, but it can be used to create material maps (modified by shape factors [1]) or to find an optimal shape in first stages of design process.

For example, to find an optimal ratio for a one end fixed beam with rectangle cross section, if load:

- bending moment $M_{b1} = 1000$ Nm
- torsional moment $M_{t1} = 2000$ Nm

There are two possible solutions – to find minimum on ζ_c indicator using numerical equation solver or draw graph of ϕ_{BT}^{f}:

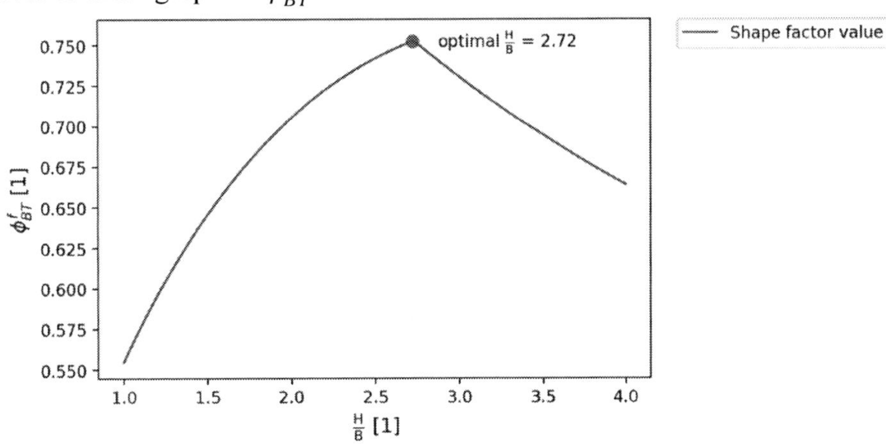

Fig. 8 Solution of the example of combined load, optimal solution marked by red point

The picture above explains that optimal cross section parameters ratio is around 2.72, the next step is to estimate cross section area size, calculate H and B dimensions and continue design process.

Conclusion

One of the goals of this paper was to show what the shape factors and materials indices are, how they works and why they are beneficial. As the next goal was to try to find out shape factor of combined stress compounded from bending and torque.

As is obvious from last part of this paper, is possible to create a shape factor for combined stress. But whole method is complicated and it takes some time. For this reason is suitable to creating complex shape factors only for special used where is possible to used it many times. For other applications is still better to use FEM analysis.

References

[1] ASHBY, Michael F. Material Selection in Mechanical Design. Oxford : Butterworth-Heinemann, 2011. ISBN 978-1-85617-663-7.

[2] CES EduPack 2013 Version 12.2.13. [DVD] Cambridge : Granta Design Limited, Granta Design Limited, 2013. 12, 2, 13, 0.

Load capacity of bolted joints of carbon fibre reinforced plastics

KRYSTEK Jan[1,a] and LOBOVSKÝ Libor[1,b]

[1]NTIS – New Technologies for the Information Society, Faculty of Applied Sciences, University of West Bohemia, Technická 8, 301 00, Pilsen, Czech Republic

[a]krystek@ntis.zcu.cz, [b]lobo@ntis.zcu.cz

Keywords: Bolted joint, carbon fibre reinforced plastic, experiment, load capacity.

Abstract. Experimental and numerical investigations of the load capacity for the bolted joints were the aims of this work. Influence of a stacking configuration was investigated. Influence of the geometric parameters of bolted joints on its load capacity was analyzed. Digital image correlation method was used for the measurement of displacements. Numerical model of bolted joints was created in finite element system *Abaqus*.

Introduction

Currently, amount of products that utilize composite materials is rapidly increasing. Usually, a whole structure is not replaced with composite materials, but only a certain part of it is replaced. Joints are often the critical parts of constructions. A bolted joint is one possibility of joining of composites with composite or metal parts [1, 2]. The bolted joint is an example of removable joints.

Load capacity of bolted joints depends on many parameters, especially on the types of fibres and resin, and geometric properties of the joint [4]. The type of the failure depends on the geometry of the composite part and on the type of composite (material of constituent, layup, etc.) [1].

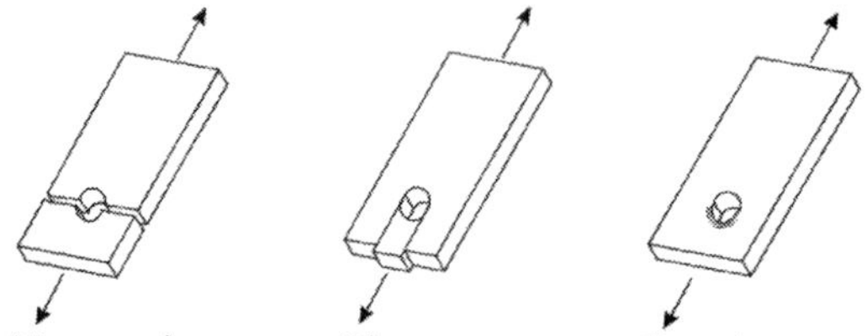

Fig. 1 Typical failure mechanisms of composite bolted joint

The experimental analysis and numerical prediction of load capacity of bolted joints is the aim of this work.

Specimens

The experimental specimens (Fig. 2) were cut using water jet from composite plate. The holes were milled. The composite plate was made from 8 prepreg layers (*HexPly 913C-HTS(12k)*). The composite material consisted of *Tenax HTS 5361* high-strength carbon fibres and epoxy resin. The specimens were cut in two mutually perpendicular directions, two types of stacking configurations were tested in this work ([0/–45/+45/90]s and [90/+45/–45/0]s).

Two types of bolts (M6 and M8) were used. The clamp load was done using a torque wrench *Atlas Copso-tensor ETV ST61-50-10*. Width of specimen was $W = 24$, 32 or 40 mm, distance from the centre of the hole to the free end was $E = 18$, 24 or 30 mm, thickness was $H = 2.4$ mm and length was $L = 50$ mm. Angle φ is orientation of the outside layer of plate relative to the longitudinal direction. Geometric parameters of specimens and tightening torque are presented in Table 1 and Table 2.

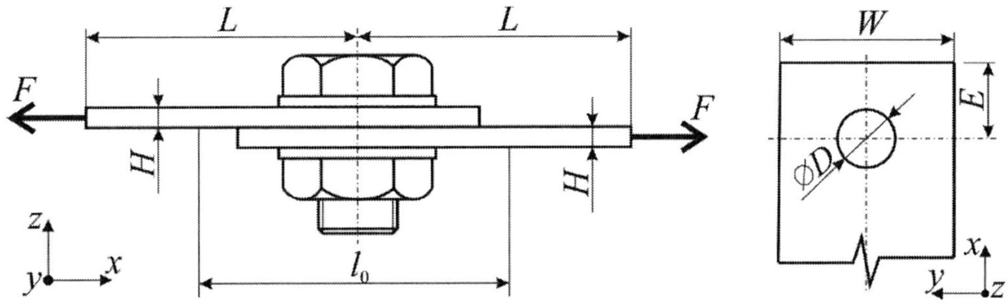

Fig. 2 Geometric parameters of bolted joint

Tab. 1 Geometric parameters of specimens, tightening torque M_t and load capacity for bolt M6

Specimen name	$2E$ [mm]	W [mm]	φ_1 [°]	φ_2 [°]	M_t [Nm]	F_{max} [N]
D6_E18_W24_0_0_1	36.7	23.9	0	0	10.6	8113
D6_E18_W24_0_0_2	36.6	23.8	0	0	10.5	8362
D6_E18_W24_0_0_3	36.4	23.9	0	0	10.6	7876
D6_E18_W24_90_90_1	36.9	23.8	90	90	10.5	7620
D6_E18_W24_90_90_2	36.3	23.9	90	90	10.6	7375
D6_E18_W24_90_90_3	36.0	23.8	90	90	10.6	7016
D6_E18_W32_0_0_1	35.8	32.8	0	0	10.6	8388
D6_E18_W32_0_0_2	36.1	32.0	0	0	10.6	8390
D6_E18_W32_0_0_3	35.9	31.8	0	0	10.7	8405
D6_E18_W32_90_90_1	36.0	31.9	90	90	10.5	8121
D6_E18_W32_90_90_2	35.9	31.9	90	90	10.6	7839

Tab. 2 Geometric parameters of specimens, tightening torque M_t and load capacity for bolt M8

Specimen name	$2E$ [mm]	W [mm]	φ_1 [°]	φ_2 [°]	M_t [Nm]	F_{max} [N]
D8_E24_W24_0_0_1*	48.9	23.8	0	0	20.6	-
D8_E24_W24_0_0_2	48.3	23.9	0	0	20.6	9744
D8_E24_W24_0_0_3	47.8	23.8	0	0	20.7	9552
D8_E24_W24_90_90_1	48.4	23.8	90	90	20.7	9168
D8_E24_W24_90_90_2*	48.2	23.8	90	90	20.6	-
D8_E24_W24_90_90_3	48.7	23.8	90	90	20.7	9631
D8_E30_W24_0_0_1	60.2	23.8	0	0	20.7	9322
D8_E30_W24_0_0_2	59.9	23.9	0	0	20.7	9723
D8_E30_W24_0_0_3	60.2	23.8	0	0	20.7	9422
D8_E30_W24_90_90_1	60.4	23.8	90	90	20.6	9654
D8_E30_W24_90_90_2	60.0	23.8	90	90	20.7	9961
D8_E30_W24_90_90_3	60.0	23.8	90	90	20.7	9084
D8_E30_W40_90_90_1	59.8	39.9	90	90	20.4	8851
D8_E30_W40_90_90_2	60.7	39.8	90	90	20.6	9011
D8_E30_W40_90_90_3	59.8	39.8	90	90	20.7	8728

* specimen slipped in the jaws

Experiments

Specimens were tested in tension in the longitudinal direction. *Zwick/Roell Z050* testing machine was used. The loading speed was $v = 3$ mm/min.

The Digital Image Correlation (DIC) method was used to analyze the plate displacements (x and y directions). A random high-contrast color pattern was applied on the side surface of the specimen. *Dantec Dynamics Q-400* system made a picture during the test. An extensometer was used for measuring the displacement (Fig. 4). The gage length was l_0 = 80 mm (Fig. 2)

The specimen during the test is shown in Fig. 3. Typical force-displacement dependencies are presented in Fig. 4. The dependencies were similar for all specimens. The first significant decrease of the force means a slippage in bolted joint. Slippage occurred about $F = 1500 \div 2000$ N in case of M6 bolt, and about $F = 2000 \div 2500$ N in case of M8 bolt. The next decrease of the force means failure of composite parts. Failure of the specimen (bearing mode) is shown in Fig. 5. This type of failure occurred for all specimens.

Fig. 3 Specimen during the test

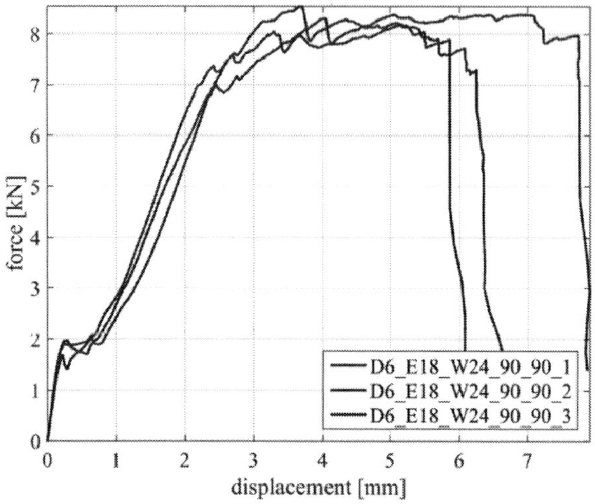

Fig. 4 Typical force-displacement dependencies for bolted joint

Fig. 5 Failure of specimen - bearing

Maximum forces (load capacities) are presented in Table 1 and Table 2. In case of M6 bolt, influence of the stacking configuration on load capacity if the bolted joint was discover. In case of geometrical parameters $E = 18$ mm and $W = 24$ mm, the load capacity for specimens with $\varphi = 0°$ was higher by 10% than for specimens with $\varphi = 90°$. In case of geometrical parameters $E = 18$ mm and $W = 32$ mm, the difference was only 5%. In case of M8 bolt, influence of stacking configuration on load capacity of bolted joints was not determined. Increasing the width increased the load capacity by 9% in case of M6 bolt and $\varphi = 90°$. In case of M8 bolt, the larger width caused a decrease of load capacity by 8% for $\varphi = 90°$.

Numerical simulation

Numerical model was created in the finite element system *Abaqus*. The composite layup solid elements were used for model of the plates. The solid elements with 8 nodes were used for model of bolt, nut and washer. Isotropic material ($E_s = 210$ GPa, $\nu_s = 0.3$) was applied in the steel parts. Transverse isotropic material (Table 3) was set in each composite layer. A friction between components was considered (composite/composite $f_{cc} = 0,12$, composite/metal composite/metal $f_{cm} = 0,12$, metal/metal $f_{mm} = 0,10$).

Tab. 3. Material properties of composite [3]

E_1 [GPa]	E_2 [GPa]	E_3 [GPa]	ν_{12} [-]	ν_{13} [-]	ν_{23} [-]	G_{12} [GPa]	G_{13} [GPa]	G_{23} [GPa]
120.0	8.0	8.0	0.337	0.337	0.33	4.6	4.6	3.0

The numerical simulation was carried out in two steps. First, the clamp load was applied in the bolt. In the second step, the specimen was loaded by means of displacement. The maximum stress criterion was used for prediction of the load capacity of the bolted joints. The strength parameters of the composite are presented in the Table 4. The maximum stress criterion predicted first failure in the composite. But this first failure is not obvious from the force-displacement dependencies. This failure occurred at about 40% of the experimental load capacity.

Tab. 4. Strength parameters of composite [3]

X_T [MPa]	X_C [MPa]	Y_T [MPa]	Y_C [MPa]	Z_T [MPa]	Z_C [MPa]	S_{12} [MPa]	S_{23} [MPa]	S_{13} [MPa]
1800	850	55	213	55	213	82	82	82

A comparison of the displacements from the experiment and numerical simulation is presented in the Fig. 6. Displacement in experiment was measured by means of DIC method.

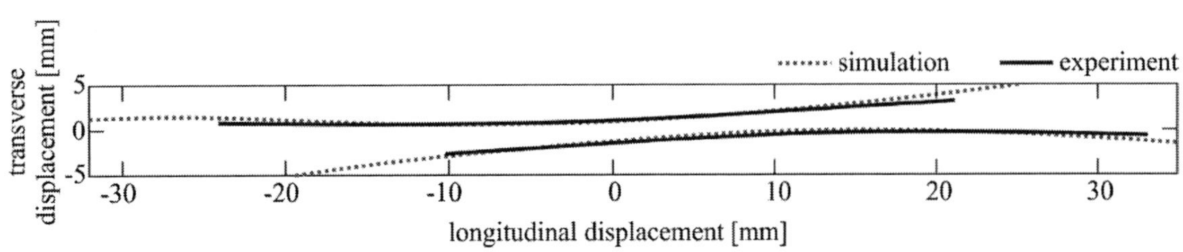

Fig. 6 Comparison of the displacements

Conclusions

Two plates were connected by bolted joint. The load clamp was applied in the bolt. Experimental analysis of the load capacity of the bolted joints was done. Displacements in x and y directions were measured by means of DIC method. The numerical model of this bolted joints was created in finite element system *Abaqus*. The maximum stress criterion was used for load capacity prediction.

Influence of stacking configuration on load capacity was determined in case of M6 bolt. This influence was not determined in case of M8 bolt. Increasing the width increased the load capacity in case of M6 bolt with outer layer having fibre orientation 90°.

Acknowledgement

This publication was supported by the project LO1506 of the Czech Ministry of Education, Youth and Sports. The authors would like to thank Robert Zemčík for helpful comments and proofreading of the manuscript. The authors would like to thank Šárka Veselá for helpful with experiments and numerical simulations.

References

[1] A. Atas, C. Soutis, Subcritical damage mechanisms of bolted joints in CFRP composite laminates, Composites: Part B, 54 (2013) 20-27.

[2] J. I. Choi, S. M. Hasheminia, H. J. Chun, J. C. Park, H. S. Chang, Failure load prediction of composite bolted joint with clamping force, Composite Structures, 189 (2018) 247-255.

[3] J. Krystek, T. Kroupa, R. Kottner, Identification of mechanical properties from tensile and compression tests of unidirectional carbon composite, Proceedings: 48th International Scientific Conference: Experimental Stress Analysis 2010, pp. 193-200.

[4] P. Lopez-Cruza, J. Laliberté, L. Lessard, Investigation of bolted/bonded composite joint behaviour using design, Composite Structures, 170 (2017) 192-201.

EAN 2018
56th conference on experimental stress analysis

Dynamic Analysis of 3D Printed and Moulded Composites with Carbon Microfibers

KULHAVÝ Petr[1,a], FLIEGEL Vítezslav[1,b], LEPŠÍK Petr[1,c] and SRB Pavel[1,d]

[1] Technical University of Liberec, Faculty of Mechanical Engineering, Studentska 2, 461 17, Liberec 1, Czech Republic

[a] Petr.Kulhavy@tul.cz, [b] Vitezslav.Fliegel@tul.cz, [c] Petr.Lepsik@tul.cz, [d] Pavel.Srb@tul.cz

Keywords: reinforced polymers, carbon fibres, DMA, 3D printing, injection moulding.

Abstract. Using of fused deposition modelling rapidly expands due to the still decreasing costs of the machines and main materials. However, the main disadvantage is still the diametrically lower strength of the printed plastic parts, especially in the direction normal to the laid plies. Therefore, the new trend in recent years was to improve the material properties by adding fibre reinforcement. In this work, two kind of samples made from the same base material (carbon reinforced ABS) but using different manufacturing technology, one made by injection moulding and second by the 3D printing have been compared. Because for real condition the basic static test is not sufficient enough, the samples were tested on the dynamic mechanical analyzer (DMA). The chosen method was 3point bending, with constant strain and various frequency and temperature. In this study, the storage and loss modulus, of the plastic samples were determined.

Introduction

3D printing is a type of additive manufacturing that can be used to fast creation of usually prototype components with highly customizable geometries. 3D printing is rapidly attracting interest the scientific, manufacturing and even public community, because of its potential for enabling low-cost, highly flexible, low- and mid-scale manufacturing [1].
A lot of worldwide companies has been producing 3D printers and printable materials enhanced for printing carbon fiber-reinforced parts for customers in almost unlimited industries [2]. A very advantageous technology for prototypes is the printing of polymers. However, the strength of this way created parts is not equivalent to that by the conventional e.g. the injection moulding processes. In the work of Nakagawa [2] and Whitwam [3] could be read about the tensile and bending strength of printed ABS samples with and without fiber reinforcement and it is stated that the strength could be significantly increased by being reinforced with carbon fibres. About dependency of the mechanical parameters on the conditions such as layer thickness, orientation, raster angle, raster width, and air gap on the tensile, flexural, and impact strengths could be read in the work of Sood [4].

Shofner [5] compounded cut nanofibres into an ABS and increased the strength by about 20%. Tensile testing of the materials shows that a material with aligned fibers have higher elastic modulus and ultimate tensile stress than with the randomly oriented fibres. Really interesting research in this field conducted Nanya [6], she studied the simultaneous printing of continuous fibre with the polymer, mutually connected in the printing head. She tested the mechanical strength and thermodynamic properties of this way prepared parts with using dynamic mechanical analyzer. Mori [7] mentioned next advantageous of the additional improving of mutual fibre/resin connection by heating – thermal bonding. He also mentioned that it is desirable to develop approaches for increasing the volume ratio of carbon fibres and the bonding force between the carbon fibres and plastic and for controlling the orientation of carbon fibres. According to Whitwam [3] the printing of thermoplastic reinforced by fibres has the potential to replace the basic alloys in industrial applications.

Materials

The 3D printing of plastic materials is relatively easy, but the resulting parts do not have sufficient durability for long-term use. Parts from the same material created by moulding injection could exhibit up to 20 times more durability and 10 times more rigidity compared to standard printed parts made from ABS. Even if the price of 3D printers is steadily declining, the possibilities of their use do not differ much from the time when their price ranged in the order of thousands of dollars. Finally, nowadays comes a solution of carbon-fiber reinforced nylon that could compete to carbon-alloys or some cheaper steels.

When replacing conventional materials by composites it is important to know the mechanical properties of the composite [8]. Whereas metal materials show one failure mode i.e. cracking, composites can exhibit one or a combination of failure modes, including fiber rupture, matrix cracking, delamination, interface debonding and void growth [9].

Thermoplastic materials, which could be repeatedly melted and formed into their shapes are nowadays one of the most commonly used materials in 3D printing. As has been already mentioned the fiber-reinforced plastics (FRP) have recently gained attention for their potential use in 3D printing to enhance the mechanical strength and elasticity of manufactured parts. Spackman [10] and Mallick [11] aimed their study to systematically studying the effect of key processing parameters e.g. fiber alignment, area coverage, and surface energy of the fiber carrier substrate, on the tensile properties and failure mechanisms is heavily influenced by the types of plastic resin selected, as well as the length and loading of the carbon fibers seen in 3D and even 4D printed polymers. In the presented work, the samples made of ABS reinforced with carbon micro fibres randomly distributed in the matrix has been used (Fig. 1).

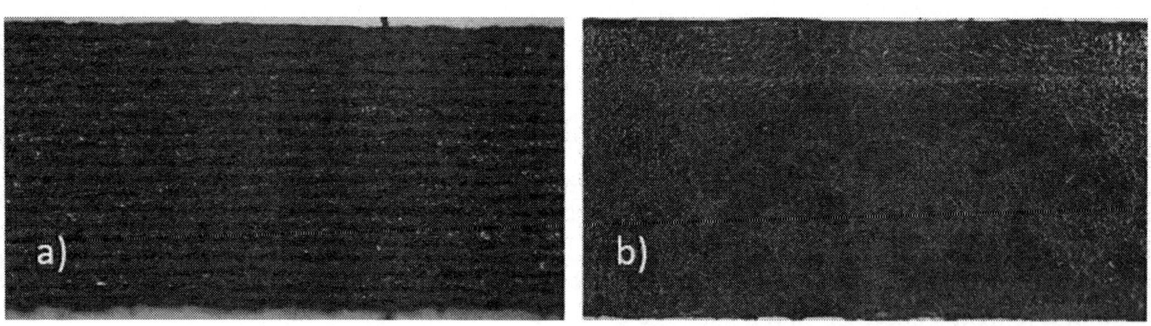

Fig.1 The base reinforced material a) printed b) moulded

Experiment

The aim of the carried experiment was using the dynamic mechanical analyzer (DMA) compare elastic behavior in dynamic loading, during simultaneous external thermal loading. The used test was the conventional 3-points bending (Fig. 2), that is also with the 4-points bend probably the most appropriate method useable for testing layered composites work [8, 9].

The created experiment consist in the various frequency and constant strain amplitude. Because, the plastic materials are really sensitive to thermal changes, simultaneously with the experiment, the temperature was also changed. A frequency sweep may be performed to evaluate the viscoelastic response of a material by the observed changes in the storage and loss moduli and tan δ. The loss tangent is the ratio of loss modulus and storage modulus, which indicate the viscosity and elastic properties of material, respectively.

In the DMA experimental device, the amplitude is defined as the maximum amount of oscillatory motion in micrometers (one half peak-to-peak). According to [6] [12] for composite samples the amplitude of 10 to 50 μm should yield good results and it is recommended to use "Logarithmically spaced" frequencies (*e.g.*, 10, 5, 2, 1) rather than "linearly spaced" frequencies (*e.g.*, 9, 7, 5, 3, 1).

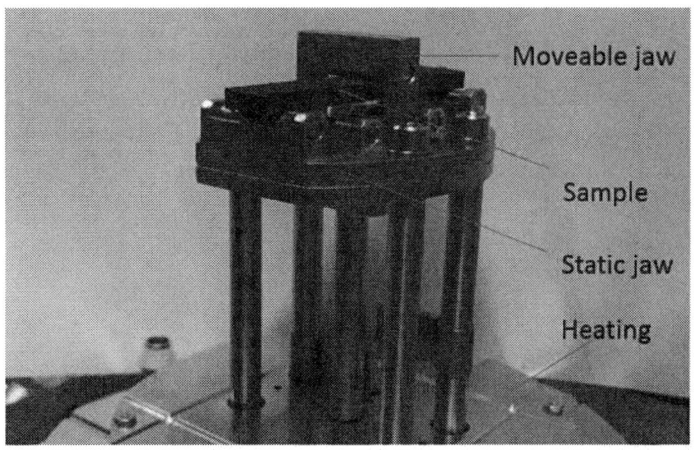

Fig.2 The experimental DMA device

For testing highly elastic material like the plastic one, the force track should be used to circumvent the deformation arising due to the thermal loading. The dynamic force is that which is required to maintain the programmed amplitude at a certain frequency [12]. When force track is activated, the static force is no longer constant, but is dependent on the dynamic force. The ratio of static to dynamic force is essentially for the force track. The following formula (1) is used to maintain the ratio between the preload force and dynamic force (force to drive amplitude).

$$F_S = F_T \times K \times A \qquad (1)$$

Where the F_S is the static force [N], F_T the force track [-], the K stiffness [N/m] and A the amplitude [m].

Before the main testing, it is appropriate to set the value of the geometry factor. For the 3-point bending clamps with rectangular samples it is possible to determine the optimal sample size according to (2):

$$GF = \frac{L^3}{48\,I}\left[1 + \frac{12}{5}(1+\mu)\left(\frac{t}{L}\right)^2\right] \qquad (2)$$

Where according the Tab. 1 the individual parameters of the tested samples are:

Tab.1 Parameters of the tested samples

Parameter:	*Mould*	*Printed*
L = sample length [mm]	50,000	50,000
I = geometric moment [mm^4]	52,267	53,867
t = sample thickness [mm]	4,000	4,000
h = sample width [mm]	9,800	10,100
ν = Poisson's ratio [-]	0,350	0,380
GF [1/ mm]	49,928	48,447

Results

In the pictures below are the result of dynamic micro 3 point bending test. It is possible to see the comparison of the printed (Fig. 3) and moulded parts (Fig. 4) in dependency on the loading frequency in the range 1 – 10 Hz.

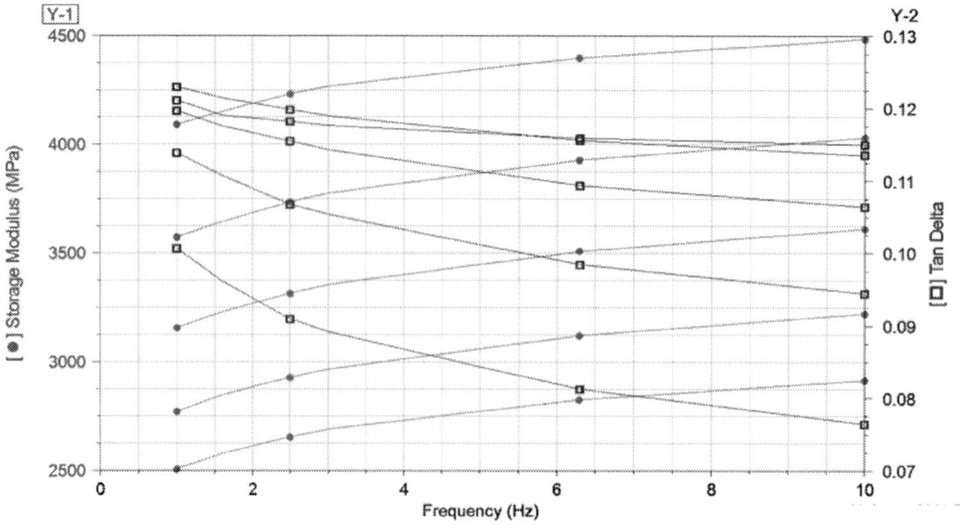

Fig.3 Experimental result of the modulus in dependency on frequency of the printed part

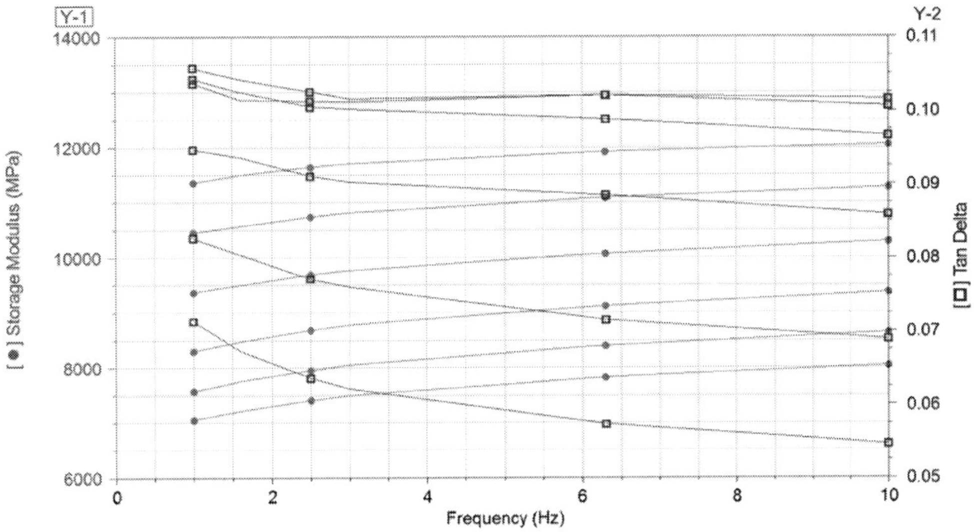

Fig.4 Experimental result of the modulus in dependency on frequency of the moulded part

Because the mechanical properties of plastic materials are really sensitive to the temperatures even within the values near their usual working conditions, it was necessary to asses also the Modulus dependency on the temperature in the range 30 – 50 °C. In the Fig. 5 there are results measured on the printed parts and in the Fig. 6, from the moulded parts.

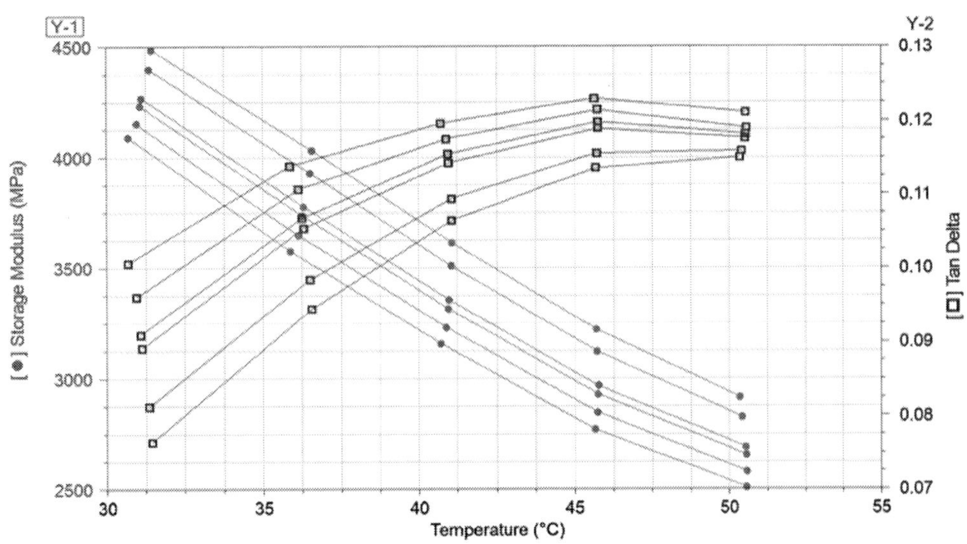

Fig.5 Experimental result of the modulus in dependency on temperature of the printed part

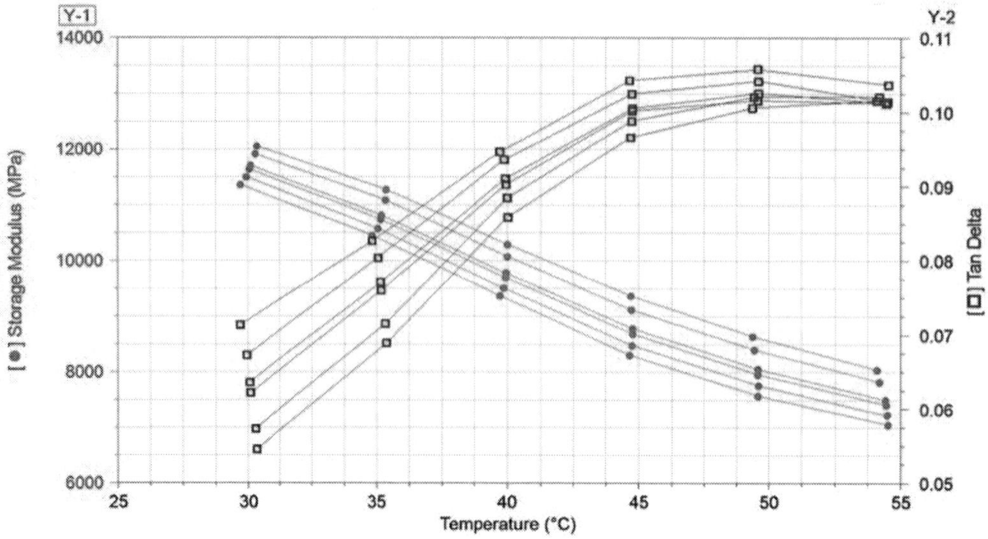

Fig.6 Experimental result of the modulus in dependency on temperature of the moulded part

Conclusions

It has been found that the resultant modulus of elasticity of the printed part reaches approximately one third of the value achieved on the injected and moulded part. Even if this value is still small, it is possible to mention significantly higher values of the material strength compared to conventional printed materials, those have the strength usually many times lower compared to the moulded parts.

Further, for the printed part, even if the increase in E is noticeable with increasing frequency of the loading, from the point of view to the thermal degradation of the material, there is no significant difference between the printed and the injected material.

As the future aim of our work it could be advantageous to study, whether the optimal layout of the printed part (angles, stacking sequence of individual plies etc.), could be optimized and computed according to the classical laminate theory.

Acknowledgement

This publication was written at the Technical University of Liberec as part of the project "Innovation of technical systems structures with the use of composite materials" with the support of the Specific University Research Grant, as provided by the Ministry of Education, Youth and Sports of the Czech Republic in the year 2018.

References

[1] B. Zhenyu, 3D Printing of Carbon Fiber-Reinforced Composites, Sigma - Aldrich, technical documentation, Milwaukee, 2017.

[2] Y. Nakagawa, K . Mori, T. Maeno, 3D printing of carbon fibre-reinforced plastic parts, The Int. Journal of Adv. Man. Technology, Vol. 91, (2017) 2811–2817.

[3] R. Whitwam, New 3D Printers Can Print Materials as Strong as Steel, Extremetech, 2017.

[4] A. K. Sood, R.K. Ohdar, S. Mahapatra, Parametric appraisal of mechanical property of fused deposition modelling processed parts. Mater Des 31(1):287, (2010).

[5] M. L. Shofner, K. Lozano, F.J. Rodriguez-Macias, E.V. Barrera, Nanofiber-reinforced polymers prepared by fused deposition modeling, J. Appl Polym Sci 89(11):3081, (2003).

[6] N. Li, Y. Li, S. Liu, Rapid prototyping of continuous carbon fiber reinforced polylacticacid composites by 3D printing, Journal of Materials Processing Technology 238 (2016) 218–225.

[7] K. Mori, T. Maeno, Y. Nakagawa, Dieless forming of carbon fibre reinforced plastic parts using 3D printer, ICTP 2014, Procedia Engineering 81, (2014) 1595 – 1600.

[8] G. Gruber, S. Wartzack, Three-point bending analyses of short fiber reinforced thermoplastic: Comparison between simulation and test results. SAS Tech Journal, (2013).

[9] P. Kulhavy, P. Lepsik, Study of bending properties of thin walled carbon and flax composite rods winded of prepreg fibers, EAN 2017, Novy Smokovec, Slovakia, (2017).

[10] C. C. Spackman, Ch. R. Frank, K. C. Picha, J. Samuel, 3D printing of fiber-reinforced soft composites: Process study and material characterization, Journal of Manufacturing Processes, 23 (2016) 296–305.

[11] P. K. Mallick, Fiber-reinforced composites: materials, manufacturing, and design; CRC press: London, 2007.

[12] QDMA Manual, TA Instruments - Waters LLC, New Castle, DE 19720, 2006.

Computation of Carbon Nanotubes Eigenfrequencies Based on Nonlocal Theory and Using Finite Element Method

LENGVARSKÝ Pavol[1,a] and BOCKO Jozef[1,b]

[1]Department of Applied Mechanics and Mechanical Engineering, Faculty of Mechanical Engineering, Technical University of Košice, Letná 9, 042 00 Košice, Slovakia

[a]pavol.lengvarsky@tuke.sk, [b]jozef.bocko@tuke.sk

Keywords: carbon nanotube, eigenfrequency, nonlocal parameter, chiral angle

Abstract. In this paper the eigenfrequencies of carbon nanotubes are investigated. Especially, effect of chirality of the carbon nanotubes is studied. The effect of the change of chiral angle is investigated for the carbon nanotubes with the same diameter, length and two different types of boundary conditions. The nonlocal continuum theory results to application of nonlocal parameter which has influence to eigenfrequencies of carbon nanotubes. All computations in the paper are made by finite element method, where standard mass matrix of finite element is corrected and updated by nonlocal parameter.

Introduction

Carbon nanotubes (CNTs) are closed graphene sheets of cylindrical shape with interesting mechanical properties. Especially, high tensile strength and stiffness in relation to very small weight attracts attention of engineers and researches. Three basic methods of nanotubes investigation are used to accomplish simulation of SWCNTs: molecular dynamic computations, atomistic-based modelling and the continuum approach. The last one is used here for description of bending vibration and accordingly it can serve as a mean tool for indirect specification of carbon nanotube properties. The continuum beam relations used in the paper are based on the theory of nonlocal elasticity published by Eringen [1]. The equations are solved by the finite element method.

In this paper the finite element modal analysis of the CNTs is performed. The effect of chirality, dimensions and nonlocal parameter on the eigenfrequencies of CNTs is investigated. The matrix formulations of the stiffness matrix and mass matrix of finite element is used for the modelling and computations of CNTs. The effect of nonlocal parameter is represented by the coefficient used in the mass matrix.

Finite element modelling of carbon nanotubes

In contrast to the classical elasticity theory Eringen takes into account the scale effect and its influence to stress at the reference point of a body. Accordingly, the stress at a point is a function of strain field at every point of body. Review of nonlocal elastic models of CNTs is given in [1]. The interesting results of CNT research can be found in literature [1]-[6]. The nonlocal finite element formulation is presented for computation of eigenfrequencies of CNTs. The mass and stiffness matrices of finite element are derived from equation

$$m_b(x)\mu\frac{\partial^2 u}{\partial t^2}+\frac{\partial^2}{\partial x^2}\left[EJ(x)\frac{\partial^2 u}{\partial x^2}\right]=0. \tag{1}$$

Here, u is the transverse displacement of the beam, m_b is the mass of beam related to the unit length, μ is nonlocal parameter, E is Young modulus, J is the second moment of cross-section area, x is a coordinate of point in beam nanotube, t is time.

As there is a problem in definition of thickness of the nanotubes, the values necessary for computations were determined according to [4] as

$$m = \rho A = 2.4 \times 10^{-24} D \quad [\text{kg/nm}], \tag{2}$$

$$EI = 428.48\,D^2 - 397.08\,D + 109.24 \quad \lfloor \text{kg}\,\text{nm}^3/\text{s}^2 \rfloor, \tag{3}$$

where D is a diameter of nanotube. The nanotube diameter can be computed from equation

$$D = 2R = a_0\sqrt{3(m^2 + n^2 + mn)} / \pi, \tag{4}$$

where $a_0 = 1.42$ Å is a carbon–carbon bond length and the integers m and n are called as chiral indices. There are known three configuration types of chiral indices called as armchair (m, m), chiral (m, n) and zigzag (m, 0). Here, for armchair nanotube we have $m = n$.

For the modelling of CNTs, we have used plane beam element with four degrees of freedom, length l_e and mass m_b. The basic FEM matrices are formulated by standard matrix of shape functions $\mathbf{N}(x)$. Then, the element stiffness and mass matrices can be defined respectively by equations

$$\mathbf{K}_e = EJ\int_0^{l_e} \frac{d^2\mathbf{N}(x)}{dx^2}\frac{d^2\mathbf{N}^T(x)}{dx^2}dx = \frac{EJ}{l_e^3}\begin{bmatrix} 12 & 6l_e & -12 & 6l_e \\ 6l_e & 4l_e^2 & -6l_e & 2l_e^2 \\ -12 & -6l_e & -12 & -6l_e \\ 6l_e & 2l_e^2 & -6l_e & 4l_e^2 \end{bmatrix}, \tag{5}$$

$$\mathbf{M}_e = m_b\mu\int_0^{l_e} \frac{d^2\mathbf{N}(x)}{dx^2}\frac{d^2\mathbf{N}^T(x)}{dx^2}dx = \frac{m_b\mu l_e}{420}\begin{bmatrix} 156 & 22l_e & 54 & -13l_e \\ 22l_e & 4l_e^2 & 13l_e & -3l_e^2 \\ 54 & 13l_e & 156 & -22l_e \\ -13l_e & -3l_e^2 & -22l_e & 4l_e^2 \end{bmatrix}. \tag{6}$$

At first, the effect of chirality to eigenfrequencies of CNTs is investigated. The computations of the first two eigenfreqeuncies are accomplished for different types of CNTs but with approximately the same diameters. The length of tubes is 10 nm and the following types of boundary conditions are used: fixed–free (F–Fr) and simply–simply supported (S–S). From computations (Table 1) we see that the chirality has effect to the results mainly due to small change of tube diameter. Under otherwise identical conditions, the effect of chirality of SWCNTs is negligible. For this reason, only armchair CNTs with chirality (m, m) are investigated in this paper.

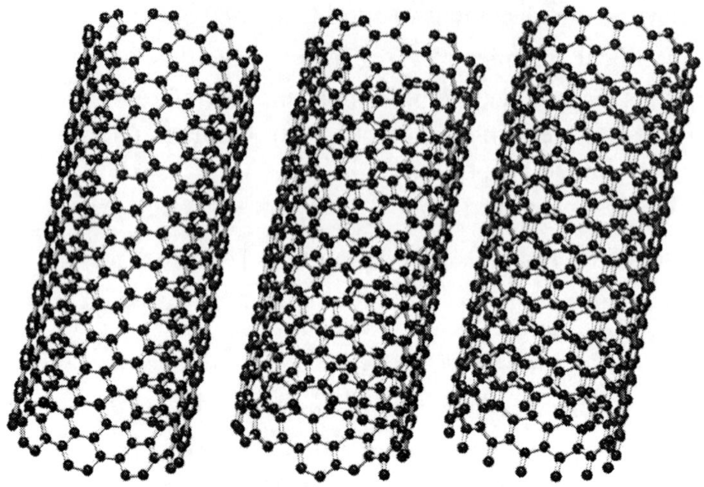

Fig.1 Types of carbon nanotubes

Tab.1 Eigenfrequencies of CNTs with different chiral angles and boundary condition

SWCNT (n, m)	Chiral angle (°)	Diameter (nm)	Frequency (THz) F-Fr (1st)	F-Fr (2nd)	S-S (1st)	S-S (2nd)
(9, 0)	0.00	0.7047	0.02795	0.17515	0.07845	0.31381
(8, 2)	10.89	0.7177	0.02858	0.17912	0.08023	0.32092
(7, 3)	17.00	0.6960	0.02753	0.17249	0.07726	0.30905
(6, 4)	23.41	0.6826	0.02689	0.16852	0.07548	0.30192
(5, 5)	30.00	0.6781	0.02668	0.16719	0.07489	0.29955

The four CNTs with the same chirality (*m*, *m*) but different diameters are modelled (Table 2). These CNTs are modelled with lengths varying from 10 nm to 100 nm and the finite element modal analysis is performed on these CNTs. Due to big amount of data, the results are given only for the first and the fifth eigenfrequency of CNTs with fixed-free boundary conditions and they are presented in Figs. 2-3. From the figures can be seen that the bigger diameter of the CNT leads to the higher eigenfrequency and the longer CNT leads to the smaller eigenfrequency.

Tab.2 Modelled carbon nanotubes

SWCNT (m, m)	Chiral angle (°)	Diameter (nm)
(5, 5)	30.00	0.6785
(10, 10)	30.00	1.357
(20, 20)	30.00	2.714
(40, 40)	30.00	5.428

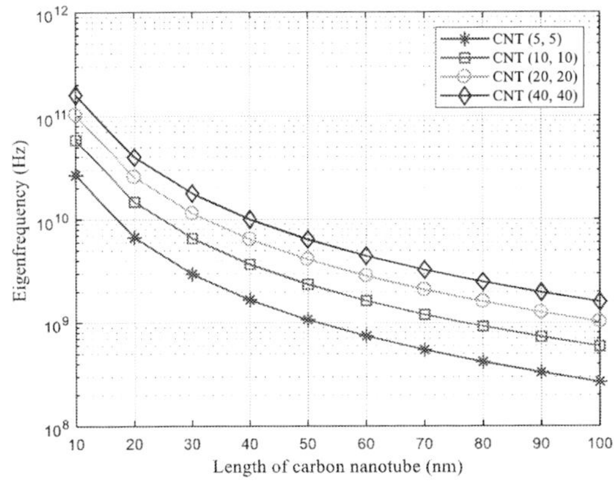

Fig. 2 The first eigenfrequencies of carbon nanotubes

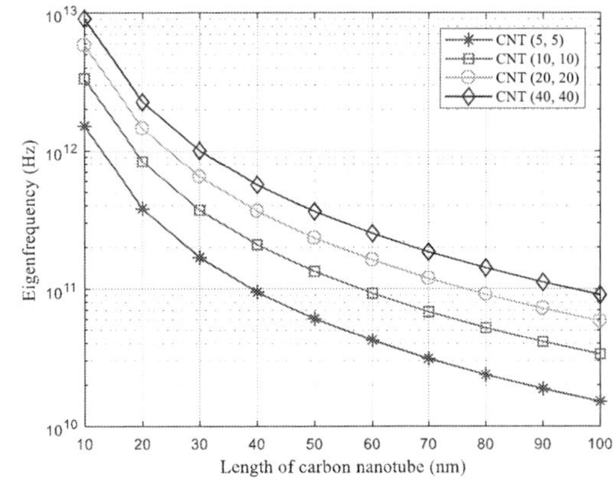

Fig. 3 The fifth eigenfrequencies of carbon nanotubes

Next, the effect of nonlocal parameter to the eigenfrequencies of CNTs is studied. The computations are accomplished for nonlocal parameters 0.01, 0.1, 0.2, 0.4, 1.2. The CNTs with the chirality (10, 10), the diameter D = 1.357 nm and the lengths from 10 to 100 nm are modelled. In the Figs. 4 – 7 are shown the first four eigenfrequencies with and without nonlocal parameter and for the fixed-free boundary conditions. From the figures can be seen that the nonlocal parameter significantly affects the eigenfrequencies of the CNTs. The nonlocal parameter smaller than one leads to higher eigenfrequencies as the eigenfrequencies computed without nonlocal parameter and the nonlocal parameter higher than one leads to smaller eigenfrequencies as the ones computed without nonlocal parameters. The nonlocal parameters smaller than one have more pronounced effect on the value of the eigenfrequencies.

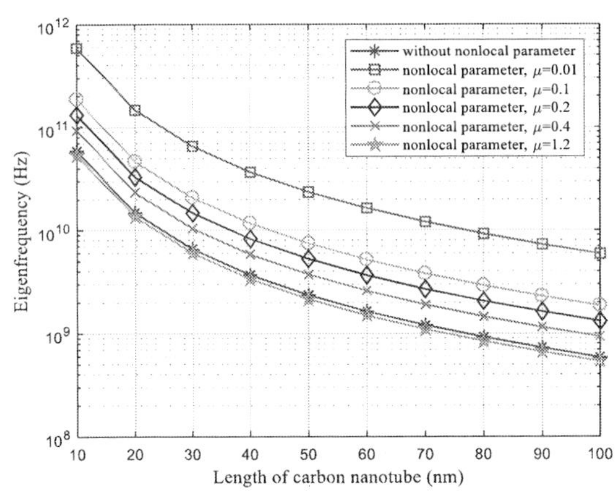

Fig. 4 The first eigenfrequencies of carbon nanotubes with and without nonlocal parameter

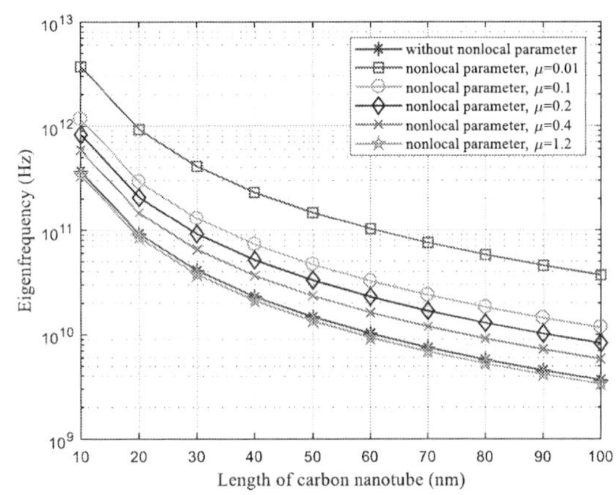

Fig. 5 The second eigenfrequencies of carbon nanotubes with and without nonlocal parameter

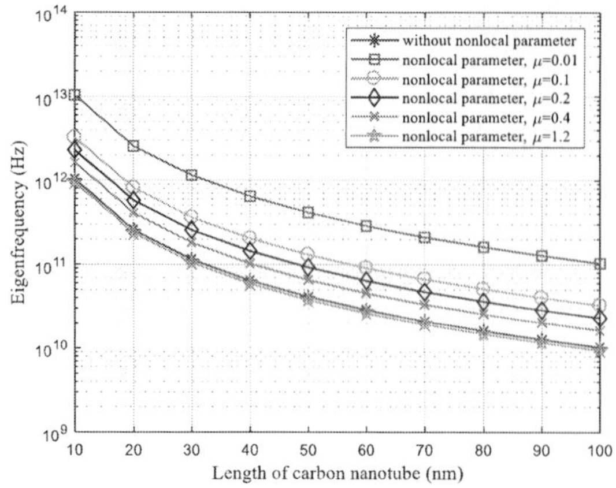

Fig. 6 The third eigenfrequencies of carbon nanotubes with and without nonlocal parameter

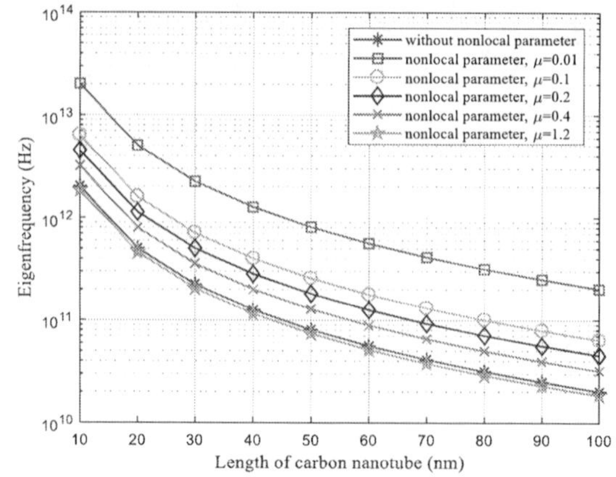

Fig. 7 The fourth eigenfrequencies of carbon nanotubes with and without nonlocal parameter

Conclusions

In the paper are described numerical computations of eigenfrequencies of carbon nanotubes accomplished for different nonlocal parameters, chirality, lengths and boundary conditions. At first, the influence of chiral angle on the values of eigenfrequencies was studied. From computations result that the eigenfrequencies of CNTs with the same diameters and lengths were not affected by the chiral angle. Next, the effect of nonlocal parameter was studied. From the results can be stated that the value of nonlocal parameter affected the eigenfrequencies. In general, the smaller value of the nonlocal parameter, the higher eigenfrequency of nanotube. Computation of eigenfrequencies of carbon nanotubes can serve as a very important tool for determination of basic material properties of homogenized nanotubes, because it is much easier to measure frequencies of nanotubes as e.g. their Young modulus, or other properties. Accordingly, the eigenfrequencies serve for indirect determination of material and geometric properties of nanotubes.

Acknowledgement

The authors appreciate financial support by grant VEGA no. 1/0731/16 - Development of Modern Numerical and Experimental Methods of Mechanical System Analysis.

References

[1] A.C. Eringen, Nonlocal Continuum Field Theories, Springer-Verlag, 2002.

[2] P. Lu, H.P. Lee, C. Lu and P.Q. Zhang, Application of nonlocal beam model for carbon nanotubes, Int. J. Solid. Struct. 44 (2007) 5289–5300.

[3] J. N. Reddy and S.D. Pang, Nonlocal continuum theories of beams for the analysis of carbon nanotubes, J. Appl. Phys. 103 (2008) 023511-1-023511-16.

[4] A. Swain, T. Roy and B.K. Nanda, Vibration behaviour of single walled carbon nanotube using finite element, Int. J. Theor. Appl. Res. Mech. Eng. 2 (2013) 129-133.

[5] C. Thongyothee, S. Chucheepsakul and T. Li, Nonlocal elasticity theory for free vibration of single-walled carbon nanotubes, Adv. Mater. Res. 747 (2013) 257-260.

[6] Q. Wang and K.M. Liew, Application of nonlocal continuum mechanics to static analysis of micro- and nano-structures, Phys. Lett. A. 363 (2007) 236–242.

EAN 2018
56th conference on
experimental stress analysis

Use of Notches in the Shafts Loaded by Torque for Improvement in Accuracy of Strain Gauge Measurements

LOPOT František[1, a], DUB Martin[1,b], JANDA Martin[1,c] and BERKA Ondřej[1]

[1]Czech Technical university, Faculty of mechanical engineering, Technicka 4, 166 07, Praha, Czech Republic

[a]frantisek.lopot@fs.cvut.cz, [b]martin.dub@fs.cvut.cz, [c]martin.janda@fs.cvut.cz

Keywords: notch, strain gauge, stress, torque, FEM, measurement, gearbox efficiency

Abstract. One of the decisive factors for the sensitivity of the strain gauge instrumentation is the sensitivity of used strain gauges. For foil strain gauges, we talk about 1 μm / m [1]. This value is reached for steel components at a stress of about 2 MPa. It can be understood as a limiting one in terms of reliable detection. Since the ideal value of strain in the practical measurements for commonly used foil strain gauges should be about +/- 0.15 % for the wire versions and circa 0.1 % for the semiconductor versions, we talk about 20 - 25 MPa on the surface of measured steel component when the final error can theoretically decrease below 1 %. However, many engineering applications are out of this range and thus the uncertainty of the results is significantly growing. Especially in direct detection of torque on the gearboxes shafts in order to determine their mechanical efficiency, this is crucial. Generally, as well as in the case of the mentioned gearboxes, a certain possibility of solving the above mentioned problem is offered by constructional and technological notches in which the mechanical stress is locally concentrated. There are many various approaches available to estimate the rate of increase in local tension against nominal values on smooth surface of stable cross-sections, unfortunately without determining of the location of calculated extreme value. This is why we work with finite element method that eliminates mentioned drawbacks and allows us to reliably apply the direct strain gauge measurement in applications with extreme demands on accuracy of results as the mentioned efficiency of gearboxes is.

Introduction

One of the decisive factors for the sensitivity of the strain gauge instrumentation is the sensitivity of used strain gauges. For foil strain gauges, we talk about 1 μm / m [1]. This value is reached for steel components at a stress of about 2 MPa. It can be understood as a limiting one in terms of reliable detection. Since the ideal value of strain in the practical measurements for commonly used foil strain gauges should be about +/- 0.15 % for the wire versions and circa 0.1 % for the semiconductor versions, we talk about 20 - 25 MPa on the surface of measured steel component when the final error can theoretically decrease below 1 %. However, many engineering applications are out of this range and thus the uncertainty of the results is significantly growing. Especially in direct detection of torque on the gearboxes shafts in order to determine their mechanical efficiency, this is crucial. It is, therefore, desirable to detect the transmitted loads with the highest possible accuracy. Generally, as well as in the case of the mentioned gearboxes, a certain possibility of solving the above mentioned problem is offered by constructional and technological notches in which the mechanical stress is locally concentrated. In literature [e.g. 2, 3], there are many various approaches available to estimate the rate of increase in local tension against nominal values on smooth surface of stable cross-sections, unfortunately without determining of the location of calculated extreme value. This is why we work with finite element method that eliminates mentioned drawbacks and allows us to reliably apply the direct strain gauge measurement in applications with extreme demands on accuracy of results as the mentioned efficiency of gearboxes is.

Motivation

In frame of preparatory conduct in a suit between a supplier and its client on reconditioning of his manufacturing line, evaluation of the true mechanical efficiency of the line driving unit became necessary. The reason for that requirement were repeated problems with motors overloading after replacement of the gearbox done during the partial manufacturing line reconditioning. The first stage of the gearbox is realized by curved bevel gearing. The second stage is planetary with three planets and fixed crown wheel with straight teeth. Power from the first stage is brought to the sun wheel. With regard to the closed setup of the line and practical impossibility to shut down the line for a longer time (already interval of a few hours was unacceptable), it was not possible to use a conventional torque sensor and direct measurement of the gearbox shaft used instead.

Measurement was performed on smooth surfaces of the in- and outgoing shafts. The finite elements method was used here merely to check that vicinity of notch and clutch hub placement either would not affect the measurement. Nevertheless, results of the FEM computations suggested that by exploitation of stress concentration in notches, measurement accuracy could be effectively increased. The presented work deals with this idea and proposes a way how to work with the FEM computation to prevent its becoming a source of uncertainty in processing of measured data and its evaluation.

Method

For the given task, an experimental shaft was made of standard construction steel (ČSN: 11523/EN: S355J2/W.Nr: 1.0579) according to Fig. 1.

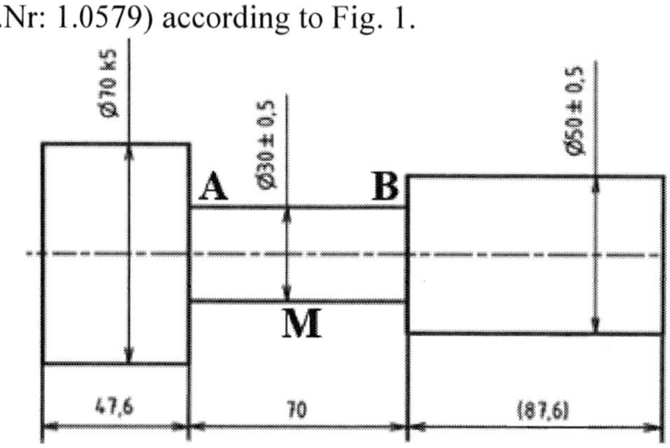

Fig. 1 Drawing of the test shaft (cut-out from the drawing)

The given shape was selected to correspond to notch on the shafts of the above described gearbox and accounted also for chucking jaws of the lab stand. Diameters of the tested shafts are four-times smaller than those of the measured gearbox. Surfaces were turned with high accuracy with roughness not exceeding Ra1,6.

Fig. 2 shows arrangement of the measurement station. The torque was induced by placing a load on the revolving segment arm of the test stand.

Fig. 2 Measurement work station setup

Standard foil strain gauges 1-XY21-1,5/120 (HBM) connected in complete Wheatstone bridge were used to avoid signal distortion by a possible effect of flexion (bending). Measurement instrumentation was set using components made by National Instruments and Dewetron connected to a standard PC. At that stage, the aim of the measurement was to verify the outcomes of the FEM calculations.

Two computational finite element models were prepared and used. Both FE-models have the same geometry corresponding to the experimental shaft and use a linear elastic isotropic homogeneous material. The calculation conditions correspond to a static linear task. The first 3D model works with A 20-node quadratic brick elements, the second 2D model uses An 8-node generalized biquadratic axisymmetric quadrilateral elements [4]. Both models are based on a method of reduced integration. Boundary conditions are defined using coupling constraints, when one end of the model is clamped and the other is loaded by torque. Fig. 3 enables quick comparative visualization of both models.

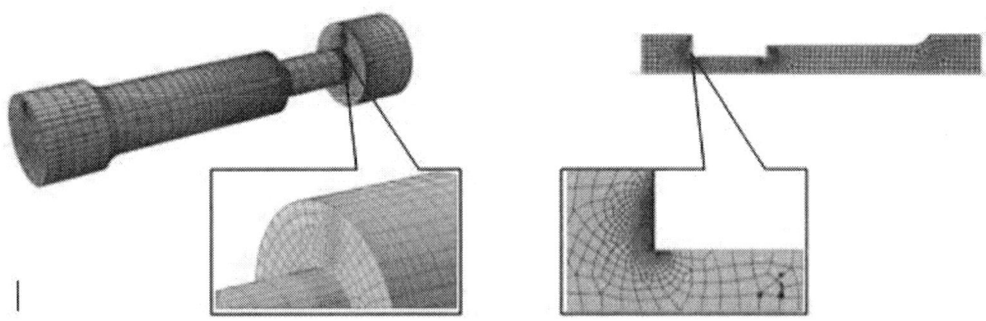

Fig. 3 3D (left) and 2D (right) model of the shaft

The 3D model was created and used mainly as a golden FEM standard for a 2D model as it corresponds very well to a real shaft without significant simplifications. However, because tuning of the 3D model has high demands on computational power and time and because the subject of investigation are shafts loaded by a torque detected by a possible bending effects eliminating instrumentation, only the 2D model was used after verification further.

The following procedure documents how to identify that the mesh and other settings correspond to the investigated situation and that the results obtained are reliable. Fig. 4 displays course of the stress caused by a given torque in dependence on mesh parameters for a site of 30 mm in diameter located between notches (see also Fig. 1).

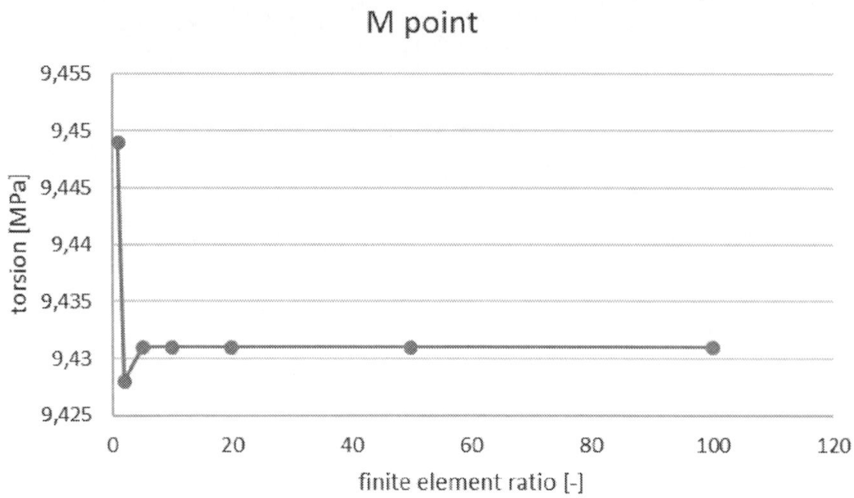

Fig. 4 Cylindrical section of 30 mm in diameter: stress vs. the element size
(finite element ratio = size of the biggest element / actual size of the local element)

From the course it is apparent that value stabilization is achieved already with the first mesh refinement.

Fig. 5. Displays the dependence for the notch A, left (according to Fig. 1), i.e. at shaft shoulder from 70 to 30 mm. Values for the plot are taken always from the same site of the model, once a fixed interface – user node - with regard to the initial geometry has been created. This user node is at a distance of 0,5 mm from the front plane of the shaft shoulder and the value corresponds to localization of the active area center of the glued-on strain gauge.

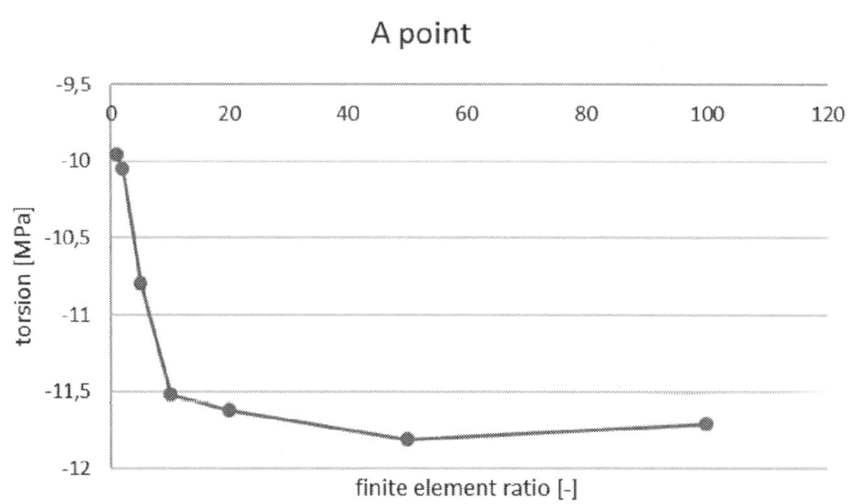

Fig. 5 Notch A, shaft shoulder 70 per 30 mm: stress vs. the element size
(finite element ratio = size of the biggest element / actual size of the local element)

Value stabilization is more gradual here – its dependence on the element size is therefore more significant than in case of a smooth cylindrical element. This effect can be expected. However, of note is "oscillation" of results around the equilibrium in dependence on the element size.

The same effect can be seen also in notch B on the right side (according to Fig.1), as is depicted in Fig. 6.

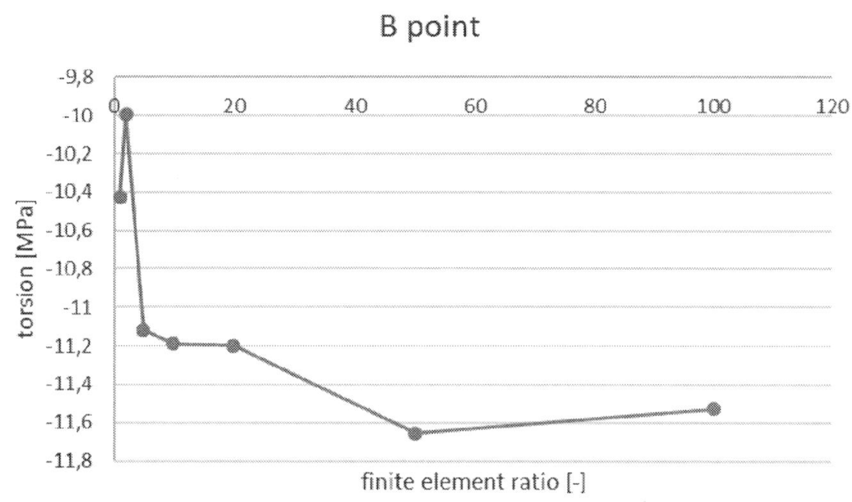

Fig. 6 Notch B, shaft shoulder 30 per 50 mm: stress vs. the element size (finite element ratio = size of the biggest element / actual size of the local element)

Results

Fig. 7 illustrates character of the obtained results. As the object of the evaluation, the course of the stress in the investigated section of the tested shaft and its FE-models with notches for selected torque loading of 50 Nm was studied.

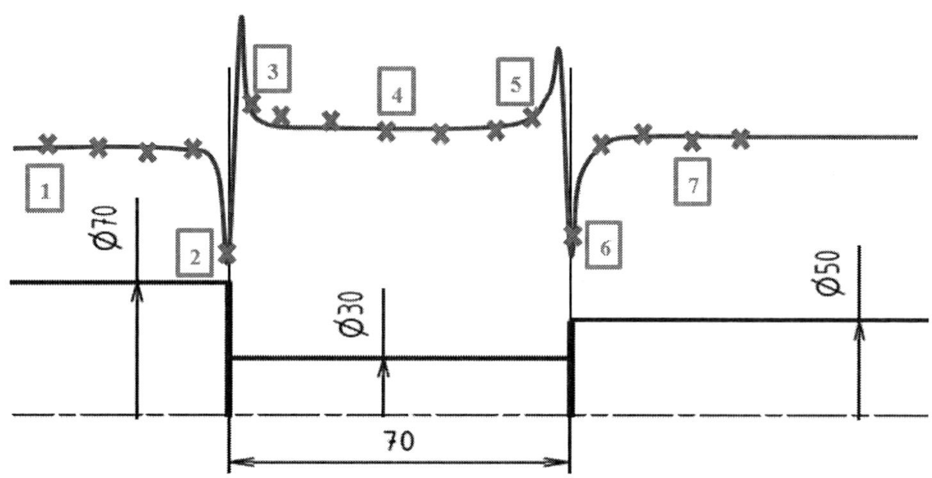

Fig. 7 Results (illustrative course of stress: x – experimental data; - - FEM)

In Tab. 1, the numerical values of the stress found by the FEM and by measurement for the selected load torque value 50 Nm are presented. For illustration of relevance of these data, both the analytically calculated values for smooth cylindrical sections of 0,742 Nmm^{-2} for Ø70, resp. 2,037 Nmm^{-2} for Ø50, resp. 9,431 Nmm^{-2} for Ø30 and Neuber's coefficient 3,16 are shown for easy comparison.

Table 1: Results (average values; position acc. to Fig. 1)

Position	1	2	3	4	5	6	7
	[N/mm^2]	[N/mm^2]	[N/mm^2]	[N/mm^2]	[N/mm^2]	[N/mm^2]	[N/mm^2]
FEM	0,744	0,000	11,740	9,431	11,636	0,000	2,039
Experiment	0,732	0,000	11,775	9,424	11,656	0,000	2,031

In the notch A and B, respectively, the stress value is 1,245- and 1,233-higher, respectively, than in a smooth cylindrical segment. Considering absolute measurement error of 2 MPa, the relative error in detection of torque, which induces the given signal, will be 1,245- and 1,233-times lower, respectively, than the relative error in measurement on a cylindrical section.

Discussion

Reaching of equilibrium of the stress value in a selected site is suggested as a sign of a sufficient refinement of the mesh to obtain relevant results applicable for placement of strain gauges for practical measurements and calculation of torque on the assessed shaft. The issue for discussion is thus mainly the oscillation of the calculated stress around the apparent equilibrium point (Fig. 5 and 6). This phenomenon can be caused by the above described definition of the user node for the site at which the evaluated stress value is to be read. The reading is taken always at the mesh node, the mentioned user node can possibly induce the mesh deformation at that site and induce some error. In 3D modeling, this problem is going to be solved by application of the so called truss elements, by means of which the length of the active strain gauge part could be effectively simulated. The read value would be an average of nods of the area defined in this way. This approach appears close to reality and should thus function well. If it does not prove to be feasible, it is still possible to adjust the condition for the mesh settings and for the recalculation of values use the equilibrium around which the calculation values would oscillate. More appropriate approach can be decided upon by an experiment.

Conclusions

The main aim of our contribution was to prove good agreement between appropriately set FEM and experimental data on basis of which a correct setting of the mesh in the model used can be defined.

On the basis of the above mentioned results obtained so far, the following conclusions can be formulated:

- Good agreement between measured data and FEM has been confirmed, for other measurements smaller strain gauges are required.
- The 3D FE-model generates usable results for the strain gauge area only after the mesh is very carefully set up.
- The 2D FE-model with axisymmetric elements generates usable results for the strain gauge area with less carefully adjusted mesh; this model is definitely the approach of choice in gearbox transmission efficiency tasks.

Despite some drawbacks, regarding the facts mentioned above, we believe that our approach extends the use of strain gauge measurements to the field of direct data detection for precise applications, like e. g. for measurement of mechanical efficiency of transmissions.

References

[1] Information on: https://www.hbm.cz/wp-content/uploads/s3996.pdf

[2] Information on: https://www.efatigue.com/constantamplitude/stressconcentration/

[3] Information on: https://babel.hathitrust.org/cgi/pt?id=uva.x000261815;view=1up;seq=1

[4] Information on: http://abaqus.software.polimi.it/v6.14/books/usb/pt06ch28s01alm01.html

EAN 2018
56th conference on
experimental stress analysis

Heat Effect on Tensile Properties of Textile Reinforced Concrete

MACHOVEC Jan[1,a] and REITERMAN Pavel[1,b]

[1] Experimental Centre, Czech Technical University in Prague, Czech Republic

[a] jan.machovec@fsv.cvut.cz , [b] pavel.reiterman@fsv.cvut.cz

Keywords: Textile reinforced concrete, high temperature, direct tension, tension, TRC.

Abstract. Experimental tests of tensile properties of textile reinforced concrete have been performed in this paper. Textile reinforced concrete (TRC) is used currently for the production of thin-walled structures, which life- and service-time are significantly affected by the presence of crack induced by the action of the environment. TRC is relatively new composite material based on textile reinforcement and high performance matrix. The most of research works was focused on the description of failure mechanism, mechanical and durability properties of used materials, however the influence of increased temperatures has not been thoroughly studied yet. Increased temperatures could occur on the surface of real structures. Dog-bone shaped specimens were heated to 40, 60 and 80°C and consequently tested in direct tension tests. Glass textile fabric have been used as a reinforcement of high performance concrete matrix. No visual defect was apparent on specimens and the concrete matrix properties decreased more significantly than properties of textile reinforcement.

Introduction

This article describes tension behavior of glass textile reinforced concrete under elevated temperature conditions. Textile concrete starts to be a popular composite material for new constructions [1,2]. Every new material needs to be tested in various extreme conditions such as acidic environment, freeze-thaw or high temperature. There have been performed several tests on the same shape of specimen with more types of reinforcement (number of layers, type of reinforcements) which showed important contribution of reinforcement mostly in after-crack part of loading [3,4,5].

Kodur in his work [6] mentions that there is not enough knowledge about effect of high temperatures on new types of concrete (such as self-consolidated concrete or fly ash concrete) comparing to normal or high strength concrete. Tested properties do not depend only on the temperature, but also on the heating rate or temperature gradient. Thorough research dealing with textile concrete was conducted by Chira et al. [3]. This work was focused on the on the cohesion of various textiles with cement based matrix. Number of experimental works was focused on the mechanical properties and properties related to the durability of TRC [4,5]. However, the knowledge of behavior under elevated temperatures induced by the sunshine is still missing.

The main aim of performed work was to document behavior of TRC concrete in elevated temperatures, which could be induced by the natural action of the environment. Any significant changes in the properties of cement based matrix are not expected, however decay of tensile strength of used TR could be crucial safe parameter. Test specimens in this work were heated on 40, 60 and 80 degrees of Celsius and tested in direct tension. Concrete mixture had properties of high strength concrete with compressive strength after 28 days 100 MPa.

Test specimens were in a shape of a "dog-bone" with a thickness of 30 mm and were loaded with loading speed 0.5 mm/min.

Experimental methods

There does not exist any code or guideline to direct tension textile reinforcement concrete testing. Direct tension tests were executed on dog-bone shaped specimens at a machine with specially shaped claws mounted on joints to the machine. This shape has been used in more previous experiments executed by authors, which gives them many data [4,7,8]. However, TRC plates could be used as well [9]. The quality of used cement based matrix was controlled in terms of flexural strength and compressive strength. The composition of used matrix followed previous research of Holcapek et al. [8]. All specimens were tested after 28 days after the production and they were cured in laboratory conditions. Next step was their conditioning in furnace to heat them on required temperature for 24 hours. Each specimen was tested immediately after its extraction from the furnace to prevent cooling down.
Direct tension tests were carried out on a MTS QTest 100 machine with special-shaped claws with installed joints at both ends to provide direct tension without bending. Loading speed was set to 0.5 mm/min and time, elongation and loading strength were recorded. Except of tensile tests, compressive strength and flexural strength was investigated at prismatic specimens 40x40x160 mm^3 according to CSN EN 1015-11 [10]. Emphasis was placed on several points in a stress diagram, namely the stress at the activation of textile reinforcement after first crack occur, maximum force at un-cracked specimen and maximum force bored by reinforcement. All values are possible to read from strength – elongation diagram.

Materials

Cementitious matrix. Composition of used matrix is shown in Tab. 1**Chyba! Nenalezen zdroj odkazů.**. Basically, it´s composition equals to high strength concrete, where the stress strength after 28 days exceeds 100 MPa [8]. These modern composites exhibited high density and durability. High resistance of commonly used HPC matrix in determined by the utilization of relatively high volume of silica fume, which reduces the content of Portlandite and simultaneously transform that on the non-soluble C-S-H phases (CaO-SiO_2-H_2O) contributing to the mechanical performance in case of cement based matrix. The development of HPC matrixes is also closely related to the application of advanced chemical admixtures, which considerably modify rheology of fresh composite and allow good workability. This matrix is suitable for TRC mostly because of the fine structure, the biggest grain is 1.2 mm and it very easily enter "eyes" in textile reinforcement.

Tab. 1 Composition of used matrix

Component	Weight ratio [-]
Cement CEM I 42.5 R	1
Silica fume	0.19
Silica sand 0.1/0.6	0.48
Silica sand 0.3/0.8	0.5
Silica sand 0.6/1.2	0.38
Silica powder	0.48
Superplasticizer	0.011
Water	0.35

Textile reinforcement. Textile reinforced concrete is relatively new composite material. In this case, E-glass textile reinforcement with styrene-butadiene coating was used in combination with high-strength concrete matrix. Glass textile reinforcement is usually used in façade systems to prevent cracks, but recently started to be used in concrete technology thanks to its easy manipulation, low price, good mechanical properties and no need of covering layer. Textile reinforcement can be made from different materials (carbon, glass, plastics, etc.), where each of them has positive and negative sides. There also exists alkali resistant glass, which does not need any surface treatment, but the price is higher. Reinforcement designated "R 585 A101" produced by Adfords Ltd. Was used in the experimental program. This code means: R – textile fabric; 585 is square weight - 585 g/m^2, which is often designated as "tex"; A101 is the code for used surface treatment, what means alkali-resistant coating on basis of styrene-butadiene. Its textile grid has dimensions 4.5 × 5.5 mm^2, what ensure suitable compaction of casted matrix and also homogeneity of produced samples.

Results

Flexural strength and compressive strength were tested on non-heated prisms to document actual properties of used matrix. Flexural strength resulted to be 10.5 MPa and compressive strength equals to 106.1 MPa. These results confirmed previous research. Even its high mechanical properties, this material is not suitable for structures because of its brittle failure mode. That is why, it is necessary to apply some kind of reinforcement.

Direct tension strength tests conducted on dog-bone specimens showed important decrease of concrete tension strength - measured as a force needed for first crack, especially for the highest applied temperature. Average value for reference samples was 5344 N, for samples heated to 40 °C the value dropped to 1629 N (30.5 %), at 60 °C to 2542 N (47.6 %) and 2998 N (56.1 %) for 80 °C heated samples. These findings is likely surprising, because such "low" temperatures are commonly used for various laboratory procedures. Otherwise, very low amount of hydrates decompose in this temperature interval.

However glass textile reinforcement did not suffer major property change, but the overall trend of strength was decreasing. Important role in this case might have the styrene-butadiene coating of glass, which is a border between glass and concrete matrix. First important value was the strength when crack in concrete matrix occurred. Average maximal strength obtained in textile reinforcement after first crack occur dropped from 1907 N (reference samples) to 1380 N (40 °C, 72.4 %), 1221 (60 °C, 64 %) and 1226 N (80 °C, 64.3%). It is obvious that the major decay take place after the exposure to 40°C, additional increasing of the temperature did not affect residual force/stress in the reinforcement.

Last important value is force/stress, when textile reinforcement was fully activated. At reference samples, the value is 1318 N and with higher temperature, the lower the strength value was, specifically 1270 N (40°C, 96.4 %), 1242 N (60°C, 94.3 %) and 830 N (80°C, 63 %). These results are well documented on the Fig. 1. It is well visible that this crucial moment in the failure of TRC composite is not significantly affected by the temperature. Nevertheless, after 80°C there is obvious loss of mechanical performance, but this value of surface temperature in relatively improbable in normal conditions.

Shape of a load – displacement curve is similar to reference samples, maximum strength values for each part varies. Heat affects the whole volume which causes changes on reinforcement. Deeper investigation with more samples and wider variety of tested temperatures is necessary for further research. No visual changes were apparent on all heated specimens.

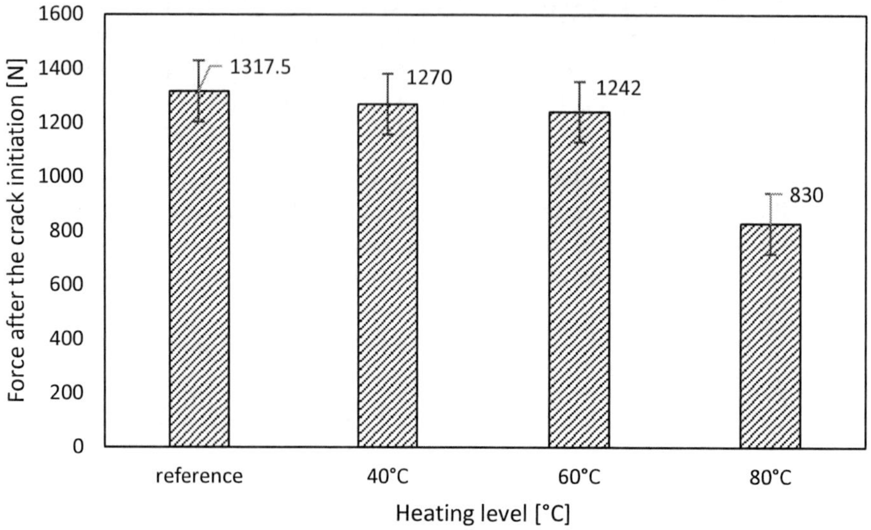

Fig. 1 Force in textile reinforcement after the crack initiation

Conclusion

Aim of this work was to experimentally assess influence of elevated temperatures to tension properties of composites with glass textile reinforcement. Studied conditions could appear in real conditions due to action of sun. For the thin-walled structures in the change of mechanical behavior crucial aspect for their design, because these changes could arise very quickly. Textile concrete was made by high-strength concrete (100 MPa after 28 days) and styrene-butadiene coated E-glass textile fabric made in Czech Republic. Specimens were preheated to 40, 60 and 80 °C and were subjected to direct tension tests afterwards. It is apparent that with rising temperature, it is mostly the concrete matrix where the biggest decrease of parameters occurs, even its high and commonly presumed good resistance. Results of test showed the slight decrease in glass textile parameters, which is probably caused in a decrease of bond strength between styrene-butadiene coating and textile fabric. However studied topic is highly interesting and should be subjected to following research, because the impact of the environment is the leading factor determining the final service-life of entire structure.

Acknowledgement

This work was supported by the CTU under the project No.: SGS18/057/OHK1/1T/11.

References

[1] R. Chudoba, E. Sharei, A. Scholzen, A strain-hardening microplane damage model for thin-walled textile-reinforced concrete shells, calibration procedure, and experimental validation, Composite Structures 152 (2016) 913-928.
DOI: 10.1016/j.compstruct.2016.06.030

[2] F. Vogel, Production and Use of the Textile Reinforced Concrete, Advanced Materials Research 982 (2014) 59-62.
DOI: 10.4028/www.scientific.net/AMR.982.59

[3] A. Chira, A. Kumar, T. Vlach, L. Laiblová, A.S. Škapin, P. Hájek, Property improvements of alkali resistant glass fibres/epoxy composite with nanosilica for textile reinforced concrete applications, Materials and Design 89 (2016) 146-155.
DOI: 10.1016/j.matdes.2015.09.122

[4] F.B. Varona, F.J. Baeza, D. Bru, S. Ivorra, Influence of high temperature on the mechanical properties of hybrid fibre reinforced normal and high strength concrete; Construction and building materials 159 (2018) 73-82.
DOI: 10.1016/j.conbuildmat.2017.10.129

[5] X. Shilang, L. Shen, J. Wang, The high-temperature resistance performance of TRC thin-plates with different cementitious materials: Experimental study, Construction and Building Materials 115 (2016) 506-519
DOI: 10.1016/j.conbuildmat.2016.04.070

[6] V. Kodur, Properties of Concrete at Elevated Temperatures; ISRN Civil Engineering, vol. 2014; Hindawi Publishing Corporation.
DOI: 10.1155/2014/468510

[7] M. Raupach, M. Cruz, Textile-reinforced concrete: Selected case studies, Textile Fibre Composites in Civil Engineering 12 (2016) 275-299.
DOI: 10.1016/B978-1-78242-446-8.00013-6

[8] O. Holčapek, F. Vogel, T. Vavřiník, M. Keppert, Time progress of compressive strength of high performance concrete, Applied Mechanics and Materials 486 (2014) 167-172.
DOI: 10.4028/www.scientific.net/AMM.486.167

[9] R. Ortlepp, S. Ortlepp, Textile reinforced concrete for strengthening of RC columns: A contribution to resource conservation through the preservation of structures, Construction and building materials 132 (201) 150-160.
DOI: 10.1016/j.conbuildmat.2016.11.133

[10] CSN 1015-11 – Testing methods for mortars – Part 11: Determination of strength of hardened mortars in bending and compression

EAN 2018
56th conference on
experimental stress analysis

Analysis of Experimental Testing of Clinched Joint

MALÝ Pavel[1,a] and LOPOT František[1,b]

[1]CTU in Prague, Department of Designing and Machine Components, Technická 4, 166 07 Prague 6, Czech Republic

[a]pavel.maly@fs.cvut.cz, [b]frantisek.lopot@fs.cvut.cz

Keywords: Clinched Joint, Experimental Testing, Strain Gauge, Measurement Analysis, FEM

Abstract. This paper describes the analysis of experimental testing of specimens that are two sheet metal plates connected with clinched joint. The strain gauge was used to obtain uniaxial stress in the vicinity of the clinched joint. The measurement results and specimen behavior were also analyzed with help of FE simulations.

Introduction

Clinching method signifies a developing technology for connection of sheet metal plates in different branches of industrial production. Therefore, there is a need to describe thoroughly both the clinching process and the resulting properties of the clinched joint, [1].

Method

The experimental method for testing clinched joint specimens with use of a strain gauge is described in [2].

Instrumentation and investigated set of specimens. The specimen with the strain gauge near the clinched joint was loaded in a special testing stand. The loading was applied by a hydraulic cylinder, measured by a force sensor and loading velocity was controlled by a proportional valve. The uniaxial stress on the surface was obtained with help of a strain gauge near the joint. Scheme of the specimen is in figure 1.

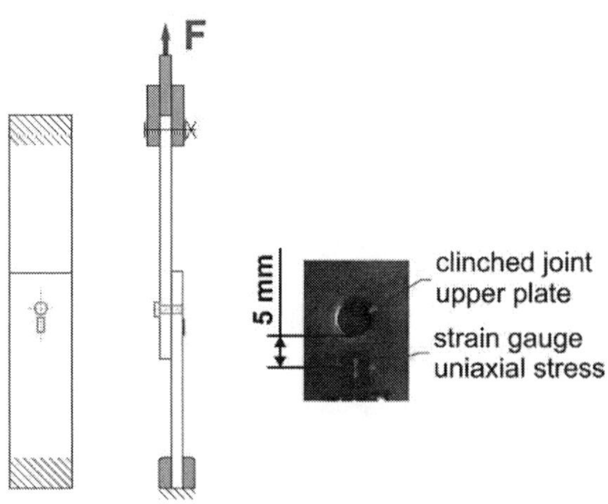

Fig.1 Scheme of the specimen with position of the strain gauge

The specimens for the testing were produced from two sheet-metal plates with thickness of 3 mm and material DX51D + Z275 that were connected with the TOX®-Clinching round point of diameter 10 mm. The shear load strength of the used clinched joint according to the specimens supplier (TOX®PRESSOTECHNIK) is 5100 N. The maximum loading during the experimental testing did not exceed this limit value of the joint shear strength.

The specimen that was subjected to the testing and analysis has the width of 40 mm, the length of both connected plates is 280 mm and the specimen is designated as 01-40.

Data analysis

Data processing was made in Matlab. The dependency of the measured data (loading force and uniaxial stress) are plotted in the graph as is shown in figure 2.

The course of the measured and processed data is specific with the loop when the loading and unloading do not have the same course. See figure 2 for the specimen 01-40 where course for different maximum loading and velocity of loading is compared. The change of slope for loading occurs in the interval of 1000 and 1500 N for testing with maximum loading more than 2000 N. The part with slope 2 during loading for testing with maximum loading of 1000 N is not so significant and occurs when reaches approximately 800 N. The testing with maximum loading of 500 N is not characteristic with the loop, the loading and unloading courses are identical.

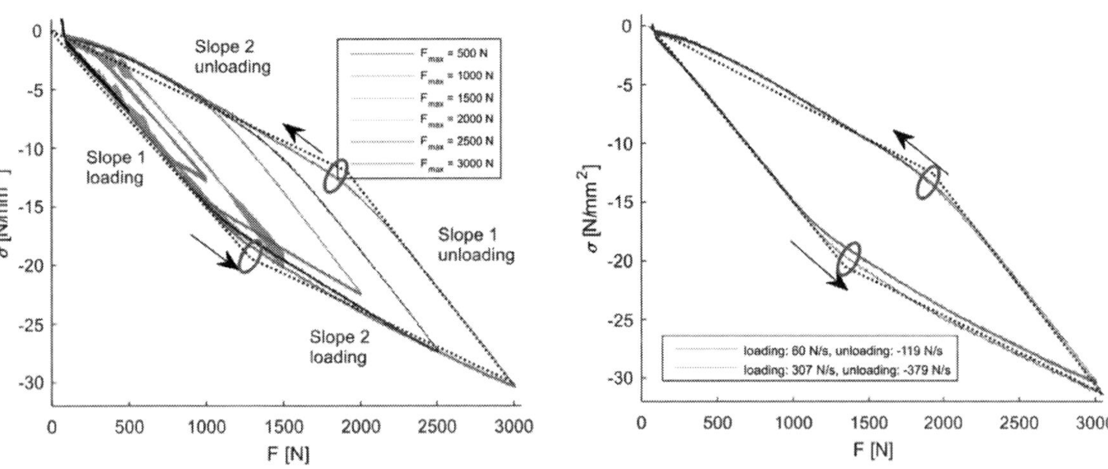

Fig.2 Processed data – comparison of magnitude and velocity of loading

The value of load when the change of the slope occurs for unloading is dependent on the maximum load. However, the sizes of slopes are identical both for loading and unloading parts of the course and therefore the loop is symmetrical and closed. This behavior can be described by the influence of the testing stand properties, additional bending of the plates or non-linear behavior of contact in the joint.

In general, the testing process was designed to be quasi-static, thus the velocity of loading was relatively small and any dynamic behavior was not expected and noticed. Anyway, measurements with different velocities of loading and unloading were conducted for the comparison of the loading velocity influence. Graphs of the results are shown in figure 2. The ratio of two velocities for loading was approx. 5 (60 N/s vs. 307 N/s) and for unloading approx. 3 (-119 N/s vs. -379 N/s). The two resulting courses of the measurement are considered to be identical, therefore the influence of the velocity of loading in the range used for experimental testing is not significant.

FEM analysis

FEM analysis in the engineering software Abaqus CAE was used to support the description of the experimental testing and measured data analysis. The complex description of the FE analysis of the clinched joint model can be found in [3]. The above mentioned bending of the specimen that is shown in figure 3 was not measured with the used testing method but it was analyzed with the help of FEM simulations

The simulation model of the specimen 01-40 uses 3D model of the clinched joint, i.e. interlock of upper and bottom plates created with use of standard hard surface-to-surface contact with friction coefficient $f = 0.1$ and tie connection. Scheme of the simulation model is shown in figure 3 where the boundary conditions and paths of the upper plate for the results analysis are specified. The starting point of the paths is marked with circle. The place of interest corresponds to the position of the strain gauge on the upper plate of the experimental specimen.

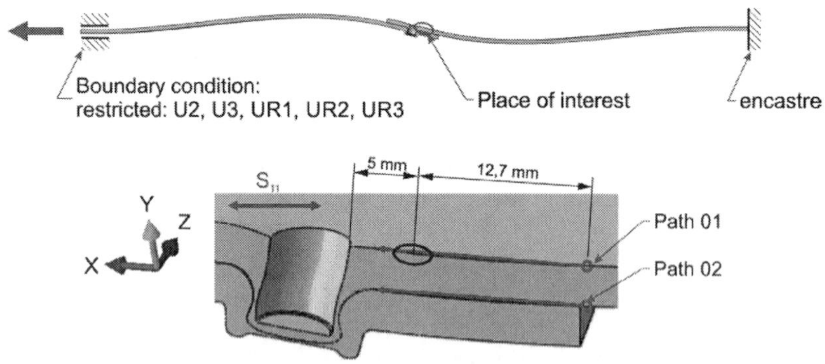

Fig.3 Simulation model with boundary conditions for FE analysis

The resultant stress in X direction (S11) along two paths of the top plate is chosen for the analysis. The detailed view on the results in the place of interest and the course of S11 value along path 01 and path 02 for corresponding loading of 3000 N are in figure 4. The unidirectional stress in the place of the strain gauge (path 01) is approximately -31 N/mm².

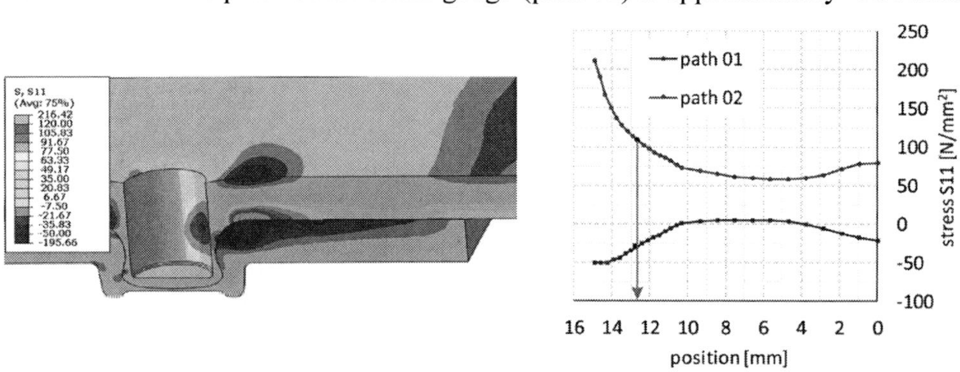

Fig.4 Results of FE simulation

The tension and bending stresses in the upper plate can be determined using the results of uniaxial stress (S11) along the two paths according to equations 1 and 2

$$S_{tension} = (S_{11(path01)} + S_{11(path02)})/2 \quad (1)$$

$$S_{bending} = (S_{11(path02)} - S_{11(path01)})/2 \quad (2)$$

The graphical representation of the tension and bending stresses is plotted in figure 5 with the marked position of the place of interest (strain gauge position). It is obvious from the FE simulation results that the stress along the upper plate is evenly distributed with the peaks near the clinched joint. The ratio of bending and tension stress ($S_{bending}$ / $S_{tension}$) changes along the upper plate in the range from 0.9 up to the 2.1. The bending stress in the upper plate is higher than tension stress at the beginning of the considered section and significant near the joint.

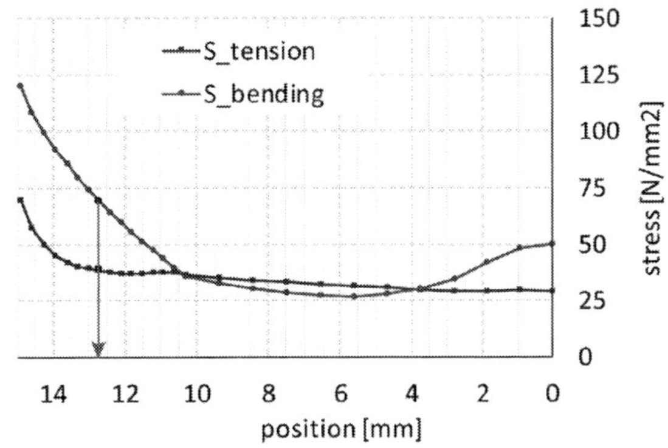

Fig.5 Tension and bending stress in the upper plate

If we focus on the FE analysis results in the specific interval in the place of interest defined with the length 2 mm (position of strain gauge ± 1 mm), the average value of tension stress is 39 N/mm² and bending stress 70 N/mm². The value of these stresses ratio is 1.8.

Conclusions

The experimental method was thoroughly described and verified by the detailed analysis of the data and with help of FE simulations that were used especially for analysis of the specimen bending. It was found out that the bending stress in the place of the strain gauge is significant, approx. 1.8 higher than tension stress. The average value of the uniaxial stress obtained by FE simulation for loading of 3000 N in the interval of 2 mm long in the place of interest (strain gauge) on the upper plate is -31 N/mm². The measured stress by the strain gauge for the same loading is approx. -29.5 N/mm². The two values correspond each other, however they are also influenced by various measurement errors and simulation configuration, e.g. boundary conditions.

The described method of experimental testing of clinched joints with uniaxial strain gauge was designed so that it could be used for further testing of complex structures connected with more clinched joints. The description of the clinched joint behavior is further subjected to the study with help of three approaches, i.e. experimental testing, FE simulations and mechanical description, [4].

References

[1] Information on http://www.tox-cz.com

[2] P. Malý, F. Lopot, F. Starý, J. Sojka, Strain gauge measurement on clinched joints specimens. EAN 2017 - 55th Conference on Experimental Stress Analysis (2017) 139-142

[3] P. Malý, F. Lopot, J. Sojka, FEM model and experimental measurement of clinched joint. IOP Conference Series: Materials Science and Engineering, volume 179, number 1 (2017)

[4] P. Malý, F. Lopot, J. Sojka, Mechanical description of the clinched joint. EAN 2016 - 54th International Conference on Experimental Stress Analysis (2016)

EAN 2018
56th conference on experimental stress analysis

Experimental Testing of Piston Lock

MALÝ Pavel[1,a] and VUILLERMET Raphaël[2,b]

[1]Institute of Physics of the Czech Academy of Sciences, Na Slovance 1999/2
182 21 Prague 8, Czech Republic

[2]CERN (European Organization for Nuclear Research), CH-1211 Geneva 23, Switzerland

[a]pavel.maly@fzu.cz, [b]Raphael.Vuillermet@cern.ch

Keywords: Piston lock, Locking System, CERN, Experimental Testing, Pullout Force

Abstract. This paper describes the experimental testing of piston lock. The purpose of the lock is to support the piston of the single acting cylinder (JF jack) during transport of New Small Wheel in CERN. The lock is a special designed locking system that utilizes the principle of rubber O-rings and conical springs expansion in the hole. The function of the lock was verified and properties were obtained with help of the experimental testing.

Introduction

The manipulation and transport of large heavy objects is very often in CERN (European Organization for Nuclear Research). Therefore, the need of different special tools and equipment is essential. New Small Wheel (nSW) is a part of the ATLAS muon subsystem which has 9,3 m in diameter and weighs over 100 tons. One of the special tooling for the transport of nSW uses single acting cylinders (JF jacks) that are placed in vertical position with piston pointing down (see figure 1). Locking system for support of the piston (JF piston lock) was designed and its function was verified with help of experimental testing.

Experimental testing on machines and newly designed structures becomes very important for the development and verification in these days. Firstly, measurements during real condition are crucial for designing of reliable products. Examples of experimental testing of different machines, elements and devices can be found in [1], [2], [3].

Fig. 1 Tested specimen (piston lock)

Method

The testing was carried out to obtain the maximum pullout force and the behavior of the piston lock during loading.

Investigated specimen. The prototype of the piston lock and test hole was produced for the test, the scheme of the specimen and the function are in figure 2. The locking force is induced by tightening of the fastening nut and compressing of the O-rings. The rings come into contact with the wall and the locking is achieved, allowing also resistance movement thanks to friction. The size of the thread is M16, the diameter of the hole is 32 mm, the length of the O-rings set is approx. 50 mm.

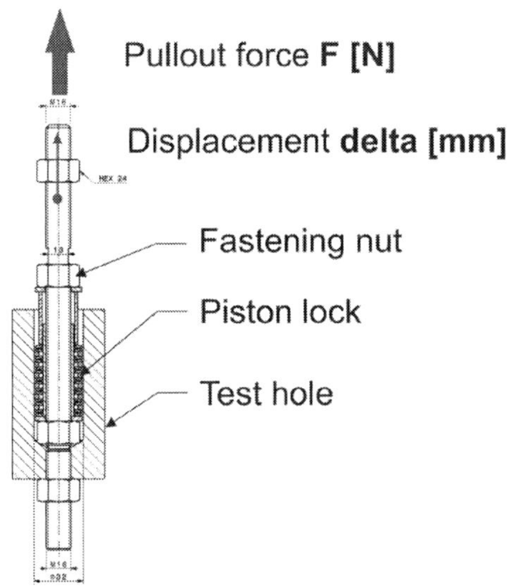

Fig. 2 Scheme of tested specimen

Measurement protocol. The specimen was subjected to the experimental testing when the piston lock was pulled out of the test hole with the constant velocity along the whole depth of the test hole. The time dependent displacement and force was measured with the constant sampling frequency of 20 Hz. The number of nut turns *n* from the initial state was the locking parameter for testing. Number of used O-rings was another parameter that influences the properties of the piston lock. The measurement was carried out several times for some parameters configuration to obtain average values.

Instrumentation. The mechanical test was carried out with a UTS™ electro-mechanical testing machine equipped of a 20 kN load cell at velocity of 20 mm/min, figure 3.

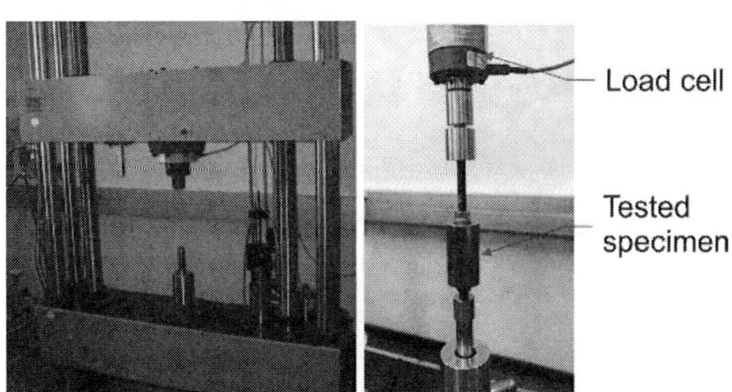

Fig. 3 Testing machine and specimen

Data acquisition and processing. The measured time dependent data of load and displacement were obtained using control software of the testing machine, further the data were processed to the graph of force dependent on the displacement using Matlab. The typical course of force (figure 4) can be divided into three parts. The first part is characteristic with the linear increase of force up to the maximum. The force in the second part is rather constant with possible fluctuation which can be explained by the friction sliding of the rings on the wall of the testing hole. The last third part is characteristic by the decrease of the force as the piston lock and O-rings are gradually extracted out of the testing hole. The maximum applicable pullout load, i.e. locking, was determined at the end of the linear increase of force course when the function of the piston lock is reliable.

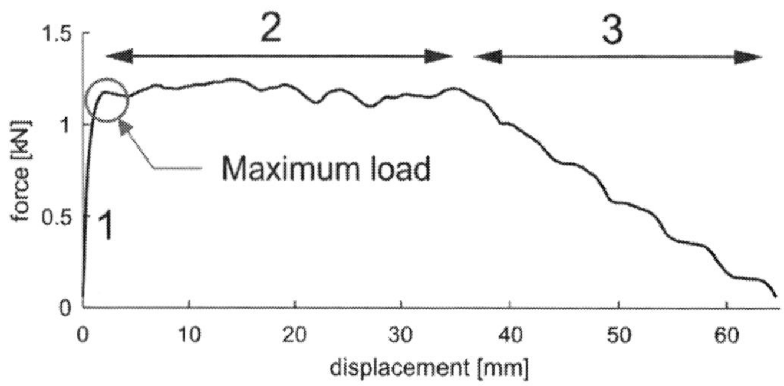

Fig. 4 Measured data (pullout force - displacement)

Experiment outcomes

The basic configuration of the piston lock was with 6 O-rings. The testing was repeated several times for locking parameter of nut turns $n = 1$ and $n = 2$. The testing for $n = 3$ was done once because large deformations of rubber rings and their crushing or displacement from the correct position occur (figure 5) with higher number of nut turns. The resultant graphs are in figure 6.

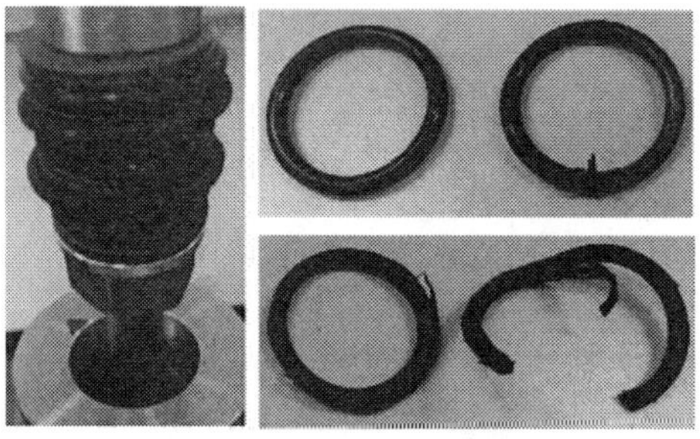

Fig. 5 Crushing and lose of O-rings

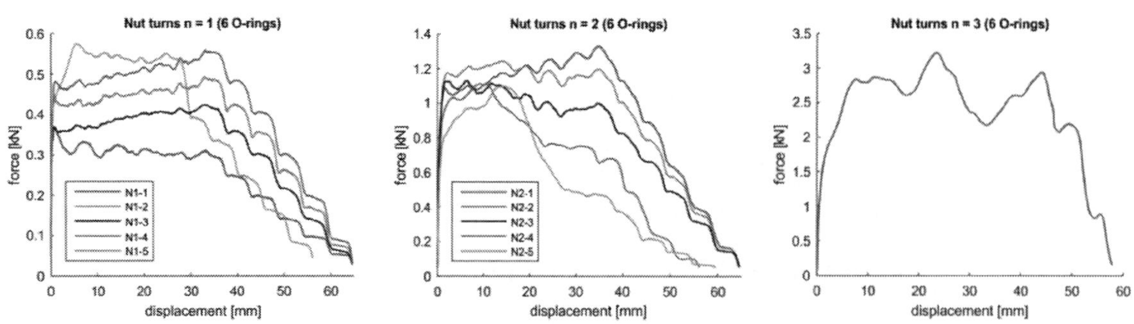

Fig. 6 Measured data for different locking parameter of nut turns

The detailed first part of the measured data for different locking parameter is in figure 7. It is obvious that the locking, i.e. maximum pullout force, increases with higher number of nut turns. Also the difference of the force course for locking parameter n = 2.5 and n = 3 is negligible. The second graph in figure 7 shows results depending on the number of O-rings. The difference between 6 and 8 O-rings for locking parameter n = 1 is evident, however is not important and crucial. The results of maximum pullout load for different piston lock configurations are summarized in table 1.

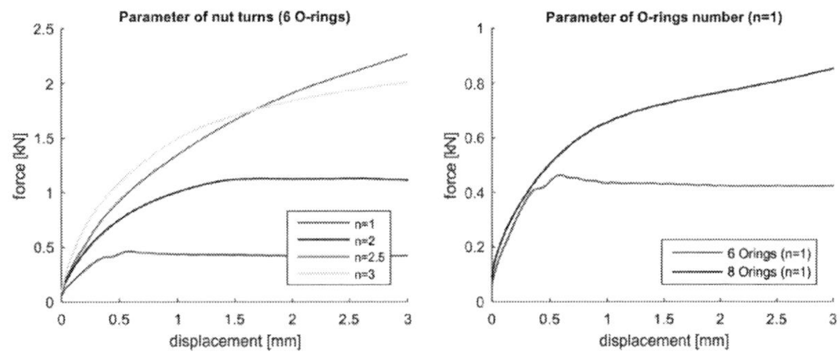

Fig. 7 Measured data for different locking parameter and number of O-rings

Tab. 1 Results of the testing

Number of turns n [1]	Number of O-rings [1]	Max. load F [kN]
1	6	0.45±0.08
1	8	0.64
2	6	1.06±0.09
2.5	6	0.95
3	6	2.84

Conclusions

The maximum pullout load for different configuration of the piston lock (number or nut turns and number of rubber O-rings) were obtained with the experimental testing. The maximum applicable load for locking parameter of 3 nut turns and six O-rings was 2.84 kN. However, crushing and lose of O-rings occurred for some tests when higher number of nut was used. Focusing on the application of the piston lock, maximum recommended nut turns is 2 when the max. load, i.e. locking, is 1.06±0.09 kN.

References

[1] M. Dub, J. Kolář, F. Lopot, V. Dynybyl, O. Berka, Experimental analysis of stress state and motion of tram gearbox hinge, EAN 2014 - 52nd International Conference on Experimental Stress Analysis (2014).

[2] O. Berka, M. Dub, F. Lopot, V. Dynybyl, Experimental analysis of stress state and motion of tram gearbox hinge, EAN 2015 - 53rd Conference on Experimental Stress Analysis (2015).

[3] P. Malý, F. Starý, V. Dynybyl, J. Hoidekr, Strain gauge measurement on the oilseed screw press. EAN 2017 - 55th Conference on Experimental Stress Analysis (2017) 625-628

EAN 2018
56th conference on experimental stress analysis

Modeling andEexperimental Measurement on Heated Lamination Cylinder for Optimization of Heat Flows

MARTINEC Tomáš[1,a], PETRŮ Michal[1,b] and NOVÁK Ondřej[1,c]

[1] Technical university of Liberec, Studentská 1402/2, Czech Republic

[a]tomas.martinec@tul.cz, [b]michal.petru@tul.cz, [c]novak.ondra1@seznam.cz

Keywords: Heat lamination, numerical modeling, experiment, heat flow

Abstract. Quality and complete distribution of temperature field on the cylinder for lamination is a critical parameter of a lamination process. Currently, this is done by using an infrared tube of the same length as a heated cylinder, heating by the circulating oil or heating segments to achieve a more even temperature distribution on the cylinder surface. However, these solutions do not offer improved energy balances for individual paper widths because heat energy is not transmitted to the surface of the cylinder steadily to the places where it is needed. Based on numerical simulations a working prototype of multi-zone heating system that would allow temperature control for the width of the paper was designed. Experimental measurements were performed on a functional prototype and compared with simulations. The results of numerical simulations, which were influenced by the boundary and initial conditions for non-stationary heat conduction, showed good correlation with experiments.

Introduction

Lamination process is a very old technology, but even older are various technical solutions for a roller heated. The oldest patent of the heated cylinder, which uses hot water as a medium, dates back to 1921 [1]. At present, the laminating rolls for a heating uses mainly resistive, infrareds and induction heating or some suitable medium (e.g. oil). Due to increasing requirements for the quality of heat and reducing of energy consumption, laminating machines constantly need the study of the thermal distribution on the rolls during lamination.

Problems of Modelling and Measurement of Heat Flows on a Heated Roller

The quality of the temperature field distribution on the cylinder is a critical parameter of the lamination process. For optimization purposes, a numerical analysis (Fig.1) of the new prototype heating [2] was performed followed by experiment on the real machine (Fig.2). This is a complex problem where the simulation and the measurement can be divided into static (the cylinder is heated without rotation and temperatures at discrete points are measured) and dynamic (the temperature is measured during production directly on the machine). A numerical model establishing of a model for the study and comparability of heated rollers must be based on the nature of phenomenon that occurs in the cylinder in terms of energy, ie. what is its energy balance. Therefore, the creation of models based on the Fourier law, which is described with parabolic partial differential equation (1).

$$c\rho \frac{\partial T(x,t)}{\partial t} = \lambda \Delta T(x,t) + Q \,. \tag{1}$$

Where: c is the specific heat capacity[J.kg^{-1}.K^{-1}], ρ is a density [kg.m^{-3}], λ is thermal conductivity [W.m^{-1}.K^{-1}], T is temperature [K], Q volume heat source [W.m^{-3}]. To solve the equation (1) it is necessary to introduce initial and boundary conditions for the spatial and time temperature distribution. The solution lies in the determination of boundary conditions (BC) at known temperature distributions at different time levels. Then it is necessary these boundary conditions to verify with the real measurements. The boundary conditions implying a characteristic heat distribution can be defined by the Dirichlet boundary condition, which determines the temperature distribution on the surface of the studied part of the heated cylinder construction, and the Newton boundary condition, which involves the exchange of heat energy between the cylinder and the surrounding environment. For the verification and calibration and experimental comparison of the model a functional prototype of a multi-zone heater was designed and constructed. An experiment on the machine simulating the course of the lamination process consists in heating of the cylinder to the desired temperature, then this temperature is maintained for a certain time period and then the heating is stopped. The heating to the working temperature was the first 2.5 minutes at full power, and then it was controlled manually. The maximum winding temperature during initial heating was 430 °C. For maintain a cylinder temperature of 120° C the power was 40 % and temperature of the winding was 220 °C. After approx. 12 minutes the heating was switched off and the temperature rose for another minute by about 1° C.

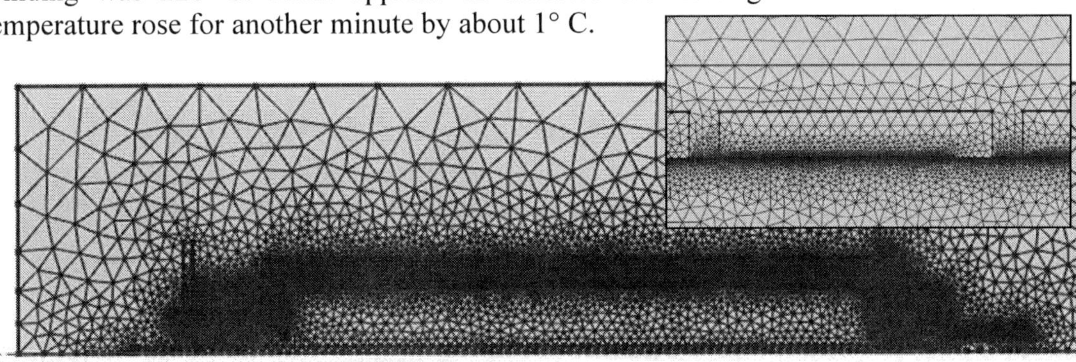

Fig.1: Numerical model of new prototype heating, detail of FE mesh (right)

Fig.2: Experimental measurement (left), measuring device (right)

Results and discussion

Static measurement is simple. The roller does not rotate; there is no need to work with the material (with paper or foil). Only the heating is switched on and the distribution of the temperature is monitored after a heating stabilization. Custom measurements can be performed, for example, by a handheld probe at arbitrarily selected points. The highest temperature is in the middle of the cylinder where the smallest heat dissipation and the accumulation of energy occurs (each point is heated and it has two heated neighbors). On the sides, the temperature decreases and the lowest is at the edge of the cylinder (heat is discharged into the flange). A completely different situation is when the device is operating. The main effect is the removal of heat by the material when the adhesive is activated on the foil (phase transformation occurs). Therefore, temperatures fall in the places that come into contact with the paper that is placed in the foil. Consumed energy must be supplied by the heater to maintain the set setpoint. Therefore, in heated places that are not in contact with the foil (where heat is not consumed) the temperature rises to undesirable values. In practice, the overheated areas of the cylinder, which are near of the working area, cause degradation of the edge of the film. Due to the rotation of the cylinder during dynamic measurements cannot be used the conventional touch sensors. Therefore, the dynamic measurement is currently uncommon and quality of heating is evaluated only indirectly by the quality of lamination. For similar fast temperature measurements in a large area, the thermal camera is very suitable solution. However, the problem is the poorly available curved surface of the cylinder, which is glossy. These are the conditions under which the thermo camera fails. Another possibility is to use the surface of the cylinder as a thermocouple material and to find the corresponding second material to form a sufficiently sensitive thermocouple. This method has been used for example in the automotive industry for the measurement of temperature fields on galvanic molds for the production of artificial skins by Slush technology [3]. The thermocouple is formed between the material of the measured surface and another material that has ideally the known properties in terms of thermoelectric properties depending on temperature. Then just is need to know the temperature and the voltage at one point on the measured object (the reference point) and the other points can be measured with only one wire (so called single-wire thermocouple).

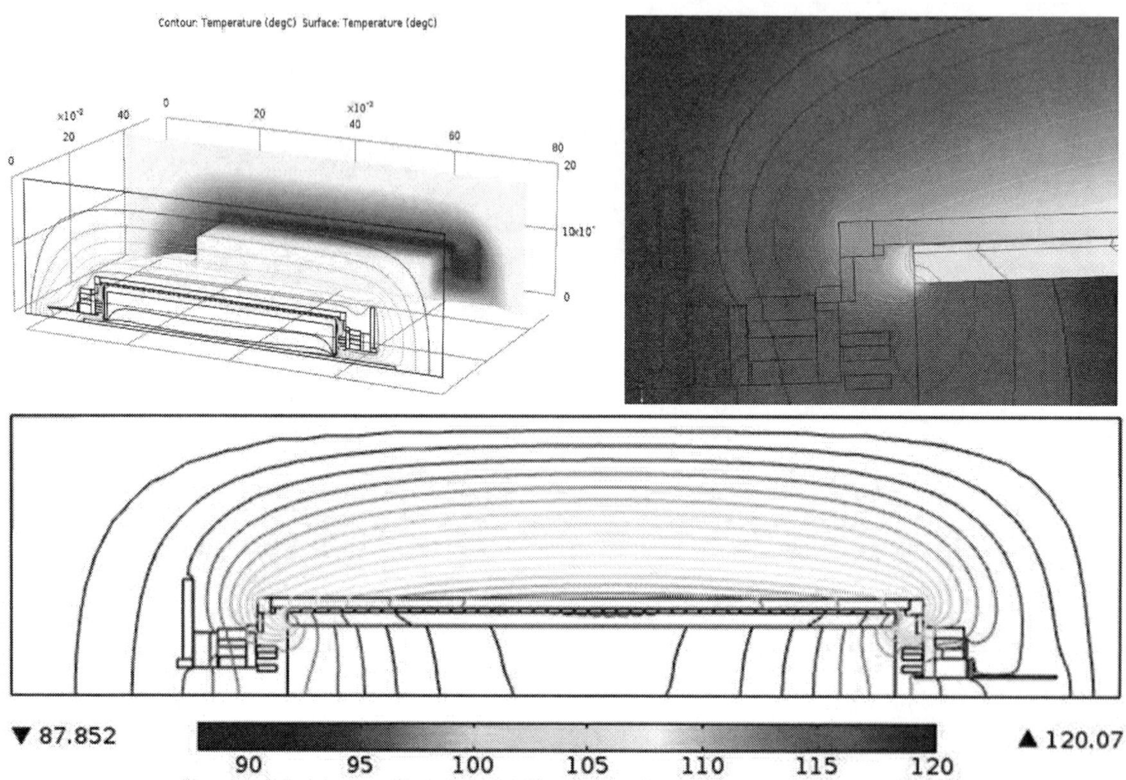

Fig. 3: Numerical model: distribution of the temperature field of a prototype construction.

Experimentally it was shown that the greatest sensitivity to the roller exhibits a negative material from J-type thermocouple. For reliable contact of the measuring wire with the rotating roller, a special caliper was made to hold the wire tensioned by the spring and at the same time with its weight presses the wire to the cylinder. The measured thermocouple voltage then corresponds to the temperature at the point of the contact between the wire and roller. A total 19 calipers were installed. Two of them are at the reference point (positive and negative J wire), other ones are in single-wire measured points. Thus, the total measured points on the cylinder are 18. The calipers are distributed most evenly along the roller, as it allows by the construction of their holders (Fig.2). The data was measured on a slowly rotating roller. The rotation of the cylinder is an interference source, therefore it was necessary to apply the filter with a time constant of 2 seconds. The simulation results are shown in Fig. 3 and the comparison of the temperatures from the numerical simulation and from the experiment are put in Fig. 4.

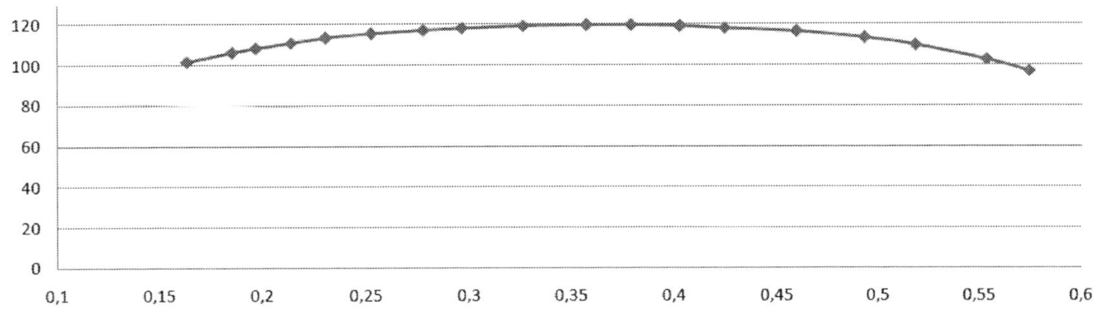

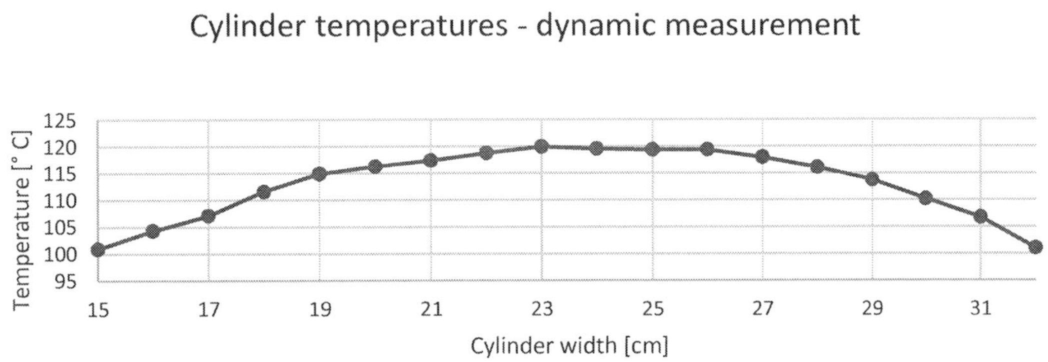

Fig. 4: Temperature distribution on the cylinder surface at dynamic measurement: Numerical model (top), Experiment (bottom).

Conclusions

In the study and modelling of the heating process in laminating machines one of the biggest problems is the measurement of temperature distribution on the cylinder. However, the knowledge of this temperature field is key for a calibrating of mathematical models and subsequently for an evaluating of optimized heating cylinder properties. This article showed one of the possible approaches for the measuring and building the relevant models. Simulation results then show a good agreement with measured data.

Acknowledge

The results of this project No. LO1201 were obtained through the financial support of the Ministry of Education, Youth and Sports, Czech Republic in the framework of the targeted support of the "National Programme for Sustainability I".

References

[1] F. V. Hadley, R. A. Carleton, Heater. US patent, US1480805A, 1921

[2] L. Sevcik, T. Martinec, M. Petru, K. Matejcek, O. Brydl, A heated laminating roller for a laminating device and a laminating device fitted with this laminating roller. Utility model n. 30982.

[3] T. Martinec, Measurement of the temperature field on the metal plate. In Proceedings of 8 th International Carpathian Control Conference ICCC'2007. Štrbské Pleso: TU Košice, 2007. s. 109. ISBN 978-80-8073-8.

Material Characteristics the CFRP for Car Seat Frame

MARTONKA Rudolf[1,a], RYVOLOVÁ Martina[2,b] and POTĚŠIL Antonín[3,c]

[1] Department of Design of Machine Elements and Mechanism, Technical University of Liberec, Studentská 1402/2, Liberec, Czech Republic

[2] Institute for Nanomaterials, Advanced Technology and Innovation, Bendlova 1407/7, Liberec, Czech Republic

[3] LENAM,s.r.o, Masarykova 457/4, Liberec, Czech Republic

[a]rudolf.martonka@tul.cz, [b]martina.ryvolova@tul.cz, [c]antonin.potesil@lenam.cz

Keywords: carbon fibers reinforced plastic, tube, flexural modulus, shear modulus, four point bending

Abstract. Material properties such as strength, stiffness and elasticity of a given composite material depend on the type of fibers used in combination with the matrix, composition and shape of the resulting product. The article deals with the measurement and evaluation of the properties of carbon fiber composite. The tubular profile is one of the basic profiles of the carrier frame of the front passenger and driver's seat. The main criterion of substitutability is to maintain the same mechanical properties of the seat frame with maximum weight reduction [1, 2]. An integral part of the realized activities is also the design of the material model of the tested composite stack. This material model will be used in follow-up FE simulations of the upgraded design of the seat frame.

Introduction

A car seat is a complicated part of the car in terms of design. It belongs to the elements of passive vehicle safety. It evolves long-term to meet the requirements of comfort, mechanical and utility properties. The weight reduction of the seat is an objective while preserving the mechanical and vibration-isolation properties of the structure. Conventional metallic materials are gradually replaced by composite materials, particularly fiber-reinforced plastics (FRP). The orientation and type of fibrous reinforcement should be adapted to the mechanical properties requirements when loading the seat design. Carbon fibers have high strength and modulus in generally [3]. Modelling of properties and behavior of composite structures is an integral part of designing parts for industrial use. Material properties, i.e. set of elastic constants of the composite structure, are the result of a virtually modelled experiment. The results of the model experiment are then used for the actual product.

Experiment

Specimens for experiment - tubes made using pultrusion from carbon fibers and epoxy resin were used as a support structure. All filaments are oriented in only one direction. This part of the composite ensures the strength of the component. Two or four layers of carbon fibers pre-impregnated with an epoxy resin with directional orientation of ± 45 ° were added to reduce vibration and to absorb deformation energies. The tubes were cured under prescribes conditions.
The tensile test is the basic method of determining static strength of composite materials. Tube testing is difficult. Tube ends crack in jaws of the tearing machine when not specially

prepared. The tensile test can be replaced by a bending test, three or four point bending test. The flexural module is calculated from measured force and deflection variables. The flexural module is recalculated to the tensile modulus. Specimens group for testing four point bending is showed Figure 1, device for four point bending is showed Figure 2 [4, 5, 6].

Fig. 1 Groups of specimens

Fig. 2 Testing device

During the test were measured this variables: the total force on the crossbar of the load medium, its displacement [mm] and the deflection in the centre of the sample w_{max} [mm]. An example of measured values of dependencies F-w_{max} is shown in Fig. 3.

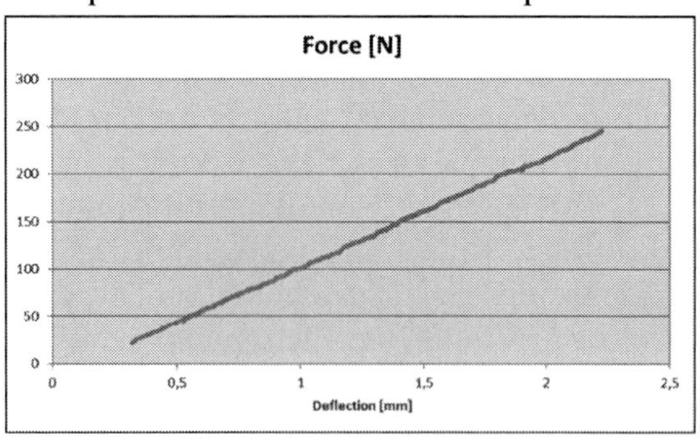

Fig. 3 Dependence F-w_{max}

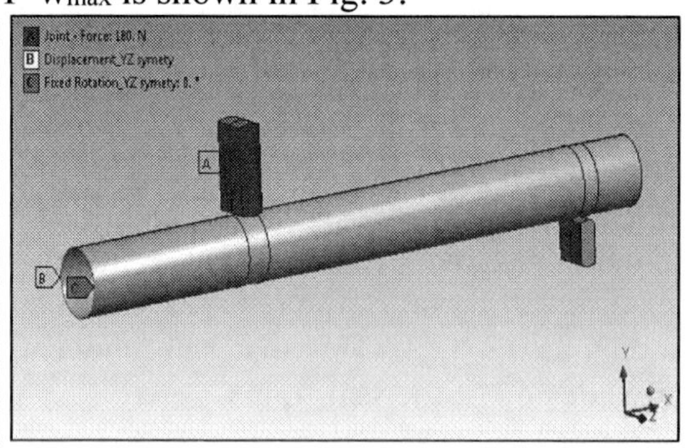

Fig. 4 Four point bending FE model

The specific stiffness of the specimens was determined from the measured values according to the equation (1).

$$Stif = \Delta F / \Delta y \qquad (1)$$

The stiffness of individual specimens and their mean values are shown in Tables 1-3.

Tab. 1 Stiffness of carbon pultruded tubes

Specimen	Stiffness [N/mm]	AVG value	standard deviation
C1	96.3	93.4	2.55
⋮	⋮		
C12	90.6		

Tab. 2 Stiffness of carbon pultruded tubes + 2Cpl

Specimen	Stiffness [N/mm]	AVG value	standard deviation
C2V1	101.3	104.1	1.89
⋮	⋮		
C2V24	107.6		

Tab. 3 Stiffness of carbon pultruded tubes + 4Cpl

Specimen	Stiffness [N/mm]	AVG value	standard deviation
C4V1	116.1	116.1	1.84
⋮	⋮		
C4V24	119.3		

The second method of finding strength modulus is based on the measurement of the velocity of the longitudinal wave propagation in a material with a defined density. The velocity of wave propagation can be measured only by the indirect method, i.e. measuring the time shift at defined distances. The time difference Δt [s] of the passage of the longitudinal wave at the tube length l [mm] was measured during the test. The statistic values of the measured values are shown in Table 4.

Tab. 4

No. specimen	AVG Δt [s]/1m	SMODCH. Δt [s]/1m
C	0.00017	0.00002
C2V	0.00086	0.00003
C4V	0.00079	0.00003

The measurement should be supplemented by measuring the shear modulus for the consistency of the knowledge about the mechanical properties of the composite material. Direct measurement of torsion is a test that gives the exact result. The method is based on the simultaneous measurement of the torque Mk [Nm] and torque angle γ [rad]. Measurements were made during the test: torque Mk [Nm] and torque angle γ [rad]. Example of measured values of Mk-φ dependence is shown in Fig. 7. Specimens group for torsion test is showed Figure 5, device for torsion test is showed Fig. 6.

Fig. 5 Specimens group for torsion test

Fig. 6 Device for torsion test

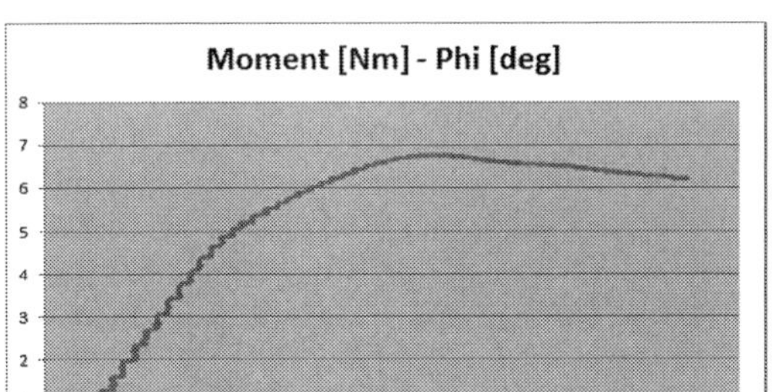

Fig.7 Dependence M_k-φ

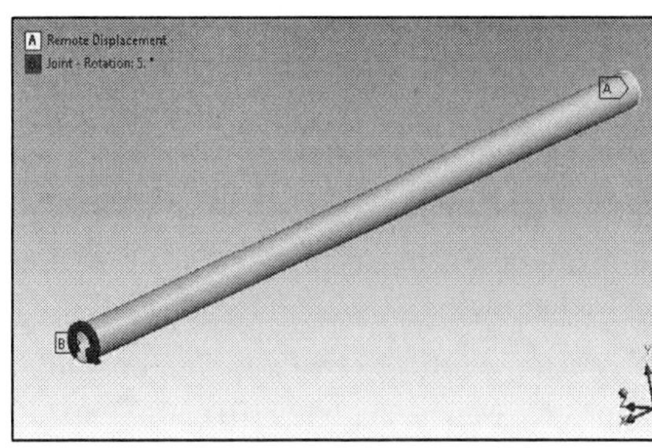

Fig. 8 Torsion FE model

The torsion stiffness was determined from the measured values according to equation (2).

$$Stif\_tor = \Delta Mk/\Delta\varphi \qquad (2)$$

The torsion stiffness of the individual specimens and their mean values are shown in Table 5.

Tab. 5 Torsion Stiffness

No. specimen	AVG Torsion Stiffness [Nm/rad]	SMODCH.
C	-	-
C2V (1 - 6)	85.37	7.10
C4V (1 - 6)	91.10	4.41

Pultruded carbon tubes was not measured by torsion test, because the specimens was destructed by low loading.

Identification of elastic properties of composite structures

The elastic material constants of the layers of the composite structure were evaluated on the basis of specific stiffness that were measured during the experiments. FE models (see Fig. 4, 8) were designed to match physical experiments. The elastic constants of the orthotropic layers of the respective composite structure were found through iterative techniques using the micromechanical theory of the Halpin-Tsai. The results are shown in Table 6.

Tab. 6 Elastic constants

E1	E22	E33	G12	G23	G13	P12	P23	P13
[MPa]	[MPa]	[MPa]	[MPa]	[MPa]	[MPa]	-	-	-
71 380	6 194	6 194	2 274	1 908	2 274	0.322	0.623	0.301

Conclusion

Based on the methodology based on the use of four point bending and torsion, together with the numerical modeling of these experiments by the finite element method and the application of one of many micromechanical theories of composite structures, the elastic constants of the tested composite compositions were determined to be used in the development of prototypes bearing frames of the seats in the means of transport.

Acknowledgment

This publication was written at the Technical University of Liberec as part of the project "Innovation of products and equipment in engineering practice" with the support of the Specific University Research Grant, as provided by the Ministry of Education, Youth and Sports of the Czech Republic in the year 2018.

References

[1] J. Holbery, D. Houston, Natural-Fiber-Reinforced Polymer Composites in Automotive Applications. JOM, 2006.

[2] Y. Bai, X. Ding, Optimal design of composite automotive seat frame based on topology optimization, Shanghai Ligong Daxue Xuebao/Journal of University of Shanghai for Science and Technology 39/1 (2017) pp. 75-80.

[3] A. Dodworth, D. Norton, Carbon multiaxial fabrics for safety-critical application JEC Composites Magazine, 54 (110) (2017) pp. 36-38.

[4] ČSN EN ISO 178 Plasty – Stanovení ohybových vlastností.

[5] ČSN EN ISO 14125 Vlákny vyztužené plastové kompozity - Stanovení ohybových vlastností.

[6] ASTM D6272-17, Standard Test Method for Flexural Properties of Unreinforced and Reinforced Plastics and Electrical Insulating Materials by Four-Point Bending, ASTM International, West Conshohocken, PA, 2017.

FEM Simulation of Elasto-Plastic Tube Indentation Comparison with Experiment

MARVALOVÁ Bohdana[1,a], HRUŠ Tomáš[1,b] and PETRÍKOVÁ Iva[1,c]

[1]Technical University of Liberec, Studentska 2, 46117 Liberec, Czech Republic

[a]bohda.marvalova@tul.cz, [b]tomas.hrus@tul.cz, [c]iva.petrikova@tul.cz

Keywords: Tube Indentation, FEA, Elasto-Plastic State of Stress, Residual Stress, Dent Assesment.

Abstract. Paper provides the summary of the FEM simulation of elasto-plastic strains and stresses in the course of the local indentation of a steel pipe by a spherical indenter. Series of 65 FE parametric studies of the elasto-plastic state of the indented tubes was performed to describe the influence of tube diameter, wall thickness, indenter diameter and the depth of indentation. The numerical results were compared with the experimental results.

Introduction

Mechanical damage is one of the most severe forms of damage that is found in pipelines. The most common defects that reduce both the static and cyclic strength of a pipeline are dents which are caused by a collision of the tube with external rigid objects. A dent in a pipeline is a permanent plastic deformation of the circular cross section of the pipe. Dent depth is defined as the maximum reduction in the diameter of the pipe compared to the original diameter. A dent causes a local stress and strain concentration. The severity of a dent depends on its size and shape and on a number of other factors, for instance whether it affects the curvature of a girth or seam weld and whether it contains other defects, such as a gouge or a crack [1]. In many cases it is necessary to repair such damage. However, in some cases, a combination of detailed inspection and assessment can result in conclusion that the damage is acceptable [2].

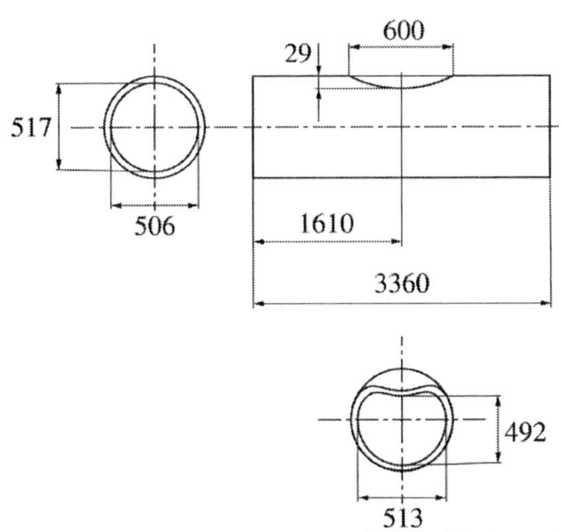

Fig. 1. Dent caused by the experimental indentation

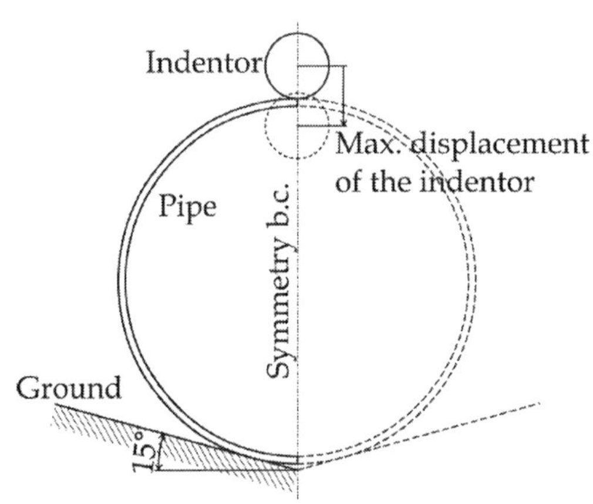

Fig. 2. FE ¼ model of indentation

FE simulation of the tube indentation

The FEM calculation was performed according to the recommendation of API 579-1 / ASME FFS-1 2007 [3] for Level 3 Dent Assessment. The configuration of the experimental loading of the indented tube has two planes of symmetry see Fig. 1. FE simulation model was adjusted as ¼ of the whole with appropriate symmetrical boundary conditions see Fig. 2. The FE model of tube was supported equally as in the experiment i.e. the support was modeled as the rigid plane with the inclination of 15 degrees. The contacts were defined between the tube and the rigid spherical indenter and between the tube and the rigid support. The loading was controlled by the displacement of the indenter. Series of 65 FE calculations of the elasto-plastic state of the indented tubes was performed. The combination of three tube diameters D=508, 920 and 1420 mm with the ratios between the diameter and the wall thickness D/h= 91, 76, 64, 51 and 37 resp. were used. The diameters of the spherical rigid indenters were 100 and 200 mm, the depth of the indent varied gradually from 75 to 150 mm. The residual stresses and strains in the vicinity of the dent after the relieving of load were determined together with the depth and length of the resulting dent. The elasto-plastic material properties were obtained from the tensile test curve of pipe material which was converted to True Stress – True (Logarithmic) Strain curve for the FE calculation purpose. FE mesh was created using 8-node hexahedron elements which allow the local adaptability of mesh that is necessary in large strains. Each single FE model from a series of 65 cases was created using a Python script which comprised the combinations of the tube diameter *D*, the wall thickness *w*, the indenter diameter *di* and the immersion depth of indenter *h*.

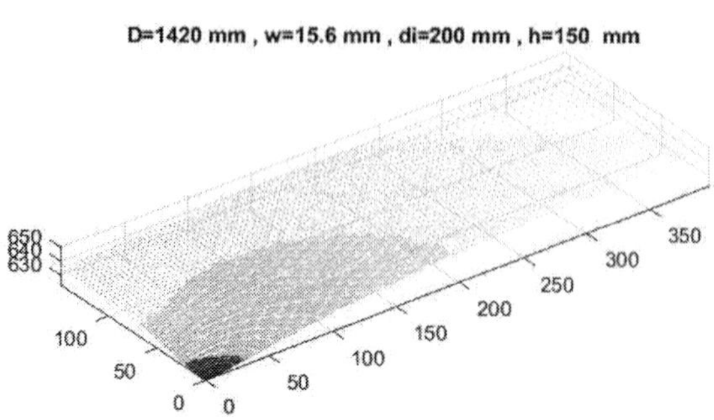

Fig. 3. Shape of tube under the indenter

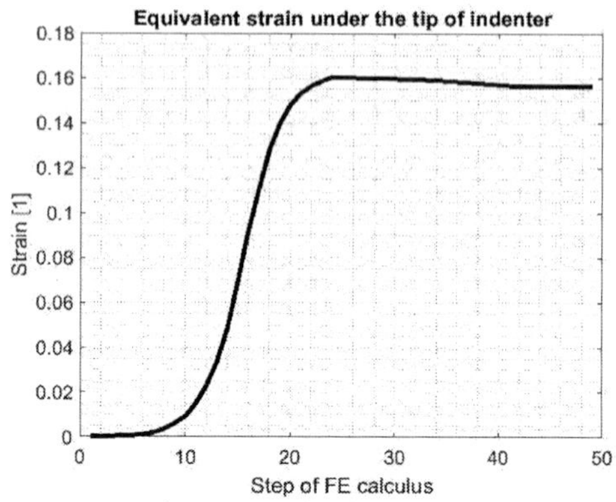

Fig. 4. Equivalent strain

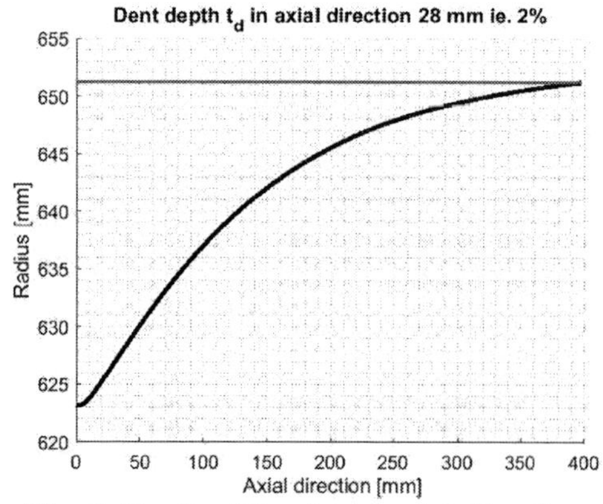

Fig. 5. Depth of the dent - axial direction

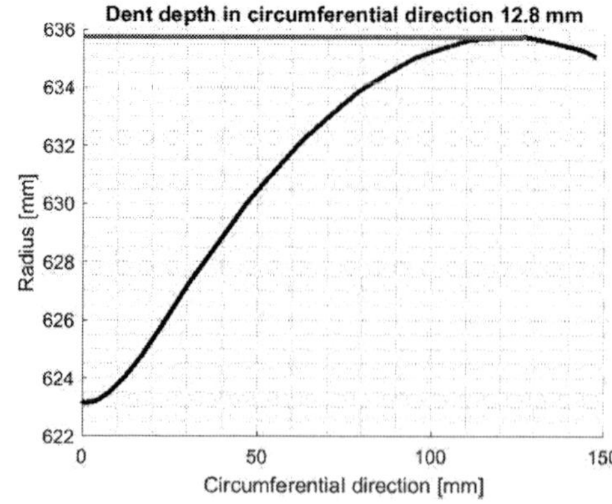

Fig. 6. Depth of the dent - circumf. direction

Results of FEA

The evaluation of the FE results and the graphic processing were performed in Matlab. Some results for a particular configuration (the tube diameter D=1420 mm, the wall thickness w=15.6 mm, the indenter diameter di=200mm and the immersion depth of indenter h=150mm) are shown in the Fig. 3 to 6 where the deformed shape of the tube is captured together with the evolution of equivalent strain inside the tube under the indenter in course of loading and the final value of the residual equivalent strain (around 15%) after the unloading. The shapes of dent in axial and circumferential directions are shown in Fig. 5 and 6. The depths of dent in these principal directions can be determined as the distances between the tangent lines (red lines) and the bottom of dent. One of the parameters characterizing the dent is the ratio between the axial depth of dent (28 mm) and the diameter of the tube (2%).

The contours of central cross-section at the different steps of loading are captured in Fig. 7 for the configuration (D=508, w=13.8, di=100 and h=25mm). The displacement of the point A located inside the tube under the indenter is in Fig. 8. The ovalization of the tube cross-section is also noticeable. The successive equivalent strain and principal stresses at this point are shown in Fig. 9 and 10. The residual deformation of the tube and the residual strain and stresses are also apparent in these figures.

It is therefore possible to establish through the FEA all the necessary parameters that US and British standards require for the assessment of safety of the pipeline with the defects of dent type.

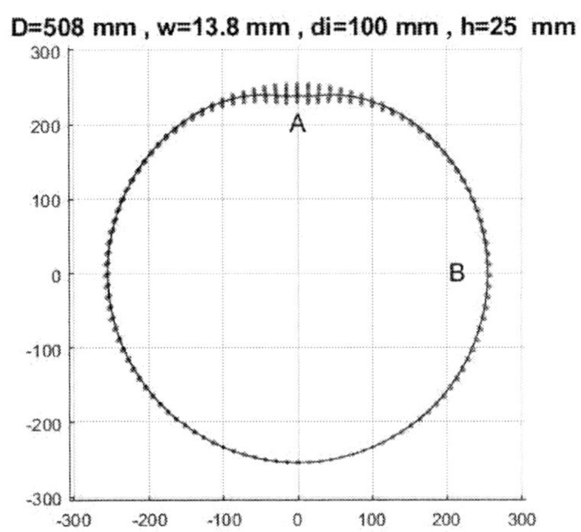

Fig. 7. Tube contour in course of indentation

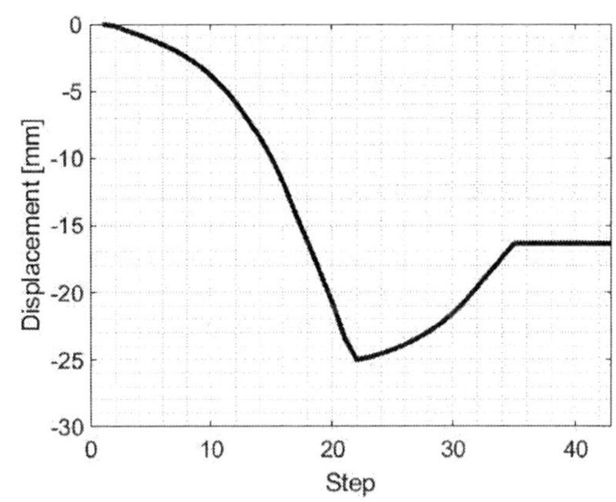

Fig. 8. Displacement of the point A

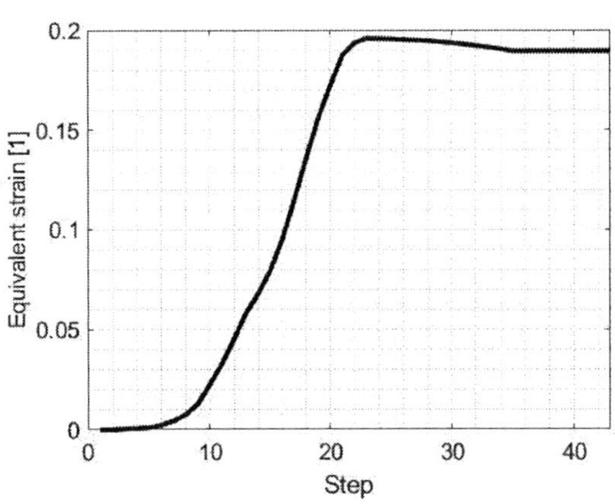

Fig. 9. Equivalent strain in point A

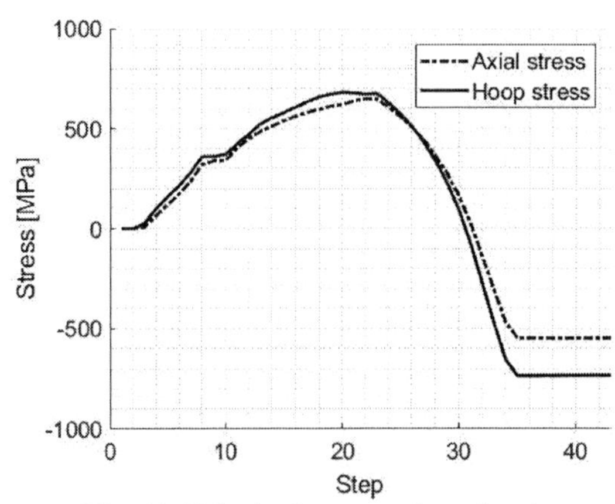

Fig. 10. Principal stresses in point A

Conclusions

In this paper, the residual stresses and strains in a series of numerical simulations of steel pipes subjected to indentation were quantitatively investigated based on nonlinear FEM. A numerical model has been developed and validated by test results, capable of predicting residual strength of dented pipe. The effects of tube and indentation variable parameters including the indentation depth, the diameters of tube and indenter and the tube thickness have been identified. The results of FEA were compared with experimental measurements [4] and they were found in good agreement.

References

[1] Allouti, M., C. Schmitt, and G. Pluvinage. Assessment of a gouge and dent defect in a pipeline by a combined criterion. Engineering Failure Analysis 36 (2014): 1-13.

[2] Macdonald, K. A., A. Cosham, C. R. Alexander, and P. Hopkins. Assessing mechanical damage in offshore pipelines–Two case studies. Engineering Failure Analysis 14, no. 8 (2007): 1667-1679.

[3] API 579-1 / ASME FFS-1 2007, Recommended Practice for Fitness-For-Service, Standard for Pressurized Equipment, ASME, 2007, http://proceedings.asmedigitalcollection.asme.org/proceeding.aspx?articleid=1634507

[4] Marvalová, B., M. Sivčák, I. Petríková. Strain Distribution in Vicinity of Dent in Tube in the Course of Indentation. Proc. of EAN 2018, 56. Intern. Conf. on Experimental stress analysis, Harrachov, Czechia, June 5th - 7th, 2018

Strain Distribution in Vicinity of Dent in Tube in the Course of Indentation

MARVALOVÁ Bohdana[1,a], SIVČÁK Michal[1,b] and PETRÍKOVÁ Iva[1,c]

[1]Technical university of Liberec, Studentska 2, 46117 Liberec, Czech Republic

[a]bohda.marvalova@tul.cz, [b]michal.sivcak@tul.cz, [c]iva.petrikova@tul.cz

Keywords: Experimental study, Steel tubes, Lateral loading, Dents, Strain gauges, Elasto-plastic strains, Stress field estimation.

Abstract. Dents are the common mode of damage which can cause failure of pipelines. The paper describes experimental and numerical investigation of the state of stress and strain in the course of the indentation of large thin walled steel tubes. The tubes resting on a V support block were laterally indented by a large spherical steel indenter. Strains in the vicinity of the dent were measured by resistance strain gauges and the elasto-plastic state of stress was evaluated from the experimental data. The results were compared with finite element simulation.

Introduction

Dents are caused by a collision of the tube with an external rigid object during the laying of the pipeline, by the lateral impact of heavy loads or by the shock caused by an excavation equipment. Considerable plastic strains are developed in the course of the formation of dents. The dent size, shape and location can significantly affect the service life of pipeline.

Experimental and numerical investigation of the indentation of tubes was a matter of interest for a number of scientists. The local dent without surface crack in a pressurized pipe was investigated by Ong and all. [1]. They were focused on the experimental gauging test and FE analysis in the elastic strain distribution. Papers of Allouti and all. deal with steel pipelines containing a gouge and dent defect [2,3]. They performed also experimental burst tests and FEA to compare the numerical and experimental results. Experimental and numerical research of Moustabchir [4] presents the stress–strain state of pressurised cylindrical shells with longitudinal external defects which includes the effects of plasticity. A fatigue analysis of dented pipeline subjected to internal pressure which is based on experimental data was presented recently by Garbatov [5]. In this paper, the different failure criteria considering the dent size and applied load are analysed and reliability as a function of number of load cycles is defined. Papers dealing with the analysis of experimentally determined strains in the elasto-plastic domain are sparse. Keil and Benning [6] described the practical procedure of the strain analysis in the elasto-plastic range based on the stress-strain curve of the material and evaluated the 2d state of stresses from the strains measured by resistance gauges. Rao [7] further developed this method. A rigorous approach founded on the flow theory of plasticity for elastic-plastic boundary value problems was presented by Sutton and Deng [8]. They suggested an numerical procedure for the determination of stresses and elastic and plastic strains from the experimental strain data measured on the surface of a structure by the numerical integration of elastic-plastic constitutive equations. The evaluation of strain gauge measurements in the elasto-plastic area was also presented by Dolhof [9].

Fig. 1. Experimental setting of the tube indentation

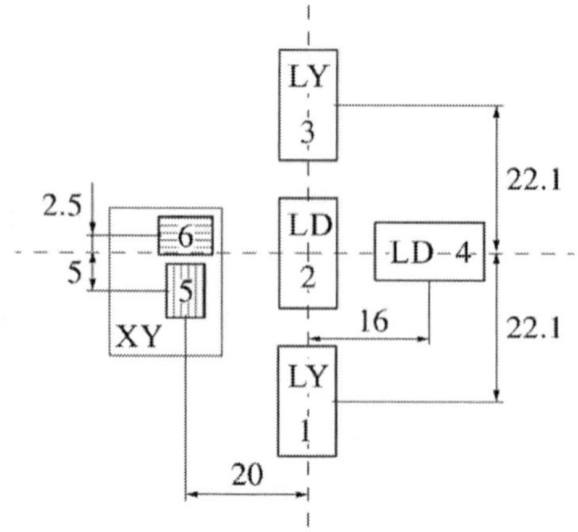

Fig. 2. Placement of the strain gauges

Experimental

The experimental work consisted of the indentation of two large steel specimens cut out of two different thin-walled tubes. The test specimens were 530 mm in diameter, 3360 mm and 2990 mm long, with corresponding D/t ratios of 56 and 66. The pipes were tested on an INOVA FU 250 kN testing machine in Fig. 1. The specimens were simply supported by a V support block at the middle-length and subjected to a concentrated transversal loading by the spherical steel indenter 200 mm in diameter. Load-displacement response and the response of resistance strain gauges as functions of dent depth are recorded for each specimen. The elasto-plastic stresses were evaluated from the records of strain gauges. HBM resistance strain gauges were fixed inside the tube under the indenter. The first experiment was performed on the tube with D/t ratio of 56. We used two linear strain gauges LY with the base 5 mm for strain up to 5%, then two LD 5 mm for large strains up to 10% and one XY rosette of 5 mm for strain up to 5%. The gauges were attached inside the tube under the center of indenter. The strain gauge layout is shown in Fig. 2. The loading process was controlled by the displacement. The indenter was plunged slowly into the tube to a depth of 60 mm then the load was relieved. The residual dent had a depth of 37 mm. The nonlinear loading curve is shown in Fig.3. The response of the strain gauges is shown in Fig.4. The strain gauges LY and the XY rosette were broken when the limit of their strain 5% was exceeded. The data of LD gauges in axial and circumferential directions were evaluated and elasto-plastic principal stresses were calculated.

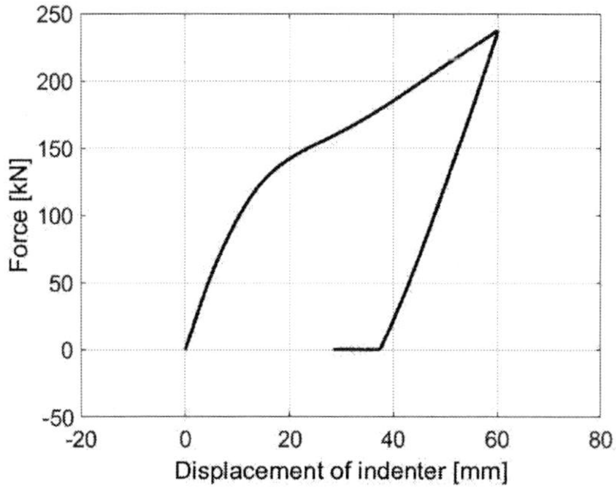

Fig. 3. Loading curve of the tube D/t = 56.

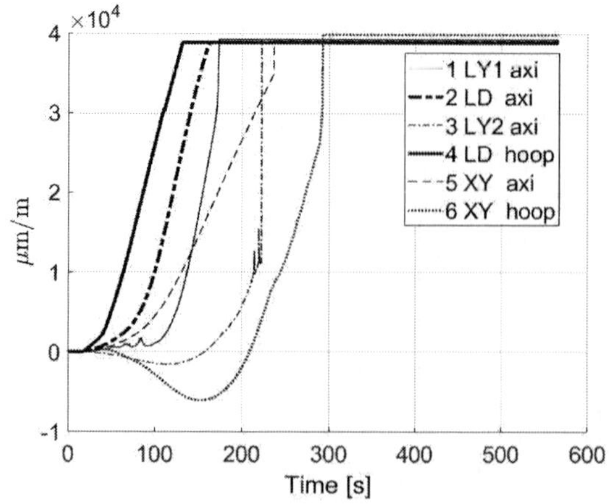

Fig. 4. The response of the strain gauges.

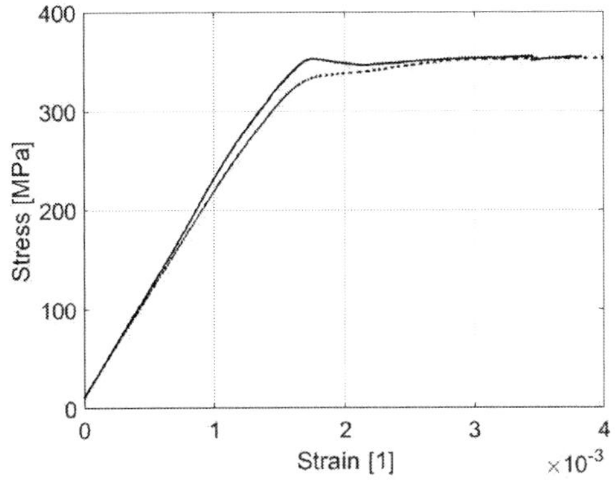

Fig. 5. Tension diagram of tube material

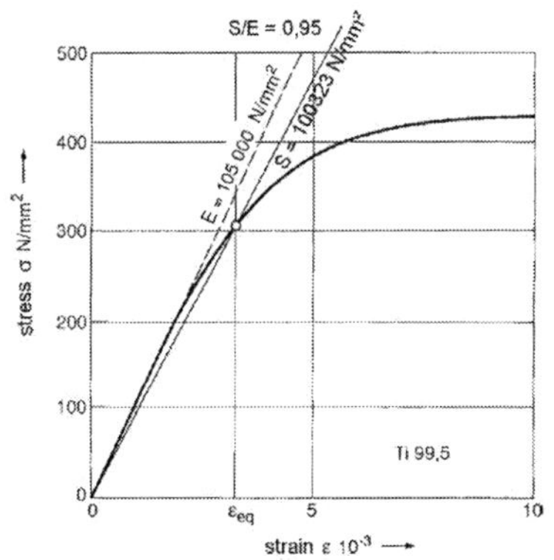

Fig. 6. Example of the evaluation of the secant modulus *S* and the ratio *S/E* from Keil [6]

Evaluation of experimental data

The tensile diagram of the pipe material in Fig.5 shows the ductile steel behavior almost without hardening with the yield stress of about 350 MPa then we can suppose the ideal elasto-plastic material. We use Von Mises yield criterion in principal stresses for 2d state of stress

$$\sigma_1^2+\sigma_2^2-\sigma_1\sigma_2=\sigma_Y^2. \tag{1}$$

The elasto-plastic stress-strain behavior of 2-d state of stress can be formally described in the same way as the linear elastic ones [6] if the variable secant modulus $S=\sigma_{eq}/\varepsilon_{eq}$ in Fig. 6 and the combined Poisson's ratio ν_g are used instead of the elastic E and ν. The combined Poisson ratio is evaluated by relation

$$\nu_g=0.5-(0.5-\nu)\frac{S}{E}, \tag{2}$$

where S is the secant modulus. Apparent Poisson ratio ν_g for our material is in Fig. 6, the value of the ratio lies between elastic ν and the plastic Poisson ratio 0.5. The elasto-plastic 2d stresses will be evaluated from the relations

$$\sigma_t=\frac{S}{1-\nu_g^2}\left(\varepsilon_t+\nu_g\varepsilon_a\right),\quad \sigma_a=\frac{S}{1-\nu_g^2}\left(\varepsilon_a+\nu_g\varepsilon_t\right), \tag{3}$$

where ε_t and ε_a are strains measured by gauges. The equivalent strain is given by relation

$$\varepsilon_{eq}=\sqrt{N_q\left(\varepsilon_1^2+\varepsilon_2^2\right)+N_g\varepsilon_1\varepsilon_2},\quad N_q=\frac{1-\nu_g\left(1-\nu_g\right)}{\left(1-\nu_g^2\right)^2},\quad N_g=\frac{-1+\nu_g\left(4-\nu_g\right)}{\left(1-\nu_g^2\right)^2}. \tag{4}$$

The elasto-plastic axial, hoop and equivalent stresses were determined from LD gauges records. The stresses satisfy the condition of plasticity as shown in the graph in Fig. 7. The evaluation through an iteration process was done in Matlab.

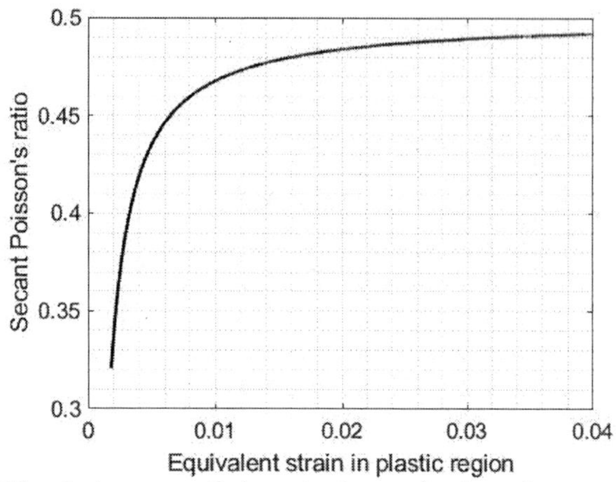

Fig. 6. Apparent Poisson's elasto-plastic ratio

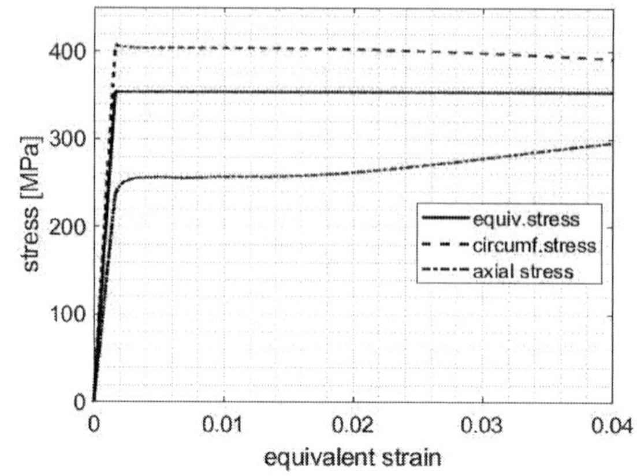

Fig. 7. Stresses calculated from LD gauges data

Conclusion

Elasto-plastic axial and circumferential stresses were determined from the principal strains measured by gauges in the course of tube indentation. The simplified method of the evaluation of strain gauges measurement above the yield limit of material was presented.

References

[1] Ong, L. S., A. K. Soh, and J. H. Ong. Experimental and finite element investigation of a local dent on a pressurized pipe. The Journal of Strain Analysis for Engineering Design 27, no. 3 (1992): 177-185.

[2] Allouti, M., C. Schmitt, and G. Pluvinage. Assessment of a gouge and dent defect in a pipeline by a combined criterion. Engineering Failure Analysis 36 (2014): 1-13.

[3] Allouti, M., C. Schmitt, G. Pluvinage, J. Gilgert, and S. Hariri. "Study of the influence of dent depth on the critical pressure of pipeline." Enging Failure Analysis 21 (2012): 40-51.

[4] Moustabchir, H., Z. Azari, S. Hariri, and I. Dmytrakh. Experimental and numerical study of stress–strain state of pressurised cylindrical shells with external defects. Engineering Failure Analysis 17, no. 2 (2010): 506-514.

[5] Garbatov, Y., and C. Guedes Soares. Fatigue reliability of dented pipeline based on limited experimental data. International Journal of Pressure Vessels and Piping 155 (2017): 15-26.

[6] S. Keil, O. Benning, On the evaluation of elasto-plastic strains measured with strain gauges, Experimental mechanics, (1979),19 (8), pp.265-270.

[7] B. N. Rao, A simplified procedure for stress analysis in the elasto-plastic range with the aid of strain gauges, Experimental techniques, (1991), pp. 38-39.

[8] Sutton, Michael A., Xiaomin Deng, J. Liu, and L. Yang. Determination of elastic-plastic stresses and strains from measured surface strain data. Experimental Mechanics 36, no. 2 (1996): 99-112.

[9] Dolhof, Vaclav. Evaluation of Strain Gauge Measurements in Elasto-Plastic Area. In Experimental Stress Analysis, pp. 429-437. Springer, Dordrecht, 1986.

EAN 2018
56th conference on experimental stress analysis

The Design of Final Drive Pinion Temperature Measurements Realized During the Real Passenger Car Operation

MAZAČ Martin[1,a], HERAJN Pavel[2] and MALÝ Miroslav[3,c]

[1,2,3]Technical University of Liberec, Liberec, Czech Republic

[a]martin.mazac@tul.cz, [c]miroslav.maly@tul.cz

Keywords: temperature, measurement, gearbox, final drive, pinion, gearing, contactless method.

Abstract. This short article describes the design of unique experimental passenger car gearbox final drive pinion teeth temperature measurement. The MQ200 (Skoda Auto a.s.) for the measurement was chosen. The special temperature measurement system in article is described. The measurement system combined the contact temperature system using the NTC sensors with contactless optical IR data transfer method. The temperature sensors parameters are contained. The hardware (used electronic parts) of the measurement system are shortly described too. The temperature sensors and other measurement system parts placing possibilities are designed.

Introduction

The temperature is one of the most important parameters for the machine operation condition analyses. The temperature can cause machine parts damage or destruction [f.e. 1]. The temperature is danger for all kinds of used technical materials. The physical material parameters are changed by temperature. For the correct and economical operation of any machine the optimal temperature is necessary. The passenger car gearbox is a typical sample of the temperature depended mechanical system. The temperature changing the mechanical wills and physical parameters of oil. The right oil temperature for the optimal viscosity and lubrication parameters is necessary. The part of the input power to heat is changed during the gearbox operation. The heat cause the temperature changes. The gearbox efficiency is depend on temperature directly. The majority of dissipated mechanical energy during the gears operation to heat is transformed. The temperature of the gears is very useful parameter shows the tooth flank friction processes consequences [f.e. 2]. The gearbox parts temperature measurements is very useful for the construction optimization. The modern simulation methods are not able to summarize all of the ongoing processes yet. The right boundary conditions for the simulation in necessary to know. The heat is not produced by the gearbox during the real passenger car operation only. The heat from the combustion engine by the clutch case to gearbox is conducted. The many processes and surroundings influence the gearbox operational condition. The temperature measurement is possible way how to find the gearbox temperature condition.

The Temperature Measurement System (TMS)

The temperature measurement system [f.e. 3] was developed by Technical University of Liberec employees. The system was tested and the functionality was verified. The system is used for the temperature measurement and data transportation primary. The temperatures are measured on a rotating shaft by the thermistors.

The miniature NTC sensor for the temperature measurement can be used. The TMS for the 2K7MCD1 sensors by Telemeter Electronic GmbH (Fig. 1) is designed. The small thermistor size is suitable for the application- the teeth are only about 5 millimetres high (teeth module about 2,5mm, circular teeth thickness about 2,5mm).

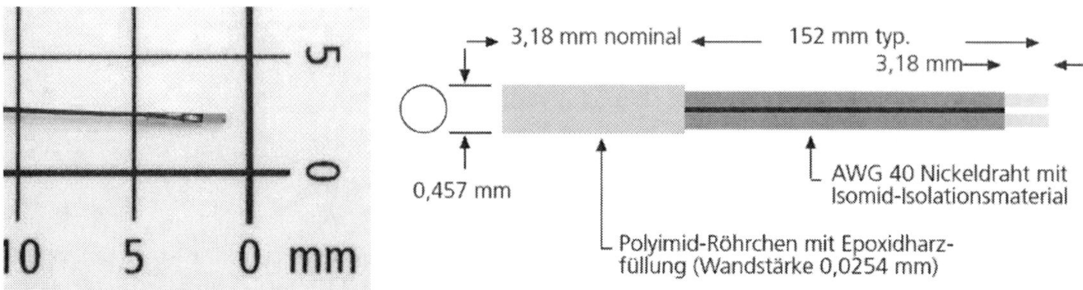

Fig. 1 Thermistor NTC 2K7MCD1

The 2K7MCD1 is one of the smallest temperature sensors. The reaction time of this sensor is 200 ms only during use in a liquid medium. The characteristic resistance of 2K7MCD1 is 2 kΩ in 25 °C/77 °F. The possible error of the measurement in 25 °C/77 °F is ± 0,2 °C. The temperature sensor is good protected by a polyamide tube cover and inner filling of the epoxy resin. Connection of the sensor is realized by the sheathed cable. The connection cable and sensor are supplied by the producer as a set.

The rotating part of the system evaluate the temperatures and coding them for transmission to static parts of the system. The data transmission by the infrared radiation technology is realized. The infrared diode in the rotation axis of a shaft is placed.

a. b.

Fig. 2 a.- TMS application, b.- TMS

The rotation parts of the system are on the metal disc out of a gearbox mounted- the last application in the MQ100 gearbox on the Fig. 2a. The static part of the system with the IR sensor in front of the rotating IR diode is placed. The static part of the system decode the IR signal and processes data. The computer for other data processing by USB to connect is possible. The TMS rotation and static parts are on Fig. 2b. The data processing by the

LabVIEW software is realized. The TMS in the [3] is deeply described. The main electronic function principle as a block scheme is on the Fig. 3.

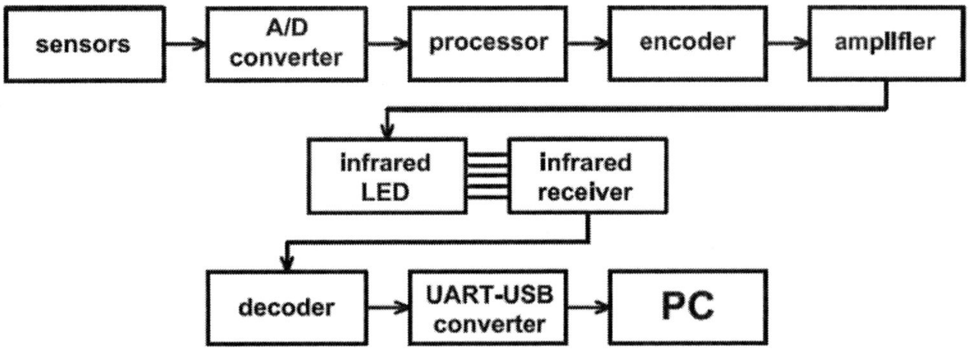

Fig. 3 TMS block scheme

The rotating part of the equipment contents two printed circuit boards with mounted electrical parts. The sensor signal evaluation and a battery are there. For the measurements of the voltage drop on the NTC sensors an A/D converter is used. There are four A/D converters type ADS1248. The ADS1248 is the 7-channel, 24-bits A/D converter with an internal amplifier and two adjustable electric current sources. The three-wire configuration for connection of the temperature sensors is used. Sixteen channels are possible to use.

The rotating part contains the processor (microcontroller) ATTiny2313 8-bit microcontroller with 2K bytes in the system programmable flash. The microcontroller is necessary for operation of the A/D converter and next sending of information to an encoder. The communication between the microcontroller and A/D converters is realized as a SPI with auxiliary signals (START, CS...).

The data from the microcontroller are sending to the encoder by the UART. The encoder transforms data to the right format for the successful sending to the static parts of the equipment. For transforming of the data the TIR1000 encoder is used. The signal transformed by the encoder is amplified by the transistor. The sending of the amplified signal out by the TSMF1020 high speed infrared light emitting diode (with double hetero technology (DH), moulded in a clear SMD package with dome lens) is realized.

The stator part is primarily designed for receiving and treatment of the IR signal from the rotor part. The TFDU4101 infrared receiver module is used here. The MCP2120 microchip for decoding of the signal from the IR receiver on the stator is used. The IR signal is decoded back to the UART.

The output data from the decoder are the UART format and the USART/USB converter for the transformation is used. The FT232RL converter for conversion of the signal is used. The output data from the converter are in the USB format and reading of them by the computer connection is easy and comfortable.

The temperature measurement system to calibrated is possible. The special calibration furnace with the cooper isothermal chamber is used. The optimal thermistor constants are computed by the special algorithm, described in [3].

The Gearbox and Measurement Places

The gearbox MQ200 for the experiment was chosen. The MQ200 is produced by Skoda Auto a.s.. The MQ200 is two shaft manual shifted gearbox. The maximal input load is

200Nm. The MQ200 with a five or six gears (without a reverse) is manufactured. The gearbox with many types of combustion engines is set. The final drive pinion gearing as a critical temperature loading place was chosen.

The termistors to mount near of a tooth flank is possible. The thermistor position with the teeth strength consideration can be chose. The termistors can be put into the holes in the teeth. The electrical discharge machining for the holes manufacturing can be used. The sample of the thermistors position (as MQ100 final drive pinion was used) on the Fig. 4.

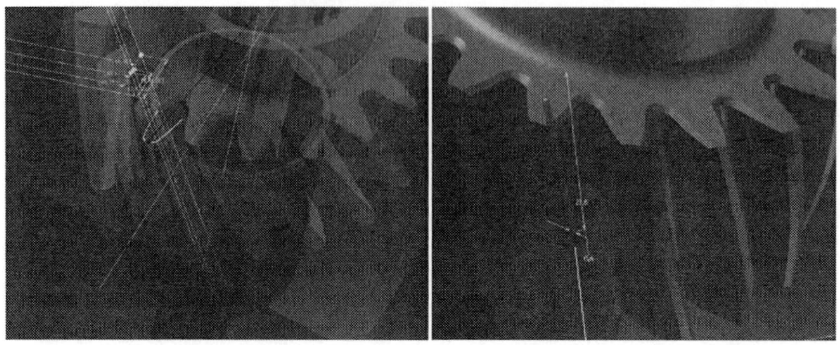

Fig.4 Temperature sensors position sample- CAD model

The Design Description

The shafts of gearbox are axially drilled for the weight reduction. The wires of sensors inside of the shaft can go through. The rotating output of the temperature measurement system out of the gearbox (near of fifth gear) can be placed. The rotating parts can be placed on a metallic disk which is mounted on a steel shaft extending part. The steel shaft extending part can be pressed to the gearbox output shaft. The shaft extending part with the case sealing element can go through.

The modified gearbox in a real car can be mounted and the final drive pinion temperature measurements during the normal operations can be realized. The design of experiment on Fig. 5.

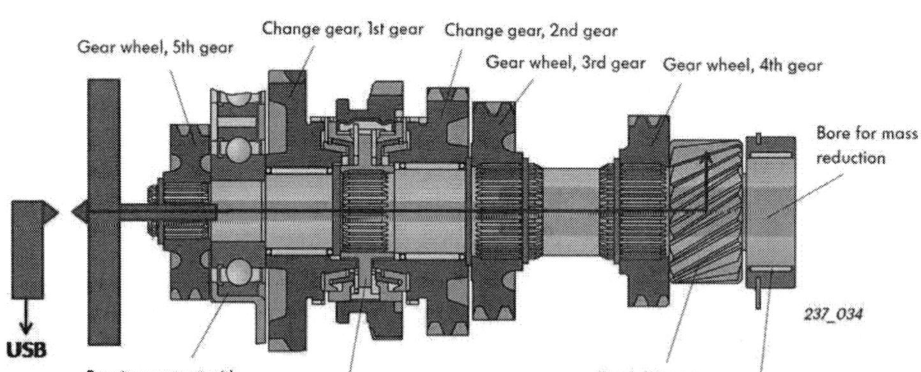

Fig. 5: The TMS application for MQ200 gearbox [4]- modified

The thermistors mounting is possible with the supporting plastic tubes. The supporting tubes to lead by the holes at first is necessary. The thermistors final mounting by the moving of supporting tubes is realized. The thermistors by the epoxy resin in the final places can be fixed- for the heath conduction the using of epoxy resin with the metallic filling is the best. The mounting of thermistors on the Fig. 6 is depict- application in MQ100 final drive pinion.

Fig. 6 The thermistor mounting

Conclusions

The short article deals with an unique design of the experimental temperature measurements of the final drive pinion gearing of the MQ 200 passenger car gearbox. The application and important technical parameters of TMS are described. The aim is design of the experimental real gearing operational temperature measurements. The unique experiment can be realized during the common passenger car operation, the parameters as a air flow or heath convection are not ignored. This short study describes the equipment for the teeth temperature measurement developed by TUL, the equipment is suitable and ready for chosen application. The designed experiment is the next important source of the operational parameters for other experiments and optimizing.

Acknowledgment

The results of this Project LO1201 were obtained with co-funding from the Ministry of Education, Youth and Sports as part of targeted support from the "Národní Program Udržitelnosti I" program.

References

[1] B.R. Höhn, K. Michaelis, H.P. Otto, Influence of immersion depth of dip lubricated gears on power loss, bulk temperature and scuffing load carrying capacity, International Journal of Mechanics and Materials in Design [online], 4.2 .2008: 145–156, DOI 10.1007/s10999-007-9045-z.

[2] S.B Rao., D.R. Mcpherson, Gear tooth temperature measurements, Gear Solutions, Pelham - Alabama: Gear Solutions Magazine, August 2009.

[3] M. Mazac, P. Herajn, M. Svoboda, The Gear Tooth Flank Temperature Measurements System. Transactions on Electrical Engineering Vol.4 [online], No.1, p.27-30, ISSN 1805-3386, Praha CTU, 2015.

[4] Volkswagen AG, The Self-study Programme 237- Manual Gearbox 02T- Design and Function, Wolfsburg, Germany, 2001.

EAN 2018
56th conference on experimental stress analysis

Experimental Verification of FEA Model

MEDŮNA Ondřej[1,a]

[1]VÚTS, a.s. Svárovská 619, 46001 Liberec, Czech Republic

[a]ondrej.meduna@vuts.cz

Keywords: FEA simulation, material test, material model, verification, experiment, 3D scan, deformation, variable test frame.

Abstract. The article deals with our procedure during steps FEA simulation, real test and their comparison which is shown on deformation tests of car bumper. These steps are supported by material tests and sophisticated scanning method.

Introduction

The output verification of the each procedure step during the R&D process is very important. Primarily, it means validation of result correctness for each single step in development cycle mostly related to the conformity of simulation and real test. In our practice we solve this issue with a comprehensive use of our current possibilities in the field of test, simulation and scanning methods. This is part of a cycle of connected activities analysis – optimization – design – prototype – test, which leads from initial analysis to final tested product.

The part of this cycle is verification of FEA simulation model via test results on the real part.

Material model

The material tests are done due to need to determine the real behavior of used materials on analyzed structure. These requirements were met by our own tensile and flexural tests. These types of test are chosen because of expected type of structure load: *finger test* that means local pushing into the bumper with controlled displacement and external load force. Testing machine for static tests, biaxial extensometer and environmental chamber were used for these material tests to get stress - strain curve and Poisson's ratio. The specimens were cut from real parts of bumper (Fig. 1).

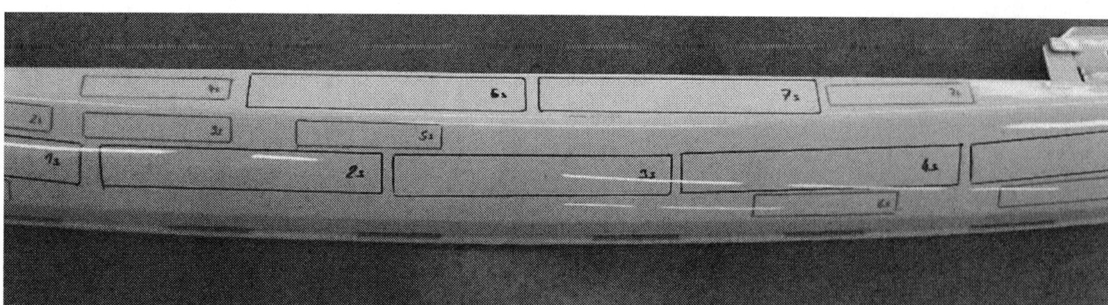

Fig. 1 Preparation of specimens

The stress – strain curve is used as input to material model. Because no applied force release is considered, a nonlinear elastic material model is used. The FEA simulation of both types of material tests is used to validate input data for the selected material model. To achieve compliance, input data of the material model is modified based on the FEA model results of the performed tests.

The results of the original (FEM - orig) and modified (FEM - final) input data for the material model compared to the real tensile and flexural test data (TEST) are shown on Fig. 2 and Fig. 3.

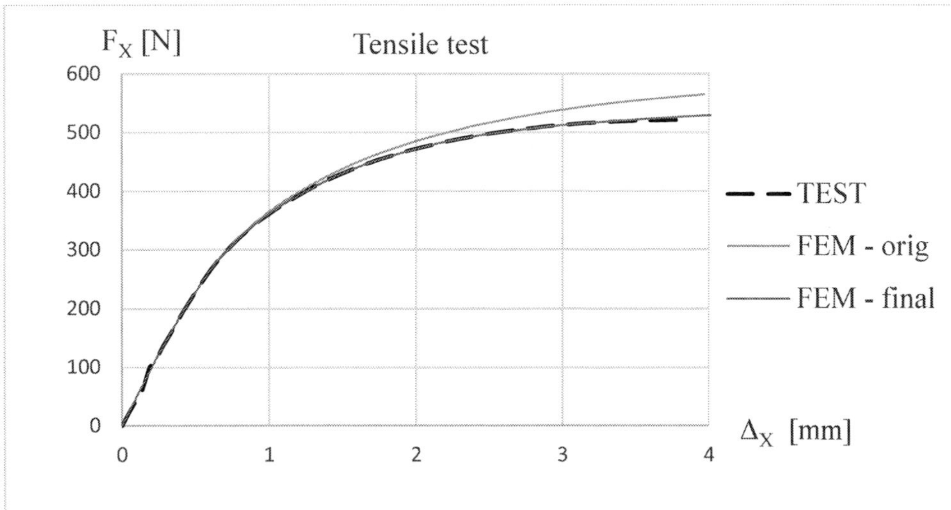

Fig. 2 Comparison of the tensile test and its FEA simulation

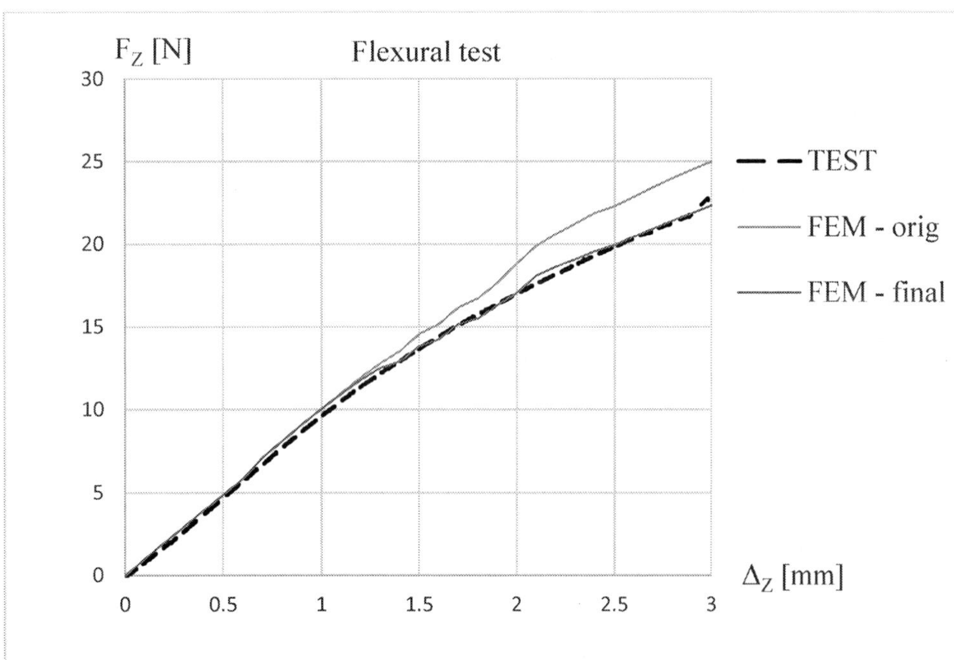

Fig. 3 Comparison of the flexural test and its FEA simulation

Test frame

For simulation and test purposes, a variable test frame [1] is designed, constructed and manufactured (Fig. 4). On the frame, there is a set of reference points for coordinate system definition that is identical to the CAD model coordinate system. It allows us to automatically input scanned data to the CAD/FEA model with the same global coordinate system. The frame consists of variable attachment points. The X, Y, Z coordinates of each one are mutually independent. This allows us to achieve high accuracy of the position of the tested part that is enabled by verification of attachment point positions by 3D scanning. In this way, the error of each attachment point positions is less than 0.5 mm (while the bumper width is over 1500 mm).

Another advantage of this concept is the use of the CAD model of the test frame to find required position of the loading system for the real test frame and define the right direction of applied force in FEA model.

In this way, compliance of the boundary conditions of the FEA model with the real test is guaranteed.

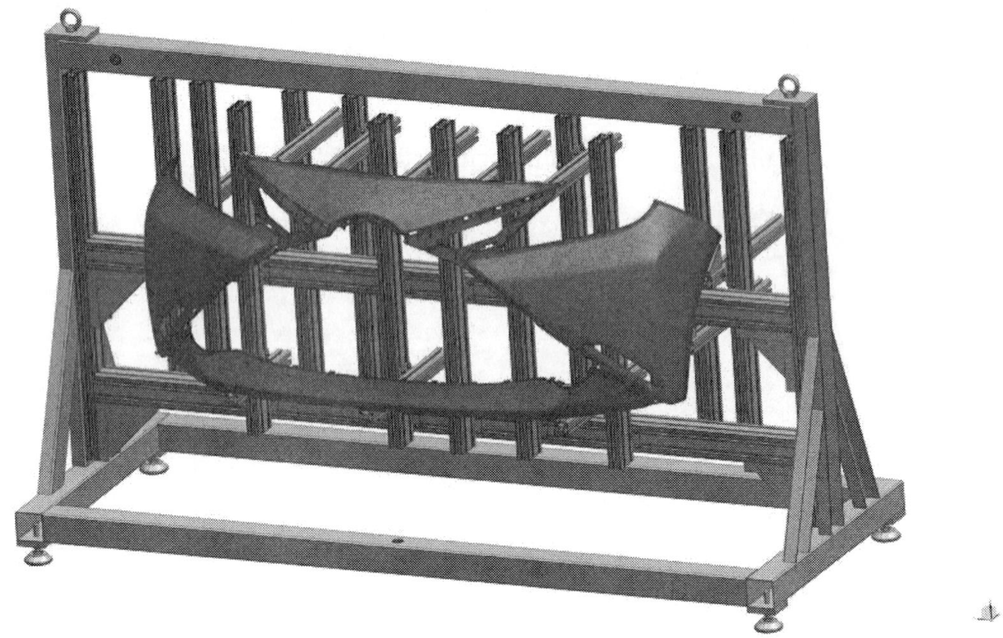

Fig. 4 Test frame with FEA mesh of the bumper

Test

A suspended weight with a pulley is used as a loading system. As mentioned above, the CAD model of test frame is used for determination of its desired position. The load force value is not determined solely by the weights used due to the influence of the friction on the pulley. Therefore, a force sensor located between the bumper and the pulley is used to determine the exact value of the load force. Such a measured load force value is then used in the FEA simulation. When a load force is applied to the bumper surface, its surface is scanned by the optical 3D scanner (Fig. 5) [2]. The STL data thus obtained are used for comparison with the results of the simulations.

Fig. 5 Scanning the bumper while the load force is being applied

Simulation

The main part of the car bumper with the auxiliary components is considered to be the deformable structure of the FEA model. This geometry is constrained to the rigid frame using preloaded bolts. 3D mesh of the whole analyzed structure is created using NX Siemens. It consists of 20-nodal hexahedral elements on large smooth faces and 10-nodal tetrahedral elements on more curved geometry. These two types of mesh are joined together by 13-node pyramidal elements. The whole FEA simulation is completed in pre/postprocessor MSC.Mentat and solved in MSC.Marc solver.

The contact between all deformable parts is considered the same as between bumper parts and rigid frame. The assembly is loaded in three stages. In the first one, the fastening of the joint is simulated by applying a pretension to each of the bolts. In the second one (after the bolts have been loaded), gravity is applied, and in the third one, the load force is applied to the top surface of the bumper to simulate finger test. Value of this force is measured in the previous real test on the bumper. The direction of this applied force is determined using CAD model of the test frame.

The STL file that represents deformed shape of FE mesh under external load is prepared from simulation results (Fig. 6). This file is then used for comparison with deformed shape of the bumper from the real test.

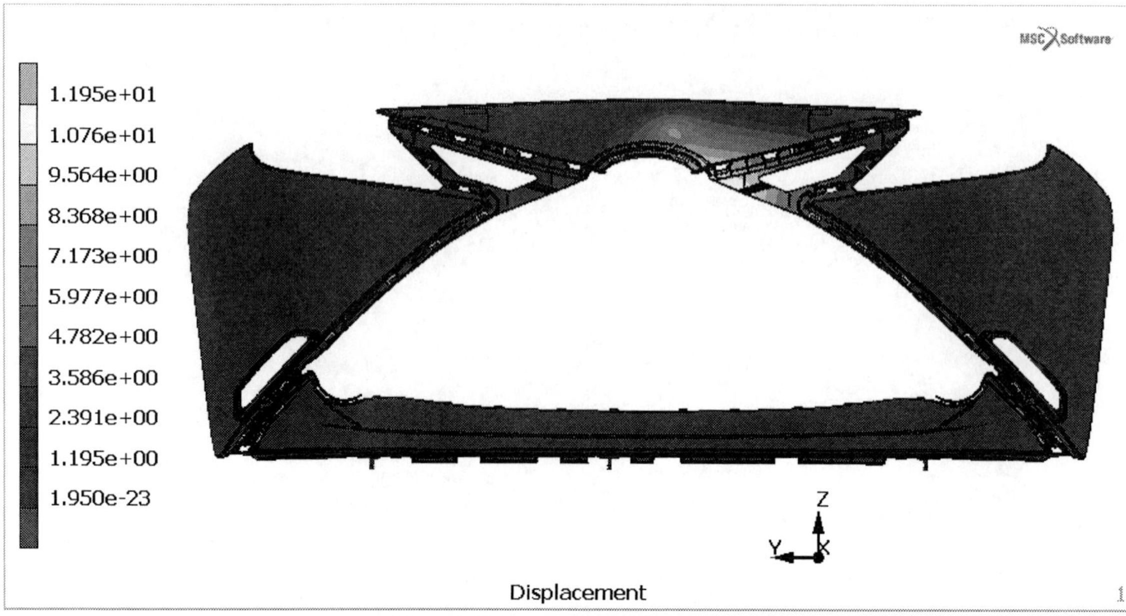

Fig. 6 FEA simulation results

Comparison of the FEA results with the real test

As mentioned above, the simulation results and the scanned deformed surface of the real bumper are compared using GOM Inspect software. Using the same coordinate system for both scanning and simulation, the deviations of both STL surfaces were computed and displayed (positive value means distance of scanned surface in positive normal direction from simulation model).

The first comparison (Fig. 7) is between the simulation results and the real test with only bolts preload and gravity load applied. There are noticeable deviations at some edges of the geometry: about +5 mm on the side areas, about -6 mm on the bottom area. The deviation in area close to the applied load force is under +1 mm.

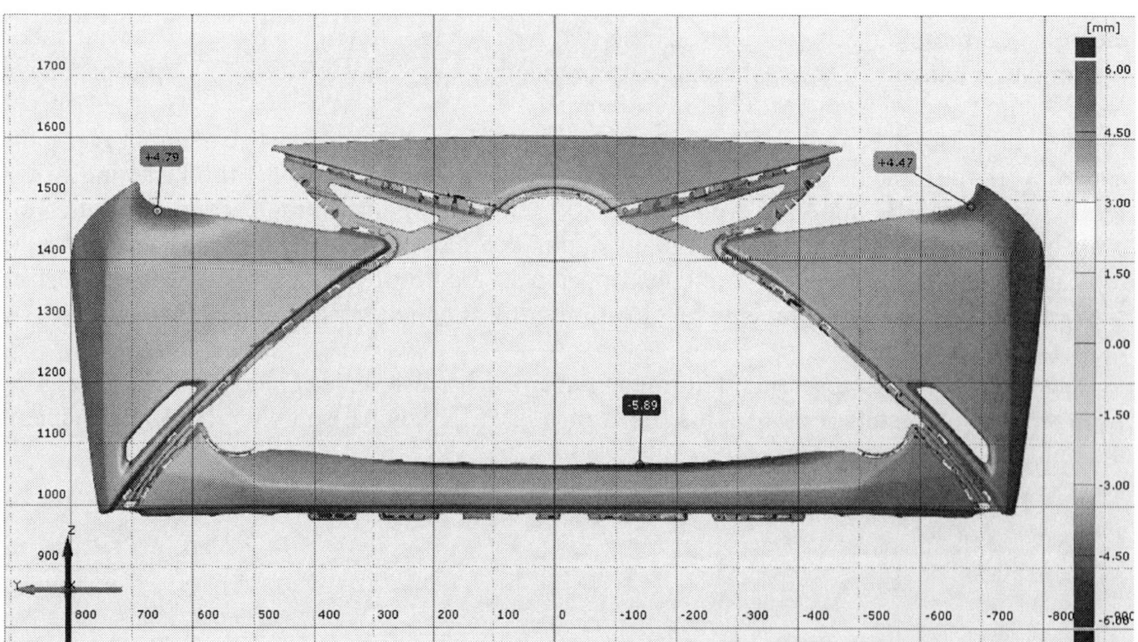

Fig. 7 FEA simulation results compared to the real deformed bumper (force load not applied)

The second comparison (Fig. 8) is between the simulation results and the real test with all loads applied (external load force included). The deviation at the point of the applied load force is now about +2 mm, the deviations on the side and bottom areas are very similar to the previous case. These appreciable deviations are mainly caused by wider bumper against its nominal CAD geometry (deformed due to its long-term storage) during mounting to the frame.

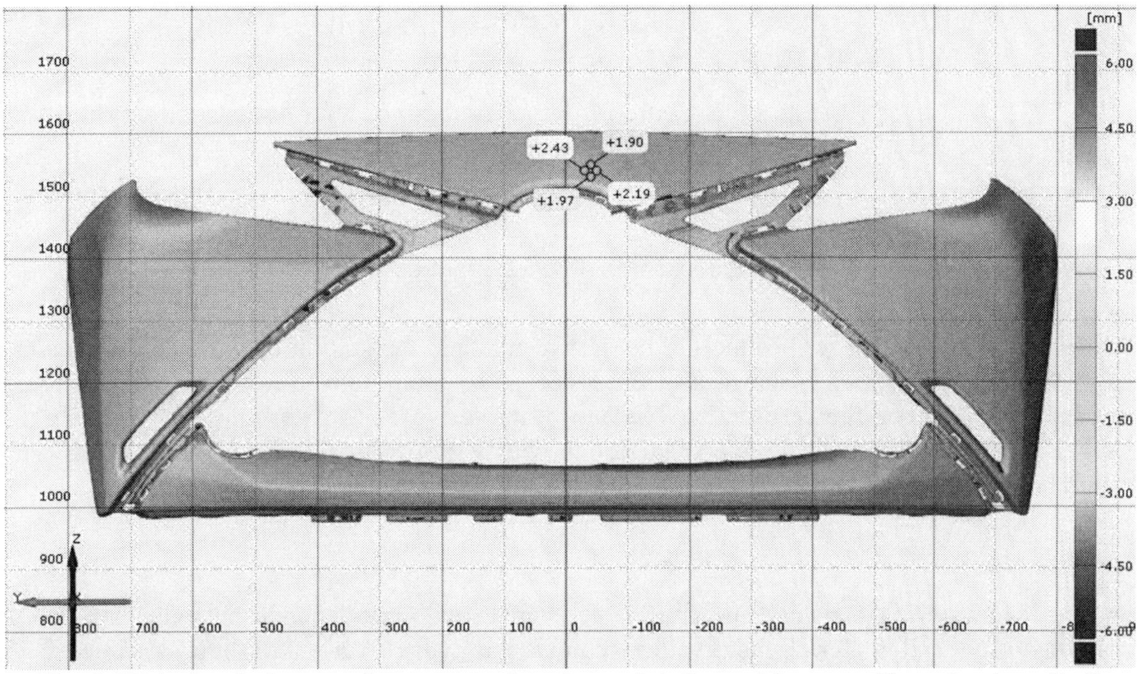

Fig. 8 FEA simulation results compared to the real deformed bumper (force load applied)

Conclusions

For validation of the simulation model of a main part of the car bumper, the deformed mesh of the FEA model with the scanned geometry of the real part with applied load force is compared under the same boundary conditions. Material tests, FEA simulation, a newly developed variable test frame and optical 3D scanning were used for this task. Resulting values of deviations may be due to various causes: manufacturing imperfections, deformation of the bumper due to long-term effect on plastic parts under gravity load, wear of contact clips due to repeated disassembling of bumper.

Acknowledgement

The publication was created with the support of the project LO1213 - Excellent Engineering Research, the grantor - MŠMT, beneficiary – VÚTS, a. s., in support of the National Sustainability Program, with the financial contribution of the Ministry of Education, Youth and Sports of the Czech Republic.

References

[1] Z. Braier, O. Medůna, P. Půlpán, M. Pustka , "Unifikovaný systém ukládání a výměny dat", VÚTS, a.s. Liberec, 2017.

[2] K. Prokeš, "Možnosti a využití měřicího ramene Romer Absolute Arm RA-7525SI-2 ve VÚTS, a.s.", VÚTS, a.s. Liberec, 2016.

EAN 2018
56th conference on experimental stress analysis

The Fatigue Strength of Various Mechanical Double Shear Rivet Joints of D16čA TV Sheets

MRŇA Tomáš[1,a], RŮŽEK Roman[1,b], HOMOLA Petr[1,c], KUCHARSKÝ Pavel[1,d] and MLCH Ivan[1,e]

[1]VZLÚ – Czech Aerospace Research Centre, Strength of Structures Department, Beranových 130, 199 05 Praha – Letňany, Czech Republic

[a]mrna@vzlu.cz, [b]ruzek@vzlu.cz, [c]homola@vzlu.cz, [d]kucharsky@vzlu.cz, [e]mlch@vzlu.cz

Keywords: aluminum alloy, double shear joint, rivet, fatigue

Abstract. This paper describes the fatigue test results and evaluation of various double strap but joints (Solid-rivets, CherryMax) of D16čA TV aluminum alloy sheets which are representative for commuter aircrafts. The aim of the tests was to determine and compare the fatigue behavior of typical double shear joints with various fasteners. The solid-rivets with round head and compensator showed the best fatigue performance.

Introduction

The load carrying capacity and fatigue resistance of joints depends on many structural, manufacturing and material factors like: connection type (overlap or butt, symmetric or asymmetric), size, rivet pitch and spacing (distance between rivets and rivet rows), sheet thickness, diameter of rivet shank and rivet type (i.e. universal, mushroom or countersunk). The most advantageous stress states occur for symmetric joints where uniaxial or biaxial tension is dominant. Unfortunately, they can be rarely used in practice. The non-symmetric joint is very unfavorable because of eccentric tension (secondary bending). The secondary bending influence is estimated by comparison of maximum bending stress in the concentration area with the nominal stress in the reduced section or with local tension stress. Additional technological factors such as riveting or holes expansion, which can also significantly affect fatigue durability, have to be taken into account [1, 2, 3].

The paper deals with a double shear joint. The typical fastener used for joining of metallic parts in commuter airframes is a rivet. The rivets are used mainly for a skin-stringer, skin-skin, skin-frame, etc. connections. In this sense the aim of presented experimental investigations was to define and compare the fatigue strength of typical riveted double strap butt joints with using of different types of fasteners. Five different joint configurations were considered. Individual joints differ in rivet row pitch, a number of rivets and type of rivet used. The methodology of fatigue tests was based on fatigue loading until failure whereas, after the failure of one joint, the specimen was split up to enable further testing of remaining joints of the specimen. The results will be applied to achieve life enhancement of principal critical areas, verification manufacturing technology and to compare calculations with the real data.

Materials and Methods

Five sets of double-riveted double strap butt joints were considered for fatigue evaluation. The overall shape of all specimens was the same – they differ in rivet row pitch, a number of rivets and type of rivets. Each set consisted of 6 pieces of test specimens. Every specimen contains the same four riveted joints in series. Scheme of the test specimen is shown in Fig. 1. After a failure of the first double lap shear joint of the specimen, the specimen was split up to enable further testing of remaining joints of the specimen. Top and bottom straps (sheet 1,3), and middle sheets (sheet 2) were made from D16čA TV aluminum alloy. The five different rivet types was used - round head (RH) solid rivet with and without compensator, countersunk head (CSH) solid rivet with compensator, CherryMax 3, and CherryMax 4 rivets. Examples of the main differences between individual rivet joints are shown in Fig. 2. Configurations of individual batches are shown in Tab.1.

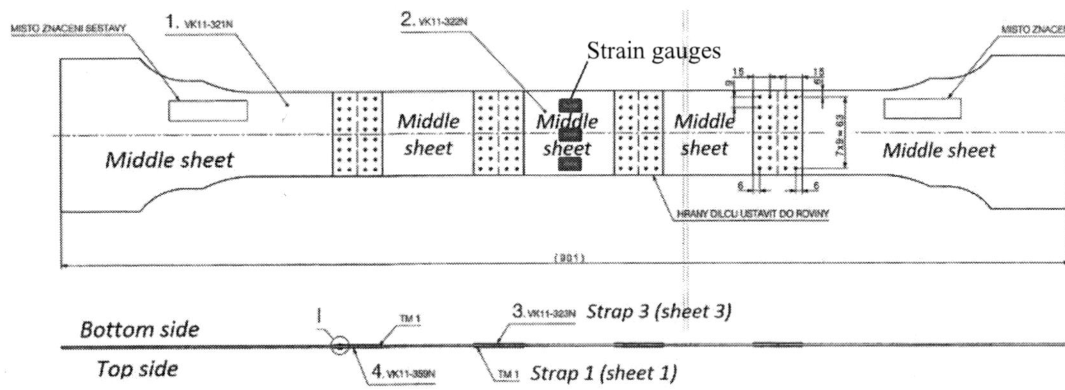

Fig. 1 Schematic drawing of double shear rivet joints

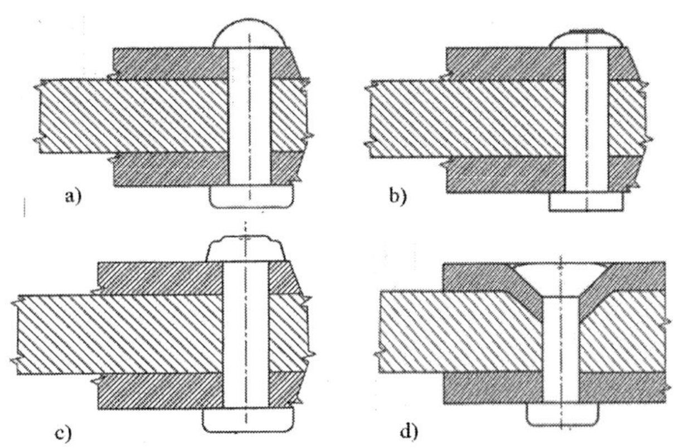

Fig. 2 Example of used rivet types (RH rivet without compensator (a), CherryMax with compensator (b), RH rivet with compensator (c) and CSH rivet with compensator (d))

One specimen from each batch was equipped with strain gauges in its middle part on the middle sheet surface for monitoring of stress uniformity during loading. The strain gauges location is shown in Fig. 1 using blue rectangles. Six strain gauges (back-to-back) were installed on the first tested specimen (from the CherryMax4 batch).

The double sheet rivet joints were loaded by a monotone loading with constant amplitude of loading force, stress ratio R = 0.05 and frequency from 3 up to 8 Hz. The tests were performed at room temperature and ambient laboratory conditions in compliancy with EN 6072 [4], ASTM E 466 [5] and MMPDS-07 [6] specifications. Tests were conducted using standard uniaxial hydraulic test machines INOVA ZUZ 100 with load cell capacity of 100 kN and MTS 250 with 250 kN load cell. The maximum stress values σ_{max} were defined based on the nominal dimensions of the specimen's middle sheet.

Tab. 1 Configurations of individual test specimen batches

Rivet type	Strap sheets			Middle sheet		Note	E_f (GPa)
	t_1 (mm)	t_3 (mm)	Material	t_2 (mm)	Material		
5DuK 3x7P	0.8	0.8	D16čA TV	1.5	D16čA TV	*Round head (RH) solid rivet*	70
CR3213-4-02-T1						*CherryMax 3 (D=3.2 mm)*	210
CR4173-4-02-T1						*CherryMax 4 (D=3.175 mm)*	70
5DuPk 3x7P						*RH solid rivet + compensator*	70
5DuZz 3x7P						*CSH solid rivet + compensator*	70

Note- E_f-Young's modulus of a rivet

Results

The strain vs force measurement was conducted up to estimated load levels corresponding to 100 000 cycles until failure. The strain gauge measurements show a strong linearity of force vs. strain data. Fig. 3 illustrates the strain distribution through the specimen width for all measured specimens. Based on comparison of front vs back strain gauges (CherryMax4 specimen), an occurrence of approximately 5% bending can be observed. Moreover, the strain gauges indicate a slight non-uniform strain distribution through specimen width. The middle-positioned strain gauges show a slightly higher strain value as compared to the side strain gauges (about 3.5%).

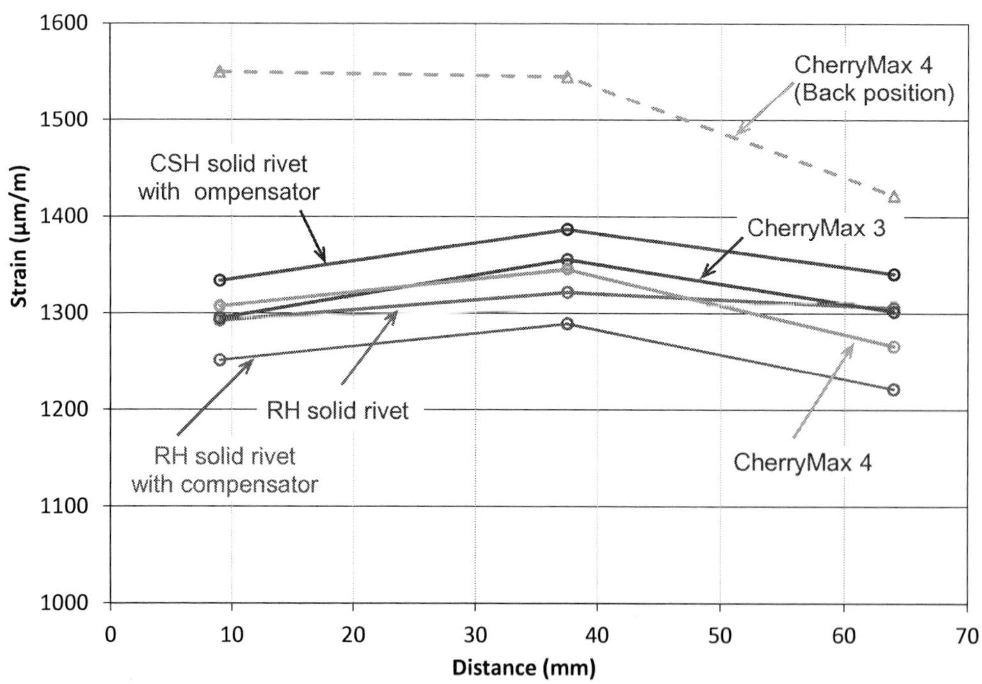

Fig. 3 Strain distribution through the width of specimens. Load level of 9.3 kN

Two main failure mode types were observed during the fatigue experiments:

1) Fracture of strap sheets 1 and 3 with a failure going through one rivet row, through two rivet rows or going through one rivet row and secondary crack initiation in different rivet row.

2) Fracture of the middle sheet with or without a crack initiation in the strap sheets. The fatigue crack initiation always started from the rivet holes.

An example of failure modes for one test specimen is shown in Fig. 4. Detailed views on failed joints with different failure mode are shown in Fig. 5.

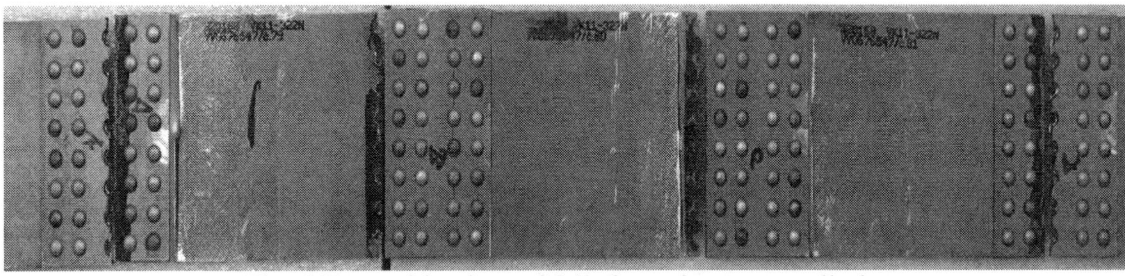

Fig. 4 Front side view of failed specimen, overall view

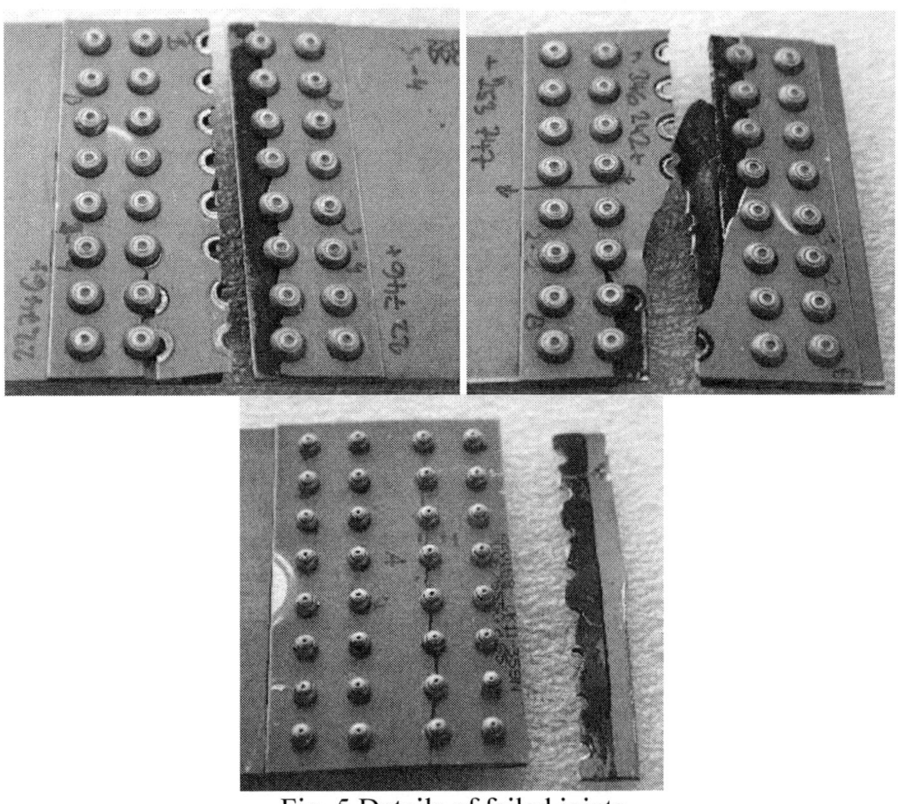

Fig. 5 Details of failed joints

The fatigue data from the each batch of specimens was evaluated using the linear regression by mean of a linear model in compliance with MMPDS-07 [6]. The linear regression model was represented by the equation:

$$\log(N_f) = A_1 + A_2 . log(\sigma_{max}), \tag{1}$$

where A_1 and A_2 are the regression constants, σ_{max} is the maximal stress level of the load cycle.

Fig. 6 shows linear regression curves for all test specimen batches. Significant fatigue strength differences were observed between individual batches differing in the rivet type. From viewpoint of the fatigue curves σ_{max} vs N_f evaluation, the longest fatigue lives were demonstrated for the 5DuPk 3x7P 1 – round head solid rivet with compensator.

Generally, the round head solid rivet types (with or without compensator) exhibit better fatigue behaviour as compared to other rivet types. The specimens with CherryMax 3 (D=3.2 mm) rivets, CherryMax 4 (D=3.175 mm) rivets, and the countersunk head solid rivets with compensator showed nearly one order shorter fatigue lives as compared to the test specimen with the round head solid rivet with compensator.

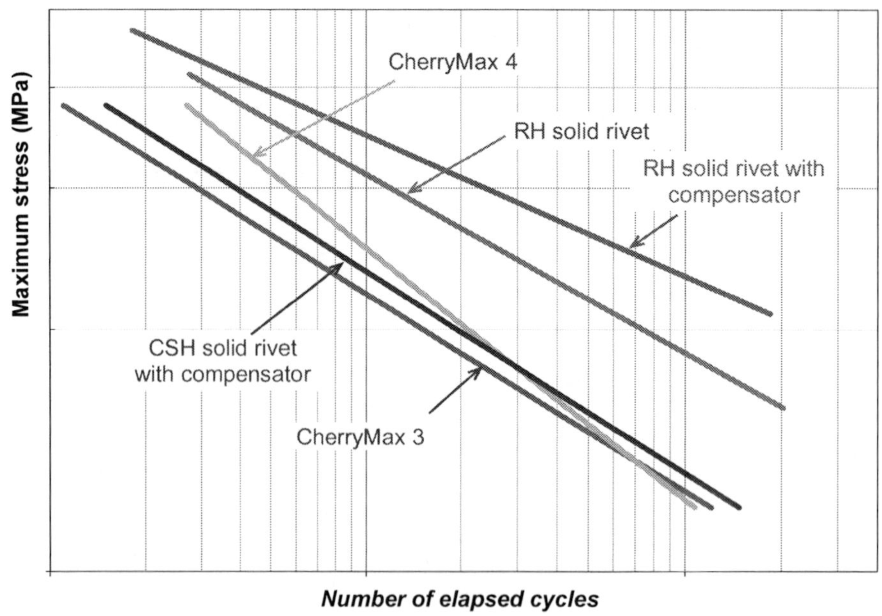

Fig. 6 Comparison of linear regression models for various mechanical double shear joints of D16č ATV sheets

Conclusions

The main goal of the fatigue tests was to determine the fatigue curves of the double strap butt rivet joints. Five batches of test specimens with different rivets were tested and evaluated. To clarify the fatigue data trends, a linear regression model was used. The best fatigue performance was showed in case of the specimens with round head solid rivet with and without compensator. For other test specimens types (CherryMax3 rivets, CherryMax4 rivets and countersunk head solid rivets with compensator), the fatigue lives were achieved nearly in one order shorter.

An additional bending moment was found in the middle part of the sheet although the double shear joints are symmetric. Approximately 5% bending was measured and a slight non-uniform strain distribution through the specimen's width (about 3.5%). The results were used as a base for inputs into numerical predictions of riveted structures used in a commuter aircraft.

Acknowledgement

The research leading to these results was performed under funding of the Technology Agency of the Czech Republic, project "The advanced aerostructure research centrum", no. TA02000032.

References

[1] M. Skorupa, T. Machniewicz, A. Skorupa, A. Korbel, Fatigue life predictions for riveted lap joints, Int. J. Fatigue. 94 (2017) 41–57. doi:10.1016/j.ijfa

[2] W. Wronicz, J. Kaniowski, The analysis of the influence of riveting parameters specified in selected riveting instructions on residual stresses. Fatigue of Aircraft Structures, 1 (2014) 63-71, doi: 10.1515/fas-2014-0005

[3] Ch. Lei, Y. Bi, J. Li, Y. Ke, Effect of riveting parameters on the quality of riveted aircraft structures with slug rivet. Advances in Mechanical engineering, 9(11) (2017) 1-12, doi: 10.1177/1687814017734710

[4] ČSN EN 6072 (312005). Aerospace series. Metallic materials. Test methods. Constant amplitude fatigue loading 2010.

[5] ASTM E 466-07. Standard Practice for Conducting Force Controlled Constant Amplitude Axial Fatigue Tests Of Metallic Materials. 2007.

[6] MMPDS-07. Metallic material properties development and standardization (MMPDS). United States: Federal Aviation Administration. 2012.

EAN 2018
56th conference on experimental stress analysis

Determining Fracture Energy of unstabilised Rammed Earth by Using Three-Point Bending Test

MUŽÍKOVÁ Barbora[1,a], OTCOVSKÁ Tereza[1,b] and PADEVĚT Pavel[1,c]

[1]Thákurova 2077, Prague, Czech Republic

[a]barbora.muzikova@fsv.cvut.cz, [b]tereza.otcovska@fsv.cvut.cz, [c]pavel.padevet@fsv.cvut.cz

Keywords: Rammed earth, fracture energy, three-point bending test, mechanical properties.

Abstract. The article is focused on measurement and evaluation of the fracture energy of unfired rammed earth. Fracture energy G_f is an important magnitude for understating of crack development in quasi-brittle materials as rammed earth. The fracture energy in mode I is observed for several types of prescriptions. It was found that adding clay and greater value of the water-clay ration increases the value of the fracture energy. It was also found that the mixture with montmorillonite clay have higher fracture energy than the mixture with illite-kaolinite clay. The maximum value is 20.57 ± 4.32 J/m^2 for *GEM II* and the minimum is 7.46 ± 0.43 J/m^2 for *S IV*.

Introduction

Nowadays utilization of the unfired earth as a building material is rising. The reason of the trend is mainly the fact that unfired earth is according to the principles of sustainable building development. It is environmentally friendly thanks to using of the final product, there is no need to produce any intermediate products. More over unfired earth building is a hundred per-cent recyclable. On the other hand, the mechanical behavior of the material is not described enough to use the data for designing in practical engineering [1, 2].

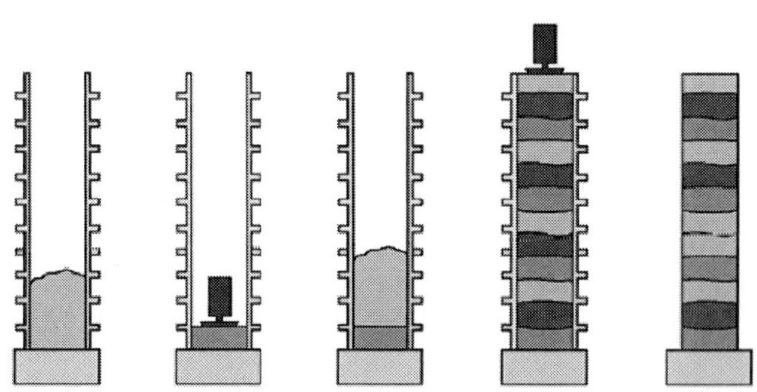

Fig.1 The principle of building with rammed earth.

Production of specimens

Building of rammed earth. The process of building rammed earth wall is as follows. First of all, the framework is built, then the first layer of rammed earth is filled in. Secondly the layer of moist earth is compressed by a tamper. Than next layers of moist earth are added and compressed up to the top of framework. Finally, the framework is removed and the result is a hard monolithic wall. The principle is shown in fig. 1.

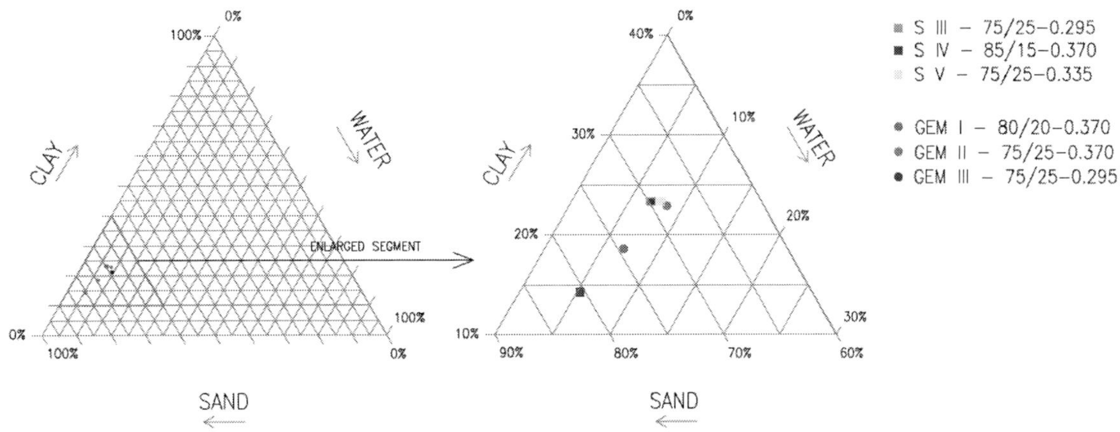

Fig.2 The used prescription in the triangular graph in the dependence on the relative procentual ratio.

Production of specimens. The material is a mixture of sand, clay and water. Six different prescriptions were tested for determining G_f. Sets *GEM* with montmorillonite clay and sets *S* with illite-kaolinite clay, sets differed in the ration of sand and clay and water-clay ration. In all there were six prescriptions of 34 specimens.

Fig.3 The produce of specimens in laboratory.

The laboratory produce of specimens is similar to building of real construction. Firstly the prescription is defined as a ration between sand (the filler) and clay (the binder) and the water-clay ratio is determined. The tested prescriptions are shown in the fig. 2, where the triangular graph illustrates the dependence on the relaive procentual ratio of each component.

Secondly the compounds were mechanically mixed together, then the moist earthen mixture was layer by layer pressed in to the mould by the drill or manually by a steel block. The specimens were removed from the mould and set in to the climatic chamber where the stable conditions were settled. The three point bending tests were made after a month after the produce. The produce and final specimens are shown in the fig. 3.

Testing of Specimens

Fracture energy. As it was written rammed earth is a quasi-brittle material. The fracture energy is the amount of energy which is needed for starting a crack growth. The methodology recommended by the RILEM – the model of fictional crack was used to determine the fracture energy. The P-δ curve from three point bending test was needed. Then the evaluation was as follows.

Firstly, the three point bending test was made and the P-δ curve was recorded. The fracture work δW could be determine as:

$$\delta W = G_f \cdot B \cdot \delta a \qquad (1)$$

Where B was the size of crack perpendicular to the direction of its growth and δa was growth of the crack length. The fracture energy could be calculated as an area A_F under the P-δ curve which was the work of the P force:

$$A_F = \int^{\delta_{max}} P\,(\delta)\; d\delta \qquad (2)$$

Then the area was divided by the crack area and B was the width of the specimen [3, 4]:

$$G_f = A_f \,/\, (B \cdot W) \qquad (3)$$

Fig.4 Specimens of rammed earth in the climatic chamber and during the three point bending test.

The three point bending tests were made to observe the P-δ curve (fig. 5) on specimen of size 40×40×160 mm. Overall 34 specimens of six different prescriptions were tested. The fracture energy was settled from the measured data.

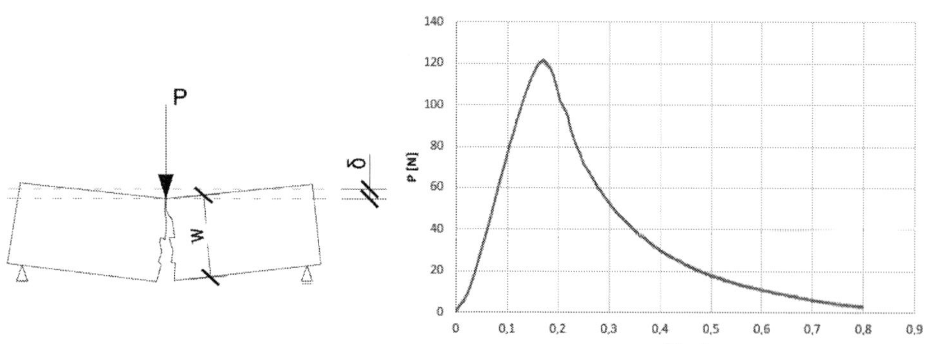

Fig.5 The three point bending test that was carried out and the P-δ curve from the test.

Evaluation of Measured Data

The evaluated data are shown in the table 1. The name of the tested set is in the first column, composition of the mixture is in the next one, then the value of fracture energy G_F, the standard deviation σ, the number of tested specimens, average bulk density of the set and type of used clay in the prescription.

Minimum content of clay. The set *S IV* (85/15 – 0.370) had the minimum content of clay (in percentage 14.2 %) and maximum content of sand (in percentage 80.5 %) and also the set *S IV* had the minimum value of G_f (7.46 ± 0.46 J/m^2).

Maximum content of clay and comparing of clay types. On the other hand the maximum content of clay (in percentage 23.3 %) and the minimum content of sand (in percentage 69.8 %) had sets *GEM III* (10.65 ± 1.85 J/m^2) nad *S III* (17.37 ± 3.75 J/m^2). For set GEM it was the minimum value but for the S set it was the maximum value. The prescriptions of these two sets were the same, they differed only in the type of used clay. As for the results the prescription with mormollionite clay had higher value of G_f than the one with illite-kaolinite clay.

Maximum water content. The maximum content of water was in the set *GEM II* (in percentage 8.2 %) and the set had the highest value of G_f of all sets (20.57 ± 4.32 J/m^2).

Tab.1 The values of evalueted fracture energy

Set	Sand/clay – water-clay ration	Gf [J/m^2]	σ [J/m^2]	Psc of tested specimens	Bulk density [kg/m^3]	Type of clay [-]
GEM I	80/20 – 0.370	18.30	1.54	6	2073	mormollionite
GEM II	75/25 – 0.370	20.57	4.32	6	2040	mormollionite
GEM III	75/25 – 0.295	10.65	1.85	4	1929	mormollionite
S III	75/25 – 0.295	17.37	3.75	6	2170	illite-kaolinite
S IV	85/15 – 0.370	7.46	0.43	6	2036	illite-kaolinite
S V	75/25 – 0.335	16.78	2.91	6	2179	illite-kaolinite

Conclusions

The article was focused on determining the fracture energy of rammed earth using the three point bending test. It has been found that adding clay and enlarging the value of the water-clay ration increased the fracture energy.

It has also been found that the mixture with montmorillonite clay had higher value of the fracture energy than the mixture with illite-kaolinite clay. The maximum value was 20.57 ± 4.32 J/m^2 for *GEM II* and the minimum is 7.46 ± 0.43 J/m^2 for *S IV*.

Acknowledgements

This research was financial supported by the Czech Science Foundation (GAČR No.18-0884S) and by the Faculty of Civil Engineering at CTU in Prague (SGS project No. 16/201OHK1/3T/11).

References

[1] Minke, G., Building with Earth - Design and Technology of Sustainable Architecture. Berlin, 2006 pp. 11-18, 158-160.

[2] Žabičková, I., Hliněné stavby. Brno, 2002. pp. 5-14.

[3] S. Shah Suredra P., S. Stuart E., O. Chenhsheng, Fracture Mechanics of Concrete, New York, John Wiley and Sons, 1995.

[4] M. Jirasek, J. Zeman, Reshape and Fracture of Materials: Reshape, Plasticity, Demage and Fracture, CTU in Prague, 2006.

X-Ray Diffraction Study of Macroscopic Residual Stresses of Al_2O_3 + TiC Oxide Cutting Ceramics after Surface Machining

NĚMEČEK Jakub[1,a], KOLAŘÍK Kamil[1,b], ČAPEK Jiří[1,c], GANEV Nikolaj[1,d], TROJAN Karel[1,e] and KUZIN Valery V.[2,f]

[1]Department of Solid State Engineering, Faculty of Nuclear Science and Physical Engineering, Czech Technical University in Prague, Czech Republic

[2]Moscow State Technological University Stankin, Vladkovsky per. 3a, Moscow, Russia

[a]nemecj17@fjfi.cvut.cz, [b]kamil.kolarik@fjfi.cvut.cz, [c]jiri.capek@fjfi.cvut.cz, [d]nikolaj.ganev@fjfi.cvut.cz, [f]karel.trojan@fjfi.cvut.cz, dr.kuzinvalery@yandex.ru

Keywords: oxide cutting ceramic, X-ray diffraction, residual stresses, machining technology.

Abstract. Currently, the extensive research in the field of oxide cutting ceramics is conducted and there are efforts to replacement cemented carbides by these materials. The reason is better availability of the source materials and lower production costs. In cooperation with the Moscow State Technological University Stankin that provides manufacturing technology, the influence of machining technologies on the value of macroscopic residual stresses in samples of Al_2O_3 + TiC oxide cutting ceramics were investigated by X-ray diffraction. Measurement was done for the both phases, the influence of the parameters of machining technology to residual stresses was studied and the resulting values were compared with each other.

Introduction

At the beginning of the 20th century, the extensive research in ceramic materials began. The first material used as ceramics as a cutting tool was Al_2O_3. Over the years, its development had moved to a level at which it was possible to replace, for example, sintered carbides using those ceramic materials. This effort has been caused by the mechanical properties of aluminium oxide and the easy availability of the starting materials.

The starting materials and their properties have been known for a long time. The problem was to make compact replaceable cutting tool with the desired mechanical properties and dimensions. The current production process consists of three parts: first, grinding the starting material to a powder with grain size of the order of 100 nm, second, sintering the powders into a compact form and third, the surface treatment of the cutting tool to the desired dimensions.

An important technological aspect of the properties of cutting ceramics is the grain size of the starting powders. If the individual grain size is approximately the same (if they have ideal spherical shape), the uniform distribution and the pore size will be gained more easily during forming and thereby for uniform compaction throughout the volume and easier reorganization of the particles during sintering. Preference is also to achieve as small as possible grain size of the sample, because it leads to faster transport of material [1, 2].

The sintering process is very similar to that used with sintered carbides (see Fig. 1). The powder is pressed below the melting point into a compact body. But the absence of a binder which during sintering formed a liquid phase is the main difference, because it would cause degradation of mechanical properties. Today, ceramic is most commonly sintered into rods of cross section and then cut to the desired dimensions. Individual plates are then variously surface treated and shaped [2].

Fig. 39 Scheme of creation technology of cutting ceramic sample:
refinement powders (wet grinding) – evaporation – pressing – sintering

Experiment

The influence of machining technologies on the state of residual stresses in samples of Al_2O_3 with approx. 30 % of TiC oxide cutting ceramics were investigated by X-ray diffraction (XRD).

The *X'Pert PRO MPD* diffractometer was used for the samples surfaces analyses. Values of residual stresses were determined in two orthogonal directions: L (grinding/feed direction) and T (transverse direction). For Al_2O_3 phase, the planes {0210} were analysed using manganese radiation ($2\theta = 109$ °), and for TiC phase, the planes {311} were analysed using chromium radiation ($2\theta = 123$ °).

X-ray elastic constant for the Al_2O_3 phase were $\frac{1}{2}s_2 = 3.18$ TPa^{-1}, $s_1 = -0.61$ TPa^{-1} and for TiC $\frac{1}{2}s_2 = 2.63$ TPa^{-1}, $s_1 = -0.42$ TPa^{-1}.

Result and discussion

Grinding by a diamond disc. After sintering, all samples were ground using a diamond wheel. An effect of the size of transverse displacement at constant cutting speed $v_c = 30$ m/s, feed rate of the workpiece in the cutting direction $f = 12$ m/min and depth of cut $a_p = 0.04$ mm was observed. Values of transverse displacement is given in Tab. 1.

Tab. 7 Parameters of grinding by diamond disc

Sample	Transverse displacement [mm/displ.]
1	0.5
2	1.0
3	1.5

Due to the different crystallographic structure, the values of compressive residual stresses are assumed to be higher for the titanium carbide phase than for the corundum phase. This is due to the different values of the modulus of elasticity and the different slip systems for the trigonal structure of Al_2O_3 and the fcc structure of TiC. This assertion was experimentally demonstrated and the resulting residual stress values are shown in Tab. 2.

Tab. 8 The values of normal σ_N and shear σ_S residual stresses after grinding by diamond disc

Al_2O_3		TiC	
L direction		L direction	
Sample	σ_N [MPa]	Sample	σ_N [MPa]
1	-216	1	-518
2	-182	2	-536
3	-165	3	-530
T direction		T direction	
Sample	σ_N [MPa]	Sample	σ_N [MPa]
1	-436	1	-776
2	-470	2	-805
3	-580	3	-938

It is also apparent that the stress values differ in the given directions. During machining, the grinding wheel moved in the direction L. Lower compressive residual stresses were analysed in the direction L. This effect is caused by moving the tool in the L-direction, due to the mechanical interaction of the cutting tool with the material, the so-called stretching of the subsurface layers occurs. In the T-direction, a higher displacement causes the higher compressive residual stresses for both phases. In the direction L, the higher displacement leads to the higher tensile stresses for the Al_2O_3 phase but not for TiC. This is due to the already mentioned different slip systems of materials, generally different mechanical properties [4].

Thermal annealing the samples after grinding by diamond disc. Next samples were after grinding thermally annealed same time at 800°C. It is evident from Tab. 3 that the values of the residual stresses decreased considerably (mainly in direction T). In contrast, the value of microstrains and grain size given by the parameter FWHM remained almost unchanged (see [5]).

Tab. 9 The values of residual stresses after thermal annealing

Al_2O_3			TiC		
L direction			L direction		
Sample	σ_N [MPa]	FWHM [° 2θ]	Sample	σ_N [MPa]	FWHM [° 2θ]
12	-185	0.666	12	-516	0.965
13	-203	0.888	13	-290	0.609
14	-199	0.889	14	-289	0.847
T direction			T direction		
Sample	σ_N [MPa]	FWHM [° 2θ]	Sample	σ_N [MPa]	FWHM [° 2θ]
12	-303	0.781	12	-599	0.782
13	-359	0.802	13	-474	0.613
14	-342	0.769	14	-473	0.908

Machining by blasting with compressed air (air abrasive machining). In this method of machining, air of high pressure with added microparticles (size approx. 100 nm) is jetted on a rotating sample (see Fig. 2). The effect of air pressure changes on the residual stress in the sample was studied (see Tab. 4) with constant time of blasting $T = 60$ s and rotate per minute $RPM = 50$.

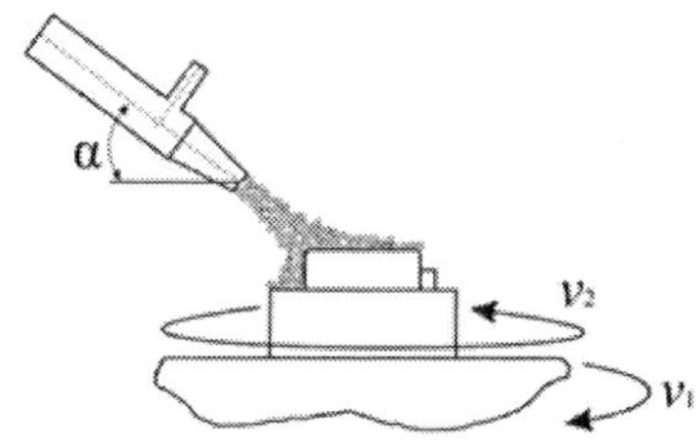

Fig. 40 Scheme of air abrasive machining

Tab. 10 Parameters of air abrasive machining

Sample	Air pressure [bar]
4	1.5
5	2.0
6	2.5

The obtained values of residual stresses (Tab. 5) show that when a pressure increases from 2 to 2.5 bars a decrease of compressive residual stresses (stress relaxation) is observed. It is probably a result of overcoming the yield strength of the material. Perhaps machining process with higher pressure values influences deeper sub-surface layers of material. The energy required for crack propagation is therefore higher for these seemingly lower compressive residual stresses. To confirm this assertion, a depth-profile analysis of residual stresses by sequential force-free etching of surface layers is needed.

Rotational movement of the specimens during the machining leads to an isotropic stress distribution i.e., the values in both the directions L and T are similar.

Tab. 11 Values of residual stresses after air abrasive machining

Al_2O_3		TiC	
L direction		L direction	
Sample	σ_N [MPa]	Sample	σ_N [MPa]
4	-352	4	-799
5	-367	5	-928
6	-242	6	-278
T direction		T direction	
Sample	σ_N [MPa]	Sample	σ_N [MPa]
4	-295	4	-737
5	-389	5	-899
6	-284	6	-283

Water Jet Machining (WJM). This technology is based on the same principle as air abrasive machining, when a focused beam of water is jetted out by varying the pressure on the rotating sample for 60 seconds (see Fig. 3 and Tab. 6)

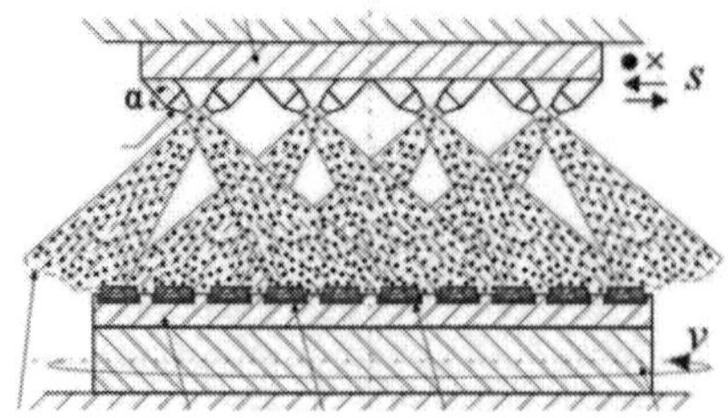

Fig. 41 Scheme of WJM

Tab. 12 Parameters of WJM

Sample	Water pressure [bar]
7	1.5
8	2.0
9	2.5

It is evident, that the yield strength value has not been overcome for applied values of pressure jetting, see Tab. 7. The values of residual compressive residual stresses increase steadily with increasing pressure in contrast to the samples after air abrasive machining. Due to the rotating of the samples an isotropic residual stress distribution is achieved again.

Tab. 13 Values of residual stresses for WJM

Al_2O_3		TiC	
L direction		L direction	
Sample	σ_N [MPa]	Sample	σ_N [MPa]
7	-99	7	-406
8	-149	8	-363
9	-255	9	-568
T direction		T direction	
Sample	σ_N [MPa]	Sample	σ_N [MPa]
7	-85	7	-374
8	-166	8	-364
9	-238	9	-484

Laser Shock Peening. By this technology the effect of power changes of pulse laser on surface residual stresses was studied (see Tab. 8). The displacement of the laser at each spot is larger than the radius of the spot. The constant parameters of machining were the frequency of laser $f = 30$ Hz and displacement at one pulse $s_x = s_y = 30$ μm, see Fig. 4.

Laser machining leads to sample heating and a subsequent decrease in residual stresses, which in this case will mainly be transformation stresses. Increasing power will result in higher temperatures and thus greater so-called stress relaxation, which correlates with a different phase representation. In addition, the influence of overlapping on the anisotropy of residual stresses distribution can be observed (see Tab. 9).

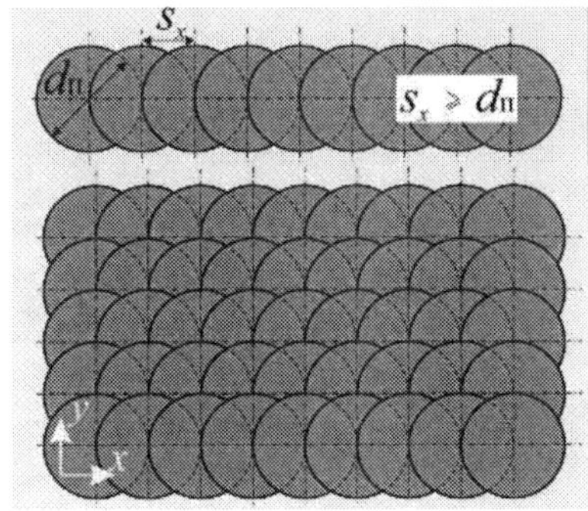

Fig. 42 Scheme of linear and surface laying of spots

Tab. 14 Variable parameter of Laser Shock Peening

Sample	Power [W]
10.1	2
10.2	5
10.3	10
10.4	15

Tab. 15 Values of residual stresses after Laser Shock Peening

Al_2O_3		TiC	
L direction		L direction	
Sample	σ_N [MPa]	Sample	σ_N [MPa]
10.1	-53	10.1	-318
10.2	31	10.2	-259
10.3	-29	10.3	-200
10.4	-62	10.4	-220
T direction		direction	
Sample	σ_N [MPa]	Sample	σ_N [MPa]
10.1	-155	10.1	-509
10.2	-125	10.2	-448
10.3	-71	10.3	-367
10.4	-48	10.4	-288

Friction tests. After grinding of samples by diamond disc their surface was loaded by rotating steel disk at different time intervals, see Tab. 10. The influence of time of friction on values of residual stresses was investigated.

Tab. 16 Parameters of friction tests

Sample	Time of loading [s]
11.1	2
11.2	6
11.3	12

Due to friction, the surface of the sample was plastically deformed and heated, which could cause thermal stress relaxation. Another possibility is that the surface could crack. Tab. 11 shows that TiC is more sensitive to loading than Al_2O_3, as changes in residual stress values for titan carbide are greater than for alumina. This can also be explained by the different structure of these phases and different thermo-mechanical properties, e.g. slip systems.

Tab. 17 Values of residual stresses after friction test

Al_2O_3		TiC	
L direction		L direction	
Sample	σ_N [MPa]	Sample	σ_N [MPa]
11.1	-42	11.1	-453
11.2	3	11.2	-129
11.3	-78	11.3	-242
T direction		T direction	
Sample	σ_N [MPa]	Sample	σ_N [MPa]
11.1	-145	11.1	-348
11.2	-15	11.2	-190
11.3	47	11.3	-240

Conclusions

Values of residual stresses of cutting ceramics Al_2O_3 + TiC were analysed in two perpendicular directions. The influence of machining technologies to residual stresses was studied in the both phases.

Anisotropy and different values of residual stresses for both the phases during grinding by diamond disc were obtained, and the thermal relaxation of macroscopic residual stresses after thermal annealing was also analysed.

For air abrasive cutting overcoming of the yield stress was found out with increasing the pressure to 2.5 MPa. Due to rotational movement during air abrasive machining and water jet machining the values of surface residual stresses are isotropic.

Samples machined by laser exhibit thermal relaxation of residual stresses. At the same time, the observed anisotropy of state of residual stresses is due to overlapping of laser-processed areas.

On the samples after the loading tests thermal relaxation of residual stresses was observed. It was shown that titan carbide is more sensitive to load than alumina.

Acknowledgement

This work was supported by the Grant Agency of the Czech Technical University in Prague, grant No. SGS16/245/OHK4/3T/14.

References

[1] A. Humár, Materiály pro řezné nástroje. MM publishing, Prague, 2008.

[2] C. B. Carter, M. G. Norton, Ceramic materials: science and engineering. Second edition, Springer Science & Business Media, New York, 2013.

[3] V. Kuzin, N. R. Portnoi, S. Yu. Fedorov,V. I. Moroz, Effect of Air-Abrasive Treatment on Oxide-Carbide Ceramic Object Operating Properties. Refractories & Industrial Ceramics, 2016, 56.5.

[4] P. Putyra, S. Skrzypek, B. Smuk, M. Posiadlo, Analysis of residual stresses using the $\sin^2\psi$ method for Al_2O_3 materials before and after grinding and heat treatment processes. Materiały Ceramiczne, 2010, 62.3, 301-306 s.

[5] J. Němeček, Diffraction study of selected characteristic of cutting ceramics, Praha, 2017, Diploma thesis, CTU in Prague, Faculty of Nuclear Science and Physical Engineering, 2017-6-15

EAN 2018
56th conference on experimental stress analysis

Numerical Modeling of Liquid Filtration Process

NOVÁK Ondřej[1,a], PETRŮ Michal[1,b] and VOSÁHLO Josef[1,c]

[1] Technical university of Liberec, Studentská 1402/2, Czech republic

[a]novak.ondra1@seznam.cz, [b]michal.petru@tul.cz, [c]josef.vosahlo@tul.cz

Keywords: Liquid flow, FE model, pressure drop, fibrous structure, SEM, microtomography.

Abstract. This paper deals with a flow modeling of liquid medium through a filter represented by a fiber structure. The CAD geometry of a microfibrous filter is based on structure and morphology observations using the microtomography and electron microscopy. The CAD model is created as two-dimensional. In the numerical model that uses real fluids, flow velocities in the pores of the filter are monitored. In conclusion, the possibilities and limitations of the model, incl. suggestions to optimize the simulation of the fiber filtration process are discussed.

Introduction

The choice of an individual filter depends on the specification of the required separation, so the specific filters differ not only in the separation properties, but also in the structure. Therefore, it is important to find the relationship among structural parameters such as size and distributions of pores, porosity, filter media thickness and separation properties. Filter manufacturers usually do not have clear information about which parameter is best suited for the performance and efficiency of the filter for a specific application. Therefore, filter performance characterization requires modeling of the morphological and structural properties of the filter unit.

Theory

For a model establishing of the liquid media flow through pores in porous system different relationships can be used. For example, the equation (1) according to the Kozeny-Carman relationship can be used. The relation is used in the field of fluid dynamics to calculate the pressure loss of fluid flowing through a filter medium that the pores are replaced with slits.

$$\frac{\Delta p}{\Delta L} = \frac{150 \cdot \mu}{{\varphi_p}^2 {d_p}^2} \frac{(1-\varepsilon)^2}{\varepsilon^3} \cdot v \qquad (1)$$

Where Δp is pressure drop, ΔL is filter thickness, φ_p a weight mean of flowing particles, d_p, v is superficial velocity, where from the continuity equation follows the relation for the mean particle velocity in the model capillary $v_p = v/\varepsilon$. In the curved model capillary apply $v_p = (v/\varepsilon) \cdot q$, where q the pore curvature factor that has a value for spherical particles $q = \sqrt{2}$. It is therefore apparent that the transport of substances (liquid and particles) in filter media depends on the properties of the flowing medium and the particular type of material, morphology and structure of the filter medium. For modelling it is important to be able to design and create geometry of the structural unit of the filter.

Experiment

A validation measurement was performed to determine the pressure drop and flow rate of the filter. A sleeve filter made of polypropylene with a declared pore size of 5 microns was used (Figure 1). The measurements were carried out on a filter line designed for this purpose (Figure 2). Measurement was carried out for different values of the flow and pressure drop was measured. The results are shown in the graph Fig. 3.

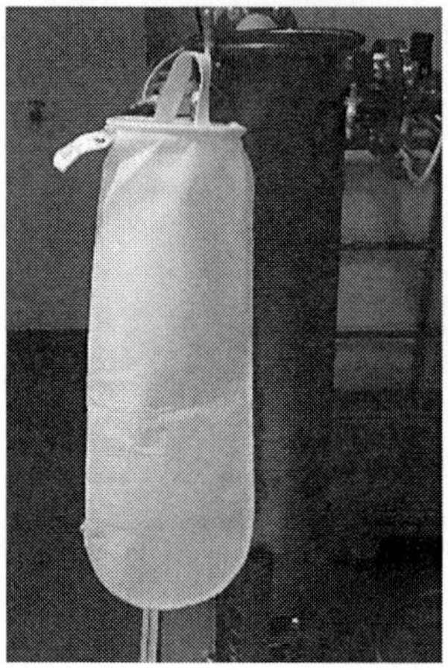

Fig. 1 and 2 Sleeve filter (left), filtration line (right)

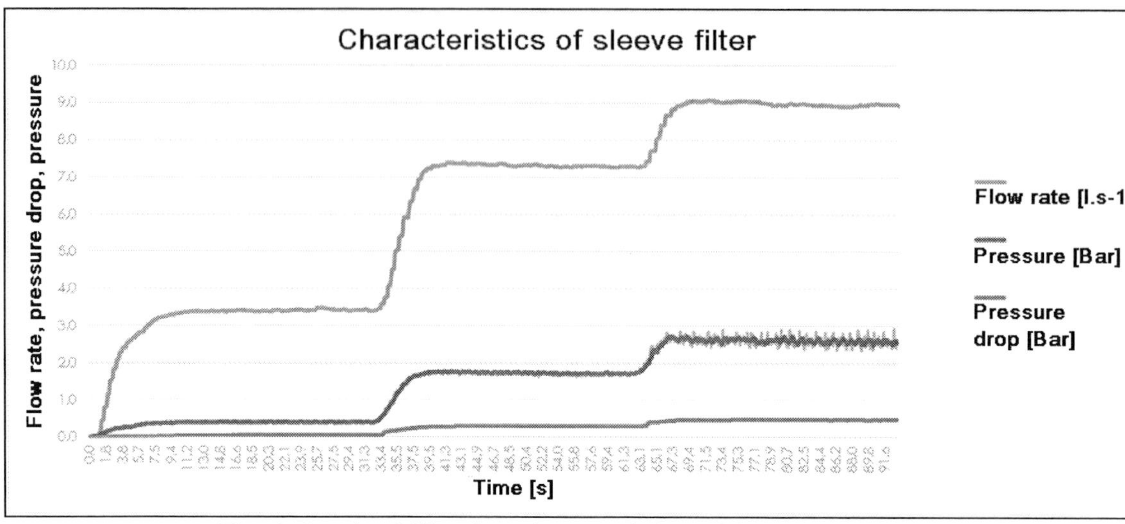

Fig. 3 Graph of filtration characteristics of the sleeve filter

From the pressure drop of the filter it is necessary to subtract the value of 0.02 MPa, which represents the pressure resistance of the filtering line. For the simulation were chosen values of the pressure gradient at a flow rate of 7.3 $l.s^{-1}$. The values applied in numerical model are put in tab. 1.

Tab. 1 Values applied in the model

Flow rate [$l.s^{-1}$]	Pressure drop [MPa]	Input pressure [MPa]
7,3	0,2 – 0,06	1,8

Establishing of the model

The design and creation of model geometry of the filter structural unit, which appropriately represents the layout of the given material structure, should be based on detailed visualization. For this purpose, a micro-computed tomography SKYSCAN 1272, using the X-ray principle, was used to capture the entire structure of the test sample. The picture of the filter structure from this device is put in Fig. 4.

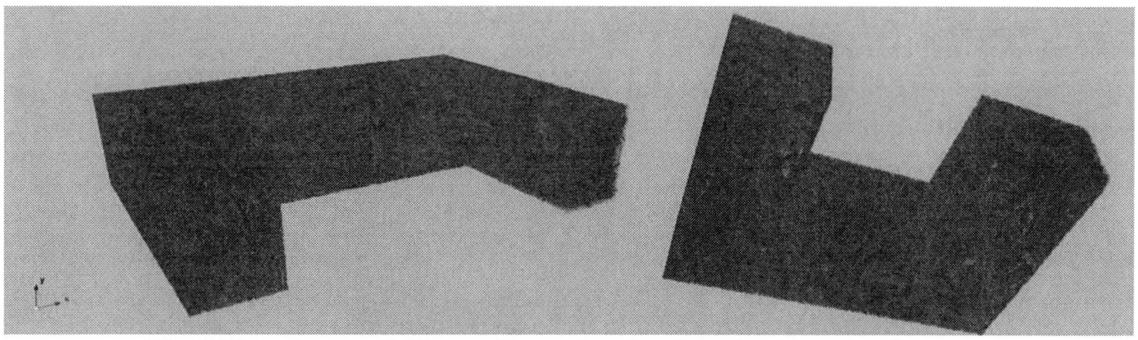

Fig. 4 The picture of fibrous filter taken by the microtomography

Electron microscopy was used to determine the morphology (Fig. 5). From these images, a 2D substrate in binary form (Fig. 6) can be obtained by image processing and used as a background for a CAD model of a structural unit of the filter media.

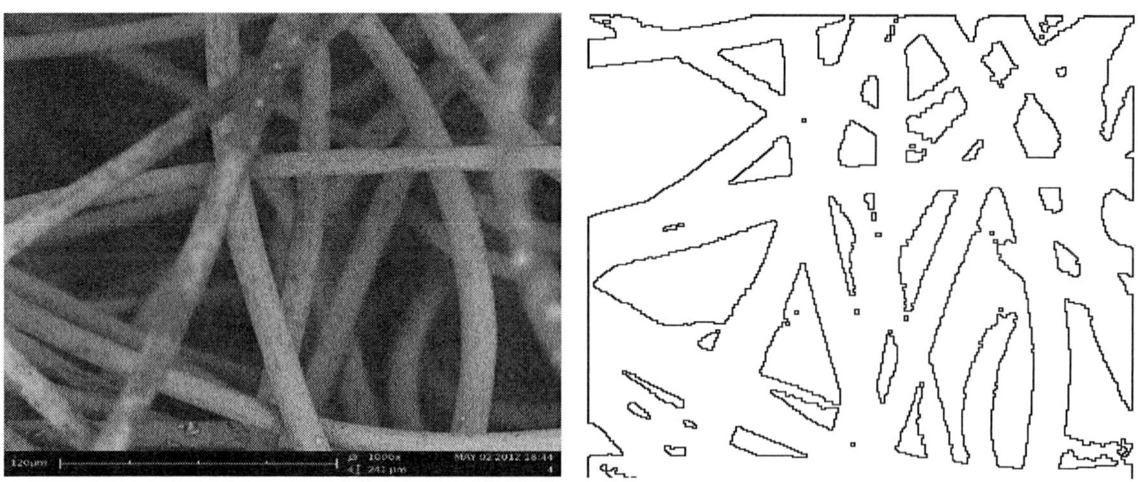

Fig. 5 and 6 SEM picture and its binary form

Based on the information about the geometry of the structure, morphology, configuration and type of fibers then can be modelled CAD model of structural units of the filter medium that can be consequently optimized. The CAD data prepared by this way can be already applied for complex model simulations of filtration media properties. The momentum equation of the fluid is described by the Navier-Stokes equation according to the relationship (2).

$$\rho(\boldsymbol{v} \cdot \nabla)\boldsymbol{v} = \nabla \cdot [-p\mathbf{I} + \mu(\nabla\boldsymbol{v} + \nabla\boldsymbol{v}^T)]. \quad (2)$$

Where **I** is identity matrix.

The geometry of a structural unit was based on CAD data. These data were based on image analysis. The material properties of the water and the fiber structure applied in numerical model were defined. To solve the equation (2) it is necessary to introduce initial and boundary conditions for space and time flow studies. The solution is to determine the boundary conditions from a known pressure distribution and velocities at different time levels. The initial conditions can be introduced in a stationary task as the initial temperature, T(x, 0) = T0 = 23 °C (initial ambient temperature) or in a non-stationary task as a function of temperature change over time. The boundary conditions enforcing a characteristic distribution of the flow through the filter media can be defined by input and output as shown in Fig. 7. The adaptive mesh refinement shows Fig. 8 and 9. The designed 2D mesh contained approximately 29077 elements.

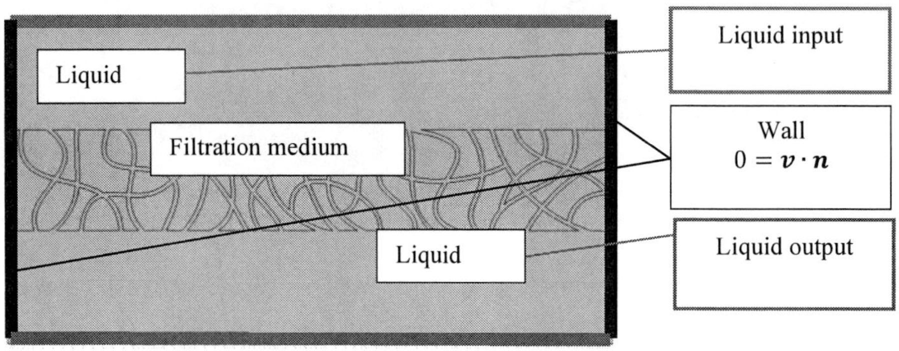

Fig. 7 Numerical model for analyzing the filter properties of the proposed media: defining of boundary conditions

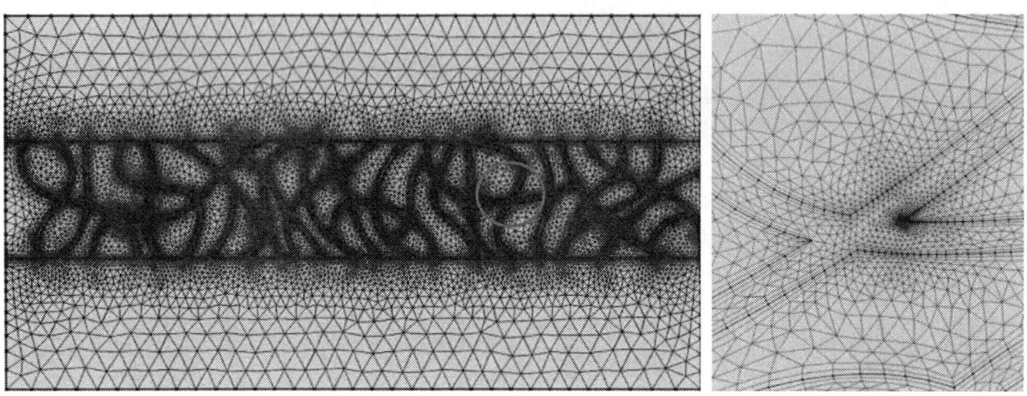

Fig. 8 and 9 Finite element mesh of the model geometry; a refiment of the mesh depends on dimension and a shape of locality

Results

From these results shown in Figures 10 and 11 it is clear that the flow of pure liquid is uneven in the individual pores of the designed filter media, because it is affected by the diameter and the curvature of the slits. It is evident that larger pore diameters and straighter slots will result in increased permeability. It is furthermore characteristic that some small pore diameters are poorly permeable compared to large pores, which is very well apparent on the model shown in Fig 10 and 11. This will at a minimum pressure$\mathbf{p_{min}} = \mathbf{2\gamma / r_{pmax}}$, where $\boldsymbol{\gamma}$ the surface tension is and $\mathbf{r_{pmax}}$ is the largest radius of the pores, that they will be impermeable. Therefore, for higher flow and small pore diameters, the operating pressure must be increased.

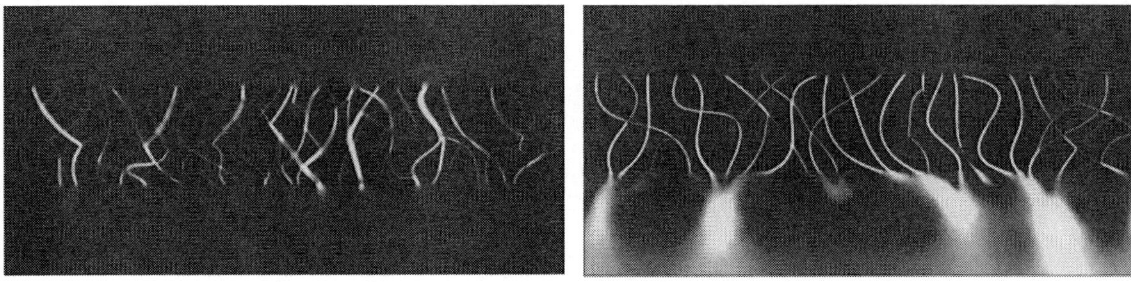

Fig. 10 and 11 The liquid flow under low and higher pressure

Conclusions

Fiber structures utilized for microfiltration do not contain pores of the same size, and therefore, the flow rate of the liquid through the filter media cannot be proportional to the increase of operating pressure. The total flow rate at a given pressure will be the sum of the contributions of all permeable pores. As mentioned above, this is a numerical model of the filter medium in two-dimensional form. This type of model can serve to better understand the processes that occur during filtration, and can also help to optimize the filter media as they allow determining the effect of the shape and the pore size of the filter media. In reality, however, these media are three-dimensional. As a result, it is impossible to obtain specific values of significant characteristics of filter media, including filtering efficiency and pressure drop. The reason is that the model does not respect the spatial shape of the pore, which in the 2D model represents only a planar projection. Information about the actual pore length, which has a significant effect on the filter characteristics due to the friction of the liquid on the pore walls, is lost. A partial solution could be a suitably introduced coefficient that would respect the length of the pores and the pressure loss and the filtering efficiency could then be determined with a certain degree of uncertainty. Thus a more appropriate model is three-dimensional. However, it is significantly more demanding on the number of elements, the quality of the mesh and the calculation time. However, it allows estimating of the pressure loss and filtering efficiency by introducing of filtered particles. Currently the authors develop such a type of the model.

Acknowledgement The results of this project No. LO1201 were obtained through the financial support of the Ministry of Education, Youth and Sports, Czech Republic in the framework of the targeted support of the "National Programme for Sustainability I"

References

[1] P.C. Carman, "Fluid flow through granular beds." Transactions, Institution of Chemical Engineers, London, 15: 150-166, 1937.J. van der Geer, J.A.J. Hanraads, R.A. Lupton, The art of writing a scientific article, J. Sci. Commun. 163 (2000) 51-59.

[2] Acheson, D. J. (1990). Elementary Fluid Dynamics. Oxford University Press. p. 205. ISBN 0-19-859679-0

The Diagnostic System of Low Speed Bearings

OELJEKLAUS Michael[1,a] and PEŠÍK Lubomír[2,b]

[1] Škoda Auto a. s. Mladá Boleslav & Faculty of Mechanical Engineering, Technical University of Liberec, Czech Republic

[2] Faculty of Mechanical Engineering, Technical University of Liberec, Czech Republic

[a] michael.oeljeklaus@skoda-auto.cz, [b] lubomir.pesik@tul.cz

Keywords: Low Speed Bearings, Bearing Diagnostics, Vibration Diagnostics.

Abstract. The subject of paper is the design of control system of slow running bearings including strength analysis and load capacity measurement with the intention to apply the results to the manufacturing operation by the chain conveyor in the paint shop.

Introduction

The diagnostics of low speed bearings is very desirable in a wide range of industrial sectors. Mainly in the transport technologies is necessary to identify the damaged bearing before a production failure and large economic losses become. Chain conveyors are used in the bodywork painting process and chain wheels are mounted on shaft, which is currently mounted on two roller bearings. Shaft speed is relative slow, ones of rpm. There is no doubt that the reliability of this construction has a main influence to the volume of production, especially since the paint shop is the bottleneck of whole production process of cars. The diagnostics of high speed bearings is based on vibration measurement that is reliably used for many years. For the diagnostics of low speed bearings there is no appropriate physical method, which has been able to identify the bearing damage. The topic of the article is focused on solving problem of diagnostics of low speed bearings with intention to apply obtained results to the manufacturing process in the paint shop.

Used Methods

The design of control system of slow running roller bearings is based on analyzation of the current state. The basis is to determine a load in all manufacturing conditions and to determine the force of the chain in the most exposed position. With knowledge of manufacturing operation force of the chain, it is possible to calculate the load of shaft and roller bearings. The results may be used for FEM analysis, calculation stresses and deformations of each part of the current state as a basis of design solution.

The design solution of a system of diagnostics of slow running roller bearings can be based on the principle of rolling resistance or vibration measurement. However effectivity of the vibration detection depends on a sufficient level of measured acceleration, which cannot be achieved in slow rotational movements. It means for the current state to dismount chain and then measure the rolling resistance or increase the rpm by external power and measure the vibration.

Pursuant to demand for the reliability of chain conveyors and the shafts and chain wheels mounting, was designed a system consists of two pairs of roller bearings - shaft and frame. These are connected to each other by the so called reference part that is the freely rotable part.

This solution allows identification of bearings damage with two methods. The first method is depending on change of rolling resistance during forced rotation of reference part. The other method uses vibration measurement by increasing the rpm with properly connected external power. This innovative solution is protected by patent.

Kinematic link between shaft and reference part allows distribution dynamic load between shaft and frame bearings, which is convenient for manufacturing operational matters. Based on this kinematic link shaft and reference part are in forced rotation. Due to kinematic link of shaft, reference part and frame, the design solution leads to integration of planetary mechanism with spur or bevel cogwheels. This solution is protected by patent as well.

After strength analysis of the designed solution, the prototype was made with the intention to experimentally measurement of deformation by nominal and maximum value of load. The fatigue test is already prepared.

Description of Current State

Current state of shaft mounting of chain wheel of conveyor is made of two spherical roller bearings mounted to the frame by bearing units. Chain wheel is connected to free end of shaft and is separated by front plate of frame, which is connected to the frame of paint shop production line.

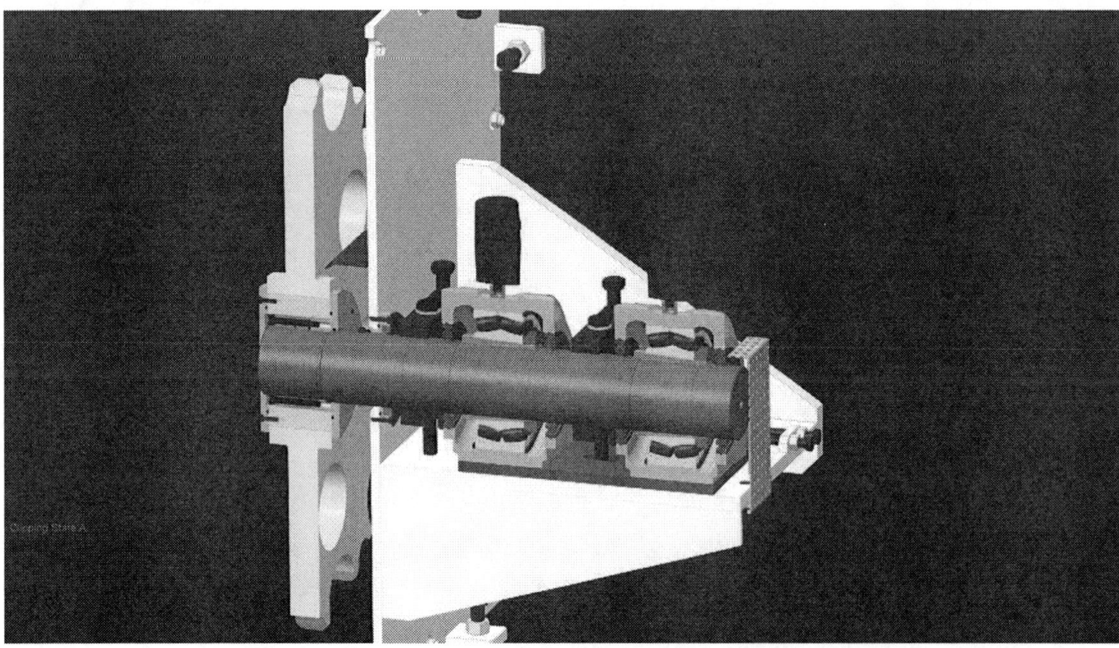

Fig.1 Current state of shaft mounting of chain wheel

The advantage of current shaft mounting of chain wheel of conveyor is its relative simplicity. Significant disadvantage is complicated dismounting and changing of bearings. Because of that, the vibrodiagnostics of bearings was implemented with using latest technologies. Sensors of low frequency vibration are used for identification of damaged bearing. The difficulties consist of insufficient intensity of vibration of slow running bearing. Its value merges with vibrational background of paint shop production line. The reliability of identification of damaged bearing is insufficient.

Design Solution of Diagnostic System

Reliable design of control system of slow running bearings can be based on identification of change the rolling resistance of bearing or its vibrations during the sufficient rpm. Both these methods are difficult to be realized during the manufacturing operation. The change of rolling resistance can be identified only with increased power load, which is usually too late. The sufficient rpm value of bearing can happen only if the standard, low speed operation is off.

The design solution of design of control system of slow running bearings, which is the topic of this paper, comes from the idea, that if one ring of the roller bearing is forced low speed, the second ring of the roller bearing is freely rotable. This diagnostic solution allows identification of change of rolling resistance of shaft bearing or sufficient rpm increase on its freely rotable ring that allows identification of damage of bearings using known methods of vibrodiagnostics. The freely rotable ring of the shaft bearing is rotably mounted to frame using frame bearing. The reference part is mounted between shaft and frame bearing (Fig. 2).

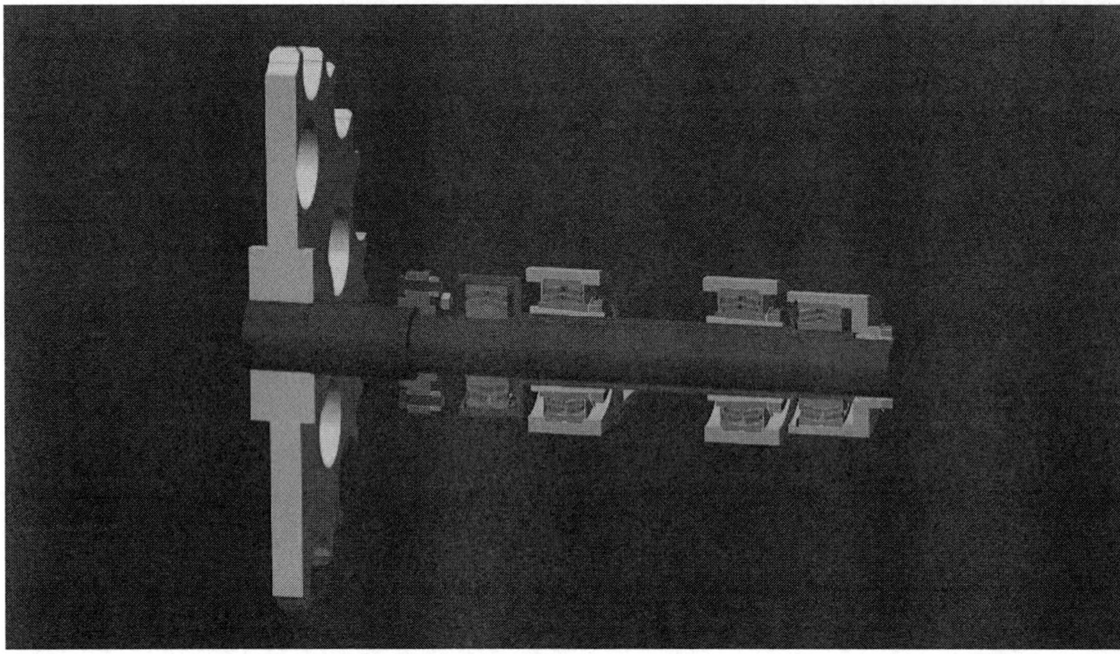

Fig.2 Drafted solution of shaft mounting of chain wheel

The change of bearings is quite a difficult operation, and because of that, the shaft is divided between chain wheel and bearing. This design allows to repair or change of bearing outside of the paint shop production.

This solution provides a number of design variants, one of which was chosen as optimal and then analyzed.

Strength Analysis

Strength analysis is an important field of every design solution. With this shaft mounting solution, the strength analysis is mainly focused on stress and deformation calculations. In all cases, the Ansys software for FEM analysis was used.

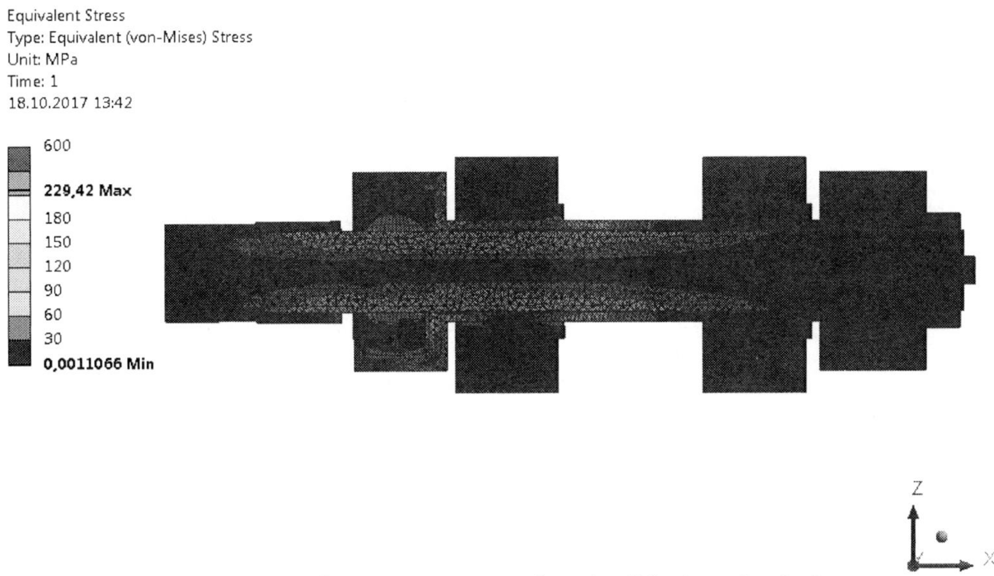

Fig.3 Equivalent stress (von Misses) of designed solution

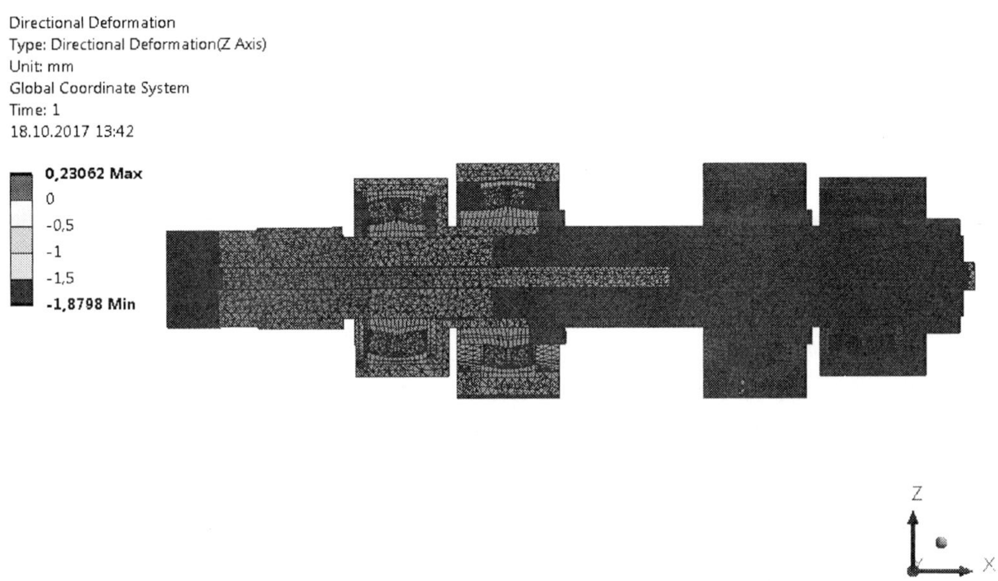

Fig.4 Deformation in z-axis direction of designed solution

Damage Identification System

The identification of damaged shaft or frame bearing can be made based on the change of rolling resistance. For this purpose, the kinematic and moment link of shaft, reference part and frame that consists of planetary gear with simple planets and bevel cogwheels, was designed. Shaft cogwheel with main axis of rotation is tightly connected to shaft. Frame cogwheel with main axis of rotation is connected to frame by adjustable frictional moment. The link between these cogwheels is made by satellites that are mounted to the pins of reference part.

The motion of reference part is determined by kinematic and moment link of planetary gear. If the rolling resistance of shaft bearing increases out of load capacity limit, the frictional moment between frame cogwheel and frame is exceeded. The frame cogwheel then starts to rotate in the same shaft direction.

If the rolling resistance of frame bearing increases out of load capacity limit, the frictional moment between frame cogwheel and frame is exceeded. The frame cogwheel then starts to rotate in the counter shaft direction.

Both rotations of frame cogwheel against the frame can be easily identified by connected sensor (Fig. 5).

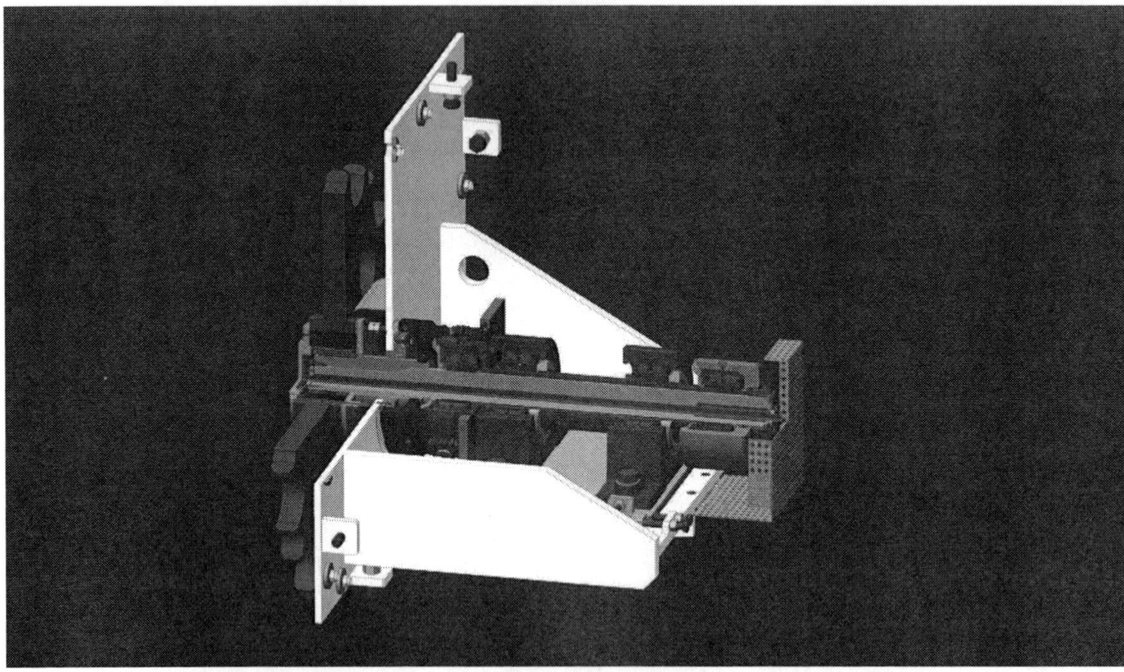

Fig.5 Designed solution of shaft mounting of chain wheel

Conclusion

The paper is focused on actual issue of diagnostic of low speed roller bearings. The designed solution uses reference part between shaft and frame bearing. Change of rolling resistance, which leads to bearing damage, can be identified using reference part. This can be realized during the manufacturing operation. The designed diagnostic system consists of planetary gear with bevel gears. Considering this is an original solution, two applications for patent were filed. One of them has been granted.

Acknowledgements

This publication was written at the Technical University of Liberec as part of the project "Innovation of products and equipment in engineering practice" with the support of the Specific University Research Grant, as provided by the Ministry of Education, Youth and Sports of the Czech Republic in the year 2018.

References

[1] Oeljeklaus, M. and Pešík, L. *Kinematic Similarity of Ball and Roller Bearings with Planetary Gears*. 58. ICMD 2017, Prag, 2017.

[2] ŠKODA AUTO a. s. Zařízení pro diagnostiku poruch pomaloběžných ložisek. M. Oeljeklaus, L. Pešík a M. Jancák. Česká Republika. Patent č. 306656. 13.3.2017.

[3] ŠKODA AUTO a. s. Vorrichtung zum Überwachen wenigstens einer Lagereinrichtung sowie Maschine mit wenigstens einer solchen Vorrichtung. M. Oeljeklaus, L. Pešík a M. Jancák. Česká Republika. Přihláška vynálezu PV 2018-86. 21. 2. 2018.

EAN 2018
56th conference on experimental stress analysis

Drying Rate of Rammed Earth with Illite-kaolinitic Clay

OTCOVSKÁ Tereza[1,a], MUŽÍKOVÁ Barbora[1,b] and PADEVĚT Pavel[1,c]

[1]CTU in Prague, Faculty of Civil Engineering, Department of Mechanics, Thakurova 7, Prague 6, Czech Republic

[a]tereza.otcovska@fsv.cvut.cz, [b]barbora.muzikova@fsv.cvut.cz, [c]pavel.padevet@fsv.cvut.cz

Keywords: Rammed earth, clay, water coefficient, draying speed, form coefficient

Abstract. This submission concerns with unfired clay as a construction material. Its main goal is to determine rammed earth drying speed. The speed of drying is crucial because it is only process with which unfired clay obtains its strength. A main part of the submission is dedicated to description of carried out experiments and their evaluation. There were two earth mixtures tested with four testing body sizes manufactured by technology of ramming.

Introduction

Scientific research of unfired earth mechanical properties was not accomplished sufficiently because the unfired earth had become minority building material in 2nd half of 20th century. But history of unfired earth is rich and long. Probably the most famous unfired clay building is The Great Wall of China [1–5].

Nowadays, there is a growing interest in unfired earth for its properties that fit into principles of a sustainable building, positive effects on human health, good adaptability of relative humidity in interior and attractive visual aspect [1, 6, 8].

The growing interest is evident from a number of scientific papers or quantity of sales of unfired bricks by HELUZ cihlářský průmysl v.o.s [9].

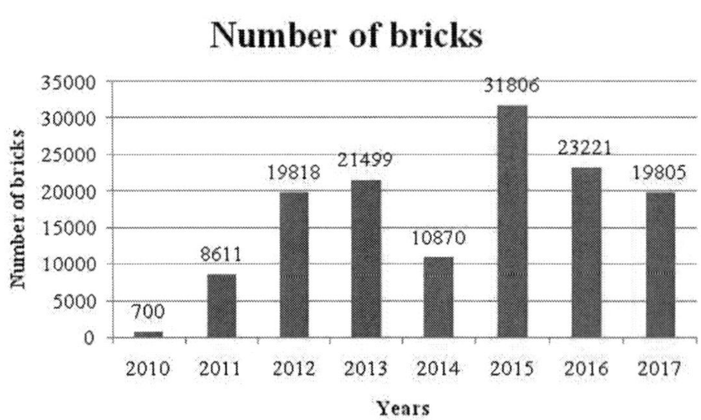

Fig. 1 Sales of unfired bricks by HELUZ cihlářský průmysl v.o.s

There is a few types of unfired earth. One of the primary types of unfired earth is rammed earth which is a topic of this paper. A final appearance of rammed earth is similar like monolithic concrete. A principle of rammed earth production is press down of earth mixture layers to formwork (see Fig. 2).

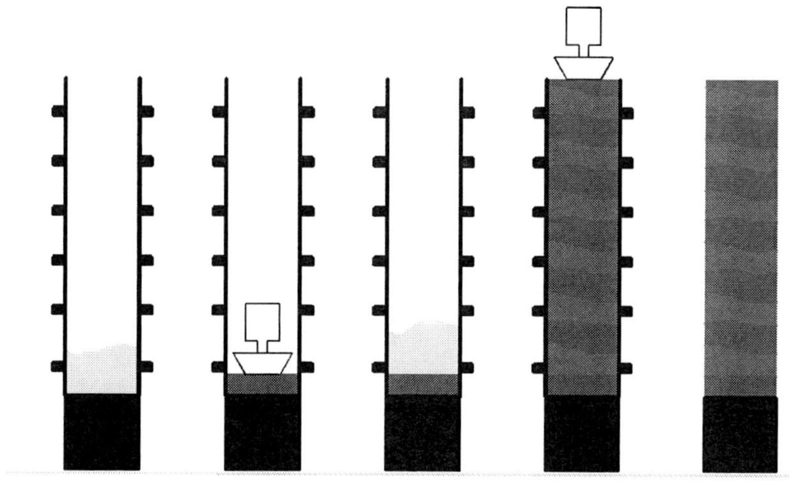

Fig. 2 Production of rammed earth

The essential topic about unfired earth is its moisture characteristics and its water resistance. There are a few parts of research about unfired earth moisture characteristics. Absorption of water, capillary storage capacity, resistance of unfired earth in standing water, weather resistance and drying rate are some of the fields of the research [1, 6].

In general, increasing moisture content in unfired earth construction could have an adverse effect on its mechanical properties.This low water resistance is the most common argument why the unfired earth is not suitable material for building construction. And that's why inorganic binders like cement or lime are added to the clay mixture often [1, 6, 7,10].

However, there are reasons why it is a good way to use of earth mixture without inorganic binder. Firstly, rammed earth with inorganic binder loses some of its good properties. Secondly, there are researches that indicate that the rammed earth can have satisfactory water resistance. And thirdly, there are a lot of unfired clay historic buildings which are in good condition by centuries. That is why we decided to investigate the rammed earth without the addition of inorganic binder [1, 5, 11].

This article is focused on drying speed of rammed earth without addition of inorganic binder. The drying speed is important because drying is the only process with witch unfired clay receives its strength. The rammed earth is usually load bearing construction. It is thus necessary to know drying speed to be able to estimate the time that must pass from a manufacture to a full construction load [1, 6].

Earth appropriate for building constructions consists from three elementary components: sand, clay and water. The clay fulfils a function of binder, the sand a function of filling agent and the water serves for activation of bonding properties of the clay and for good earth mixture processing [1, 2, 12].

However, the earth in nature is usually composed of several kinds of clay and contains another components like stones, gravel, etc. The natural earth composition varies and its composition is unique for each locality. The properties of unfired earth are dependent on composition of the earth mixture. These two fact are the reason why it is very difficult to describe the properties of the final unfired earth construction [1, 2, 12].

Knowledge of mechanical properties of rammed earth dependence on earth mixture composition is necessary for creation of such methodology. That is why we have decided to produce our own clay mixtures from the basic components.

Manufacturing Method

The earth mixtures for testing bodies manufacture were proposed and produced in our laboratory. The earth mixture was prepared from sand, illite-kaolinite clay and mixture water.A grain curve of the used sand can be seen in the Figure 3. The clay was used in powdery form (see Fig. 4). The amount of mixture water was determined by water coefficient. The water coefficient is defined by Equation 1.

$$W = m_W/m_C \qquad (1)$$

W - water coefficient [-]
m_W - weight of water [g]
m_W - weight of clay [g]

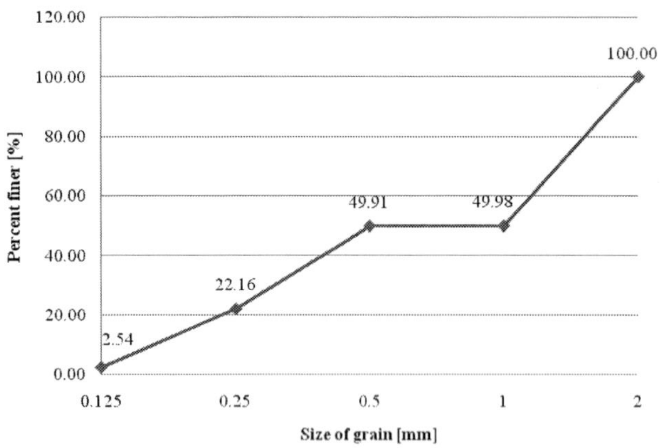

Fig. 3 Grain curve of used sand

Fig. 4Illite-kaolinite clay

There were produced two sets of clay mixtures (sets KRVII and KRIX).The sand to clay ratio was 80/20 for both of the sets and water to clay coefficient (water coefficient) was 0.25 (set KRVII) and 0.45 (set KRIX). The components were mixed and mechanically and manually kneaded until homogenous clay mixture was obtained.

Fig. 5 Ramming of the earth mixture to the steel moulds

The clay mixture was rammed to steel moulds and the testing bodies were created (see Fig. 5). It was manufactured for both of the sets 4 test bodies of size 4/4/16 cm, 4 testing bodies of size 2/2/10 cm, 4 testing bodies of size 2/2/5 cm and 4 testing bodies of size 4/4/8 cm. Ready-made testing bodies were right after unmolding inserted into air-conditioned chamber.

The environment properties in air-conditioned chamber were set to 20 °C and 50 % relative air humidity. The composition of the earth mixtures and number of the test bodies can be seen in the Table 1.

Tab.1 Composition and number of the test bodies

Set	Sand/clay ratio [-]	Water coefficient [-]	Number of the test bodies			
			2×2×10 cm	2×2×5 cm	4×4×16 cm	4×4×8 cm
KRVII	80/20	0.25		4		4
KRIX	80/20	0.45		4		4

Principle of measurement

The test bodies were inserted into air-conditioned chamber and were regularly weighed.The decrease of test bodies mass due to water drying in air-conditioned chamber was measured. Moisture content was calculated from value of measured weight (see Equation 2).

$w = m_{wet} - m_{dry}/m_{dry} \,.\, 100\ \%$ (2)

w - moisture content [%]

m_{wet} - weight of wet material [g]

m_{dry} - weight of dry material [g]

A form coefficient was determined from shape and size of tested bodies. The form coefficient was defined like volume to surface ratio (see Equation 3). There is a hypothesis that the smaller value of the form coefficient means quicker decrease of the moisture content. The question is what dependence between shape and drying rate is.

$F_{coeff} = V/S$ (3)

F_{coeff} - form coefficient [m^3/m^2]

V - volume [m^3]

S - surface [m^2]

Results

The moisture contents of the sets KRVII and KRIX in time are showed in the Table 2 and 3. The experiments show drying speed dependence on the form coefficient (see Fig. 6, 7).

Tab.2 Moisturecontentofset KRIX over time

Size [cm]	**Form coefficient [-]**	**Moisture content over time [%]**						
		0 h	**0.5 h**	**3.25 h**	**4.25 h**	**5.5 h**	**22.25 h**	**47.25 h**
4x4x16	0.89	9	8.2	6.6	5.9	5.1	1.4	0.9
2x2x10	0.45	9	7.6	5	3.9	2.9	1.3	1.2
4x4x8	0.80	9	7.4	4.5	3.6	2.8	1.1	0.8
2x2x5	0.42	9	6.5	1.8	1.2	1	0.2	0.2

Tab.3 Moisturecontentofset KRVIIover time

Size [cm]	**Form coefficient [-]**	**Moisture content over time [%]**					
		0 h	**0.75 h**	**2 h**	**3 h**	**5 h**	**23 h**
4x4x16	0.89	5	4.3	3.4	2.7	1.9	0.6
2x2x10	0.45	5	2.8	0.9	0.8	0.1	-
4x4x8	0.80	5	4	3.4	3.1	2.6	1.6
2x2x5	0.42	5	2.4	1	0.3	-	-

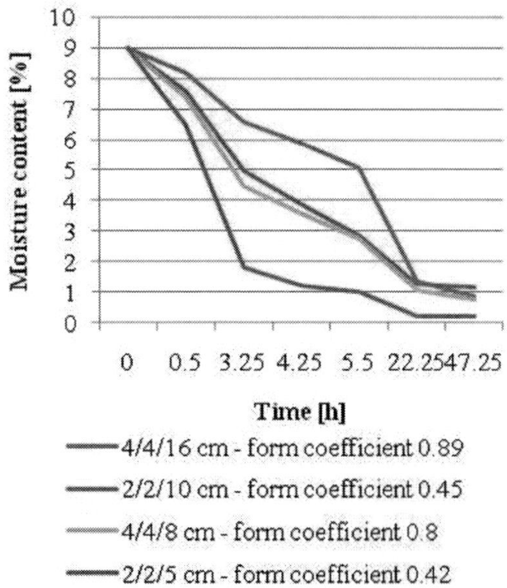

Fig. 6 Moisture content of set KRIX in time

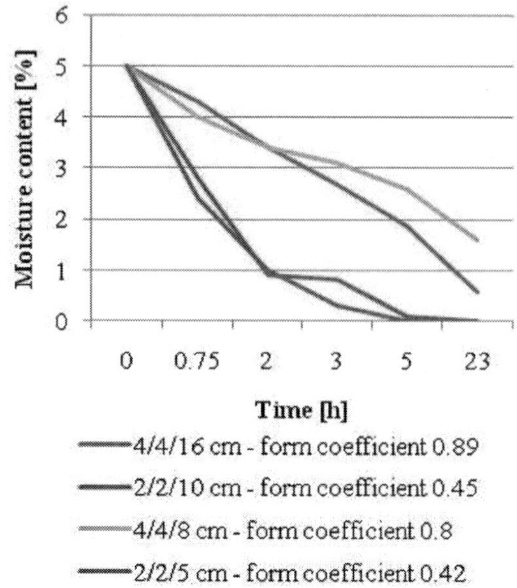

Fig. 7 Moisture content of set KRVII in time

One day after the production of the test bodies equilibrium humidity occurred (state of constant moisture content of the test bodies).

The slowest drying rate had the test bodies with the highest value of the form coefficient and the fastest drying rate had the test bodies with the lowest value of the form coefficient.It corresponds to our hypothesis. But, shape of the curves of graphs are not sufficiently convincing about our hypotheses (see Fig. 6, 7).

In particular Figure 6 shows that drying rate of the test bodies with different form coefficient (form coefficient 0.45 and 0.8) were very similar.This discrepancy could have been caused by induced deviation in geometry of the test bodies.

If we focus on the extreme values of the form coefficient we can see that the moisture content was determined approximately 4-9 times lower after three hours from the production of the test bodies (see Fig. 8).

Because of the inaccuracy of measurement it is not possible to determine the dependence the drying speed on the form coefficient from a narrow interval.

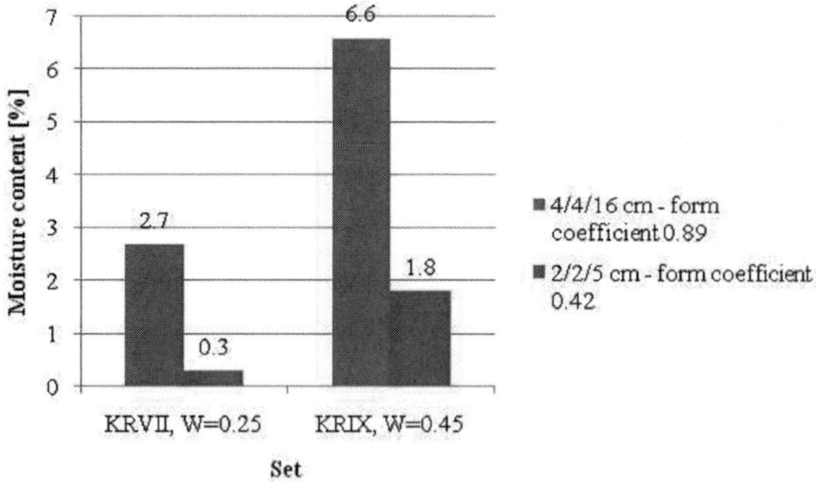

Fig. 8 Moisture content of set KRVII in time

Conclusions

The conclusions from this part of research are not satisfactory. The method for determination of the drying speed was not suitable, because the form coefficient of the test bodies were selected on the narrow interval. The next fault was to produce too small test bodies. The grains of sand that crumbled from the test bodies had significant proportion in the weight of a test body. After these experiences we found a good solution to produce special moulds for it. The rammed earth would be in these moulds whole drying time and it would be weight with these moulds. The grains of sand would not fall off and we will be able to obtain more exact value of the form coefficient.

Acknowledgement

The financial support of this experiment by the Czech Science Foundation (GAČR project NO. 18-10884S) and Faculty of Civil Engineering, Czech Technical University in Prague (SGS project No. SGS16/201/OHK1/3T/11) is gratefully acknowledged.

We would like to express our thanks to LB MINERALS, s.r.o. company for free supply of material necessary for experimental measurement and we would also like to express thanks to HELUZ cihlářský průmysl v.o.s. for giving statistical data and information.

References

[1] MINKE, Gernot. Building With Earth [online]. B.m.: ÖkobuchVerlag, Staufen, 2006 [vid. 2015-08-18]. ISBN 978-3-0346-0822-0. Dostupné z: http://archive.org/details/Gernot_Minke-Building_With_Earth

[2] P. Walker. Rammed Earth: Design and Construction Guidelines, vol. 2010. IHS BRE Press, Watford, 2010.

[3] NORMA. ČSN 1168-1939 - PODMÍNKY PRO ZEDNICKÉ A PŘIDRUŽENÉ PRÁCE POZEMNÍCH STAVEB. 1951.

[4] JAQUIN, Paul a Charles AUGARDE. Earth building : history, science and conservation [online]. Report. B.m.: Bracknell : IHS BRE Press. 2012 [vid. 2017-02-16]. Dostupné z: http://www.envia.bl.uk//handle/123456789/4131

[5] P. Jaquin, C. Augarde. Earth building : History, science and conservation. Report, Bracknell : IHS BRE Press, 2012.

[6] ŽABIČKOVÁ, Ivana. Hliněnéstavby [online]. Brno: Era 21, 2002 [vid. 2015-08-18]. ISBN 80-86517-21-7. Dostupné z: http://www.kosmas.cz/knihy/107760/hlinene-stavby/

[7] SIREWALL USA | STABILIZED INSULATED RAMMED EARTH, 2018.

[8] Agenda 21 - United Nations Environment Programme (UNEP). http://www.unep.org/Documents. Multilingual/Default.asp?documentid=52

[9] Scopus - Analyze search results.

https://www.scopus.com/term/analyzer.uri?sid=8EE866586B66D91CE67A3F34F5AF5EFE. wsnAw8kcdt7IPYLO0V48gA%3a70&origin=resultslist&src=s&s=TITLE-ABS-KEY%28rammed+clay%29+AND+PUBYEAR+%3e+1999&sort=plf-f&sdt=b&sot=b&sl=26&count=70&analyzeResults=Analyze+results&txGid=8EE866586B66D91CE67A3F34F5AF5EFE.wsnAw8kcdt7IPYLO0V48gA%3a49.

[10] V. B. Reddy, P. P. Kumar. Structural Behavior of Story-High Cement-Stabilized Rammed-Earth Walls under Compression. Journal of Materials in Civil Engineering 23(3):240–247, 2011.

[11] Q. B. Bui, J. C. Morel, B. V. Venkatarama Reddy, W. Ghayad. Durability of rammed earth walls exposedfor 20 years to natural weathering. Building and Environment 44(5):912–919, 2009. doi:10.1016/j.buildenv.2008.07.001.

[12] E. Nemecz. Clay Minerals. AkadémiaiKiadó,, 1st edn., 1981.

EAN 2018
56th conference on experimental stress analysis

Shrinkage of Rammed Earth Based on Illitic Clay

PADEVĚT Pavel[1,a], BITTNAR Petr[1,b] and OTCOVSKÁ Tereza[1,c]

[1] Czech Technical University in Prague, Faculty of Civil Engineering, Thákurova 7, Prague 6, 166 29, Czech republic

[a]pavel.padevet@fsv.cvut.cz, [b]petr.bittnar@fsv.cvut.cz, [c]tereza.otcovska@fsv.cvut.cz

Keywords: Rammed Clay, Illitic Clay, Shrinkage, Creep, Compression Strength, Lever Mechanism.

Abstract. The paper is focused on testing rammed earth, prepared from illitic clay, sand and water. The rammed earth was used for testing in creep and in shrinkage test. Six specimens were used for preparation of tests. Temperature and weight of specimens were observed during testing, which was 100 days long. Finally, curves of measured parameters were established from measured data. The comparison of influence between creep and shrinkage on the deformation was realized.

Introduction

The lack of material properties of unfired and rammed clay is at the same time one of the main causes of its unpopularity in the building industry in Europe [1]. Rammed clay has its undeniable advantages over modern building materials [2]. The advantage is its ecological and energy modesty during construction [3]. Of course, this building material has its disadvantages. The unavailability of material characteristics and thus considerable design uncertainty is one of the drawbacks.

The paper seeks to explain the behavior of one type of rammed clay in terms of material properties. Attention is focused on material shrinkage and creep [4]. In parallel with the rheological properties, mechanical properties of the clay are needed, because deformation of construction made from unfired clay can be relatively large. The process of shrinkage of unfired clay is coupled with changes of content of the water in material [5]. On the other hand creep of material depends on the loading of material.

Fig.1 The specimens used for measurement material properties, prepared from illitic clay

About of material

The experiment was realized with earth prepared form illitic clay, sand and water. The sand was composed of two fractions, 0-1 and 1-2 mm [6]. Smooth line granularity was created using of both fractions in the ratio of 60 : 40 (fine and coarse part). The clay was in a mixture represented 20 %, and sand accounted for 80 % of the dry mixture. The water ratio, defined as the relation between clay and water weight had size 0.37.

Preparation of mixture is basic step for production of quality material [7]. Firstly, sand was mixed with little content of water. Next step was in addition of full plenty of clay to sand. After then the sand and clay was blended by stirring attachment. During blending the remainder of water was added to the mixture. After a mixing process the mixture was knead in hands. Finally, the mixture was mixed by stirring attachment for 1 minute.

The mixture was placed into the moulds and rammed in several layers. The dimension of small specimens used for testing of creep was 20 × 20 × 70 mm. The specimens for compression testing had length 100 mm and same cross section area. Testing of bending stress was realized on specimens with dimension 40 × 40 × 160 mm. Compaction of small specimens was realized around in 5 layers. Bigger specimens included approximately 6 layers of compacted material.

The specimens were removed from moulds in same day like their production. All specimens were placed in climatic chamber to condition 20°C and relative humidity 50 %. Drying of material is process for increasing of its material properties. The weight changes of specimens were observed [4]. Creep of rammed specimens was started 6 hours after their production.

Test of material

Material tests were realized on the 12 specimens of different age, see Fig. 1. Creep test was carried out on the 3 specimens and shrinkage was measured on the 3 specimen, too. Loading area of specimens was octane with distance of opposite edges approximately 20 mm. Length of specimens for measurement was 70 mm. Length of measurement of rheological properties was 100 days. Temperature and humidity were measured in laboratory room in whole length of measure, see Fig.2.

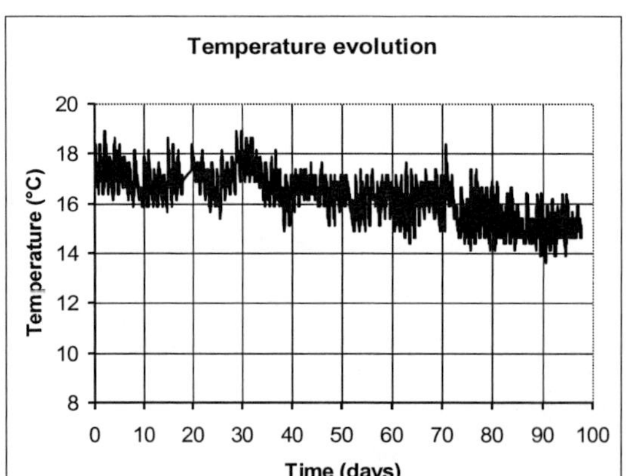

Fig.2 The curve of evolution of temperature during the test

Principle of measurement is in the placing of specimens into the levers mechanisms and measurement its axial deformation [8]. One lever mechanism is used for measurement of the one specimen. Deformation of whole length of specimen is measured by three deformation sensors with accuracy 0.0001 mm. The measurement of shrinkage is realized in the lever mechanism, too, but without loading of mechanism. The specimens are able respond on changes of temperature and their humidity, it is decreased during measurement. Maturation process is related with drying of material. Compression strength depends on the content of free water in the rammed unfired earth.

The specimens were not covered for ability of change of humidity. Weight of same specimens placed in area of measurement was measured, because weight of measured specimens is not possible measuring. These changes of weight were assigned to changes of weight of creep and shrinkage tested specimens. The deformation caused by creep contains deformation from shrinkage. If we want know pure impact of load on the deformation, than is necessary separate impact of shrinkage. Effect of separation of shrinkage is possible express by Eq.1, where δ_{cr} is pure creep, δ_{all} is deformation included creep and shrinkage and δ_{sh} is shrinkage of material.

$$\delta_{cr} = \delta_{all} - \delta_{sh}. \quad (1)$$

Results

Results are summarized in the graphs. Specimens No. 1 – 3 represents loaded state and specimens 4 – 6 were used for measurement of shrinkage, see Fig.3. As is possible see, for the shrinkage is important humidity and temperature condition. The main change took place in first ten days, see Fig.4. The specimen No.4 had expansion 0.015×10^{-3} milistrain. The specimens No. 5 and 6 had expansion near to 0.019 and 0.022×10^{-3} milistrain. The specimen No. 4 had different course of development to specimens No. 5 and 6 on beginning part of measurement.

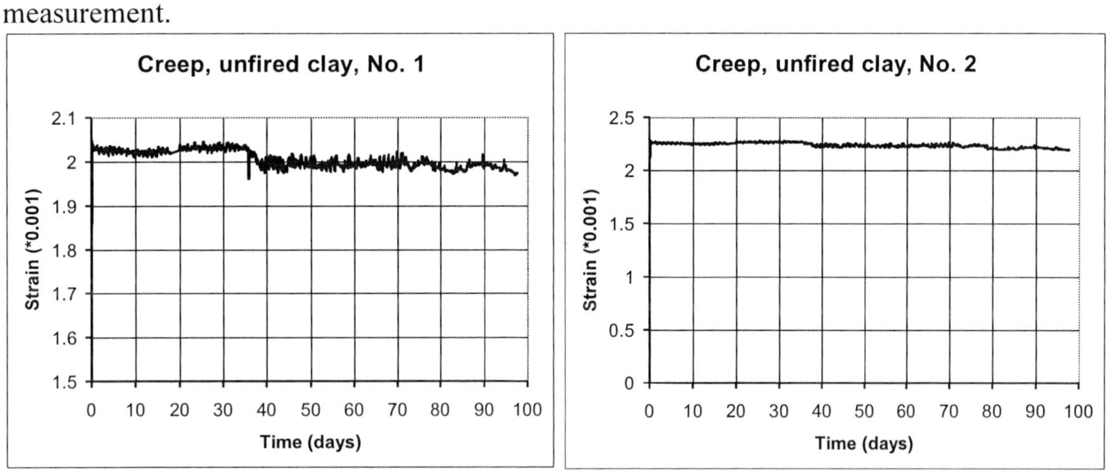

Fig.3 The curves of the creep included shrinkage of specimens 1 and 2

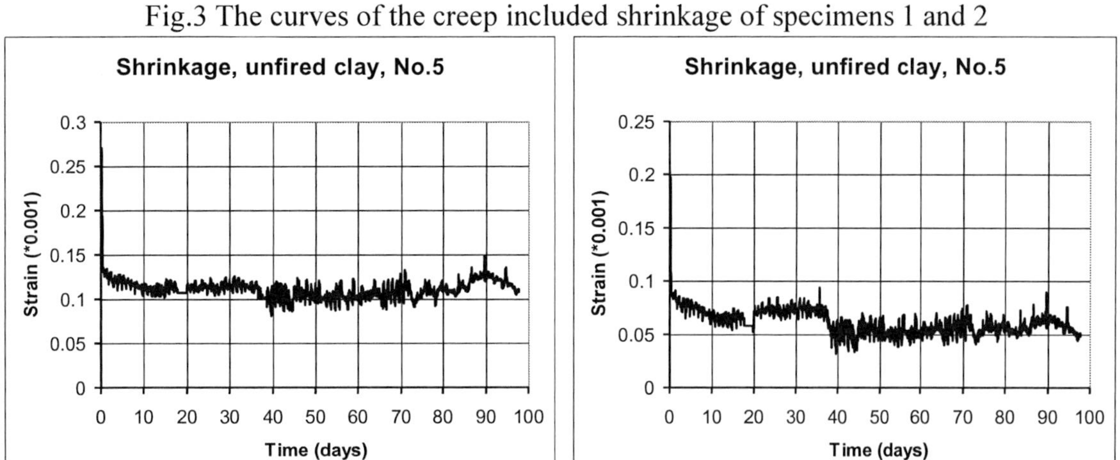

Fig.4 The curves of the shrinkage of specimens 5 and 6

The creep included shrinkage (δ_{all}) had size of strain stable during whole time of loading. Little changes relates with temperature changes, which are is possible see in graph of temperature evolution. The change of interior temperature in laboratory had occurred at 37 and 77 day of measurement, as is possible see on Fig.2. The temperature decreased about 1 °C in both cases of change. Of course, these sizes of creep included the shrinkage and do not reflect only influence of creep.

Using the Eq.1 is possible achieve pure creep. It is displayed in graphs in Fig.5 expressed creep without shrinkage for specimens 1 – 3. Influence of shrinkage was calculated by measurement on specimen No. 5. The deformation of specimens firstly increases due to loading in compression direction. Next, during one day the deformation is decreasing near to half level of initial deformation after the loading. Further development of deformation is possible characterized as a significant slowdown of deformation changes. The process change of the creep deformation ceases after 10 days from start of measurement. The size of this small change is near to 0.03×10^{-3} milistrain. Next evolution of deformation caused from pure creep is stopped.

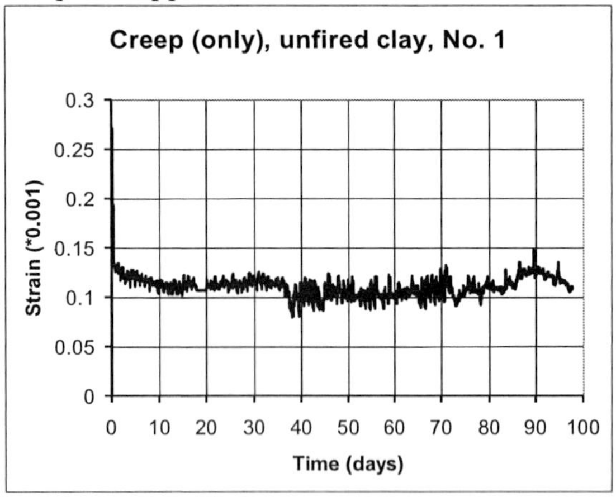

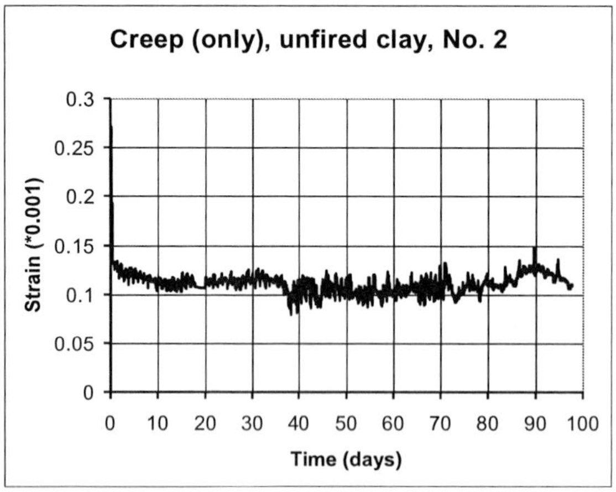

Fig.5 The curves of the creep without influence of shrinkage of specimens 1 and 2

Conclusions

Rammed earth has a very interesting perspective of creep properties. Totally interesting result is the size of shrinkage and creep after 100 days. This reached a value of 0.03×10^{-3} milistrain in the direction of compression for shrinkage, and approximately 2×10^{-3} for creep in same direction, see Fig. 2. But this value is influenced by shrinkage. The creep without shrinkage achieved value of strain 0.11×10^{-3} milistrain from initial deformation after loading. The evolution of creep and are very influenced of the reaching of compression strength. Iimportant influence on the size of deformation of rammed unfired clay has a shrinkage of material as is possible see from graphs and measured date. The end of creep is related with increasing of strength of material. The strength of material depends on rate of drying unfired clay. The curves of shrinkage are influenced by changes of humidity in first days of measurement. Next changes of shrinkage depends on the increased or decreased temperature of environment [9]. Results are comparable with material prepared from same ingredients but in another one relation [10].

Acknowledgement

This research has been supported by project SGS project No.16/201/OHK1/3T/11 Advanced from design of materials to the constructions and from project GACR No. 18-10884S Unfired and rammed clay for construction.

References

[1] G. Minke, Building with Earth [online], available from:

http://archive.org/details/Gernot_Minke-Building_With_Earth

[2] F. Pacheco-Torgal, S. Jalali, Earth constructions: Lessons from the past for future eco-efficient constructions, Constructions and Building Materials, Vol. 29, April 2012, pp. 512-519.

[3] S. Ferrari, A.F. Gualtieri, The use of illitic clays in the production of stoneware tile ceramics, Applied Clay Science, Vol. 32, Issue1-2, April 2006, pp. 73-81.

[4] T. Otcovská, B. Mužíková, P. Padevět, Drying Characteristics of Rammed Earth with Illitic-Kaolinitic Clay Content, NMM 2017 – Nano & Macro Mechanics 2017, Acta Polytechnica CTU Proceedings, Vol 13 (2017), 89 – 92,

DOI: https://doi.org/10.14311/APP.2017.13.0089

[5] M. Maddison, T. Mauring, K.Kirsimäe, U. Mander, The humidity buffer capacity of clay-sand plaster filler with phytomass from treatment wetlands, Building and Environment, Vol. 44, Issue 9, September 2009, pp. 1864-1868.

[6] P. Padevět, P. Bittnar, Creep of Unfired Clay Mortar Based on Illitic Clay, 55th Conference on Experimental Stress Analysis EAN 2017, Nový Smokovec, Slovakia, 2017.

[7] J.M. Kinuthia, J.E. Oti, Designed nom-fired clay mixes for sustainable and low carbon use, Applied Clay Science, Vol. 59-60, May 2012, pp 131-139.

[8] P. Padevět, P. Bittnar, Measuring the creep and material properties of cement paste specimens, WSEAS Transaction on Applied and Theoretical Mechanics pp. 81-90, 2010.

[9] M Jankula, T. Hulan, I. Štubňa, J. Ondruška, R. Podoba, P. Šín, P. Bačík, A. Trnik, The influence of heat on elastic properties of illitic clay Radobica, Nippon Seramikkusu Kyokai Gakujutsu Ronbunshi/Journal of the Ceramic Society of Japan, Vol. 123, Issue 1441, 2015, pp. 874-879.

[10]P. Padevět, P. Bittnar, B. Mužíková, Smrštění nepálené hlíny připravené z montmorillonitu, Proceedings of the 14th International Conference on New Trends in Statics and Dynamics of Buildings October 13-14, 2016 Bratislava, Slovakia (in Czech).

EAN 2018
56th conference on
experimental stress analysis

Evaluation of Acoustic Emission Signals Obtained during the Three-Point Bending Loading of Thermally Degraded Concrete Samples

PAZDERA Lubos [1,a], TOPOLAR Libor [1,b], MIKULASEK Karel [1,c], HODULAKOVA Michaela [1,d], DVORAK Richard [1,e], SMUTNY Jaroslav [1,f] and CHOBOLA Zdenek [1,g]

[1] Brno University of Technology, Faculty of Civil Engineering, Czech Republic

[a]pazdera.l@fce.vutbr.cz, [b]Libor.Topolar@vutbr.cz, [c]mikulasek.k@fce.vutbr.cz, [d]hodulakova.m@fce.vutbr.cz, [e]dvorak.r1@fce.vutbr.cz, [f]smutny.j@fce.vutbr.cz [g]chobola.z@fce.vutbr.cz

Keywords: acoustic emission, concrete, temperature, bending test, degradation

Abstract. Due to fires of building structures, there has been a growing interest in investigating the properties of concrete at high temperatures. In a fire, a temperature of up to 1400 °C may be measured. Such high temperatures will cause changes in the structures of the entire construction. Ultrasonic methods such as acoustic emission method can be used to monitor changes in the structure caused by a fire.
Heated to selected temperatures of 200 °C, 400 °C, 600 °C, 800 °C, 1000 °C, and 1200 °C, the concrete specimens were maintained for 60 minutes. Then, they were dried for another 48 hours at 110 °C and, after that, cooled down to a laboratory temperature. Beam concrete specimens were loaded under a three-point bending test. The evaluations of the experiments imply the possibility of using the acoustic emission method to describe the changes in a concrete structure after it is fire-degraded at the three-point bending loading.

Introduction

Due to fires of building structures, there has been a growing interest in investigating the properties of concrete at high temperatures. In a fire, a temperature of up to 1400 °C may be measured. Such high temperatures will cause changes in the structures of the entire construction. Ultrasonic methods such as acoustic emission method can be used to monitor changes in the structure caused by a fire [1].

The service-life of concrete structures depends on the behaviour of concrete in its environment and on the concrete mixture. Concrete behaviour depends mainly on the cement type, the aggregate type, the sand and the water-to-cement ratio [2].
Some studies have been made to identify the presumed causes and predict the degradation of concrete under fire conditions.

The thermal mismatch between the aggregate particles and the hydrated cement paste binder leads to a loss of bond among the constituents in addition to the loss of stiffness and strength. The increase in temperature results in water evaporation, C-S-H gel dehydration, calcium hydroxide and calcium aluminates decomposition, etc. Along with the increase in temperature, changes in the aggregate take place [3].

The following three phrases are important for the behaviour of concrete. (1) phase transformations (e.g., loss of free water at about 100 °C, decomposition of calcium hydroxide at about 450 °C, and crystal transformation of quartz at 573 °C from the form), (2) pore structure evolution (e.g., volume and surfaces of pores increase up to a temperature of about 500 °C and then decrease with further temperature increase, and (3) coupled thermo-hygro-chemo-mechanical processes (e.g., temperature gradients leading to thermal stresses, multiphase transport of water, and chemical changes that affect pore pressure and structure) [4]. Therefore, when the temperature exceeds 100°C, the capillary pore water changes into water vapour and the cement hydrates begin to desiccate. At higher temperature values, over 200°C, the dehydration of cement hydrates takes place which leads to the cracks in the cement paste being nucleated [5].

A simple three-point bending test can determine the behaviours of the tested specimens. The crack propagation path should be perpendicular to the central force [6].

The acoustic emission method is a tool of a group of the non-destructive testing methods. Due to local plastic deformations, elastic or acoustic waves spread throughout the structure of the material. The acceleration sensors then detect and record these elastic waves [7].

Experiment

Experimental concrete samples with dimensions of 100 mm x 100 mm x 400 mm were prepared from 15 % of Portland cement CEM I, 0.1 % of superplasticizer, 1 % of water and amount sand and gravel aggregates with Table 1 in the laboratory of the Institute of Technology of Building Materials and Components, Faculty of Civil Engineering, Brno University of Technology. Each specimen was immersed in a water bath for 28 days. Then, they were dried for another 48 hours. Heated in a programmable laboratory furnace at a heating rate of 5 °C/min to the selected temperatures of 200 °C, 400 °C, 600 °C, 800 °C, 1000 °C, and 1200 °C, the concrete specimens were maintained for 60 minutes. Then, they were dried for another 48 hours at 110 °C. Then, the specimens were cooled down to a laboratory temperature.

Table 1 Percentage of sand and gravel aggregates in monitored mixtures

Mixture	A	B	C
Sand 0/4 mm	15	38	35
Gravel aggregate 4/8 mm			43
Gravel aggregate 8/16 mm	42	22	
Gravel aggregate 11/22 mm		16	

Beam concrete specimens were loaded under a three-point bending test using the displacement-controlled method [8]. The fracture tests were carried out using a Heckert FPZ 100/1 testing machine within the range of 0 to 10 kN. The notch up to one third of the cross section determined unambiguously the fracture process zone [9]. The depths of notches were 2 mm. The acoustic emission sensors were placed on the surface of the concrete beam specimens. The XEDO acoustic emission system made by Dakel (Czech) recorded acoustic emission hits.

Figure 1 Experimental set-up

Results

The graphs in Figure 2 suggest the following conclusions of the comparison between the formulas (Table 1) in terms of the acoustic emission at a three-point bending test:

For samples at the normal temperature of 20°C - formula B seems to be the best one for which it is evident that, in concrete with three aggregate fractions, a minimum of number of cracks created during the total destruction of a sample. When formula C is loaded, an activity of the acoustic emission appears at the beginning caused probably by the supports being pushed into the sample and the samples settling down.

For samples tested at the temperature of 400 °C, containing no more freely-bound water, the greatest force has been achieved with formula C, which is likely to be caused by the closeness of the aggregate fractions. Although for evaluating the acoustic emission activity during the bending test, the behaviours of the formula A and C samples are similar, a slightly higher activity has been measured for formula A samples. The formula B samples show huge acoustic emission activity immediately before the break, which is caused by the samples being broken in a more abrupt manner as compared to the A and C formula samples.

In the samples tested at the load temperature of 600 °C, the crystalline phase of the aggregate quartz changes from alpha to beta (hexagonal configuration) at a temperature of 573 °C under normal pressure. At this temperature, also the Portlandite begins to disintegrate. The results of the acoustic emission show that the formula A samples showed lowest strength, probably because of a larger space between the aggregate grains filled with cement, which started to disintegrate. For the B and C formula samples, however, the situation is slightly different with the aggregate grains being closer to each other so that there is less cement in between and the samples show an activity typical for acoustic emission before breaking.

By a temperature of 800 °C, the Portlandite has already been totally disintegrated. For this reason, samples of all formulas break by very small forces. The acoustic emission activity is caused mostly by the cement being crushed and crumbled out and, thus, the grains moving in the freed space as proved by the results of the formula B samples containing the largest number of aggregate fractions.

By a temperature of 1000 °C, the samples of all formulas have become almost unmovable with very strong crumpling at the surface caused even by very slight manual handling, which is also corroborated by the results of the maximum force achieved and the acoustic emission activity.

The samples degraded by the temperature of 1200 °C then show completely different properties. The question is then, whether this is still a concrete.

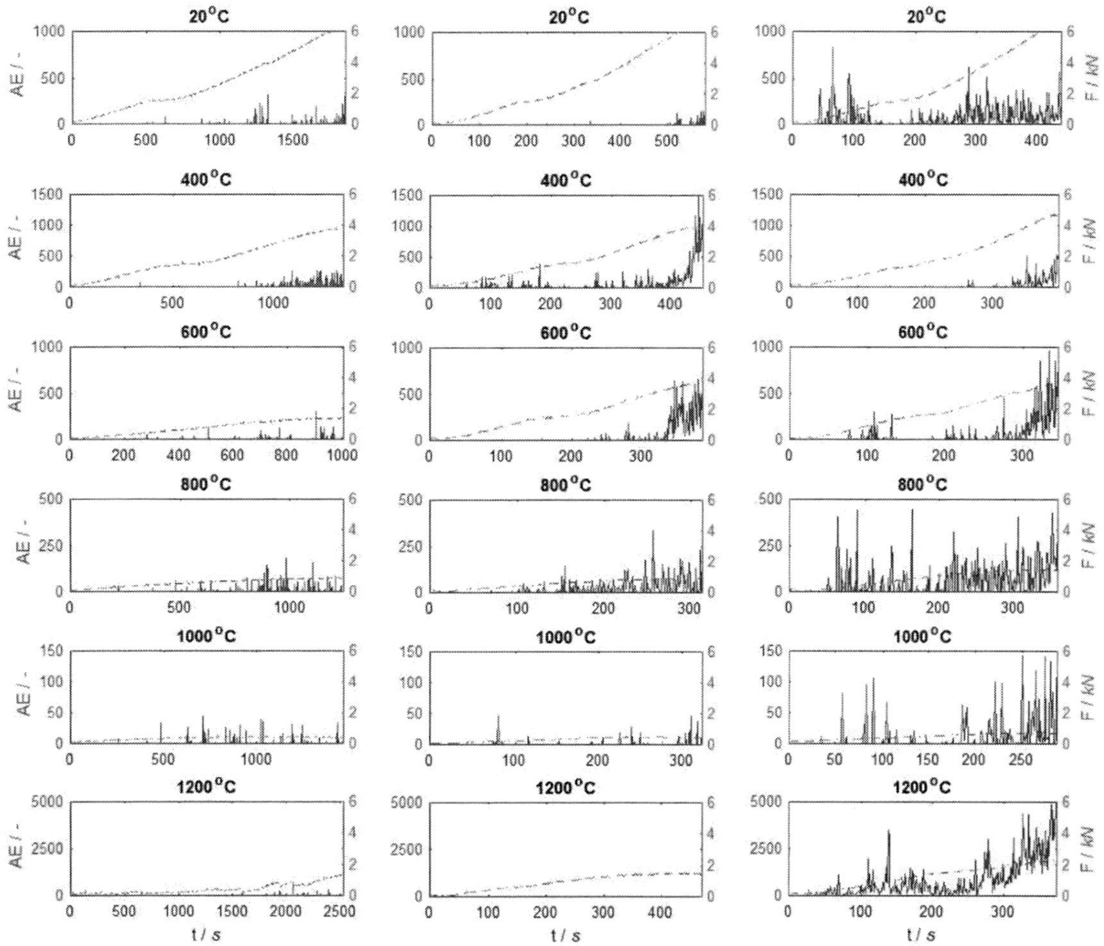

Figure 2 Dependence of acoustic emission activity *AE* (left axis) and load *F* (right axis) on the loading time *t*, of three mixtures A, B and C (diagrams from left to right).

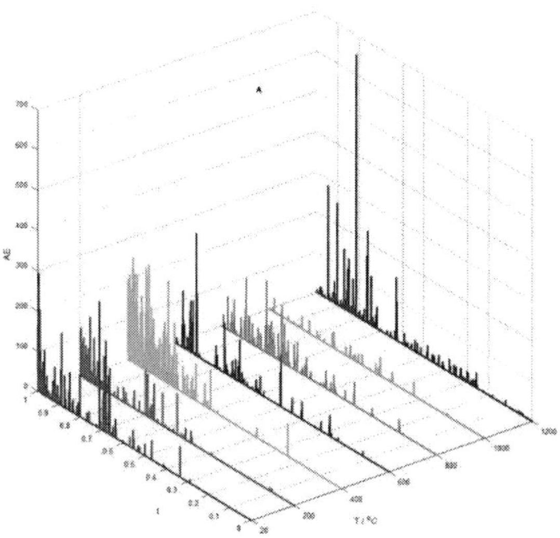

Figure 3 Mixture A

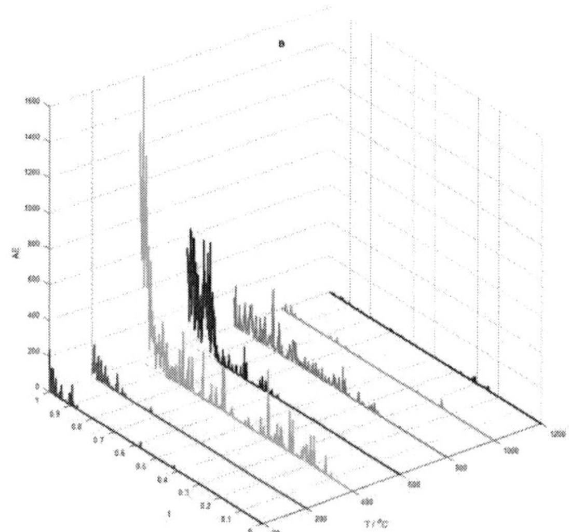

Figure 4 Mixture B

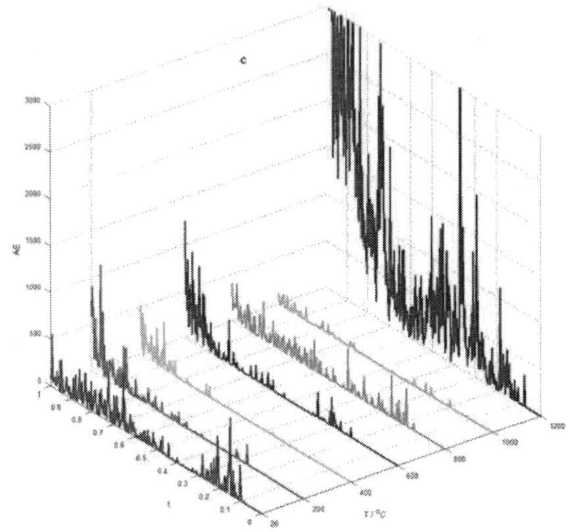

Figure 5 Mixture C

The behaviours of the acoustic emission activity for samples at different temperatures are summarized by diagrams in Figures 3 to 5. The time axis show the time elapsed from the beginning of loading to the sample destruction. Particularly from the diagram in Figure 5, it is clear that the behaviour of the sample after a temperature load over 1000 °C is unexpected due to its disintegration.

Conclusions

The results of an analysis of the acoustic emission signal show that, in terms of heat resistance, it is suitable to use a mixture of more aggregate types, which will be the subject of the next part of the research.

The paper describes an evaluation of the acoustic emission tests carried out on temperature degraded concrete specimens. The evaluations of the experiments imply the possibility of using the acoustic emission method to describe the changes in the concrete structure after its degradation by fire at three-point bending loading. The outcomes of the acoustic emission method can help provide the numerical models with more accurate input parameters.

Acknowledgement

This paper has been worked out under the project Czech Science Foundation GACR No.16-02261S supported by Czech Science Foundation and the project No. LO1408 "AdMaS UP - Advanced Materials, Structures and Technologies", supported by Ministry of Education, Youth and Sports under the „National Sustainability Programme I" and the internal project FAST-J-18-5275 supported by Faculty of Civil Engineering of Brno University of Technology.

References

[1] L. Pazdera, L. Topolář, M. Hoduláková, K. Mikulášek, J. Smutný, Impact of the heat load of concrete on the propagation of ultrasound waves. In Experimental Stress Analysis 2017. Košice (2017) 392-397.

[2] D. Gawin, C. Alonso, C. Andrade, C.E. Majorana, F. Pesavento, F Pesavento, Effect of damage on permeability and hygro-thermal behaviour of HPCs at elevated temperatures, Computers and Concrete, 2,3 (2005) 189-202

[3] I. Hager, Behaviour of cement concrete at high temperature, Bulletin of the polish academy of sciences technical sciences, 61, 1 (2013) 1-10.

[4] Z.P. Bazant, M.F. Kaplan, Concrete at High Temperatures, Longman, (1996)

[5] X.K. Li, R.T. Li, B.A. Schrefler, A coupled chemo-thermo-hygro-mechanical model of concrete at high temperature and failure analysis, International Journal for Numerical and Analytical Methods in Geomechanics, 30,7 (2006) 635-681

[6] G. Wardeh, E. Ghorbel, Prediction of fracture parameters and strain-softening behaviour of concrete: effect of frost action, Materials and Structures, 48,1-2 (2015) 123-138

[7] D.G. Aggelis, N.M. Barkoula, T.E. Matikas, A.S. Paipetis, Acoustic emission as a tool for damage identification and characterization in glass reinforced cross ply laminates, Applied Composite Materials, 20,4 (2013) 489-503

[8] I. Havlikova, V. Bilek, L. Topolar, H. Simonova, P. Schmid, Z. Kersner, Modified cement-based mortars: crack initiation and volume changes, Materiali in tehnologije, 49, 4 (2015) 557-561

[9] F.H. Wittmann, X. Hu, Fracture process zone in cementitious materials, International Journal of Fracture, 51,1 (1991) 3-18

EAN 2018
56th conference on experimental stress analysis

Measurement of Response Distribution on Mass Crossbeams in Dependence on Load Position

PETR Karel[1,a], STARÝ František[1,b], DUB Martin[1,c], DYNYBYL Vojtěch[1,d] and ČEŠPÍRO Zdeněk[1,e]

[1]Czech Technical University in Prague, Faculty of Mechanical Engineering, Technická 4, 166 07 Prague 6, Czech Republic

[a]karel.petr@fs.cvut.cz, [b]frantisek.stary@fs.cvut.cz, [c]martin.dub@fs.cvut.cz, [d]vojtech.dynybyl@fs.cvut.cz, [e]zdenek.cespiro@fs.cvut.cz

Keywords: measurement, distribution, experimental, mast crossbeams, strain gauges, construction hoist

Abstract. The article deals with experimental measurement of three design variants of construction hoist masts. There is described methodology of measurement and marking of measured places. All three masts were experimentally tested by means of strain gauge measurement and results are discussed.

Introduction

There is a demand on higher load capacity of construction hoists due to increasing productivity of construction work. Manufacturers of hoists want to satisfy this demand by innovation of masts, but they are limited by connection dimensions of nowadays used masts so the old and new masts could work together. Usage of construction hoists is shown on Fig. 1.

Fig.1 Photos of construction hoists [1]

Variants Description

In earlier published article [1] is described FEM analysis of original mast (variant A) and modified masts (variant B and C). Selection of measuring points was made on base of this analysis. Reason for this measurement was getting information about distribution and magnitude of deformation of mast crossbeams in different operational modes.

The variant marked as A is original construction which started to lose its reliability due to increased load capacity of construction hoist (see Fig. 2). The variant marked as B is based on variant A. Middle crossbeam shape was changed so half of its length is C profile and half is L profile and it is made from bent sheet metal. Top and bottom crossbeam is closed by a second L profile from bent sheet metal and welded to original rolled L profile. The variant marked as C is based on variant B. Top and bottom crossbeam has been changed from closed shape to C profile from bent sheet metal. Middle crossbeam is same as at variant B [1].

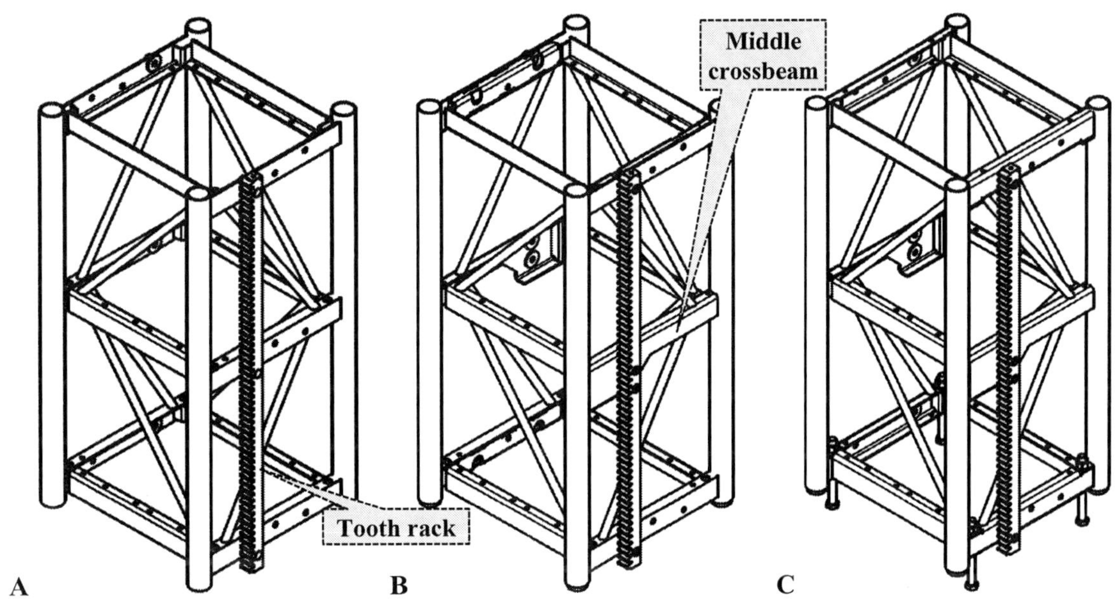

Fig.2 Mast variants (A – original, B – closed profile, C – C-profile) [1]

Measurement methodology

Relative deformation ε was chosen as measured magnitude to evaluate reaction of crossbeams on load.

Measurement was carried out via strain gages 1-LY11-6/120 (HBM). Position of load (position of cage pulleys) was sensed with strain gages placed on tooth rack for all measurements (see Fig. 3).

Measurement was carried out for these load levels: empty cage, cage with set of weight 3 200 kg and cage with set of weight 3 900 kg. Sample rate was set at 10 kHz. Measurement was carried out during following operational modes: going up and down, braking via instantaneous safety gear, loosen tooth rack and not adjusted pinions. There were used 6 strain gages on each variant of mast on crossbeams and 3 strain gages at tooth rack near tooth root.

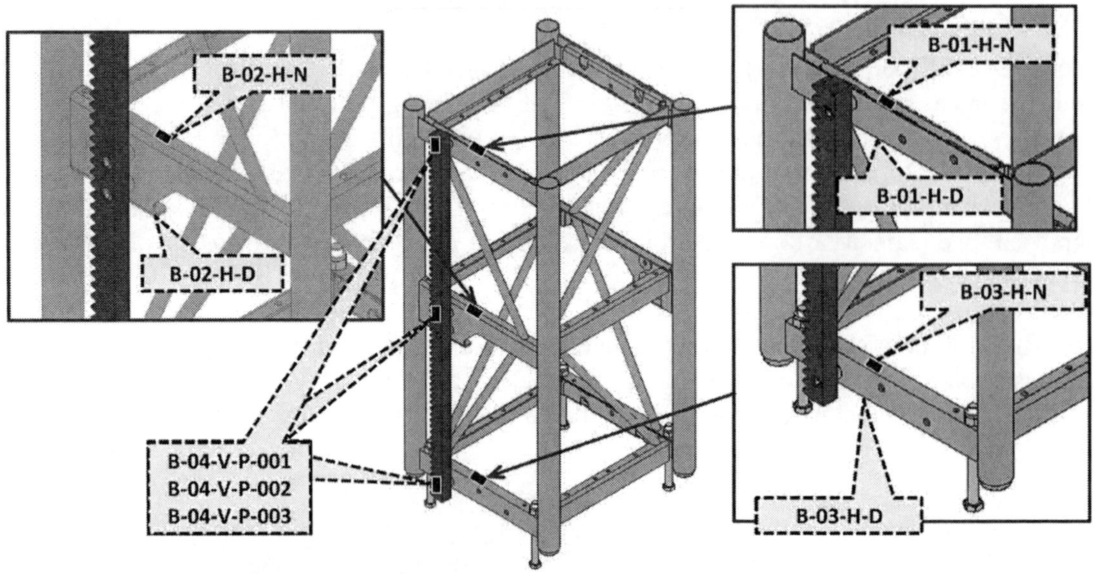

Fig.3 Indication of strain gages placed on tooth rack and on hoist variant B

There are in bellow mentioned examples represented measured data from middle crossbeams, specifically from upper strain gages marked as (A-02-H-N; B-02-H-N and C-02-H-N). First letter determines hoist variant, two digits determines position of crossbeam, letters H and V determines orientation (H – horizontal or V – vertical) and last letter determines upper (N) or lower (D) flange.

Measurement example 1 – Going up and down with full speed and maximal set of weight 3 900 kg

There is shown time behavior of relative deformation at middle crossbeam (Fig. 2) in Fig. 4 for variants A, B and C. Measured values of individual variant were for going up and down following: $\varepsilon_A = 288.5 / 139.0$ μm/m; $\varepsilon_B = 380.2 / 319.5$ μm/m; $\varepsilon_C = 212.9 / 253.6$ μm/m.

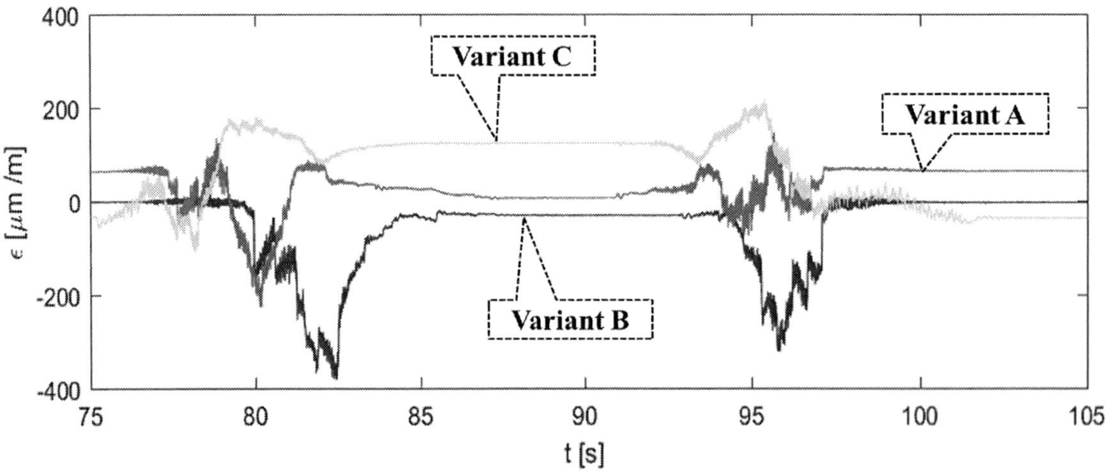

Fig.4 Deformation behavior at middle crossbeam for variants A, B and C – Going up and down with full speed and maximal set of weight 3 900 kg

Measurement example 2 – Going up and down with full speed and empty cage

There is shown time behavior of relative deformation at middle crossbeam in Fig. 5 for variants A, B and C. Measured values of individual variant were for going up and down following: ε_A = 166.6 / 81.8 μm/m; ε_B = 191.8 / 157.2 μm/m; ε_C = 126.2 / 81.8 μm/m. Behavior is of a same shape as with maximal set of weight (Fig. 4), but with lower amplitude values.

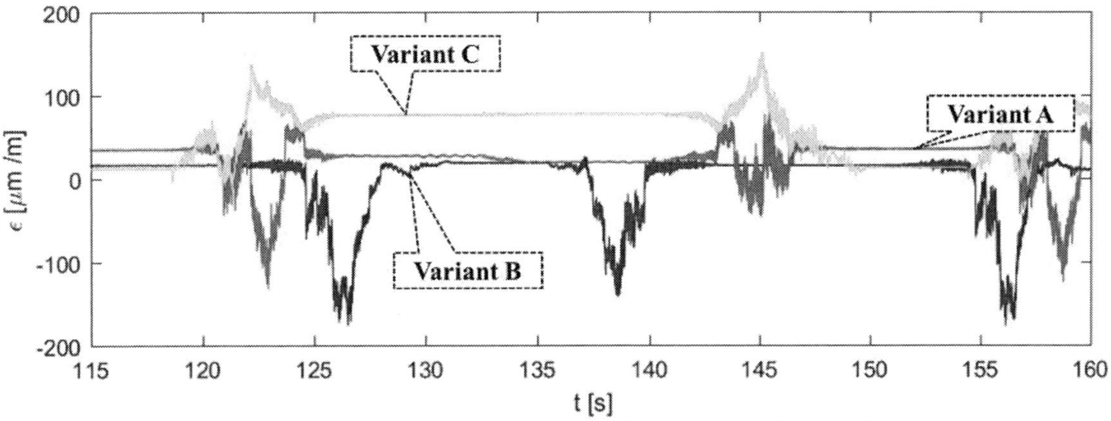

Fig.5 Deformation behavior at middle crossbeam for variants A, B and C – Going up and down with full speed and empty cage

Measurement example 3 – Going up and down with full speed and maximal set of weight 3 900 kg, loosened tooth rack

There is shown time behavior of relative deformation at middle crossbeam in Fig. 6 for variants A, B and C. From comparison of Fig. 4, Fig. 5 and Fig. 6 follows, that if there is change (not significant) of load distribution among individual crossbeams, it is given by lowering of bend stiffness of tooth rack and crossbeam. Not enough measuring channels were available during measurement (they were used for measurement of other strain gages), so data for middle crossbeam of hoist variant B is missing. Measured values of individual variant were for going up and down following: ε_A = 336.9 / 135.8 μm/m; ε_C = 208.6 / 262.9 μm/m.

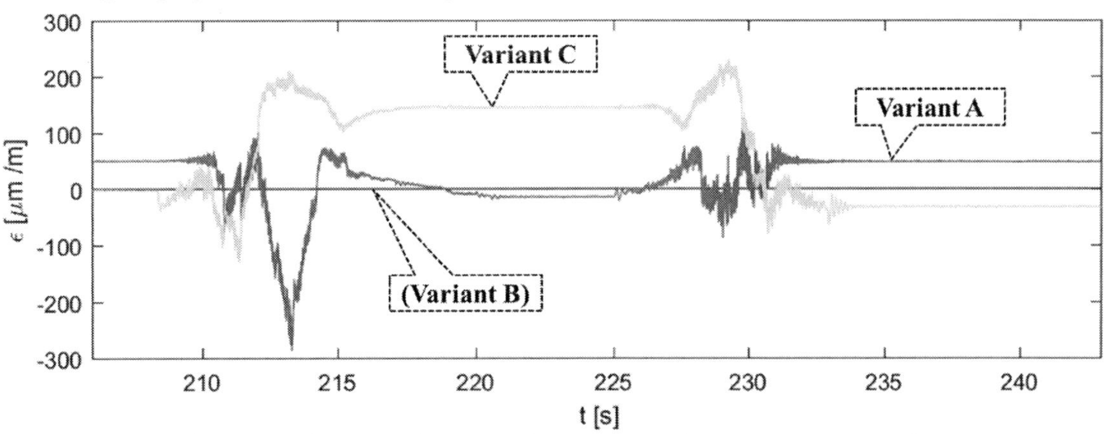

Fig.6 Deformation behavior at middle crossbeam for variants A and C – Going up and down with full speed and maximal set of weight 3 900 kg, loosened tooth rack

Measurement example 4 – Activation of instantaneous safety gear in upper half of hoist A

This measurement was carried out for mapping transient performance, which can came during instantaneous safety gear activation. Hoist variants were placed at each other in order from top: B, A, C. Fig. 7 shows as expected swing of measured strain behavior with following

damping to static load from cage weight. This dynamic impulse caused crossbeams of hoist B to oscillate. Hoist C has minimal oscillation.
Measured values were following: $\varepsilon_A = 488.7$ μm/m; $\varepsilon_B = 247.1$ μm/m; $\varepsilon_C = 101.3$μm/m.

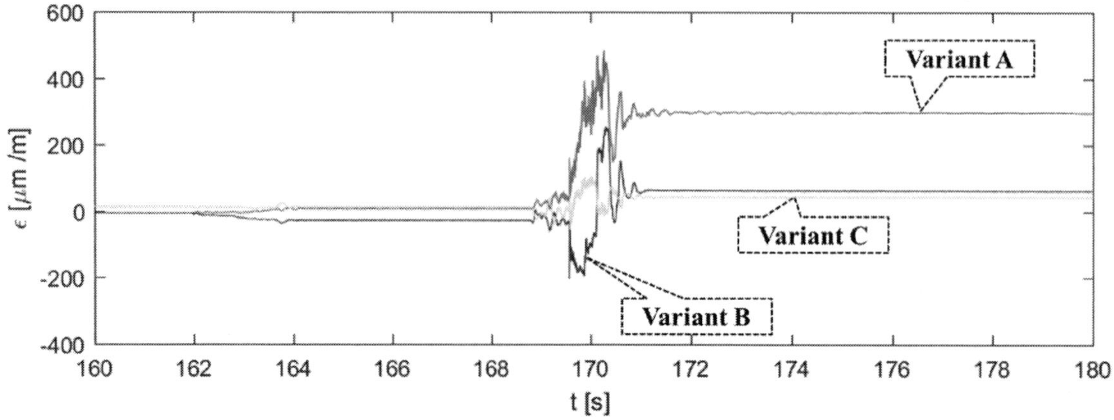

Fig.7 Deformation behavior at crossbeam for variants A, B and C – Activation of instantaneous safety gear in upper half of hoist

Discussion

Fig. 4 shows maximal values of amplitudes and time behavior of relative deformation during run. It is obvious that crossbeams reacts on cage passing differently (different crossbeams design and different tooth rack interaction with crossbeams).

Important fact is oscillation of some signals around zero value. This is not detected at crossbeam variant B (see Fig. 4 – 6), that has top and bottom crossbeam from closed profile. These profiles shows stability and do not vibrate. Moreover the stepped load from 3 pinions during cage passing is very well visible. It is most important for dynamical analysis evaluation of crossbeams, that load has pulsation cycle character i.e. from zero to maximum. When there is oscillation of signal around zero value, it is usually asymmetrical oscillating cycle.

Table 1. contains maximal values of relative deformations, which were measured with strain gages during different load levels (Example 1 – 4).

Tab. 1 Comparison of measurement results of individual variants

Variant	Example 1		Example 2		Example 3		Example 4
	up	down	up	down	up	down	
A	288.5	139.0	166.6	81.8	336.9	135.8	488.7
B	380.2	319.5	191.8	157.2	-	-	247.1
C	212.9	253.6	126.2	139.4	208.6	262.9	101.3

Strain gauges shows deformation in longitudinal direction of flange, which most corresponds with bend of crossbeams. Diagrams in Fig. 8 and Fig. 9 shows, that lowest bending stiffness shows hoist A. The highest bending stiffness shows hoist C.

Influence of loosened tooth racks and not adjusted pinions in not significant. Loosened tooth racks fit closely to bolts at all places similar, so the load is distributed equally analogously as with properly tightened tooth racks.

Important information from measurement with activation of instantaneous safety gear is that dynamical load to hoist is approximately twice as much as during service run. However activation of safety gear is not common during service run.

Diagrams in Fig. 8 and Fig.9 show proportion of relative deformations among hoists A, B and C.

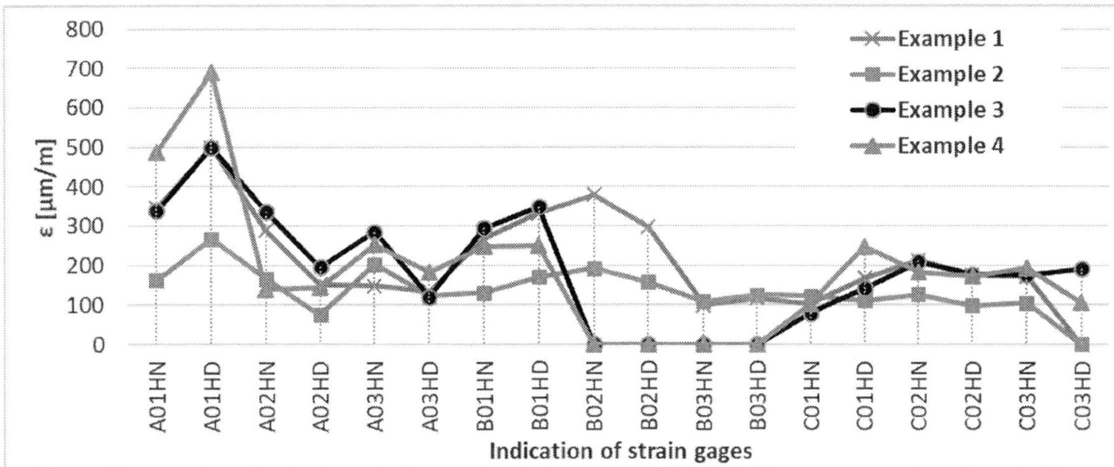

Fig.8 Review of measured relative deformations during different service runs – going up

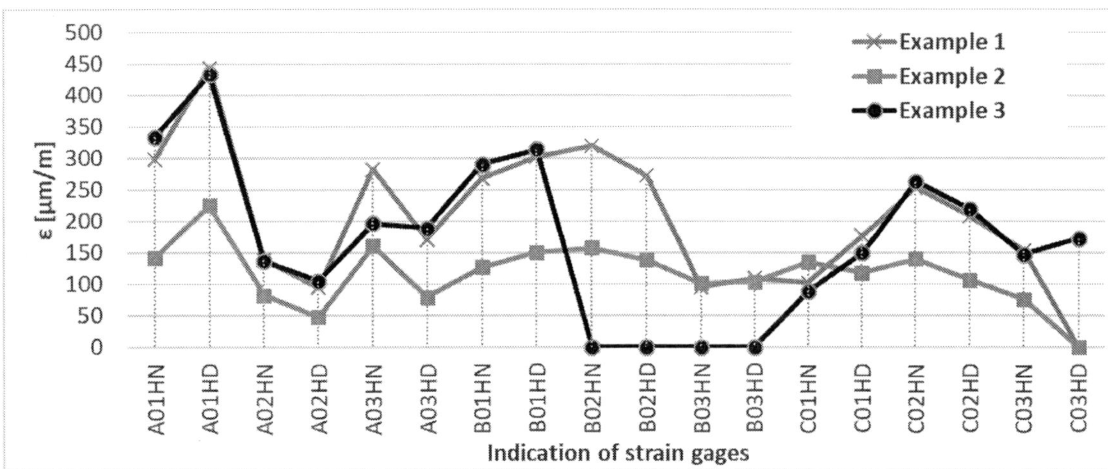

Fig.9 Review of measured relative deformations during different service runs – going down

Conclusions

One of important crossbeam property was their attenuation behavior after cage passing. Best attenuation results had crossbeam B, which has closed profiles. It´s top and bottom crossbeams reacts only by increasing of relative deformation during cage passing and it´s dropping to zero after unloading – it had character of pulsating cycle. Other variants of crossbeams oscillates around zero values – they had character of oscillating cycle.

Simulated assembly mistakes (loosened tooth racks and not adjusted pinions) did not influence load of hoist significantly. Load is distributed in similar way as when the assembly is correctly done.

References

[1] STARÝ, F., PETR, K., DUB, M., et al. Comparison of Different Construction Hoist Masts by FEM Analysis. In: Proceedings of 58th International Conference of Machine Design Departments – ICMD 2017. 58th International Conference of Machine Design Departments. Prague, Horoměřice, 06.09.2017 - 08.09.2017. Prague-Suchdol: Faculty of Engineering. 2017, s. 344-347. ISBN 978-80-213-2769-6.

EAN 2018
56th conference on experimental stress analysis

Investigation of Mechanical Properties of Cement Pastes with Fine Recycled Concrete as Partial Replacement for Portland Cement: Kind of Method High-Speed Milling

PROŠEK Zdeněk[1,a], TESÁREK Pavel[1,b], ĎUREJE Jakub[1,c], TREJBAL Jan[1,d] and ZOBAL Ondřej[1,e]

[1] Faculty of Civil Engineering, Czech Technical University in Prague, Thákurova 7, 166 29 Prague, Czech Republic

[a]zdenek.prosek@fsv.cvut.cz, [b]tesarek@fsv.cvut.cz, [c]jakub.dureje@fsv.cvut.cz, [d]jan.trejbal@fsv.cvut.cz, [e]ondrej.zobal@fsv.cvut.cz

Keywords: Cement pastes, Recycled concrete powder, Old concrete, Mechanical activation, Mechanical properties.

Abstract. This article deals with mechanical activation of old concrete by high-speed mills and using recycled concrete powder as a partial replacement for cement. In particular, we focus on mechanical properties of cement pastes containing different kinds of fineness of micronized recycled concrete powder. As recycled materials, we used the monolithic construction of a column from a construction of supporting structures. Recycled material was milled by company LAVARIS Ltd. at high-speed milling process with different types of milling. Mechanical properties were obtained by non-destructive (resonance method) and destructive method. Investigated mechanical properties were the dynamic modulus of elasticity and the dynamic shear modulus, flexural strength and compressive strength. Results obtained from these materials were compared with the reference material.

Introduction

Only the fraction larger than 1 mm is used in this recycling. However, the problem arises when processing a very fine fraction (<1 mm) of recycled concrete. One option is thermal activation of an old cement paste [1, 2]. Due to high demands on energy consumption and CO_2 production, it is particularly suitable for sustainable and environmentally friendly methods where the very fine fraction of recycled concrete is used in its raw form [3, 4].

Unlike mentioned researches, we focused directly on mechanical activation of old cement pastes. The article deals directly with the use of possible hydration potential of micronized concrete. The micronizing of the old cement paste into fine particle results in uncovered unhydrated clinker minerals, which can further hydrate upon contact with water [5]. In addition, the size of the grains of inert portions such as filler and the old cement matrix will be reduced. Small inert grains in the future composite can create nucleation centers for Portland growth and thus positively affect the hydration of cement. In addition, small grains may have a micro-filler function [6].

Recycling through high-speed milling is a highly energy-intensive method. For this reason, its optimization is very important. Company LAVARIS ltd. has patented grinding elements and a new generation of high-speed mills that can grind recycled material in two levels and have several possible sampling points [7]. Depending on the size of the particle size, individual sampling points can be used. Using these mills, you can optimize the entire course of recycling of materials so that it is economical and environmentally friendly. A similar way of optimization has already been used to recycle waste marble sludge [8].

Materials and Specimens

Cement pastes composed of 70 wt. % Portland cement CEM I 42.5R from Radotin and 30 wt. % non- / micronized recycled concrete has been prepared for investigating the impact of micronizing. Recycled concrete was obtained from an old construction of a column. The difference between individual specimens was in the mode of high-speed milling and sampling of the micronized recycled concrete. Testing was performed on beams of dimensions equal to 40 × 40 × 160 mm.

Micronized recycled concrete without milling (REC) had average grain size of 304 microns. Micronized recycled concrete I-level milling removed from the front of the mill (REC I-A) had average grain size of 48 microns. Micronized recycled concrete I-level milling removed from the middle of the mill (REC I-B) had average grain size of 6.17 microns. Micronized recycled concrete I-level milling removed from behind of the mill (REC I-C) had average grain size of 5.51 microns. Micronized recycled concrete II-level milling removed from the front of the mill (REC I-A/II-A) had average grain size of 38.7 microns. Micronized recycled concrete II-level milling removed from behind of the mill (REC I-A/II-B) had average grain size of 3.04 microns.

Experimental methods

Investigated mechanical parameters were the dynamic modulus of elasticity, the dynamic shear modulus, flexural strength and compressive strength. Values of the dynamic shear modulus, the dynamic modulus of elasticity, flexural strength and compressive strength of the specimens were determined after 28 days.

Resonance method was used for non-destructive testing; its principle was explained in the work of Topič et al. [9]. Measurement of the dynamic modulus of elasticity was carried out using the measurement line from Brüel & Kjær, consisting of a measuring unit type 3560-B-X12 impact hammer type 8206 and an acceleration sensor type 4519-003 53422 assemblies were connected to the control computer. The dynamic modulus of elasticity was achieved by using the basic longitudinal natural frequency of specimens and the dynamic shear modulus was achieved by using the basic torsional natural frequency of the specimen. The mass of all specimens was measured and the dimensions were recorded before the resonance method was applied. Bulk density was calculated from these values.

The flexural and compressive strength were determined using the Heckert device, model FP100. The testing was displacement controlled at a constant rate of 0.1 mm/s in the case of three-point bending and 0.3 mm/s for the compressive test. The distance between supports for three-point bending test was equal to 100 mm. The uniaxial compressive test was performed on the broken halves of the specimens with effective dimensions of 40 × 40 × 80 mm.

Results and Discussion

Results (Fig. 2) show the effect of fineness of micronizing recycled concrete on mechanical properties of cement pastes, namely dynamic modulus of elasticity and dynamic shear modulus. Results do not show a very positive effect of milling on dynamic modulus of elasticity (Fig. 2). The greatest effect can be seen in the non-milled recycled material where the dynamic modulus of elasticity is higher than in the reference specimen. The average value of dynamic modulus of elasticity of references specimens was 27.1 ± 0.1 GPa. Specimens with non-milled recycled materials had the value of dynamic modulus of elasticity equal to 28.7 ± 0.3 GPa and the specimens with finest milled recycled concrete 22.6 ± 0.3 GPa. The same effect can be seen in the case of the dynamic shear modulus and compressive strength (Fig. 2 and Fig. 3). The average values of dynamic shear modulus were equal to 10.8 ± 0.04

GPa in the case of reference specimens, 11.5 ± 0.12 GPa in the case of specimens with non-milled recycled concrete and 9.1 ± 0.24 GPa in the case of specimens with finest milled recycled concrete. The average values of compressive strength were equal to 110.7 ± 3.2 MPa in the case of reference specimens 101.0 ± 4.6 MPa in the case of specimens with non-milled recycled concrete and 79.9 ± 5.6 MPa in the case of specimens with finest milled recycled concrete.

This effect was caused by decrease values of the bulk density. Average values of the bulk density were 2113 ± 8 kg/m^3 in the case of reference specimens, 2122 ± 12 kg/m^3 in the case of specimens with non-milled recycled concrete and 1986 ± 24 kg/m^3 in case specimens with finest milled recycled concrete. The reduction in bulk density indicates an increase in porosity that has a negative effect on these mechanical properties. One of the possible ways to eliminate this effect is to use plasticizers.

Flexural strength is shown in figure 2. The higher value of flexural strength had the specimen with milled recycled concrete (REC I-A), namely 4.1 ± 0.12 MPa. The references specimen had the average value of flexural strength equal to 2.5 ± 0.48 MPa. The results showed that average grain size around 40 microns had the positive effect on flexural strength. This effect is due to the fact that recycled concrete forms a supporting structure in the cement composite. The best way of micronizing is a one-level milling with sampling at the front of the milling process.

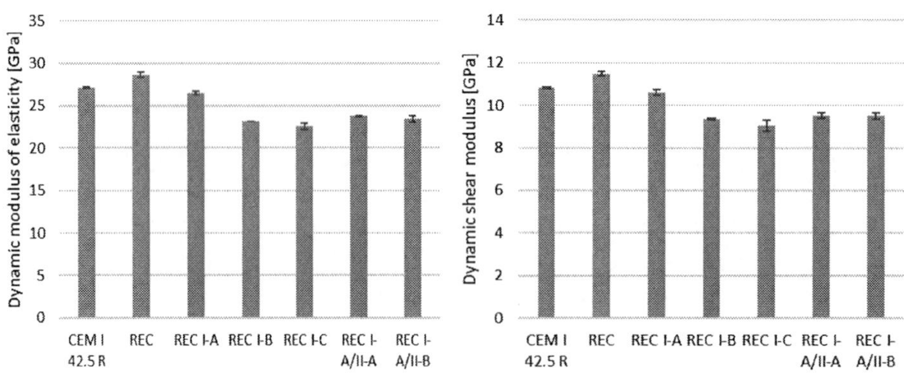

Fig.1 Comparison of dynamic modulus of elasticity and dynamic shear modulus (with standard deviations)

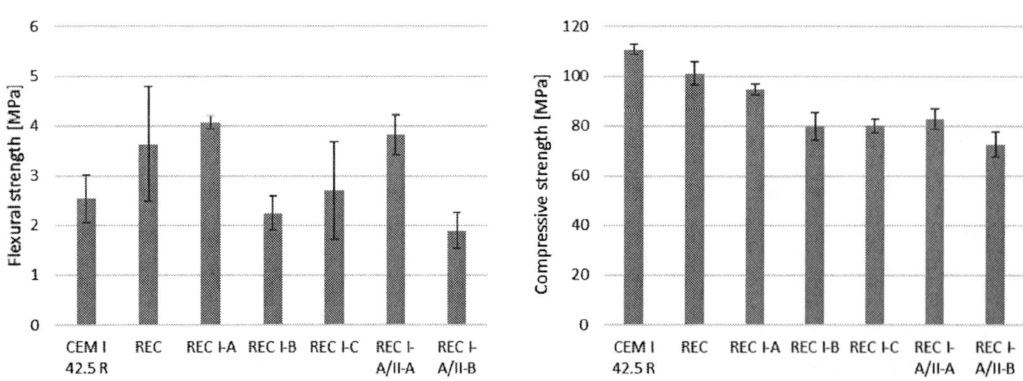

Fig.2 Comparison of flexural strength and compressive strength (with standard deviations)

Conclusions

Results does not show a very positive effect of milling on mechanical properties. It can be seen that the non-milled recycled concrete had the higher value of dynamic modulus of elasticity than references specimens. The greatest effect of milling process can be seen in the recycled concrete I-level milling removed from the front of the mill, where flexural strength is higher than in the reference specimen and the value of compressive strength was lower about 15 MPa. This is due to the fact that the recycled material had negative effect on porosity of cement composites. One of the possible ways to eliminate this effect is to use plasticizers. Future work will be dealing with using plasticizer with micromilled recycled concrete in cement composites.

Acknowledgement

This paper was financially supported by the GA ČR research a project under the number 17-06771S and by Czech Technical University in Prague under No. SGS project SGS16/201/OHK1/3T/11. The authors also thank company LAVARIS Ltd, where a high speed mill from was used for micronizing of the recycled concrete.

References

[1] Z. Shui et al., Rehydration reactivity of recycled mortar from concrete waste experienced to thermal treatment, Construction and Building Materials 22 (2008) 1723-1729.

[2] J. Shoon et al., Waste fibrecement: An interesting alternative raw material for a sustainable Portland clinker production, Construction and Building Materials 36 (2012) 391-403.

[3] V. Corinaldesi et al., Behaviour of cementious mortars containing different kinds of recycled aggregate, Construction and Building Materials 23 (2009) 289-294.

[4] S. Ahmari et al., Production of geopolymeric binder from blended waste concrete powder and fly ash, Construction and Building Materials 35 (2012) 718-729.

[5] J. Hrůza et al., Mechanical properties of the cement paste with fine ground recycled concrete depending on the microstructure, Advanced Materials Research 1144 (2017) 22–27.

[6] V. Bílek, The use of concrete filler as a mineral admixture in concrete. Proc. of 8 th int. Conf. Concrete in the Low Carbon Era (2012) 667–675.

[7] Information on http://www.lavaris.cz

[8] Z. Prošek et al., Effect of particle size distribution waste micronized marble powder on the mechanical properties of the cement composite, Advanced Materials Research 1144 (2017) 54–58.

[9] J. Topič et al., Effect of pva modification on properties of cement composites, Acta Polytechnica 55 (2015) 64–75.

Method For Vacuum Conductivity Estimation For Engineering Use

RACHAČ Marek[1,a], HEDBÁVNÝ Pavel[2,b], HOŠEK Jan[1,c] and LOPOT František[1,3,d]

[1]CTU Praha, Faculty of mechanical engineering, Technická 4, 16607, Praha, Czech Republic

[2]VAKUUM PRAHA, V Holešovičkách 2, 180 00, Praha, Czech Republic

[3]Charles University, Lab. of extrem loading, José Martího 31, 16252, Praha, Czech Republic

[a]marek.rachac@fs.cvut.cz, [b]hedbavny@vakuum.cz, [c]jan.hosek@fs.cvut.cz, [d]frantisek.lopot@fs.cvut.cz

Keywords: Vacuum, vacuum conductivity, vacuum conductance, computational model, pumping speed, effective pumping speed

Abstract. The paper focuses on the vacuum conductivity of high vacuum elements. The aim of the paper is to find a fast-use calculation method for determination of vacuum conduction in molecular flow mode for basic elements of vacuum chambers.

Work on the described computational model can be divided into two steps. The first step is to create your own computational model on the basis of literature. The second step is a series of measurements that verify the accuracy of the calculation approach and allow it to determine its error (uncertainty).

The proposed method is also tested on model examples using the Monte-Carlo method in MolFlow+ software. In addition, the verification process continues, however, for some geometries and vacuum parameters, the results are already experimentally validated and excellent conformity can be noted (deviation has not yet exceeded 10% of the measured values) allowing for seamless use of the method in engineering work.

Introduction

The paper focuses on the vacuum conductivity of high vacuum elements used in the CERN, the join institute of nuclear research JINR and other workplaces of a similar nature, because in these conditions its significance excels. Achieving the highest quality vacuum (if possible XHV) plays a major role in particle accelerators where collisions of the accelerated beam with particles of residual atmosphere occur, which has a number of adverse effects. In various methods of surface analysis, particles of residual atmosphere pollute the samples under investigation. However, the problems of various industrial plants, from relatively crude applications in the food industry, to physical vapor deposition, for example in the automotive industry, can not be overlooked. For the efficient design of the vacuum system, the knowledge of the vacuum conductivity of the elements used is of great importance and there is therefore a long-term effort to create practical background for its determination based on knowledge of the parameters of the required vacuum and geometry of the given elements of the apparatus. The aim of the paper is to find a fast-use calculation method for determination of vacuum conduction in molecular flow mode for basic elements of vacuum chambers.

Theoretical background

Vacuum (from Latin: Empty space) is defined as a state of gas whose pressure is lower than atmospheric pressure at room temperature [1]. Decreasing pressure results in significant changes in gas behavior. Therefore, three basic types of gas mode are distinguished. These are viscous flow mode, molecular flow mode and transitional area. In a molecular mode, gas changes its behavior from continuous to chaotic movement of individual atoms and molecules. A more detailed division of the vacuum is shown in Table 1.

Tab. 8 Pressure ranges in vacuum technology

Pressure range	Pressure [hPa]	Mode type
Low vacuum (LV)	300...1	viscous mode
Medium vacuum (MV)	$1...10^{-3}$	transitional area
High vacuum (HV)	$10^{-3}...10^{-7}$	molecular mode
Ultra-high vacuum (UHV)	$10^{-7}...10^{-12}$	molecular mode
Extremely high vacuum (XHV)	$<10^{-12}$	molecular mode

Achieving and maintaining a high vacuum is very economically demanding. This is the reason to deal with the details of the elements used for the construction of the vacuum chamber. One of the essential parameters is vacuum conductivity. Vacuum conductivity plays an important role in designing the vacuum pump on the apparatus as it reduces the pumping speed of used vacuum pumps.

Vacuum conductivity is characterized the flow of gas through the vacuum system [l/s, m^3/h]. With the pressure drop in the chamber and the transition between flow modes, the vacuum conductivity of the same component decreases by several orders of magnitude. This is due to the fact that in the viscous mode the vacuum conductivity is directly proportional to the fourth power of a characteristic dimension, while in the molecular mode it is only the third power. For this reason, it is much more important to study the vacuum conductivity for the molecular state.

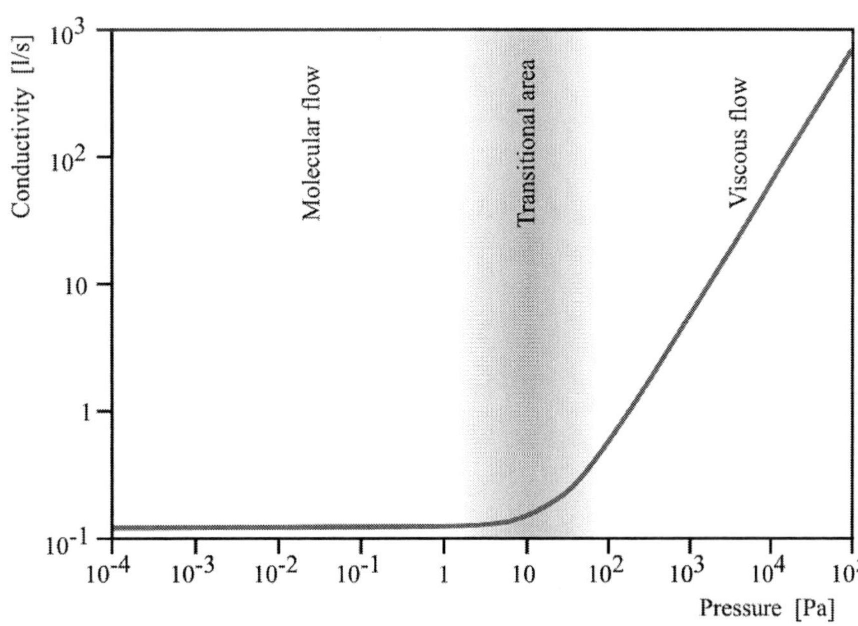

Fig. 43 Conductance of a smooth round pipe as a function of the

In addition to the characteristic dimension, the length of the tube, the mean gas pressure in the tube (viscous mode), the gas type and its temperature influence the conductivity (Eq.1, Eq.2).

$$C_{circle.lam} = C(d, l, \eta, \bar{p}) \tag{1}$$

$$C_{circle.mol} = C(d, l, \bar{c}) \tag{2}$$

C...vacuum conductance
d... tube average
l... tube length
η...viscosity (depends on temperature)
$\bar{p}$...mean pressure
$\bar{c}$...mean thermal velocity (depends on the type of gas)

The total vacuum conductivity of a system composed of individual elements can be calculated analogously to the conductivity of the electrical circuits (Eq. 3 - Parallel connection conductance, Eq. 4 - Series connection conductivities) [2].

$$C_\Sigma = C_1 + C_2 + \cdots + C_n \tag{3}$$

$$\frac{1}{C_\Sigma} = \frac{1}{C_1} + \frac{1}{C_2} + \cdots + \frac{1}{C_n} \tag{4}$$

The effect of total vacuum conductivity on the final pumping speed is described in Eq. 5. It will be appreciated that the effective pumping speed of the vacuum pump connected to the vacuum chamber by the system with total vacuum conductivity is lower than the indicated pumping speed. When designing a vacuum system, it is advisable to count this phenomenon. In the extreme, there is a risk of failure to achieve the desired degree of vacuum, which brings a number of complications.

$$\frac{1}{S_E} = \frac{1}{S} + \frac{1}{C_\Sigma} \tag{5}$$

S_E...effective pumping speed
S... pumping speed
C_Σ... total vacuum conductivity

Method

Work on the described computational model can be divided into two partially parallel and partly serial steps. The first step is to create your own computational model on the basis of literature. Literature lists either data for totally specific cases or data for idealized models. Their processing creates a basis for setting the calculation using the inter and extrapolation functions.

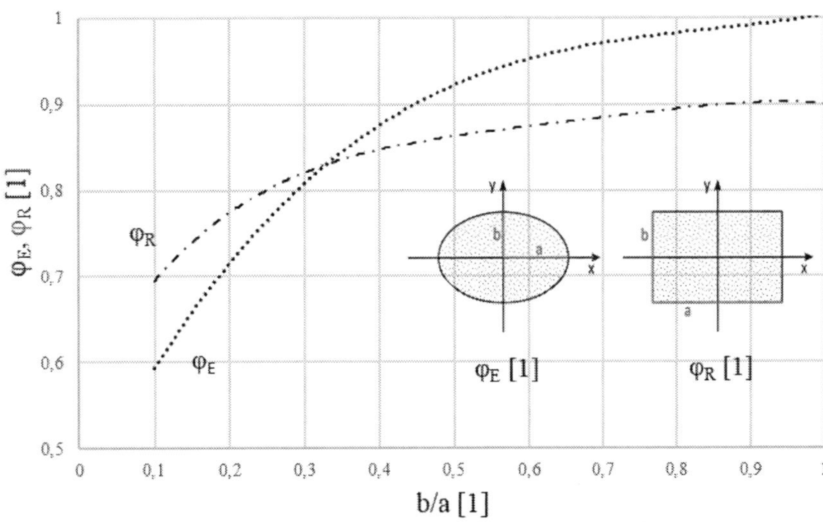

$\varphi_E = 0{,}7653x^3 - 1{,}9598x^2 + 1{,}7615x + 0{,}4357$

$\varphi_R = -1{,}2165x^4 + 3{,}2456x^3 - 3{,}2521x^2 + 1{,}5537x + 0{,}5694$

Fig. 44 φ_E, φ_R koef

The second step is a series of measurements that verify the accuracy of the calculation approach and allow it to determine its error (uncertainty). This phase is now in co-operation with the leader in the field of development and production of ultravacuum technique, spol. VAKUUM PRAHA and it will continue to be continuously, in fact, because, given the cost of creating ultra-vacuum assemblies, assemblies that are already in use or produced and prepared for the customer are used. There is also cooperation with CERN and JINR, which can help with testing. On Fig.3 there is a pumping system for NICA booster, designed by JINR and VAKUUM PRAHA. This pumping system is manufactured and assembled in VAKUUM PRAHA. A number of measurements are being carried out on the pumping system in cooperation with JINR and an adjustment is planned for the future in order to verify the proposed computational model.

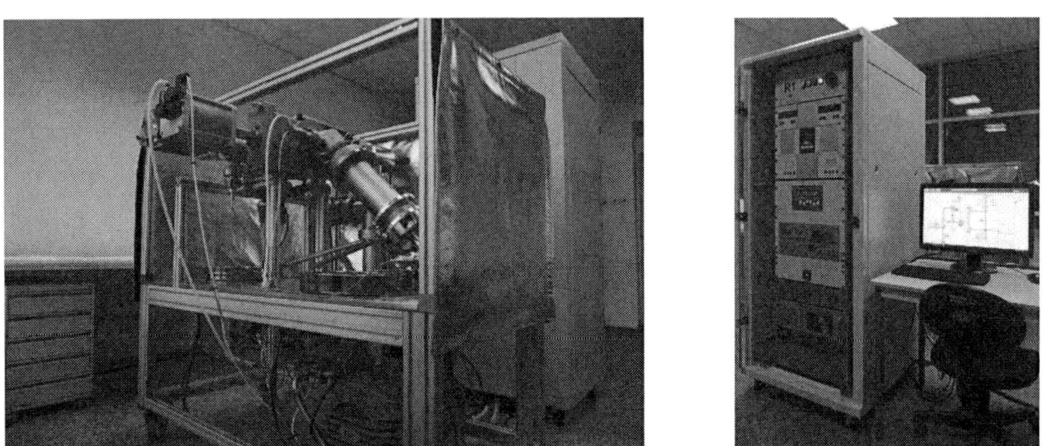

Fig. 45 Pumping system for NICA booster with control system

Conclusions

In the current state of the work, a computational model (method) is prepared for the rapid calculation of the vacuum conductivity. The proposed method is tested on model examples using the Monte-Carlo method in MolFlow+ software, which was developed in CERN and is freely available [3]. In addition, the verification process continues, however, for some geometries and vacuum parameters, the results are already experimentally validated and excellent conformity can be noted (deviation has not yet exceeded 10% of the measured values) allowing for seamless use of the method in engineering work.

Acknowledgement

The project and the paper were supported by grant no. SGS17/076/OHK2/1T/12 and SGS17/176/OHK2/3T/12.

References

[1] L. Holland, J. Yarwood, Vacuum Manual, Spon, London, 1974.

[2] J. Groszkowski, Technika vysokého vakua, Nakladatelství techn. lit, Praha, 1981.

[3] R.Kersevan, M. Ady. MolFlow+ user guide: For version 2. 4.1 [online]. In:. 3 June 2014 [cit. 2018-04-18]. https://molflow.web.cern.ch/sites/molflow.web.cern.ch/files/molflow_user_guide.pdf.

EAN 2018
56th conference on experimental stress analysis

Experimental Verification of Thickness Effect on Fatigue Crack Growth in AA 2124 Plate

RŮŽEK Roman[1,a], HOMOLA Petr[1,b] and KAKOS Martin[1,c]

[1]VZLÚ – Czech Aerospace Research Centre, Strength of Structures Department, Beranových 130, 199 05 Praha – Letňany, Czech Republic

[a]ruzek@vzlu.cz, [b]homola@vzlu.cz, [c]kakos@vzlu.cz

Keywords: thickness, crack growth, aluminum alloy, constant amplitude, variable amplitude.

Abstract. Both predictions and experimental results indicate for some materials significant discrepancies between crack propagation data of machined parts from a thick plate to different thinner thicknesses. The different fatigue crack growth behavior under constant and variable amplitude loading can be recognized. The thickness influence could be one of the main possible reasons for this unexpected behavior. The paper deals with a plate thickness of 50 mm made from 2124 T851 aluminum alloy. Crack propagation was investigated for various specimen thicknesses and different loading conditions. No significant thickness influence on the crack propagation behavior occurred both for constant and variable amplitude loading typical for a commuter aircraft loading spectra. It implies that present prediction models are able to predict the crack propagation in this material only using basic material characteristics defined under the constant amplitude loading.

Introduction

Fatigue crack propagation in metals is generally quantified in terms of stress intensity factor. Based on previous knowledge, different analytic and numerical models were derived. Various prediction models are based on the crack closure phenomenon and/or stress-strain conditions development in the surrounding area of the crack tip. All the models consider the material characteristics evaluated under the constant amplitude loading. Parameters of the prediction models are usually defined by means of experiments under constant and variable amplitude loading. Thin-walled structures made from sheets and various profiles without additional significant thickness reduction had been used in the aerospace industry in the past. Then the thickness influence is not so critical for these structures. The prediction models in this case could evaluate the fatigue crack propagation trends very well. In case of a significant thickness reduction of a plate material, standard prediction models could lead to misguided results or huge experimental work have to be performed ([1]-[4]).

Several works focused on damage tolerance investigation of metallic airframe structures were carried out in the framework of „The advanced aerostructure research centrum“, project no. TA0200003. Within this project, the crack propagation predictions based on analytical models indicated some significant discrepancies as compared with experimental results based on variable amplitude loading using randomized flight-by-flight loading sequences. These unexpected results invoked the request to perform more detailed investigations to experimentally clarify the real crack growth behavior for different thicknesses of final parts manufactured from a plate material.

The thickness influence on fatigue crack growth under constant and variable amplitude loading of AA (aluminum alloy) 7475 plate material was proved previously [5]. Significant differences between constant and variable amplitude crack growth data were found out. While crack propagation data under constant amplitude loading did not indicate any potential thickness influence, the fatigue crack propagation data obtained under variable amplitude loading showed a significant thickness influence. Thickness of 8 mm had significantly shorter crack propagation lifetime as compared with thickness of 2 mm. Thickness of 2 mm showed in about 260% longer lifetime as compare to thickness of 8 mm.

The paper documents experimental research work focused on the thickness influence investigation on crack propagation behavior of AA 2124-T851 aluminum alloy plate material under constant and variable amplitude loading spectrum representing a commuter aircraft service.

Material and specimen configuration

2124 T851 plate material (manufacturer KUMZ in Russia) was experimentally investigated as a typical airframe structural material. A plate semi product with nominal thickness of 80 mm was used. During a structure part manufacturing the plate is machined into the final varying thickness (typically from 2 up to 8 mm). Therefore, the paper consider just the data within this typical thickness range. Two extreme thicknesses of 8 and 2 mm were investigated in case of the constant amplitude loading and three nominal thicknesses of 2 mm, 4 mm and 8 mm were investigated in case of the variable amplitude loading.

Typical M(T) specimens shown in Fig. 1 were used for the crack propagation investigation. Specimen width of 100 mm was selected. The same notch shape in the critical part (middle part with the crack starter) was made in all the specimens, see Det. A in Fig. 1.

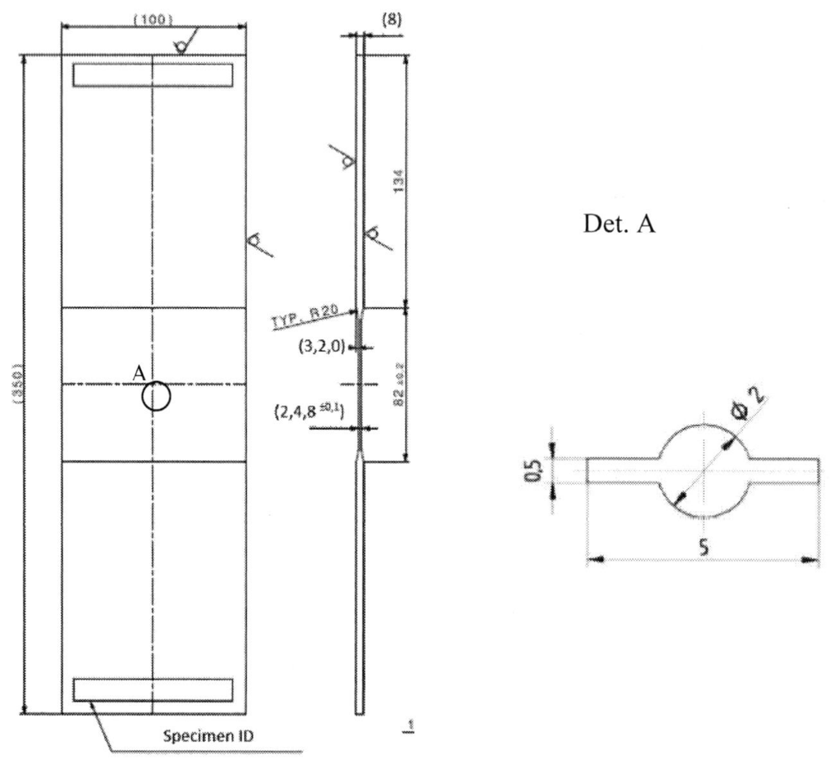

Fig.1 M(T) specimens geometry

Test methods

Fatigue crack growth tests were conducted using hydraulic SCHENCK load frame with capacity of 250 kN. The SCHENCK load frame was controlled by the INSTRON FastTrack 8800 test control system. Specimens were clamped into the hydraulic test frame through mechanical grips. Test set-up, i.e., M(T) specimen, load frame, mechanical grips and microscopes, is shown in Fig. 2.

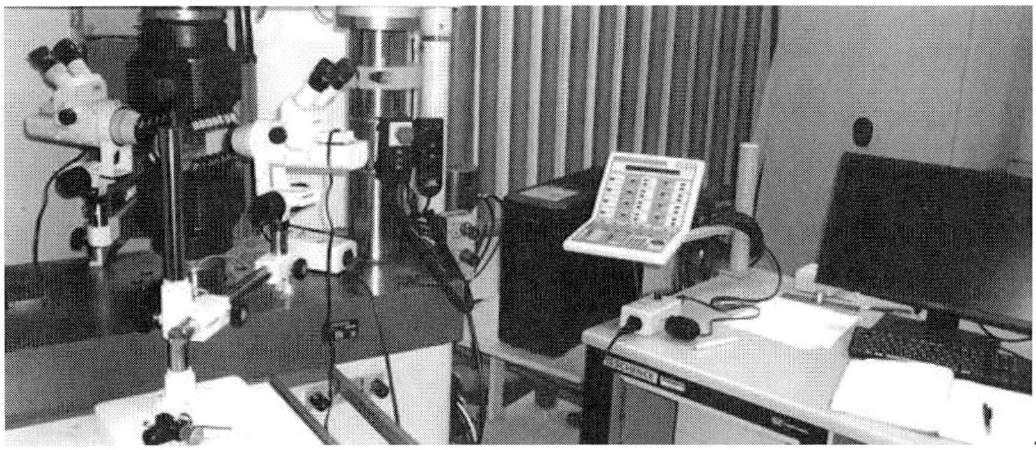

Fig.2 M(T) fatigue crack growth test configuration

Test procedure was divided into two steps. Main goal of the first step was crack initiation. The aim of the second one was the crack propagation monitoring. The pre-cracking procedure was always conducted under constant amplitude loading; the crack propagation phase was conducted under either constant or variable amplitude loading. The test procedure was performed in agreement with ASTM E-647 standard requirements [6].

Fatigue crack initiation and propagation was monitored by visual method (VT) using light stereomicroscopes Olympus SZ40 with maximum magnification of 40x. The crack length measurement was carried out on both surfaces and both sides regarding to the longitudinal axis of the specimens.

A harmonic loading (sinusoidal load cycle) with constant force amplitude was used in the crack initiation phase. The thickness influence was investigated for both constant and variable (randomized flight-by-flight sequence) amplitude loading. A typical randomized flight-by-flight sequence with variable amplitude representative for a wing of small commuter aircraft called SQ_0058 was used. The sequence consists of 3000 flights and represents stress spectra of a bottom wing panel skin surface. The SQ_0058 loading sequence contains total number of 166.302 cycles. Fig. 3 illustrates a part of the sequence in a graphical form.

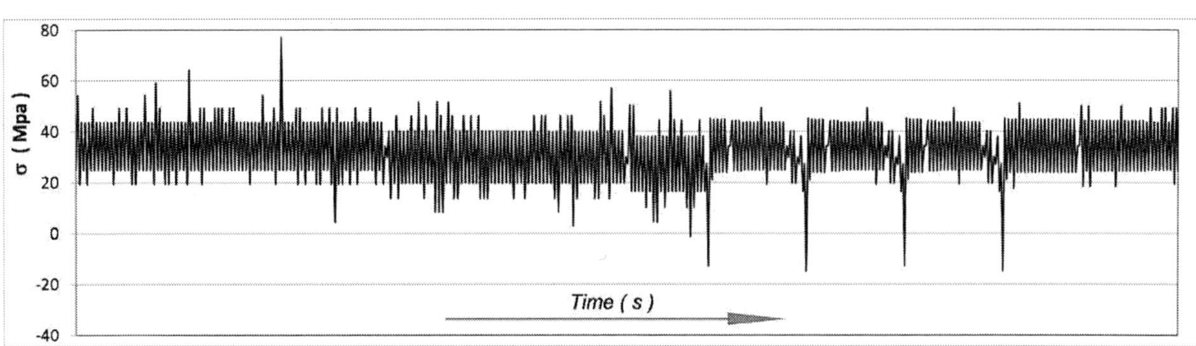

Fig.3 Visualization of a typical segment from flight-by-flight loading sequence

Experimental results and discussion

Basic fatigue crack growth material characteristics of the plate material of 50 mm in thickness were defined using specimens with thickness of 8 and 2 mm. No significant difference was found out. Maximal difference between crack growth lifetimes under similar stress conditions was ±13% of elapsed numbers of cycles. This difference is in very good compliance with the previous experience [5], [7], [8]. Fig. 4 compares data obtained under the constant amplitude loading (specimen thickness of 8 vs. 2 mm) including data from previous experiments. No significant thickness and plate material influence was found out.

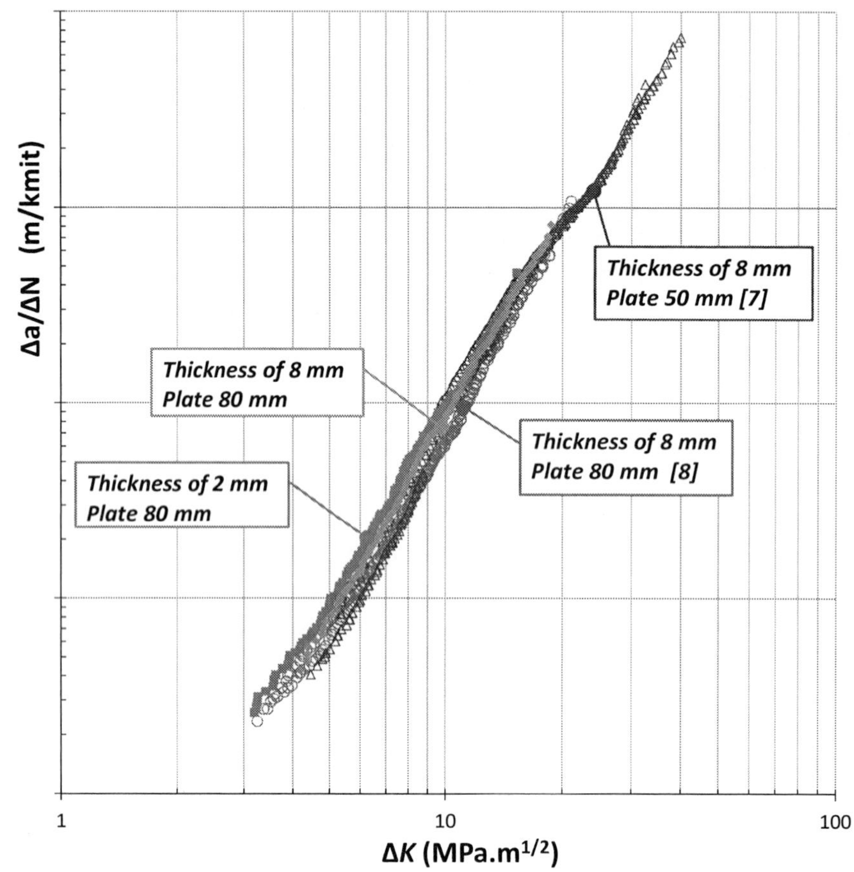

Fig.4 2124-T851 crack rate vs. effective stress intensity factor data – thickness influence

Fig. 5 documents fatigue crack propagation data measured for different thicknesses (2, 4 and 8 mm) under variable amplitude loading. All the specimens were loaded using the same stress loading parameters – loading sequence of SQ_0058 with maximum stress value in the sequence of 79.6 MPa. No thickness influence on the crack propagation was observed.

The results imply the conclusion that present prediction models are able to predict the crack propagation in investigated 2124 T851 plate only using the material characteristics defined under the constant amplitude loading without definition of additional parameters into the models. The results confirmed that various aluminium alloys have a significantly different behaviour from damage tolerance properties viewpoint ([1-8]). The selection of parameters in prediction models have to be a critical point of inspection intervals designing.

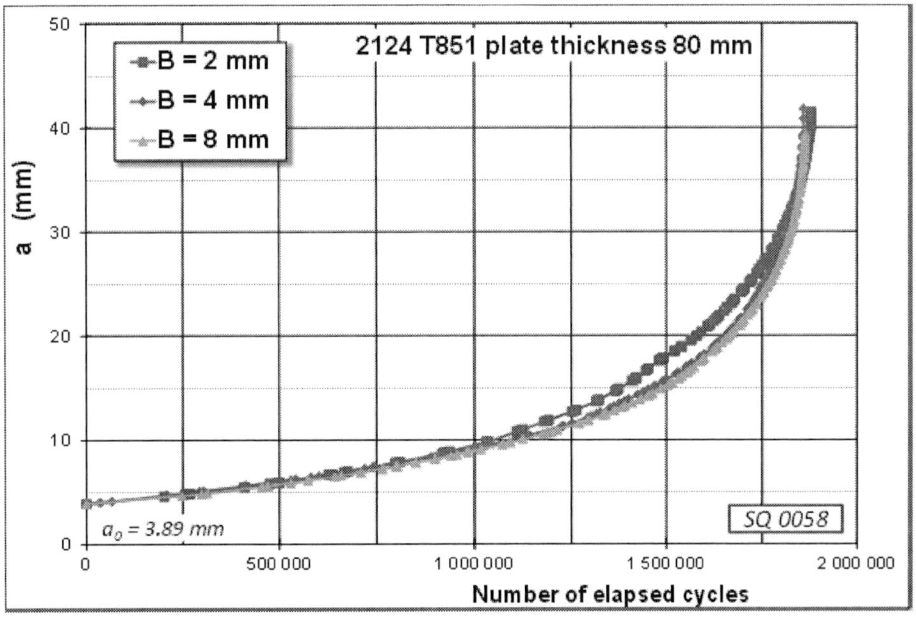

Fig.5 2124-T851 – crack propagation data comparison under variable amplitude loading (σ_{max} = 79.57 MPa)

Conclusions

Constant amplitude loading:

- All the data in range of thicknesses from 2 mm up to 8 mm lies in the same group of data represented by using of stress intensity factor.
- The data of the tested specimens shows difference scatter of ± 13 % between total crack propagation lives (development from the same initial crack length up to specimen failure). It can be considered as a good agreement from viewpoint of previous experience with similar characteristics.
- Data scatter ± 13 % is very similar to data scatter ± 15 % for 7475 T7351 data documented in [5] for the same thickness range (2 - 8 mm) data.

Variable amplitude loading:

- No thickness influence on the crack propagation behavior was proved.

AA 2124 and 7475 plate materials show different crack growth behavior, hence, it is very desirable to verify crack propagation behavior also for other aluminum alloys used in commuter airframe structures.

Acknowledgement

The research leading to these results was performed under funding of the Technology Agency of the Czech Republic, project "The advanced aerostructure research centrum", no. TA0200003.

References

[1] J. Šedek, R. Růžek, J. Raška, J. Běhal: Comparative study of prediction methods for fatigue life evaluation of an integral skin-stringer panel under variable amplitude loading. Procedia Engineering, 114 (2015) pp.124 – 131, doi: 10.1016/j.proeng.2015.08.050

[2] J. Běhal: Comparison of crack growth prediction models implemented into the AFGROW software. Technical report R-6592, VZLÚ, Praha (2016).

[3] J. Běhal, P. Homola, R. Růžek: Effect fracture resistance on retardation of crack growth. International Journal of Structural Integrity, Vol. 6 No. 4 (2015) pp.1-13, doi: 10.1108/IJSI-10-2014-0053

[4] J. Běhal, P. Homola: Simulation of Operational Loading Spectrum and its Effect on Fatigue Crack Growth, Procedia Engineering 101 (2015) pp.26-33, doi: 10.1016/j.proeng.2015.02.005

[5] R. Růžek, P. Homola, M. Kakos: Experimental verification of thickness effect on fatigue crack growth in AA 7475 plate. EAN 2017 - 55th Conference on Experimental Stress Analysis 2017 (2017), pp. 513-518, Conference proceedings, Novy Smokovec; Slovakia; 30 May 2017 through 1 June 2017 ISBN: 978-805533167-6, Scopus,2-s2.0-85026303106, Code 128696, https://www.scopus.com/inward/record.uri?eid=2-s2.0-85026303106&partnerID=40&md5=6fa73be973075d20b0cfc8b9fcc924f3

[6] ASTM E 647-13e1 Standard Test Method for Measurement of Fatigue Crack Growth Rates. 100 Barr Harbor Drive, PO Box C700, West Conshohocken, PA 19428-2959, United States, ASTM International (2013)

[7] R. Růžek, at all: Material characteristics of the plate thickness of 50 mm made of 2124-T851 alloy in L-T direction (in Czech). Praha, VZLU, Technical report MOSTA.0484.A.U.PD (2011).

[8] P. Homola, M. Kadlec, P. Kucharský, M. Kakos: Fatigue crack growth tests at low values of ΔK – D15čATV sheet (1.5 mm) and 7475-T7351 (76 mm) and 2124-T351 (50 mm and 80 mm) plates (in Czech). Praha, VZLÚ, Technical report R-6531 (2016).

EAN 2018
56th conference on experimental stress analysis

Crash Test of Carbon Composite

RYVOLOVÁ Martina[1,a] and VOSÁHLO Josef[1,b]

[1]**Institute for Nanomaterials, Advanced Technology and Innovation,** Technical University of Liberec, Studentská 1402/2, Czech Republic

[a]martina.ryvolova@tul.cz, [b]josef.vosahlo@tul.cz

Keywords: carbon unidirectional prepreg, crash test, energy absorption, FEM model.

Abstract. The article summarizes the work of authors about a carbon unidirectional prepreg. The current results are based on the previous measurement of force and deformation obtained during crash tests. The amount of impact force was determined for different variants of carbon plates. The dependence of the value of the energy absorbed by the impact on the composition of the composite was monitored and all cases were modelled under the same conditions. Results of measurement and mathematical model were compared.

Introduction

Generally, components made of fibre reinforced plastics have a higher energy absorption than similar components made of metals. High strength and low weight are typical properties of structural laminates. The behaviour of the laminates depends on the type of fibre, the type of resin, the thickness of the laminate - the number of layers and their orientation, and production technology of fibre reinforcement. A number of works show that the appropriate choice of material and construction can affect the rate of impact destruction - the amount of energy absorbed by the impact. There are many methods and approaches to determine the resistance of composite material to impact. Using numerical methods, these processes can be modelled. Thin CFRPs have a higher energy absorption efficiency (EAE) at a higher speed impact. It is deduced from the results of experiments and mathematical simulations [1]. CFRP laminates with higher thickness have a higher EAE at a lower speed impact. There are differences in CFRP response depending on the type of reinforcement used. [2] described the difference energy absorption for laminates from woven fabrics (plain and satin weave were compared) and from woven reinforcement consolidated with kevlar sewing at high impact velocity. Different test method is used for composite energy absorbers. Composite materials absorb energy through controlled gradual material failure. Total absorbed energy (AE) and specific absorbed energy (SAE) are determined. The total absorbed energy is the area under the force-displacement curve. Series of tests on shape-defined absorbers was realized, which had a different composition and were made of different materials. The SAE value was dependent on the orientation of the reinforcement. Specimens with reinforcement oriented in the load direction have achieved the best results [3]. Comparison of absorbers produced by different technologies was made. Specimens were made from pressed composite boards absorb more energy than specimens were made from filament wound composite tubes [4].

Material and methods

The experiment was performed on specimens made from unidirectional carbon prepreg [5]. It is semi finished product formed high strength carbon fibres that are pre-impregnated epoxy matrix (matrix constitutes 38% by weight of prepreg). The fibres are oriented in one direction only. Three groups of specimens was prepared for experiment, groups was designated by composition L, T, W. Specimens were made with different composition, orientation of the reinforcement and with different number of plies. Table 1 contains composition and properties of specimens.

Tab. 1 Properties of specimens

specimens	composition	thickness [mm]	areal weight [gm^{-2}]
L	1 pl;	0.16	250
T	3 pl; 0°/90°/45°; non crimp	0.38	600
W2	2 pl, plain weave	0.29	425

For crash test was used hydraulic motor for high speed uniaxial loading with hemispherical impactor with a radius of 20 mm. The speed of the impactor for the duration of the sample penetration was constant 10 ms^{-1}. The course of the test is shown in Fig. 1. A graph for each composite specimen was created from measured force and displacement values. An example of a force dependence on displacement for a W marked specimen is shown in Fig. 2.

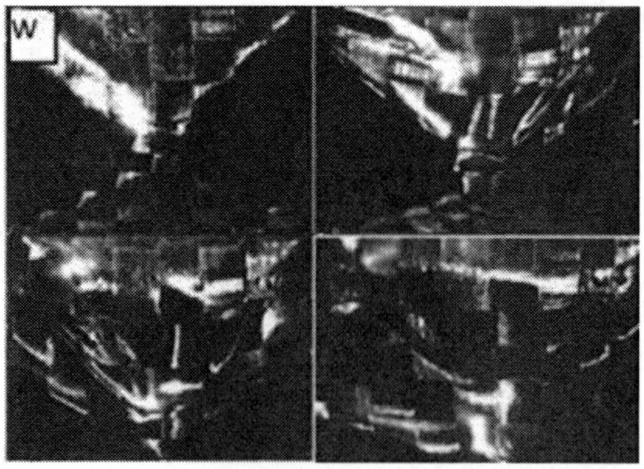

Fig. 1 The course of crash test

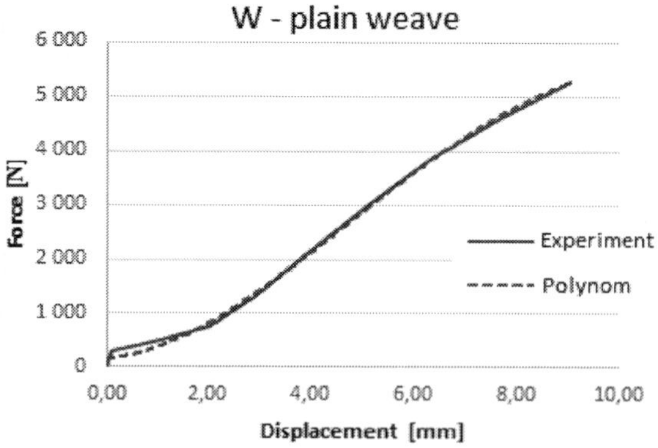

Fig. 2 Graph of W specimen

The curve of the graph was spaced by a 4th degree polynomial that approaches the experimental curve (Fig. 2). The polynomial data was used to calculate value of impact energy – total absorbed energy (Equation 1).

$$U = \int_{X1}^{X2} F(x)dx \qquad (1)$$

In the preceding formula represents: U - mechanical energy [J], F(x) - function of the force-deformation of the test piece [N], x - deformation of test [m].

Verification of the correctness of the calculation was performed by combining the start and end points of the curve (minimum and maximum force) with the tangential. Results of measurements shows Table 2.

Tab. 2 Results of experiment

Specimens	Avr. of force [kN]	Avr. of displacement [mm]	Average of energy [J]
L (L1-L3)	0.63	7.41	3.25
T (T1-T3)	4.48	10.55	25.19
W (W1-W3)	4.83	19.96	39.26

FEM model

The geometry of the three test specimens was created in the Solidworks CAD program. Sample parameters were based on real values, size 100x100mm and thickness 0.16, 0.38 and 0.29 mm. A 20 mm diameter ball end impactor was positioned perpendicular to the centre of the sample. Sample assemblies were saved in * .step format and imported into ANSYS software. The module for explicit dynamics was used to simulation. The new material model was created in the library. The measured mechanical properties were inserted into the model. The linear elastic orthotropic properties of the material were used to create the material modulus [6]. The elastic modulus, the Poisson number, and the shear modulus were also calculated from the measured values. Material properties of the steel were introduced into the model according to Table 3.

Tab. 3 Material properties of model

Material	Density [kg.m^{-3}]	Elastic modulus [GPa]		Shear modulus [GPa]		Poisson ratio [-]		Tensile strength [Gpa]
		$E_{11}^{f,m}$	$E_{22}^{f,m}$	$G_{12}^{f,m}$	$G_{23}^{f,m}$	$v_{12}^{f,m}$	$v_{23}^{f,m}$	
Carbon fiber	1318	12.33	77.80	5	3.08	0.27	0.42	1.332
Structural steel	7850	200		76.9		0.3		0.25

Describe of destruction

Fixation boundary conditions were inserted into the edges of each sample. Conditions prevent rotation and displacement in all directions. The motion boundary condition was inserted into the impactor model that moved perpendicularly to the model of composite plate. The displacement corresponded to the maximum measured values of the experiment. The result of each simulation was the value of the impact energy [7, 8].

Specimens L: single layer of prepreg, fibres with directional orientation 0 °. The matrix was broken during impact, the fibres were broken too (Fig.3). The direction of deformation in the model is in accordance with the orientation of the fibres, the maximum energy is concentrated at the point of impact (Fig. 3)

Specimens T: three layers of prepreg, 0 ° / 90 ° / 45 ° directionally oriented. The impact energy is concentrated at the point of contact, the fibres are damaged in this area (Fig. 4). The model shows instead of the maximum energy concentration and its propagation in the plate (Fig. 4)

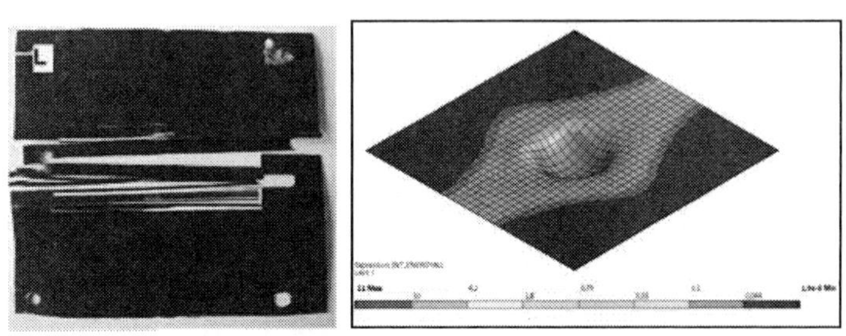

Fig 3. Specimen L after test and model

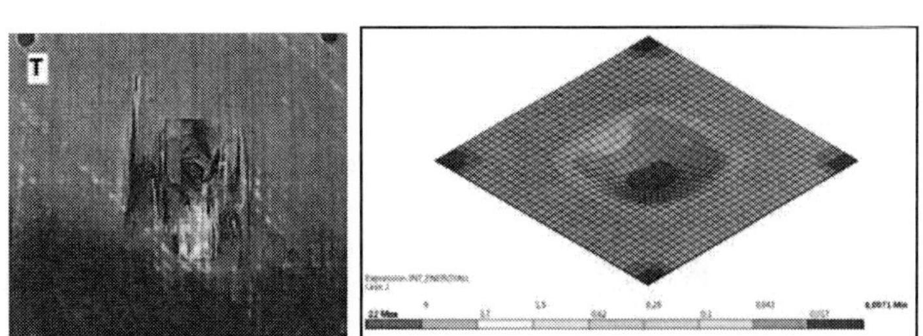

Fig 4. Specimen T after test and model

Specimens W: The square shows the binding point (warp or weft, change of fibre orientation 0°/90°and 90°/0°). The specimen area is composed of binding elements. In the event of an impact, the energy is concentrated to the point of impact and continues to be limited (Fig. 5). The model shows instead of the maximum concentration of energy. The binding elements absorb energy using alternation the directional orientation of the layers and limit crack propagation.

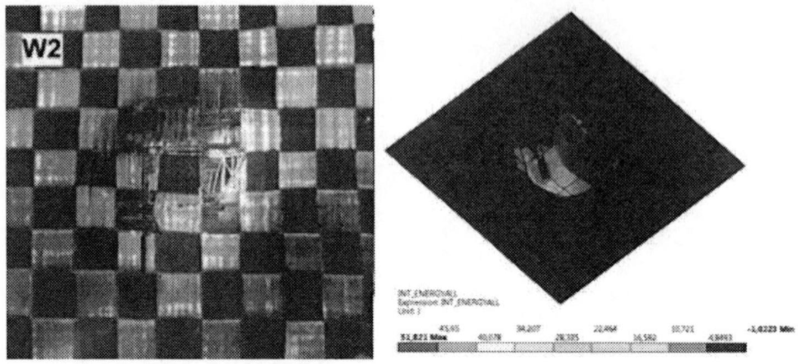

Fig 5. Specimen W after test and model

Conclusion

Crash tests were performed on three variants of carbon prepreg specimens. Measurement results (force and deformation) were used to calculate the energy required to break individual specimens. Calculated values were verified using a mathematical model. Results is showed Table 4.

Tab. 4 Comparison of energy

	L	T	W
Energy (calculation) [J]	3.25	25.20	39.26
Energy (simulation) [J]	6	22.1	51.8

The energy value for the L series specimens is 50% higher than the energy obtained from the measurement. Values are almost equivalent for T-series specimens. The mathematical model assumes a higher energy value for W series specimens by 32% than measured. Conclusions of the experiment are difficult to generalize due to the low number of specimens.

Acknowledgment

The results of this project LO1201 were obtained with co-funding from the Ministry of Education, Youth and Sports as part of targeted support from the "Národní program udržitelnosti I" program.

References

[1] B. Wang and col., Energy absorption efficiency of carbon fiber reinforced polymer laminates under high velocity impact, Materials and Design 50 (2013) 140–148.

[2] M.V. Hosur, U.K. Vaidya, C. Ulven, S. Jeelani, Performance of stitched/unstitched woven carbon/epoxy composites under high velocity impact loading, Composite Struct.64 (2004) 455–66.

[3] M. Mališ, Měření specifické absorbované energie kompozitních materiálů, Transfer 29, 2017.

[4] M. Růžička, V. Kulíšek, S. Bogomolov, V. Shánël, Development of composite energy absorber, Procedia Engineering 96 (2014) 392 – 399.

[5] M. Syrovátková, M. Petrů, Experimental and numerical analysis of impact test of carbon prepreg composite, EAN 2016 – 54th International Conference on Experimental Stress Analysis, 1. vyd. Západočeská univerzita v Plzni (2016) ISBN 978-802610624-1.

[6] M. Jirásek, Z. Bažant, Inelastic analysis of structures,Wiley, 2002, ISBN 0-471-98716-6.

[7] M. Talia, H. Lankarani, J.E. Talia, New experimental technique for the study and analysis of solid particle erosion mechanisms, Wear pp. 225-229 (1999) pp.1070–1077.

[8] G. Zavarise, P. Wriggers, B.A. Schrefler, On augmented Lagrangian algorithms for thermomechanical contact problems with friction, Int. Journal for Numerical Methods in Engineering (1995) 38(17):2929–2949.

Developing the Optimal Geometry of the Trabecular Structure

ŘEHOUNEK Luboš[1,a] and JÍRA Aleš[1,b]

[1] Czech Technical University in Prague, Faculty of Civil Engineering, Department of Mechanics, Thákurova 7, 166 29, Prague, Czech Republic

[a]lubos.rehounek@fsv.cvut.cz, [b]jira@fsv.cvut.cz

Keywords: Dental Implant, Optimization, Porous, 3D Printing, SLM.

Abstract. This research deals with the problem of geometrical solutions of porous dental implants. The parameters associated with optimizing the geometry of the structure are usually overall porosity, cell size, wall thickness and mechanical parameters such as modulus of elasticity and hardness. We have developed a few variants of these structures according to the research of other authors. These structures were then assessed based on the difficulties associated with their manufacturing process and subsequently numerically analysed. The variant Diamond 30% relative density was evaluated as optimal among our current structures.

Introduction

The long-term stability of implants is greatly influenced by their surface treatment. These treatments are usually done either by sand-blasting, acid-etching the surface or applying a plasmatic Ti coating onto the surface of the implant. The shortcomings of these methods is mainly their cost, demanding technological procedures and high surface stiffness of the implant [1].

With the advancement of additive manufacturing (AM) technologies, we are able to produce morphologically complex structures that resemble biological tissues, from both geometrical and mechanical standpoint [2]. These structures are usually 3D-printed using the Selective Laser Melting (SLM) or Selective Laser Sintering (SLS) technology and various metal powders (e.g. Ti-6Al-4V). The main goal of this work is to create and subsequently test and compare these structures.

The main reason behind creating new biostructures is increasing the implant success rates and decreasing the number of reoperations. This can be done by optimizing the structure so as to provide better stability during the early stages following implantation, where most failures occur due to the loss of stability [3]. The most common reasons for reoperation are mechanical release of the implant, infection and aseptic loss of stability or its dislocation [4].

Trabecular Pore Size. The process of optimizing the geometry of the trabecular structure is dependent on multiple factors. First, mechanical and biological factors come to mind. The latter are far more intriguing – no research to our knowledge has actually decisively determined the optimal size of the pores of the structure. There are, however, experiments that allude to a certain range of dimensions [5]. These dimensions should not, however, in our opinion, be taken as a definitive answer to the problem of pore size as they are performed under very special conditions that never truly resemble real live organisms. They should rather be taken as a recommendation towards other experiments to create more data.

The main properties to consider are overall porosity, wall/beam thickness, pore size and trabeculi orientation inside the 3D matrix. The problem of determining the optimal pore size has been investigated before – as long ago as 1980 [6], authors have tried to come to a conclusion on optimal pore size. This particular research concluded that the greatest value of fixation strength was achieved using pore sizes in the range of 50-400 μm (17 MPa at 8 weeks following the implantation). Other important research, too lengthy to describe in this short paper, can be found in the following references [7, 8].

Authors [5] investigated the effect of pore size of collagen-glycosaminoglycan scaffolds on proliferation of osteoblast MC3T3-E1 cells. Three time intervals were tested (24 hours, 48 hours, 7 days). The pore size that exhibited the greatest amount of cell proliferation across all time intervals is 325 μm (62% of cells in contact with the structure) [5]. We consider this a viable starting information as there is not much research performed in this area.

The problem also gets more complicated as it is vital to observe *how* the bone material attaches to the structure (or the implant) and in what form. The process of healing and new bone formation is associated with cartilage formation and subsequent ossification and formation of osteons – the Haversian system [9]. During experiments, it is important to measure the volume of bone attached to the implant. This way, we can determine how well does a certain structure perform inside the living tissue as we need to distinguish between the bearing beams of the osteons and the low-mineralized fibrous tissue.

Experiment. We have previously conducted in-vivo tests on laboratory pigs with focus on bone contact percentage after 6 months following implantation (Fig. 1) [10]. This experiment is similar to the experiments carried out by authors mentioned above, however, it better and truly represents the conditions inside a living organism and the outcome we want to achieve. The pore size in this experiment was around 900 μm and all specimens were 3D-printed using the Ti-6Al-4V metal powder alloy. We want to further investigate other, different structures with different pore sizes (Fig. 2), apply them onto real implants and redo the tests with pore sizes more corresponding to the findings presented by M. M. Ciara et al. [5].

It is important to note that the value of pore size of 325 μm should not be considered final as it comes only from a single series of experiments and does not truly represent the nature of an in-vivo experiment. The authors of this work plan to take it as a base value and try to develop different structures and observe how they perform inside the aggressive surroundings of a living organism.

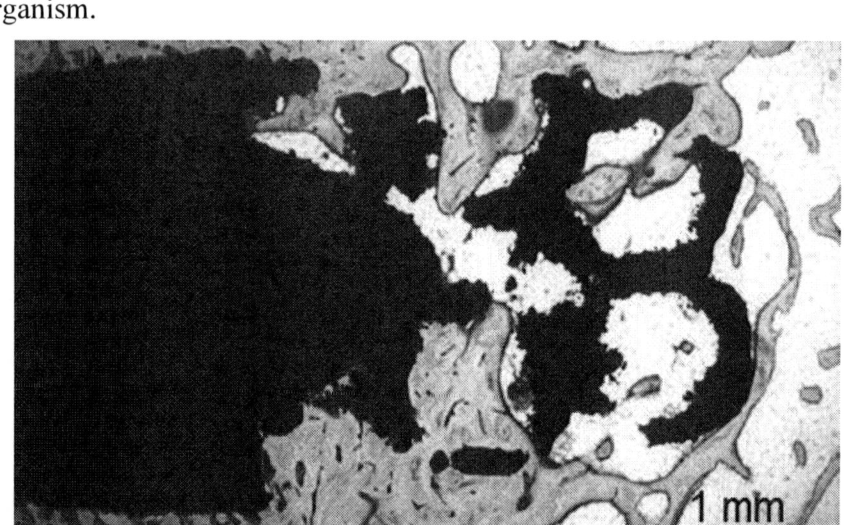

Fig.1 A longitudinal section of the trab. specimens extracted 6 months after implantation. White and light yellow color represents fibrous tissue, orange represents beams of newly formed bone tissue and black color represents the trabecular implant. In-vivo experiment performed at the Inst. of Animal Phys. and Gen., v. v. i., Liběchov

Geometrical Solutions. The main goal of this work is to create an optimal structure that will at the same time endure the mechanical loads and allow for the greatest amount of bone ingrowth and ongrowth. We have considered a total of 28 variants and decided to manufacture 3 variants shown below in Fig. 2. The pore sizes of all structures vary from 200 μm to 450 μm. The most promising structure seems to be the n.2 Diamond 30% relative density that has a pore size of approximately 340 μm and does not exhibit any difficulties during manufacturing.

We have also created a few other basic models of other structures for comparison in numerical simulations. These structures will serve to give a better overview of possible variants and show possible flaws of design and places of stress concentrations (Fig. 3).

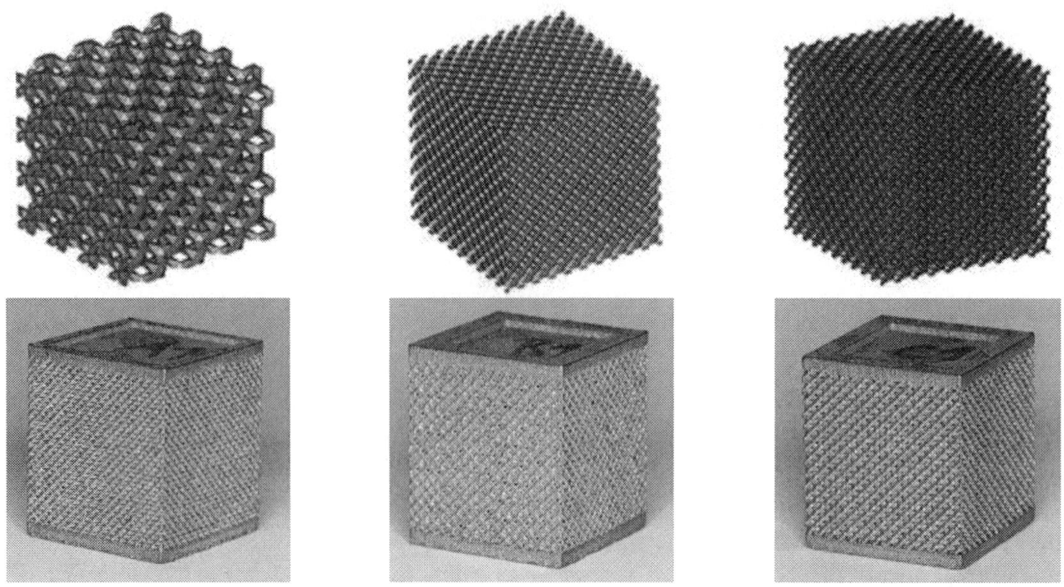

1) Dode Thick [MSG] 2) Diamond 30% relative density 3) Rhombic dodecahedron, relative density 30%

Fig.2 Upper row – figures of the .STL model files used to create the geometries of the final specimens. Bottom row – corresponding 3D-printed specimens for use in the uniaxial compression mechanical test. Used material is Ti-6Al-4V.

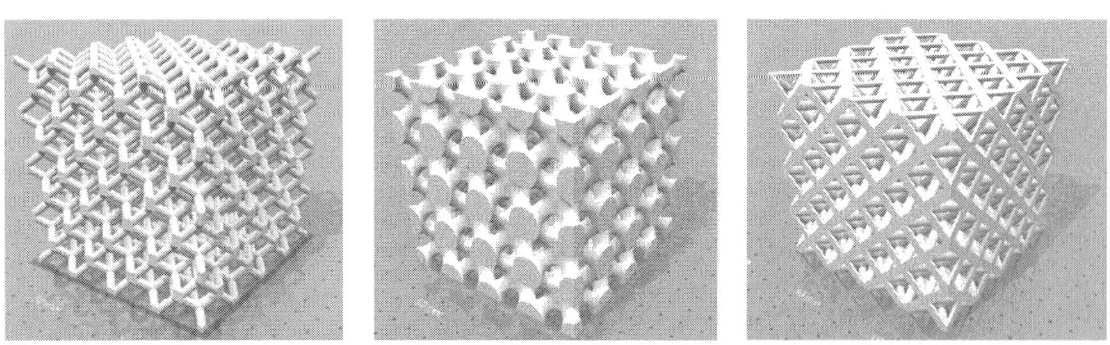

1) Box structure 2) Body diagonal 3) Octet truss 20%

Fig.3 Other geometrical solutions of structures that will be used for comparison in the process of numerical simulations.

Mechanical Tests. We performed a series of macromechanical and micromechanical tests to determine the behavior of different trabecular structures. For this purpose, we performed uniaxial tensile and compression tests and nanoindentation test. The nanoindentation test proved the values of micromechanical properties dependent on the depth of the indent (the trend of reduced modulus E_r and hardness H_{it} is constant in the range of contact depth 470 nm or greater, see Fig. 4). The mean values of micromechanical properties of all specimens (not dependent on geometrical solution) are E_r = 118 GPa and H_{it} = 5,187 GPa. The micromechanical properties correspond with the values given by the manufacturer.

Heat treatment. The values of reduced modulus (Fig. 4 bottom) generally correspond with the values provided by other researchers [11], but the values of microhardness significantly vary. We have concluded that in the case of 3D-printed specimens, post-production heat treatment will play a significant role in the hardness of the final product. The values of hardness provided by our experiment (Fig. 4 top) are higher than most values provided by other research because they undergone a special annealing process that increased hardness significantly. Determining the right method of post-production heat treatment will be the next task in order to fully explore the capabilities of 3D-printed structures.

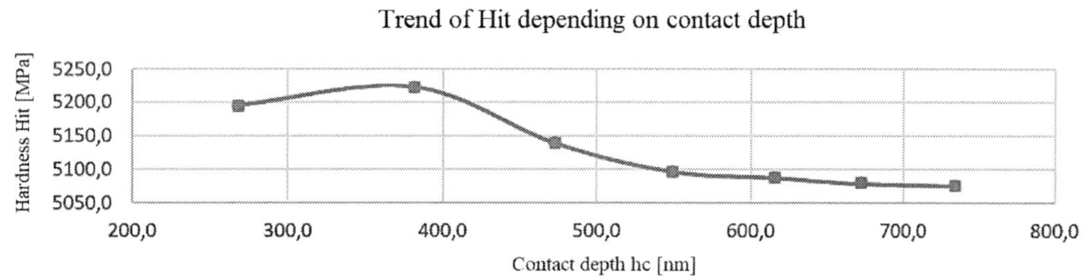

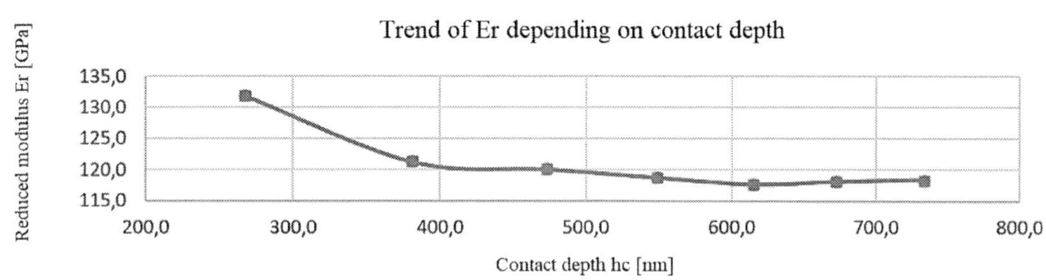

Fig.4 Two graphs showing the dependency of micromechanical properties on contact depth.

Conclusions

Multiple geometrical solutions of the porous trabecular structure have been created and tested. The challenge to create a structure that is optimal in terms of new osteon formation inside the structure is met with mechanical challenges, namely values of modulus, ultimate strength of the structure and its ability to resist fragmentation.

We have evaluated the conditions for optimal bone ingrowth and decided to create our own specimens based on these rough values of pore sizes and observe which structure yields the best results.

Micromechanical tests proved the values of E_r and H_{it} of 3D-printed specimens (Fig. 4) are in accordance with the values of conventionally machined specimens. The macromechanical tests coupled together with a numerical model of the individual structures will show further insight into the behavior of the trabecular structures. The optimal structure also has to abide by the conditions for osseointegration of the implant, which is yet another factor to account for when optimizing the geometry of the implant.

Acknowledgment

The financial support by the Faculty of Civil Engineering, Czech Technical University in Prague (SGS project No. SGS17/168/OHK1/3T/11) and by the Technology Agency of the Czech Republic, project no. TJ01000328 is gratefully acknowledged.

References

[1] L. Řehounek, F. Denk, A. Jíra, Trabecular structures of dental implants formed by 3D printing and their mechanical properties, EAN 2016 - 54th International Conference on Experimental Stress Analysis, Czech Republic (2016).

[2] L. Le Guéhennec et al., Surface treatments of titanium dental implants for rapid osseointegration, Dental Materials 23 (2007), 844-854.

[3] R. B. Osman and M. V. Swain, A Critical Review of Dental Implant Materials with an Emphasis on Titanium versus Zirconia, Materials 8 (2015), 932-958.

[4] R. Wauthle et al., Additively manufactured porous tantalum implants, Acta Biomaterialia 14 (2015), 217-225.

[5] M. M. Ciara et al., The effect of mean pore size on cell attachment, proliferation and migration in collagen–glycosaminoglycan scaffolds for bone tissue engineering, Biomaterials 31 (2010), 461-466.

[6] J. D. Bobyn et al., The optimum pore size for the fixation of porous-surfaced metal implants by the ingrowth of bone, Clinical Orthopaedics and Related Research 150 (1980), 263-270.

[7] G. Campoli et al., Mechanical properties of open-cell metallic biomaterials manufactured using additive manufacturing, Materials & Design 49 (2013), 957-965.

[8] S. A. Yavari, Bone regeneration performance of surface-treated porous titanium, Biomaterials 35 (2014), 6172-6181.

[9] A. M. Phillips, Overview of the fracture healing cascade, Injury 36 (2005), 5-7.

[10] L. Řehounek and A. Jíra, Numerical and Mechanical Analyses of a 3D-printed titanium trabecular dental implant, Acta Polytechnica 57 (2017), 218-228.

[11] S. Q. Wu, Microstructural evolution and microhardness of a selective-laser-melted Ti–6Al–4V alloy after post heat treatments, Journal of Alloys and Compounds (2016), 643-652.

EAN 2018
56th conference on experimental stress analysis

The Chloride Ion Penetration Resistance of Fine Recycled Aggregate Concrete

SEFFLOVA Magdalena[1,2,a]

[1]CTU in Prague, University Centre of Energy Efficient Buildings, Trinecka 1024, 273 43, Bustehrad, The Czech Republic

[2]CTU in Prague, Faculty of Civil Engineering, Department of Building Structures, Thakurova 7, 160 00, Prague, The Czech Republic

[a]magdalena.sefflova@fsv.cvut.cz

Keywords: Chloride ion penetration, Fine recycled aggregate, Durability.

Abstract. This paper is focused on chloride ion penetration of fine recycled aggregate (FRA) concrete. The durability of FRA concrete is connected with doubts. A total four concrete mixtures were designed. The first concrete mixture was reference, only with natural aggregate. In other concrete mixtures, natural aggregate was replaced by FRA in varying replacement ratio. Chloride ion penetration resistance of FRA aggregate was tested. The test results showed that the use FRA in concrete influences chloride ion penetration of concrete.

Introduction

The chloride ion penetration resistance is one of the most important durability concrete properties. Primarily the question of durability is very important for recycled aggregate concrete. The use coarse fraction of recycled aggregate is already accepted but with specific restrictions [1]. After literature review it is possible to say that many researchers obtained totally different results. One of many studies observed that chloride ion penetration of coarse recycled aggregate concrete is not significantly different from concrete without recycled aggregate [2]. Evangelista [3] showed contrary results. The chloride ion penetration resistance is reduced by adding of recycled aggregate in concrete mixtures.

The problem is with the use the FRA. A little researcher studies of FRA concrete durability were published. Therefore, this paper is focused on the chloride ion penetration of FRA concrete.

Experimental Research Program

Materials. Portland cement CEM I 42.5 R, natural sand, water and the FRA were used in the experimental program. The FRA originated from crushed concrete waste. Concrete waste was crushed to fraction 32/64 mm in the recycling plant and to fraction 0/4 mm by jaw crusher in the laboratory. There were tested geometrical and physical properties of FRA. Testing results were comparison with natural sand. The test results of physical properties of aggregate are given in Tab. 1. The geometrical properties are given in Fig. 1.

Tab. 1 Physical properties of FRA and natural sand

Types of aggregate	Fraction [mm]	Density[kg/m^3]	Water absorption capacity [%]
FRA	0 – 4	2090	8,29
Natural sand	0 – 4	2600	2,00

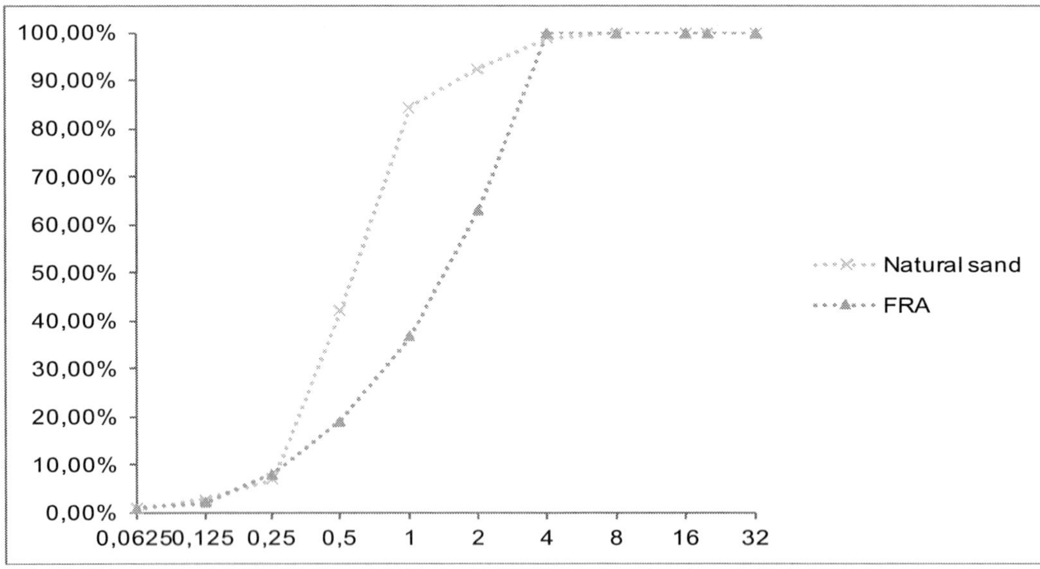

Fig. 1 Geometrical properties of natural sand and FRA

Concrete mixtures. A total four concrete mixtures were designed. All concrete mixtures were designated with same amount of cement and the same w/c ratio for clear comparison. The first concrete mixture was prepared only with natural sand (REF). In other mixtures, natural sand was replaced by FRA in varying replacement ratio (specifically 50 %, 60 % and 70 %). All concrete mixtures are given in Tab. 2.

Tab. 2 Concrete mixtures

Designation	REF	REC50	REC60	REC70
cement [kg/m^3]	486	486	486	486
water [kg/m^3]	243	243	243	243
natural sand [kg/m^3]	1458	729	583	438
FRA [kg/m^3]	0	729	875	1020

Testing. Beams of dimensions 40x40x160 mm were used for the testing. Chloride ion penetration was tested according to CSN 73 1326 [4]. The surface of concrete samples (40 x 160 mm) was plunged in 3 % NaCl solution. Concrete samples were exposed a total 100 freeze – thaw cycles. One freeze – thaw cycle took around 2 hours and was composed of freezing on temperature -15 °C (minimal duration 15 minutes) and thawing on temperature 20 °C (minimal duration 15 minutes). A pass was made at depth to grind the concrete samples into dust after each 25 cycles, which were then collected.

Results. Fig. 2 shows a concrete pass after 25, 50, 75 and 100 cycles. It is evident that concrete samples REC50 show the worse chloride ion penetration resistance. The concrete pass was almost 1800 g/m^2 after 100 cycles. The reference concrete samples show also poor chloride ion penetration resistance. The pass of reference concrete was 1319 g/m^2. In comparison with this, the best chloride ion resistance showed concrete samples with 60 % replacement ratio. The pass of concrete REC60 was 967 g/m^2 after 100 cycles. Fig. 3 and Fig. 4 show concrete surface disruption after 100 CHLR cycles.

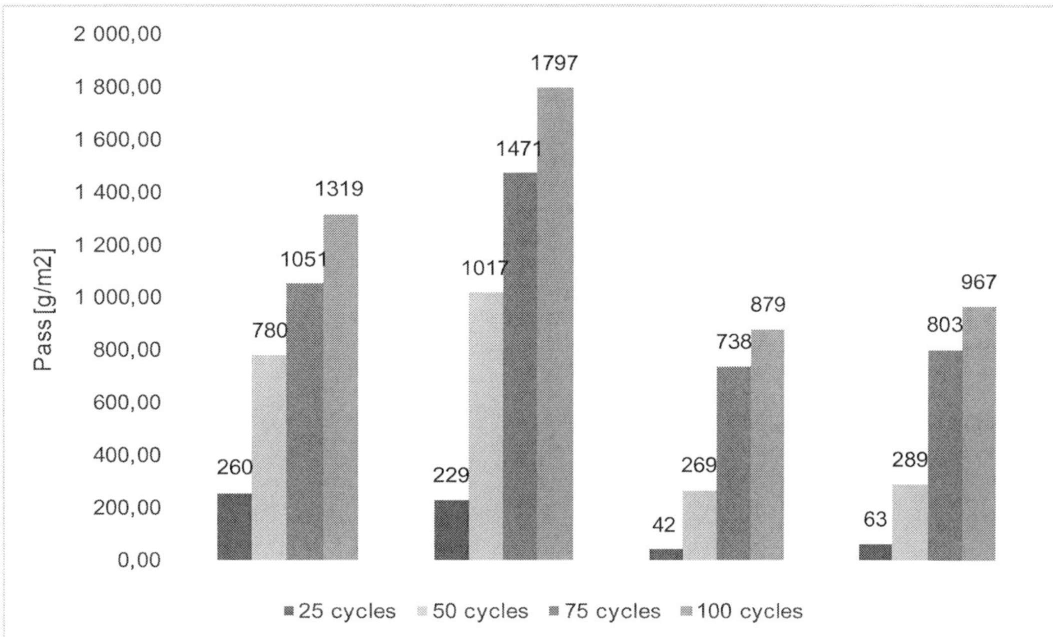

Fig. 2 Concrete pass after 25, 50, 75 and 100 CHLR cycles

Fig. 3 Concrete surface disruption after 100 CHLR cycles – REF concrete

Fig. 4 Concrete surface disruption after 100 CHLR cycles – REC60 concrete

Conclusions

In conclusion, it is possible to say that the use the FRA in concrete as partial replacement of natural sand influences chloride ion penetration of concrete. The test results are very satisfactory and confirm assumption of the possible use of the FRA even with higher replacement ratio. However, it is necessary to verify this results in future research.

Acknowledgement

This work has been supported by the Ministry of Education, Youth and Sports within National Sustainability Programme I, project No. LO1605 – Energy efficient construction and demolition waste into building structures and by TH02030649 Environmental efficiency building waste into building structures and by SGS18/108/OHK1/2T/11 Environmental aspects of high performance cement composites and concrete with recycled aggregate including their durability and service life.

References

[1] CSN EN 206. Concrete – Specification, performance, production and conformity. Prague: Czech standards institute, 2014 (in Czech).

[2] Gomes, M., 009. Structural concrete with incorporation of coarse recycled concrete and ceramic aggregates: durability performance. Materials and Structures, 42 (5), 2009, pp. 663 – 675.

[3] Evangelista L., 2010. Durability performance of concrete made with fine recycled concrete aggregates. Cement and Concrete Composites, 32 (1), 2010, pp. 9 – 14.

[4] CSN 73 1226. Determination of cement concrete surface resistance to water and chemical defrosting materials. Prague: Czech standards institute, 1984 (in Czech)

EAN 2018
56th conference on experimental stress analysis

Monitoring of Delamination Growth on the MMB Specimens using FBG Sensors and Electrical Resistance Measurement Method

SCHMIDOVÁ Nikola[1,a], DVOŘÁK Milan[1,b], KADLEC Martin[2,b] and RŮŽIČKA Milan[1,d]

[1]Department of Mechanics, Biomechanics and Mechatronics, Faculty of Mechanical Engineering, Czech Technical University in Prague. Technická 4, Praha 6, Czech Republic.

[2]VZLU – Czech Aerospace Research Centre, Beranovych 130, Prague–Letnany, Czech Republic

[a]nikola.schmidova@fs.cvut.cz, [b]milan.dvorak@fs.cvut.cz, [c]kadlec@vzlu.cz, [d]milan.ruzicka@fs.cvut.cz

Keywords: Carbon Fibre-Reinforced Polymer (CFRP) composites, FBG sensors, MMB specimens, electrical resistance measurement.

Abstract. This paper report results of damage monitoring on carbon fiber-reinforced polymer composite specimens loaded simultaneously by mode I and mode II loading (Mixed-Mode Bending - MMB specimens). The delamination propagation during fatigue loading was monitored by the embedded FBG sensor and by electrical resistance measurement method. The measurement of the electric response was carried out in the through-thickness direction of the specimen. Presented data confirm possibility to detect delamination by mixed-mode loading using embedded FBG sensor and also using the presented electrical measurement.

Introduction

Carbon Fibre-Reinforced Polymer (CFRP) composites are widely used in the aerospace industry, automotive and civil engineering because of their outstanding ratio of high strength to low density. For a reliable long-term operation, it is necessary to develop procedures of a damage monitoring. The damage often appears inside the material without visible signs on the impacted surface unlike to conventional materials such as steel and aluminum. A delamination is one type of such a damage. It often appears after the impact loading and then it can growth during the shear loading. Ultrasonic testing is the most common inspection method to reveal such damages, but it is costly and time demanding. Other methods are also under investigation [1] such as the acoustic emission [2], the electro-magnetic interference, the monitoring using FBG (Fiber Bragg Grating) sensors [3] or the comparative vacuum monitoring.

Methods based on the electrical potential or resistance measurement [4, 5] are also intensively investigated because they have a great potential to be developed to the form of automated systems. The electrical potential/resistance measurement can be used for an impact monitoring or a delamination growth monitoring.

Extensive tests with a large number of samples must be carried out to evaluate a method for the structural health monitoring. For this reason, we decided to study two methods on the same specimens. Specimens loaded simultaneously by mode I and mode II loading (Mixed-Mode Bending - MMB specimens) were used for the evaluating purpose. Optical fibers with FBG sensors have already proven its ability for the delamination detection under a pure mode I loading (tested on Double Cantilever Beam - DCB specimens) [3]. The possibility to detect delamination crack in advance also under a mixed mode I and II was examined in this study. The second studied method was a less known method based on the electrical resistance measurement. The work by Zappalorto et al. [5] described change in measured electrical resistance on DCB specimen for delamination propagation when delamination cracks already passed the area of electrical contacts. On the contrary, in presented work, electrical resistance was measured in the area of electrical contacts to bring more details about electrical response in this area. We also used a slightly different configuration of the electrical contacts in order to avoid possible influence of the contact resistance on the measurement. So called voltage change method was used instead of the potential method [6].

Fig. 46 Configuration of the test

Experimental Procedure

Material and Specimen Preparation. The MMB specimens were extracted from CFRP composite made of 16 layers of the material Hexply AGP 193PW/8552S RC40. The lay-up $[45/0]_{4S}$ was used. Dimensions of the specimens are 185 x 25 x 3.3 mm. A 25 mm long non-adhesive plastic foil sheet was inserted to the plate edge during manufacturing to the neutral axis to initiate the delamination growth between the 8th and 9th layer. Total number of five MMB specimens was tested.
A pair of aluminum alloy blocks with dimensions of 25 mm x 20 mm x 10 mm was bonded using R 21 MP-2 instant cyanoacrylate adhesive to the end of each specimen on the side with non-adhesive insert. These blocks are used to apply the load to the specimens (Fig. 1). Then a white paint marker was applied on the specimen edges to reveal the crack front during loading. Both edges of the specimen were marked using black lines with an interval of 25 mm.

The optical fiber with two FBG sensors was embedded also between the 8th and 9th layer during manufacturing. Distance of the first FBG sensor from the specimen edge was 50 mm and second FBG sensors was 75 mm from the specimen edge. The positions of the FBG sensors are shown in Fig. 47.

Electrical contacts were manufactured after the curing of the specimens. Surface of the specimens was first slightly sanded and carefully cleaned with a degreasing agent. The electrical contacts were prepared using a conductive epoxy and thin copper strips on the top and bottom surfaces of the specimens. The configuration of the electrical contacts used for electrical resistance measurement are shown in Fig. 2. Two electrical contacts served for injecting duty current of 80 mA. The voltage difference was measured on the second pair of electrodes.

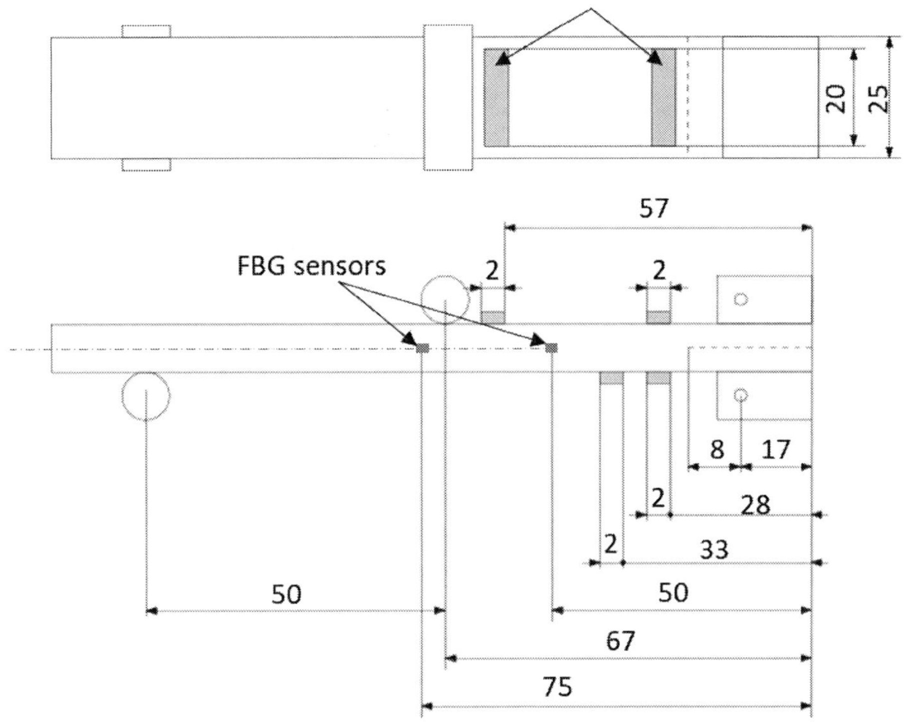

Fig. 47 Configuration of the electrical contacts and FBG sensors location

Test Configuration. The specimens were loaded by mixed-mode bending according to ASTM D6671M-06 standard. The configuration of the loading is shown in Fig. 48. MMB specimens were clamped using bonded aluminum blocks into INOVA ZUZ 50 loading machine, with a LTC-116-0.1 load cell, which has a capacity of 1 kN. The crack propagation during the loading was monitored using the cameras at 12.5 frames per second. Sinusoidal loading with displacement control set to δ_{max} = 7 mm and δ_{min} = 0.7 mm, and a frequency of 5 Hz was applied. The cycling was stopped for each measurement.

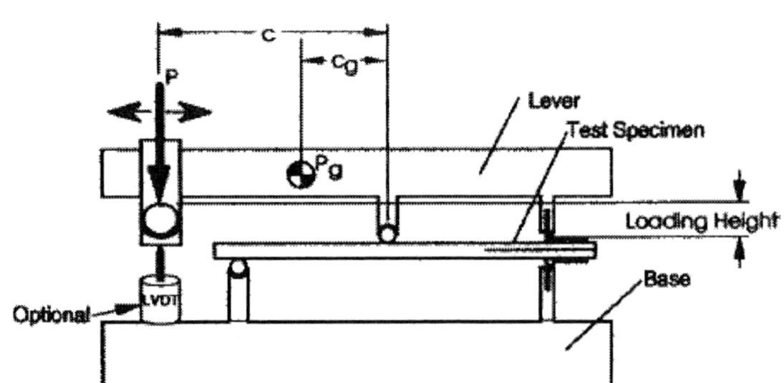

Fig. 48 MMB test configuration [7]

Measurement procedure. The Safibra FBGuard 1550 DAQ device was used to capture the signal from FBG sensors. Bragg spectra during two static load cases (loaded / unloaded specimen) were recorded after each cyclic loading block. Mechanical strain from FBG sensors versus time was also measured with data sampling frequency of 50 Hz.

Electrical resistance measurement was performed using the HP E3631A current source and two Agilent 34461A multimeters (Fig. 49). The change in electrical resistance was determined using the series comparison measurement method according to the following equation:

$$R_X = \frac{U_X}{U_N} \cdot R_N \tag{1}$$

The electrical measurement was performed also in two static load cases (loaded /unloaded).

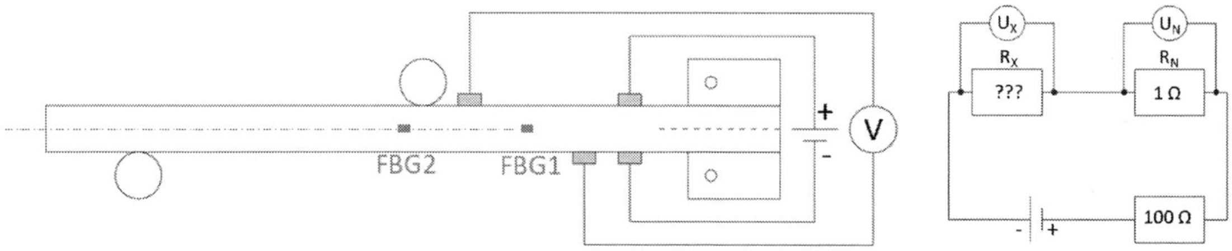

(a) (b)

Fig. 49 Measurement procedure of the electrical resistance measurement - configuration of duty current injection and voltage change measurement on the specimen (a), schematic of the series comparison measurement method (b)

Experimental Results

FBG sensors. Experimental data from FBG sensors were evaluated using the procedure, which is described in Ref. [3]. Delamination evaluation method was based on the peak distortion detection in measured spectrum of each FBG sensor (see Fig. 5). For the number of cycles, where FBG sensor has detected damage, the actual damage length from camera was obtained. Damage detection advance was then calculated as a difference between the position of the FBG sensor (50 or 75 mm, measured from the specimen's front edge) and the value from camera reading. Results are shown in the Table 1. It can be seen that in cases of MMB02-05 the first FBG sensor (50 mm from specimen edge) was able to detect damage in composite before it reaches sensor's position, with advance in range from 0.9 to 6.9 mm. In a single case of specimen MMB01, where only small crack growth was found until the last 5000 cycles, damage was detected with more than 3 mm latency after the visual method. Second FBG sensor wasn't able to detect any damage in all cases.

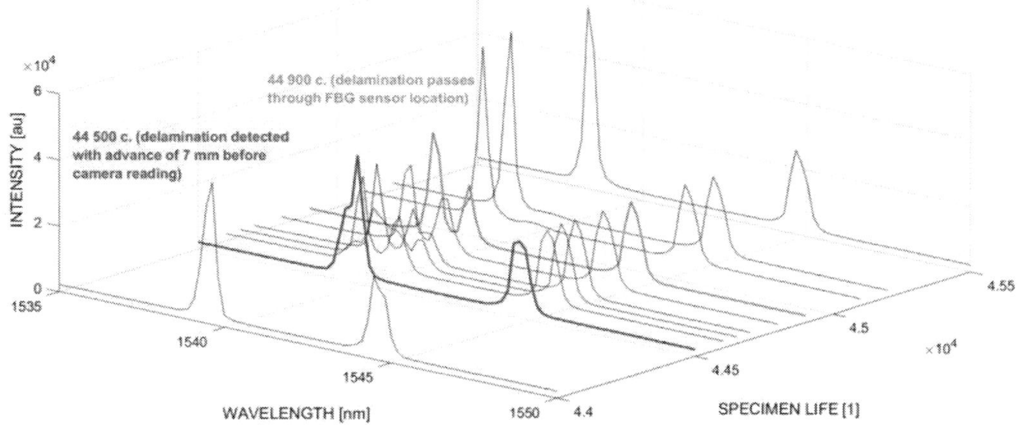

Fig. 50 Specimen MMB04 Bragg spectra reading, loaded state

Tab.1 Summary of results for specimens MMB01-05.

	distance from specimen edge [mm]	50
MMB01	damage detection advance (FBG sensor vs. camera reading) [mm], loaded / unloaded condition	-3,3
MMB02		0,9
MMB03		1,0
MMB04		6.9 / 4.6
MMB05		6.3 / 1.8

Electrical resistance change measurement. The results for electrical resistance change measurement are given in Fig. 51 and Fig. 52 for all five specimens. The ΔR value in the graph is measured electrical resistance normalized by its initial value R_0. Electrical resistance was measured in two states – loaded/unloaded – opened crack/closed crack. Measured values for both states are given in Fig. 51.

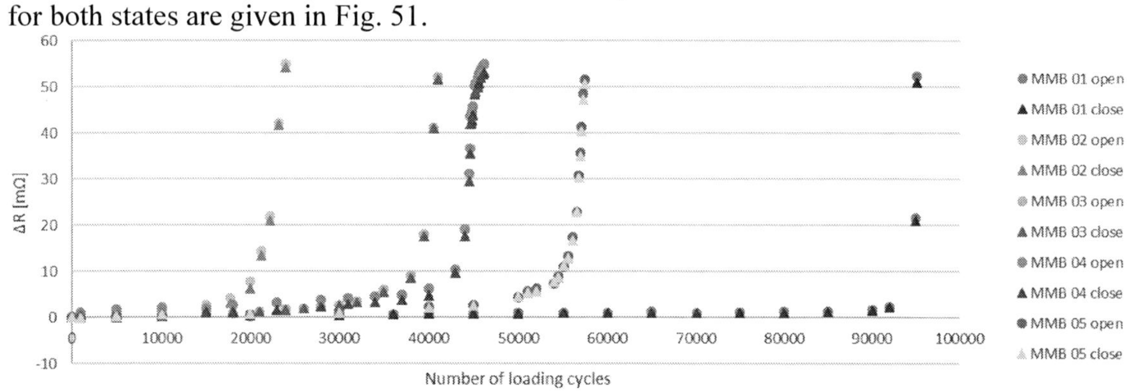

Fig. 51 Change in measured electrical resistance vs. number of cycles of all tested MMB specimens for states loaded/unloaded – opened/closed crack

The differences between measured values for loaded/unloaded state are small, the maximal difference between electrical resistance measured for open and close state was 1 mΩ for all tested specimens.

Relative change of measured electrical resistance was calculated according the following equation

$$\frac{\Delta R}{R} = \frac{R - R_0}{R_0} \cdot 100\ [\%]. \tag{2}$$

R is the actual electrical resistance and R_0 is the initial electrical resistance measured before cyclic loading of the specimen. The relative electrical resistance as a function of measured delamination length is plotted in Fig. 52. Delamination length was determined using the video-cameras. The initial length of delamination was given by the inserted 25 mm long non-adhesive plastic foil sheet.

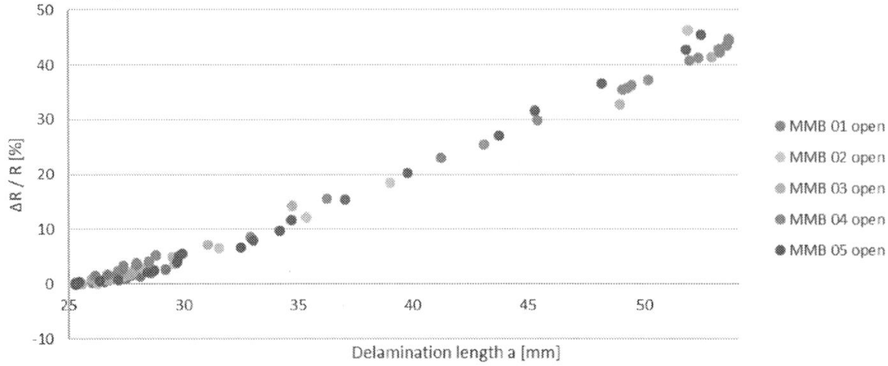

Fig. 52 Relative electrical resistance change vs. delamination length

Conclusions

In this work two methods for damage monitoring of CFRP composite were studied on the MMB specimens during fatigue loading. Possibility to delamination monitoring of such specimens was confirmed by means of the embedded FBG sensors. The electrical resistance measurement in the area of electrical contacts can be also used for delamination monitoring. Further work will be done on in the area of finite element analyses and the temperature influence evaluation.

Acknowledgement

The authors would like to thank the Technology Agency of the Czech Republic for supporting this research with project no. TE02000032 and Grant Agency of the Czech Technical University in Prague for supporting this research with grant No. SGS18/175/OHK2/3T/12.

References

[1] Gardiner, G. (2015). Structural health monitoring: NDT-integrated aerostructures enter service. CompositesWorld, http://www.compositesworld.com/articles/structural-health-monitoring-ndt-integrated-aerostructures-enter-service.

[2] L. Michalcová, M. Kadlec, Carbon/epoxy composite delamination analysis by acoustic emission method under various environmental conditions, Engineering Failure Analysis. 69 (2016) 88–96.

[3] M. Dvorak, N. Schmidova, M. Kadlec, M. Ruzicka, Use of FBG Sensors for Delamination Growth Measurement under Mode I Loading, EAN 2017 - 55th Conference on Experimental Stress Analysis 2017. (2017) 102-107.

[4] A. Todoroki, New analytical method for electric current and multiple delamination cracks for thin CFRP cross-ply laminates using equivalent electric conductance, Advanced Composite Materials, 25 (2016), 87-101, DOI:10.1080/09243046.2014.985420.

[5] M. Zappalorto, F. Panozzo, P.A. Carraro, M. Quaresimin, Electrical response of laminate with a delamination: modelling and experiments, Composite Science and Technology. 143 (2017) 31-45.

[6] L. Shen, J. Li, B.M. Liaw, F. Delale, J.H. Chung, Modeling and analysis of the electrical resistance measurement of carbon fiber polymer-matrix composites, Composite Science and Technology. 67 (2007) 2513-2520.

[7] ASTM Standard D6671M-06. Mixed Mode I-Mode II Interlaminar Fracture Toughness of Unidirectional Fiber Reinforced Polymer Matrix Composites. ASTM International, West Conshohocken, PA, 2006.

Dynamic Analysis of PUR Foam Reinforced with Coconut Fibres in Various Thermal Conditions

SRB Pavel[1,a], FLIEGEL Vítězslav[1,b], KULHAVÝ Petr[1,c] and MARTONKA Rudolf[1,d]

[1]Technical University of Liberec, Faculty of Mechanical Engineering, Studentská 2, 461 17 Liberec 1, Czech Republic

[a] pavel.srb@tul.cz, [b] vitezslav.fliegel@tul.cz,[c] petr.kulhavy@tul.cz, [d] rudolf.martonka@tul.cz

Keywords: polyurethane foam, composite, coconut fiber, DMA

Abstract. The reduction of emissions and using natural materials are the current trends. One of the ways how to reduce emissions is the vehicles weight reduction. The requirement on the weight reduction is also applied to car seats cushions. Polyurethane foam is still the most often used material for cushion of car seats. The weight saving could be achieved by manufacturing and using composite polyurethane foams with lower density and improved mechanical properties provided by the fibres reinforcement. Samples of polyurethane foam with coconut fibers were made and their inner structure was scanned by means of an electron microscope (SEM). Beside the static properties for the car parts (seats) it is important to know especially the dynamic properties. Therefore, the samples were tested on the dynamic mechanical analyzer (DMA). The chosen method was compression with constant strain, sweeping frequencies and various temperatures. This way, the storage and loss modulus of the reinforced foam samples were determined. The significant dependency of those main elastic constants on the using of reinforcement and also the varying ambient temperature has been found.

Introduction

Polyurethanes were developed in 1950s. Since then has been done many researches, and polyurethane production has become a multi-billion dollar business in the world today [1]. Polyurethanes have a wide variety of commercial use. Polyurethane foams are produced in three forms as flexible, semi-rigid and rigid. Flexible foams are widely used in different applications mostly in automotive, furniture and packaging. Nowadays car manufacturers have high demands on car seat development. The main limiting factor is the constant effort to reduce weight and price while keeping or even improving the main mechanical properties. One way to achieve a reduction of price of car seat cushion is using cheap natural material as a filler in polyurethane foam. Important features for seat evaluation are comfort and vibration isolation. In order for these parameters to be met, it is necessary to check the replacement material by a number of mechanical tests. Many studies describe the possibility of using natural fibers as reinforcement in polyurethane foams. Gu [2] describes the effect of adding wood fibers as a filler into several rigid and flexible polyurethane foams. The addition of fibers resulted in an increase in tensile and compressive strength of the rigid foam and increased stiffness of flexible foam. Banik [3] determined the effects of different cellulosic fibrous materials on the formation of the foam structure. The addition of few percent of fibers had a significant effect on the structure and the resulting properties of the polyurethane foam.

Combining coconut fibers with polyurethane foam could also possitively affects the damping of arising vibration. After the addition of 2.5% coconut fibers to the flexible polyurethane foam, an increase in the damping properties of the polyurethane foam was found and the resonance frequency was shifted to higher values compared to pure foam [4]. The dynamic properties of car seat cushions are very important in terms of long-term seating comfort [5]. Long-term exposure of human body to low frequency vibration is associated with health problems, especially, lower back pain. The vertical vibrations with frequency around 3 Hz increases fatigue and drowsiness [6]. A dynamical mechanical analyzer is a useful tool for measuring the mechanical properties of different materials [7].

Materials

For this study foam samples with 5 % of coconut fibers were created. In Fig. 1_a the microscopic photo of a wad of coconut fibers could be seen. There it is possible to see the varying diameters of the individual fibers. Fig. 1_b shows the SEM photo of one fiber in polyurethane foam, there can be seen the damage of the fiber where the part of its surface is torn off. On the torn off layer good coupling polyurethane with fiber is evident. That means the suitable composite material has been created. In addition, improvement of mechanical properties of this way prepared composite foam compared to the pure polyurethane foam could be expected.

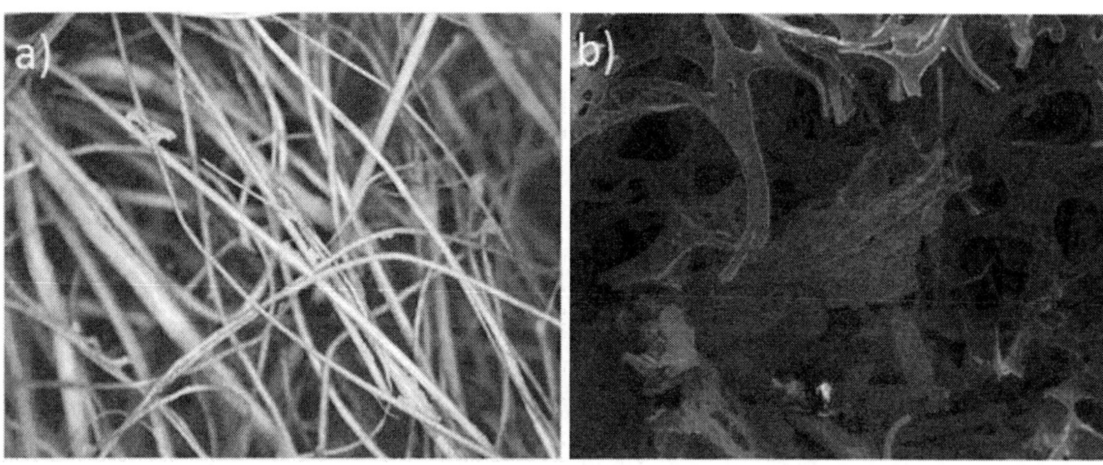

Fig 1. a) wad of coconut fibers b) SEM photo of one fiber in polyurethane foam

Fig. 2 Samples for DMA compression testing

Experiment

The aim of the carried experiment was to compare elastic behavior of the neat polyurethane foam samples and the reinforced one by means of the dynamic mechanical analyzer (DMA) as is schematically illustrated in the Fig 3. DMA compression is used to measure the properties of low and medium tough materials, including gels and weak elastomers [8]. The sample must support a static preload force during testing. The sample should have the ratio between the thickness and diameter as high as is possible, depending on the sample preparation and instrument limits. The carried experiment consists in various frequency with constant strain amplitude. Because the properties of the polymer material are temperature dependent, the measurements were carried out for different temperatures. A frequency sweep may be performed to evaluate the viscoelastic response of a material by the observed changes in the storage and loss moduli and tan δ. The loss tangent is the ratio of loss modulus and storage modulus, which indicate the viscosity and elastic properties of material, respectively. According to [8] is recommended to use logarithmically space frequencies (e.g. 10, 5, 2, 1) rather than linearly spaced for composite samples.

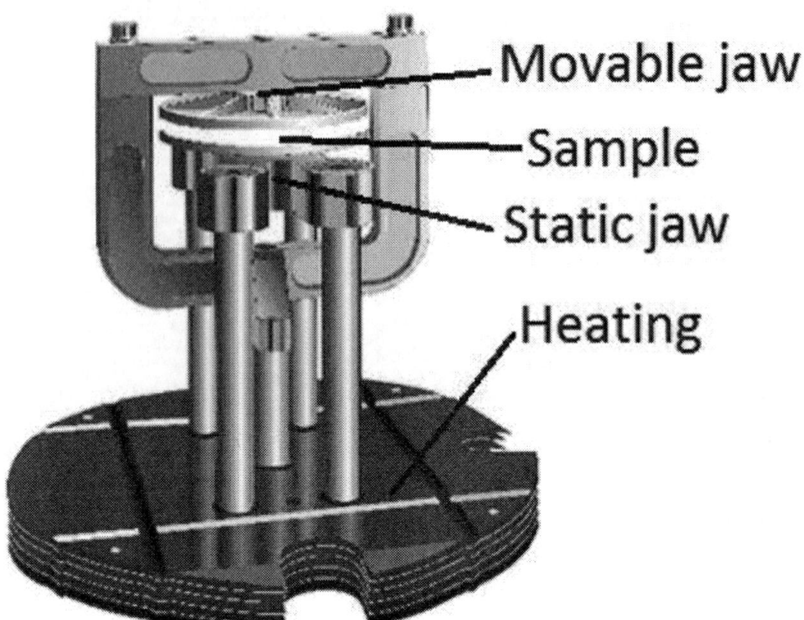

Fig. 3 The experimental DMA device

Before the main testing, it is appropriate to set the value of the geometry factor. For the compression clams it is possible to determine the optimal sample size according to (1)

$$\mathbf{GF} = \mathbf{Fe}\,\frac{\mathbf{T}}{\mathbf{A}} \tag{1}$$

where T is sample thickness [mm], A is sample cross sectional area [mm^2], Fe is sample shear factor. This factor is a strong function of Poisson's ratio and the thickness of range 0.1 to 1. The value was determined from table in device manual [8]. Tab. 1 summarizes sample values and measurement settings.

Tab. 1 Sample values and measurement settings

	Neat PU foam	Composite foam
Sample thickness T [mm]	7.8	8.1
Sample diameter D [mm]	12.8	12.7
Clamping factor Fe [-]	0.61	0.61
Geometry factor [1/mm]	0.0369	0.039
Amplitude [mm]	0.5	
Preload [N]	0.6	
Force track [%]	125 %	
Frequency range [Hz]	1 - 50	
Temperature range [°C]	30 - 60	

Results

In the following figures are the results of dynamic compression. It is possible to see the comparison of the neat polyurethane foam Fig. 4 and composite foam with coconut fibers Fig. 5 in dependency on the loading frequency in the range of 1 – 50 Hz.

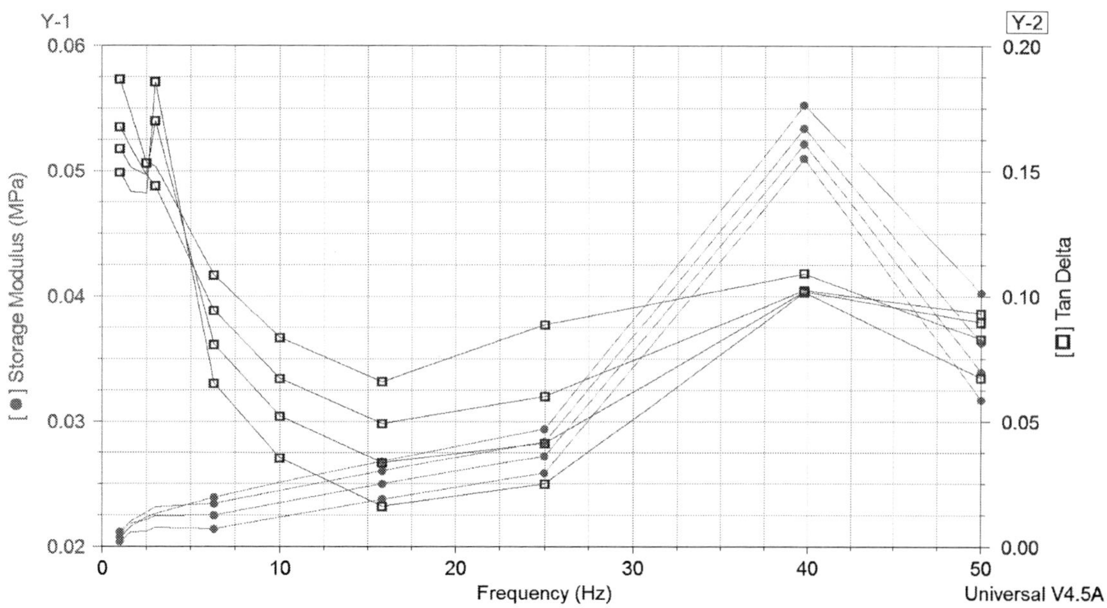

Fig. 4 Experimental results of the modulus in dependency on frequency of neat foam

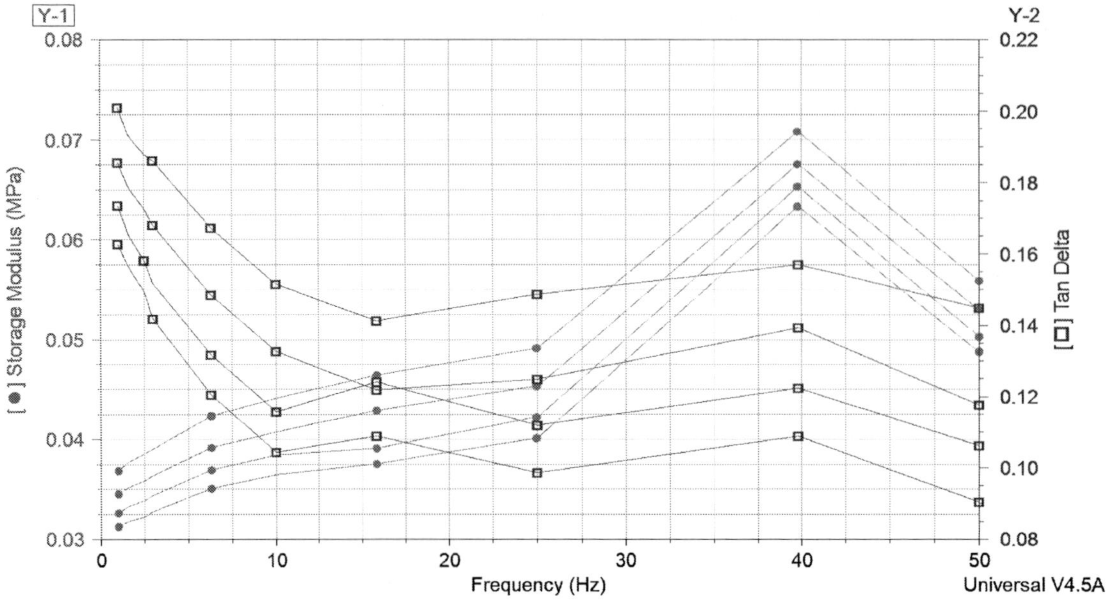

Fig. 5 Experimental results of the modulus in dependency on frequency of composite foam

Because the temperature in car could achieve relatively high values, it is appropriate asses the dependency of properties also with various ambient temperatures. The result of dynamic compression in dependency on the temperature could be seen in the following figures. It is possible to see the comparison of the neat polyurethane foam Fig. 6 and composite foam with coconut fibers Fig. 7 in dependency on the temperature in the range 30 – 60°C.

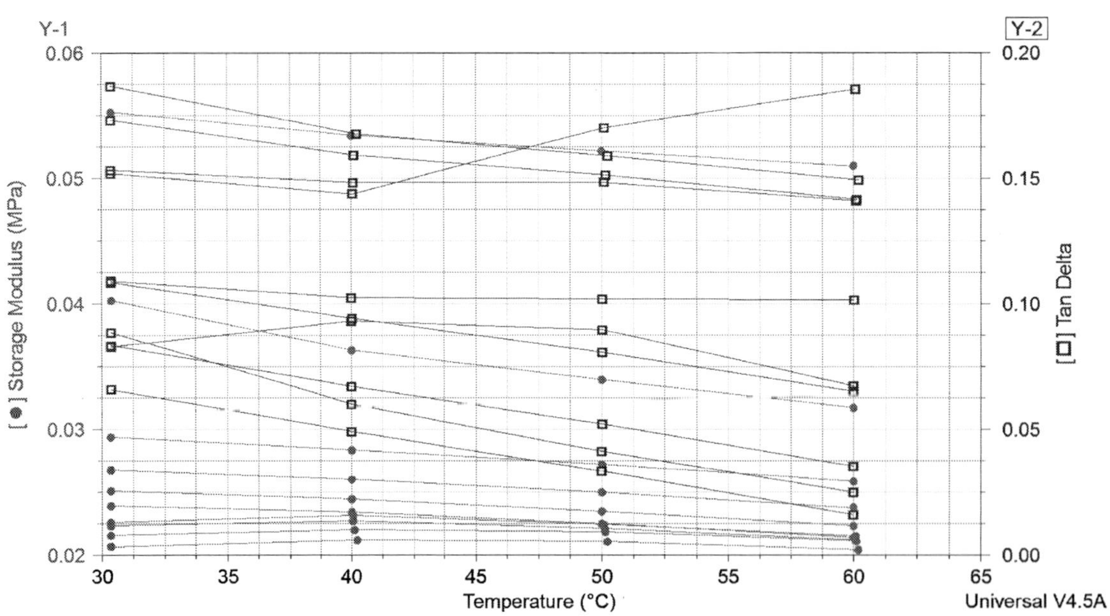

Fig. 6 Experimental results of the modulus in dependency on temperature of neat foam

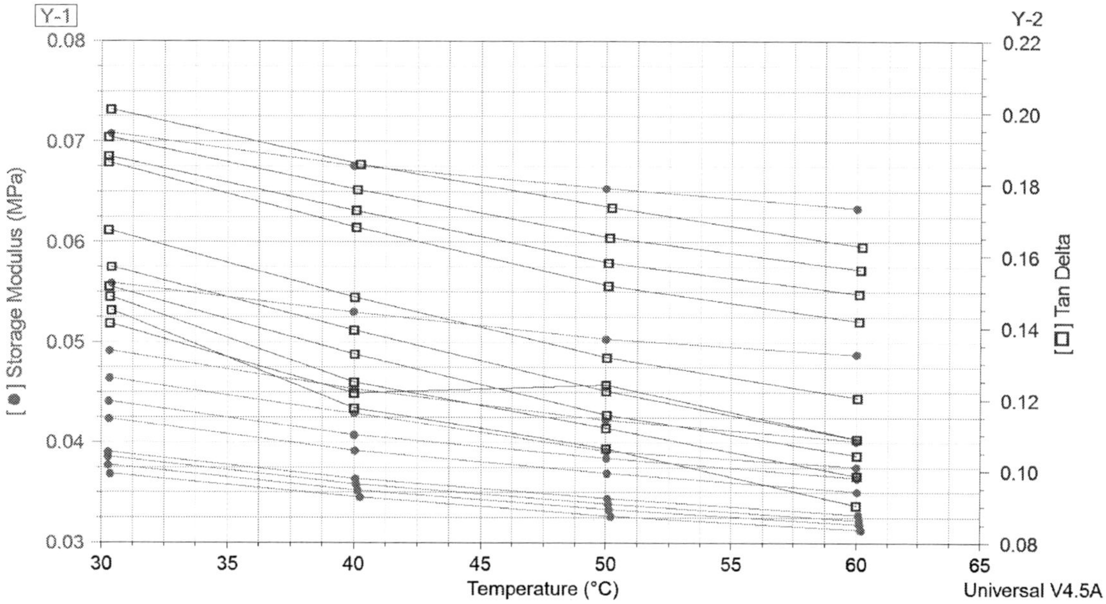

Fig. 7 Experimental results of the modulus in dependency on temperature of composite foam

Conclusions

Composite PU foam reinforced by coconut fibres was created. Mechanical properties of neat foam and composite foam were compared using the dynamic mechanical analyzer. It has been found that the resultant modulus of elasticity of composite polyurethane foam with coconut fibres is higher compared to neat polyurethane foam. Highest storage modulus was found at frequency of 40 Hz for both neat foam and composite foam. With increasing temperature storage modulus drops slightly in neat foam, for composite foam is this drop more pronounced. The results have shown that polyurethane foam with coconut fibers can be used for a car seat cushion. The further aim of the research should be targeted in order to find out the optimal volume ration of the fibre reinforcement and the best disposition.

Acknowledgement

This publication was written at the Technical University of Liberec as part of the project "Innovation of products and equipment in engineering practice" with the support of the Specific University Research Grant, as provided by the Ministry of Education, Youth and Sports of the Czech Republic in the year 2018.

References

[1] M. Khazabi, Investigation of biopolyol spray foam Insulation modified with natural fibers, Doctoral Thesis, University of Toronto, 2015

[2] R. Gu, M. Sain, S. Konar, A feasibility study of polyurethane composite foam

with added hardwood pulp. Ind Crop Prod 2013; 42: 273–279.

[3] I. Banik and M. Sain. Water-blown soy polyol based polyurethane foams modified by cellulosic materials obtained from different sources. J Appl Polym Sci 2009; 112: 1974–1987.

[4] C. Shan, M. Ghazali and M. Idris, Improved Vibration Characteristics Of Flexible Polyurethane Foam Via Composite Formation, International Journal of Automotive and Mechanical Engineering, 2013

[5] D. Ning, S. Sun, L. Wei, B. Zhang, H. Du, W. Li, Vibration reduction of seat suspension using observer based terminal sliding mode control with acceleration data fusion, Mechatronics, Volume 44, 2017, Pages 71-83, ISSN 0957-4158,

[6] N. Sombatsompop, Dynamic Mechanical Properties of Ground Flexible Polyurethane Foam Particles and Carbon-Black-Filled Rubber Vulcanisates, Polymer-Plastics Technology and Engineering, (2006), 37:1, 1-18, DOI: 10.1080/03602559808006909

[7] J. Zhou, Z. Yao, Y. Chen, D. Wei, Y. Wu, Thermomechanical analyses of phenolic foam reinforced with glass fiber mat, Materials & Design, Volume 51, 2013, p. 131-135, ISSN 0261-3069, https://doi.org/10.1016/j.matdes.2013.04.030.

[8] QDMA Manual, TA Instruments - Waters LLC, New Castle, DE 19720, 2006

Numerical Simulation of Reinforced Polyurethane Foam

SRB Pavel[1,a] and PETRŮ Michal[1,b]

[1]Technical University of Liberec, Institute for Nanomaterials Advanced Technology and Innovation, Studentská 2, 461 17 Liberec 1, Czech Republic

[a] pavel.srb@tul.cz, [b]michal.petru@tul.cz

Keywords: natural fibers, polyurethane foam, composite, FEM

Abstract. Nowadays there is a significant demand of use eco-friendly technologies in industry included automotive. Legislation is constantly increasing the pressure to reduce emissions, especially carbon dioxide and nitrogen oxides. One of ways to reduce car emissions is reduce fuel consumption, which can be achieved by reducing the weight of the car, which means reducing weight of all parts including car seat. A possible solution of weight reduction is the use of new low-density materials with better properties instead of commonly use materials. The usage of natural composite materials is currently widespread in all fields of technology. A foam composite material has been designed to use in car seat cushions. Numerical simulations are a very powerful tool of mechanical parts design. A composite material model has been designed and verified.

Introduction

The car seat consists of many parts. One of them is a comfortable cushion, despite its low density, there is also the effort to reduce weight and price with the same comfort [1, 2, 3]. One way to get a lower weight cushion can be using of a composite low-density foam with good damping and comfort properties. For combination with polyurethane foam cheap natural fibers (cellulose, coir, hay etc.) can be used [4, 5]. The addition of natural fibers increases a tensile and compressive strength of flexible polyurethane foam. Many studies describe using reinforced polyurethane foams. Banik [5] has found that adding of only few percent of the fibers to the polyurethane foam significantly improves its mechanical properties. Shan [6] states that adding 2.5 % of coir to the flexible polyurethane foam resulted an improvement of damping properties. A useful tool for designing and optimizing mechanical parts are numerical simulations [7,8]. Simulation software usually contains a number of material models of homogeneous materials [9]. For simulating of composite materials, there are usually tools for creating of laminated materials [10]. For other composite materials is a possible way simulation on samples of the microstructure. In that case, the notion of representative volume element (RVE) is of paramount importance [11, 12]. The RVE is usually regarded as a volume of heterogeneous material that is sufficiently large to be statistically representative of the composite, i.e., to effectively include a sampling of all microstructural heterogeneities that occur in the composite. RVE must include a large number of heterogeneities (grains, voids, fibers, etc.) It must however remain small enough to be considered as a volume element of continuum mechanics.

Materials and methods

The aim of this work was to create a numerical model of a sample of composite polyurethane foam with a fiber reinforcement and to verify its behavior as compared to experimentally tested samples. The numerical model was created for a brick shape sample with dimensions 100x100x100 mm. First was created model of neat polyurethane foam without reinforcement. The numerical model was created according to real compression test Fig.1 a). The numerical model consist two rigid plates between them was the test sample Fig. 1 b). The bottom plate is fixed - motion is prevented in all directions, the top plate is only allowed to move in the vertical axis where the compression is applied at a speed of 100 mm.min^{-1}, compression was carried out to 70 % of sample height. A friction coefficient of 0.1 was applied to the contact surfaces. The polyurethane foam was defined as hyperfoam material in Abaqus. The abaqus hyprefoam [11, 12] model is a nonlinear, isotropic material model that is valid for cellular solids with porosity that permits large volumetric changes and is suitable for hyperelastic foams. The model is based on the strain energy function of the form

$$\mathbf{U} = \sum_{i=1}^{N} \frac{2\mu_i}{\alpha_i^2}\left[\hat{\lambda}_1^{\alpha_i} + \hat{\lambda}_2^{\alpha_i} + \hat{\lambda}_3^{\alpha_i} - 3\frac{1}{\beta_i}\left(\left(J^{el}\right)^{-\alpha_i\beta_i} - 1\right)\right] \tag{1}$$

where N is a material parameter; μ_i, α_i and β_i are temperature dependent material parameters. The independent variables λ_i are principal stretches and are related to strain in a continuum. The term J^{el} is the elastic volume ratio and is a function of the principal stretches. Parameter β_i is related to the to the Poisson's ratio ν_i by the expression

$$\beta_i = \frac{\nu_i}{1 - 2\nu_i} \tag{2}$$

The model order N = (1÷6), must be chosen and the material parameters α_i, β_i and μ_i can either be specified, or computed by Abaqus using a least square fit that minimizes error in the computed stress when given experimental data. The Poisson's ratio can also be specified or computed. The acceptable experimental data for this model are uniaxial, biaxial and volumetric.

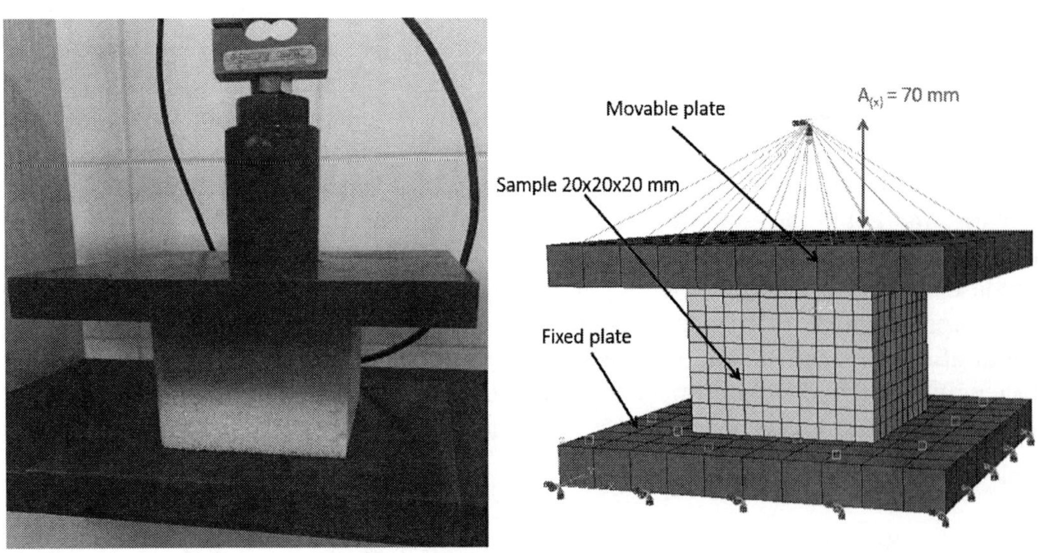

Fig. 1 a) arrangement of experiment b) numerical simulation of compression test

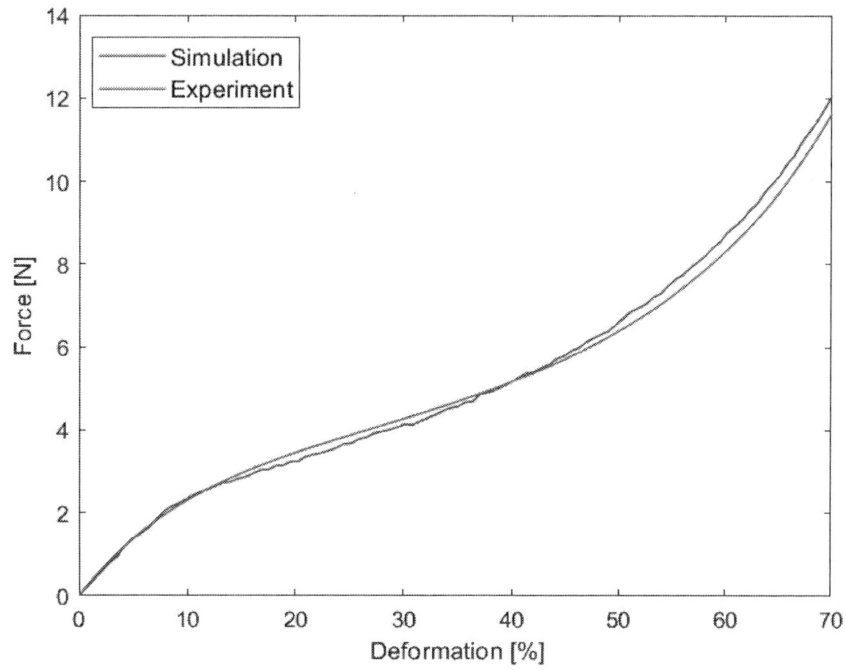

Fig. 2 Comparison of compression curves of neat polyurethane foam.

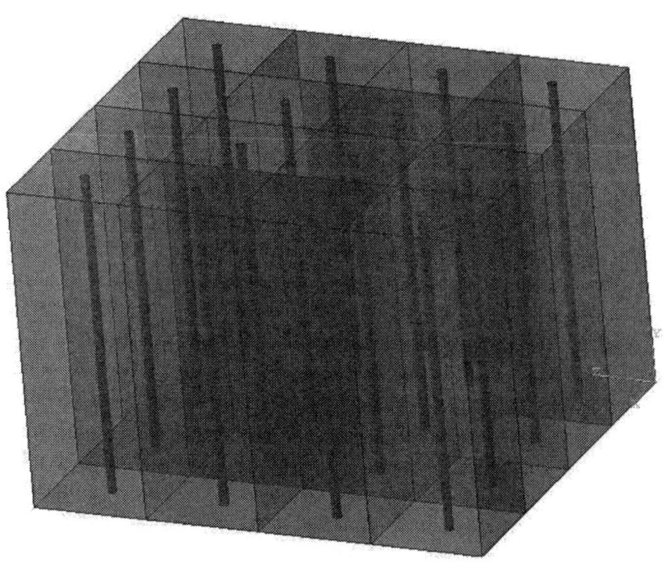

Fig. 3 Composite foam with vertical fibers.

Material parameters were determined using curves obtained by uniaxial compression of foam samples. Fig. 2 shows the force course of experimental compression of the neat polyurethane foam sample compared to the numerical model, the deviation is at most 5%. The first model of reinforced foam was designed as brick of edge length of 20 mm with 16 holes of 0.4 mm diameter. Vertical fibers of 0.4 mm diameter were inserted into these holes and connected to the foam by tie constraint fig. 3. Tie constraint connect the nearest nodes of the selected surfaces and prevents their mutual movement. An elastic material model has been selected for fibers. At the beginning of loading fibers behave like beams and greatly increase stiffness of composite. With increasing deformation, the buckling occurs the fibers will deflect and their effect on the composite stiffness will be greatly reduced. In real experiments, the buckling appears immediately after compression starts due to foam and fiber in homogeneities and uneven fiber configurations. In the simulation, the buckling may not occur at all because of the idealization of many aspects. In this simulation a buckling occurred, which is due to the contact setting when certain shocks that initiate deformation. Fig. 4 shows deformed composite foam with vertical fibers. This high stiffness at the beginning of compression is not suitable for flexible foams which are significantly deformed during use. Second was created geometry resemble the shape of the real fibers o 0.4 mm diameter fig. 5 a. shows fiber dimensions. Then a geometry of the foam segment (5x5x20 mm) was created with a cavity corresponding to the dimensions of the fiber. The fibers were connected to foam segments by tie constraint. The individual segments of foam with fiber were assembled into a block of 20x20x20 mm. Each segment was rotated at a different angle according to the longitudinal axis. The arrangement of the fibers in foam is shown in fig. 5 b.

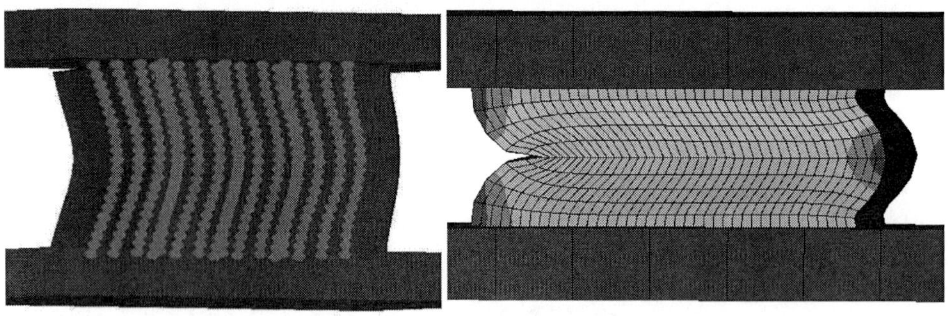

Fig. 4 Deformation of polyurethane foam with vertical fibers.

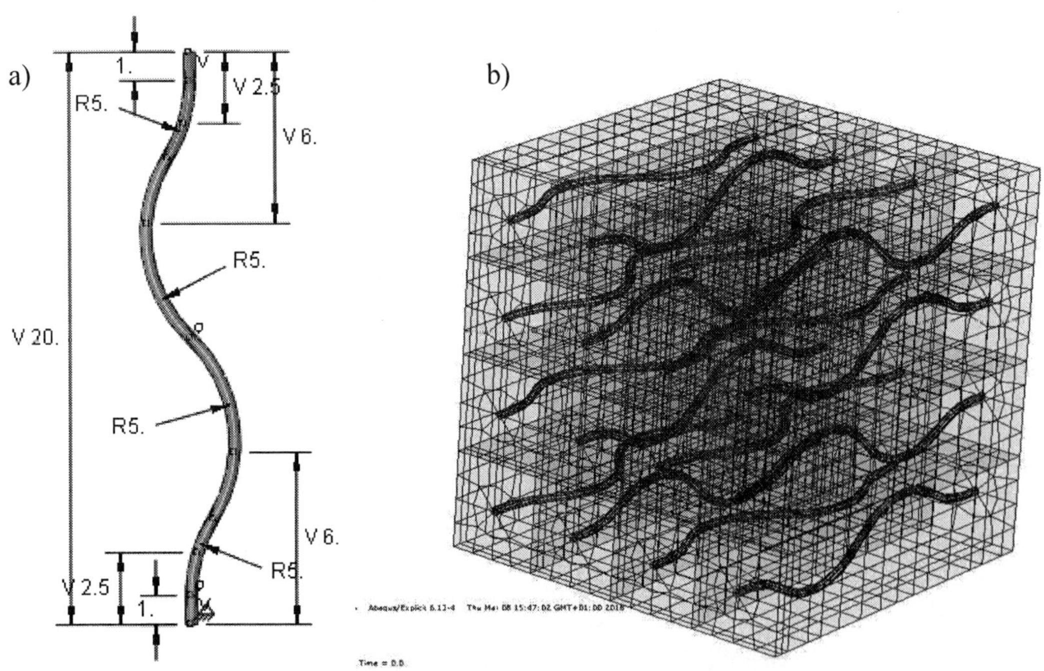

Fig. 5 a) dimensions of fiber, b) arrangement of fibers

Results

Numerical simulation of compression was performed on the assembled model of composite foam. The compression was performed both in the fiber direction and perpendicular to the fibers. Fig. 6 shows strain stress curves of individual simulations compared to the experiment. At the beginning of the compression in the fibers direction was found a significant increase of stiffness. During further compression, there was a significant uneven distortion of the fibers and whole composite sample. After compression of 18 % deformation, the simulation crashed due to too short time step. Next, was simulated compression in direction perpendicular to the fibers. At the beginning of compression, only polyurethane foam is deformed and the effect of the fibers is minimal. Between individual fibers there are spaces filled with polyurethane foam. At compression 20 %, the fibers get closer and their effect on the stiffness is beginning to show. With further compression this effect continues fibers are stressed by bending and torsion. Based on these simulations was created a real sample of composite foam with longitudinally oriented fibers. The compression test shows very good agreement with numerical model fig. 6.

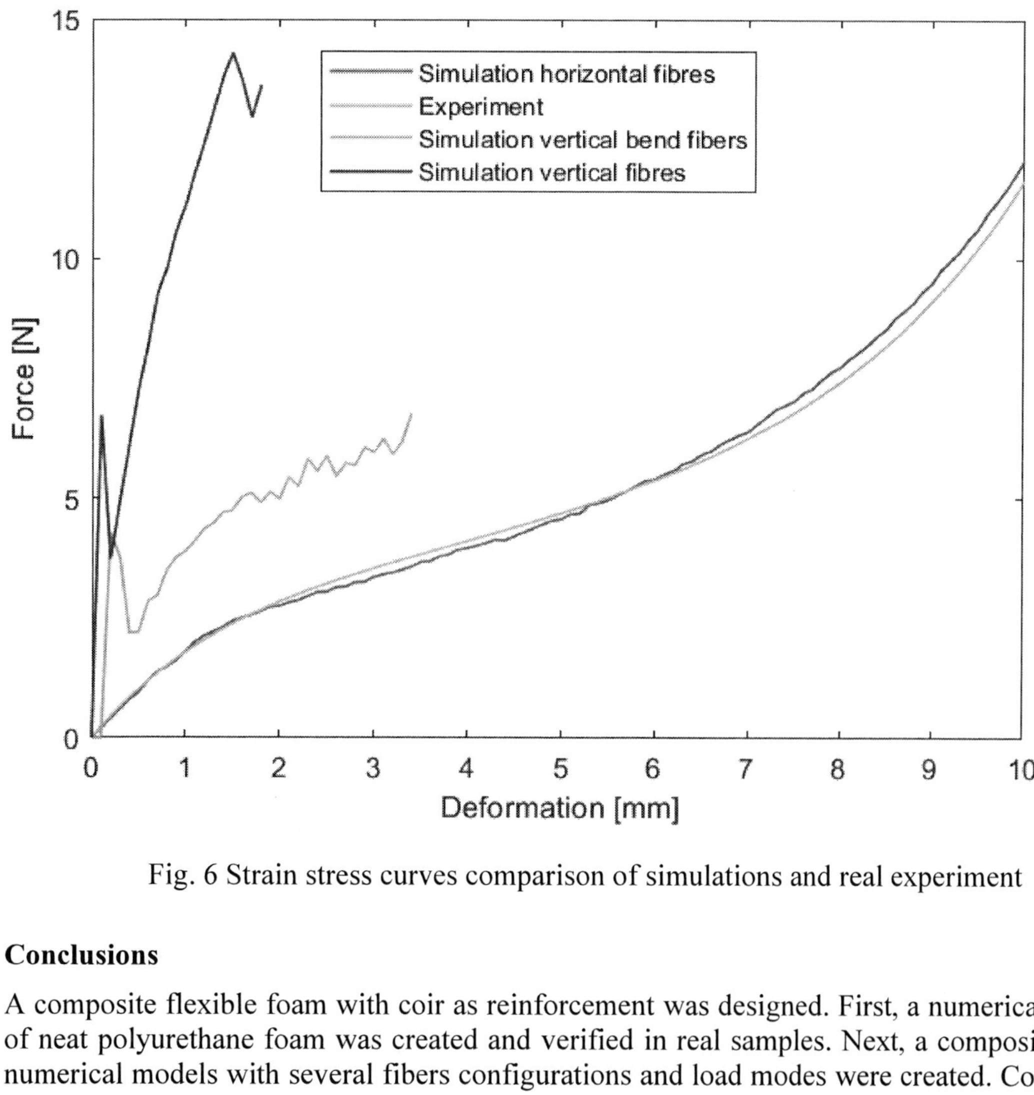

Fig. 6 Strain stress curves comparison of simulations and real experiment

Conclusions

A composite flexible foam with coir as reinforcement was designed. First, a numerical model of neat polyurethane foam was created and verified in real samples. Next, a composite foam numerical models with several fibers configurations and load modes were created. Composite foam with longitudinally orientation fibers was chosen for its good properties for use in car seat cushion. Real composite foam samples were then created to test the mechanical properties and numerical model verification. It was proved good agreement of experiments and numerical simulation. The results of numerical simulations can lead to design of whole composite car seat cushion with specific mechanical properties.

Acknowledgement

The results of this project No. LO1201 were obtained through the financial support of the Ministry of Education, Youth and Sports, Czech Republic in the framework of the targeted support of the "National Programme for Sustainability I"

References

[1] I. Kamp, The influence of car-seat design on its character experience, Applied Ergonomics, Volume 43, Issue 2, 2012, Pages 329-335, ISSN 0003-6870

[2] Petru, M., Novak, O., Lepsik, P., Mysakova, D. Numerical and experimental analysis of interphase of new environmental low-energy composites. 52nd International Conference on Experimental Stress Analysis, EAN 2014; 2014.

[3] Petru, M., Novak, O., Lufinka, A. Experimental and numerical analysis of crack propagation in light composite materials under dynamic fracturing, Communications - Scientific Letters of the University of Zilina, Volume 16, Issue 3a, 2014, Pages 82-89

[4] Gu R, Sain MM and Konar SK. A feasibility study of polyurethane composite foam with added hardwood pulp. Ind Crop Prod 2013; 42: 273–279

[5] L. Chang, M. Sain, M. Kortschot, Effect of mixing conditions on the morphology and performance of fiber-reinforced polyurethane foam, Journal of Cellular Plastics, 2015

[6] I. Banik and M. Sain. Water-blown soy polyol based polyurethane foams modified by cellulosic materials obtained from different sources. J Appl Polym Sci 2009; 112: 1974–1987.

[7] C. Shan, M. Ghazali and M. Idris, Improved Vibration Characteristics Of Flexible Polyurethane Foam Via Composite Formation, International Journal of Automotive and Mechanical Engineering, 2013

[8] Petru, M., Novak, O., Lufinka, A. Study and analysis of transmissibility of car seat with non polyurethane material. 50th Annual Conference on Experimental Stress Analysis, EAN 2012; 2012.

[9] T. Kanit, S. Forest, I. Galliet, V. Mounoury, D. Jeulin, Determination of the size of the representative volume element for random composites: statistical and numerical approach, International Journal of Solids and Structures, Volume 40, Issues 13–14, 2003, ISSN 0020-7683,

[10]Miehe, C., Schroder, J., Becker, M., 2002. Computational homogenization analysis in finite elasticity: material and structural instabilities on the micro- and macro-scales of periodic composites and their interaction. Comput. Method Appl. Mech. Engrg. 191, 4971–5005.

[11]A. Siranosian, R. Stevens, Developing an Abaqus *HYPERFOAM Model for M9747 (4003047) Cellular Silicone Foam, 2012

[12] Hibbitt, Karlsson and Sorensen, ABAQUS/CAE 6.13 User's Manual, Dassault Systemes, 2013

EAN 2018
56th conference on experimental stress analysis

Development of the Plastic Zone at the Crack Tip Under Cyclic Loading in the Middle-cracked Tension Specimen

ŠEDEK Jakub[1,a]

[1] VZLÚ – Czech Aerospace Research Centre, Beranových 130, Prague 9, 199 05, Czech Republic

[a]sedek@vzlu.cz

Keywords: Crack, Cyclic loading, FE, Plastic zone size

Abstract. The paper deals with the development of the plastic zone size during cyclic loading. The analysis was carried out using finite element method. The three-dimensional model of M(T) test specimen with central crack was discretized by hexahedral elements. In the area of crack tip the mesh was refined. The elastic/perfectly-plastic material with isotropic hardening was employed. Cyclic loading with coefficient of the cycle asymmetry $R = 0$ was applied up to 100 cycles. The plastic zone area in cross-sections and the size in the crack plane was monitored. The cyclic plastic zone size remains approximately constant in the centre of the specimen, where plane strain prevails. Whereas near the free sides, the size is reduced with increasing number of cycles and tends to form the shape and size of plane strain. The development of the cyclic plastic zone in upward part of the cycle can be described by interpolation function in dependence on the monotonic plastic zone.

Introduction

Loading the part with a crack leads to developing the plastic zone at the crack tip in most metal engineering materials. The plastic zone increases in hollow material according to power law. At the sides of the body, the crack tip is in the conditions of plane stress, while toward to the centre, the condition of plane strain prevails, provided the small plastic zone size in comparison with the characteristic dimension of the body. The condition of plane strain is often introduced in the form of the ratio $r_p/B = 0.0637$, where B denotes the characteristic dimension of the body and r_p is the plastic zone size [1]. The FE analysis is widely applied when studying three-dimensional part. The plastic zone and the strain constraint at the crack tip during monotonic loading were studied on laboratory specimens in [2]. The constraint factor α was analysed in order to describe the out-of-plane constraint as well. The plastic zone under cyclic loading is analysed in this work for the ratio of $B/r_{p\,max} = 79$ (in the reversed form for $r_p/B = 0.0127$) using the same FE-model of M(T) specimen. The ratio is expressed at the maximum load level of the cycle. This indicates that the material at the crack front is mostly in the plane strain conditions. The dimensions of the specimen were 100x80x20 mm with half crack length of 20 mm. The material was described as ideally elasto-plastic with isotropic hardening. The characteristic mesh size at the crack tip was 0.003 mm and the hexahedral elements with reduced integration were used. The mesh size was increased in locations away from the crack tip in order to sustain the analysis price reasonable. Six layers of elements gradually tightening to the sides were used in 1/8 model by applying the symmetry planes. The pre/post processing and the solution using implicit solver with the step increments of 0.1 was carried out in Abaqus 2017 software.

Development of plastic zone during cyclic loading

Cyclic loading creates the cyclic plastic zone at the crack tip and the residual plastic zone is present even in the unloaded state. The elasto-plastic behaviour and strain hardening results in blunting of the crack tip during cyclic loading and the plastic zone is forced to change. The cyclic loading with the coefficient of cycle asymmetry $R = 0$ is analysed. The shape of the plastic zone is shown for 1st cycle in Fig. 1 a), after 10 cycles in b) and after 100 cycles in c). The reduction of size of plastic zone is clearly visible. The plastic zone under plane stress splits apart by application of increasing number of cycles. Nevertheless, only tightening of the plane stress zone might be the reason and the finite element is not thin enough to describe this change.

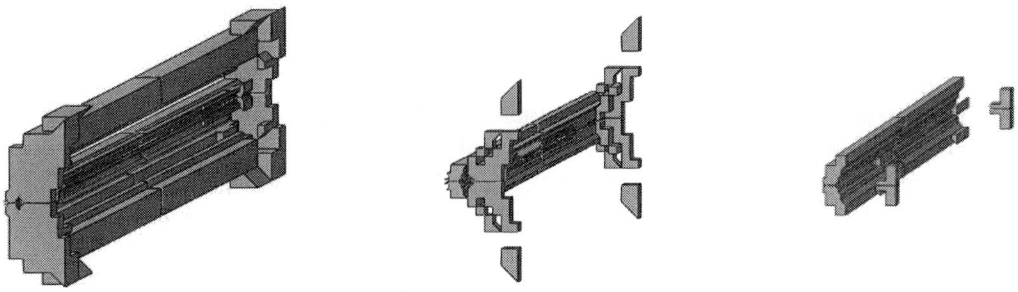

a) b) c)

Fig.1 Plastic zone after a) 1 cycle, b) 10 cycles and c) 100 cycles; FE analysis ($B / r_{p\,\max} = 79$).

The plastic zone size measured in units of area. In Fig. 2 a) there is shown the development of the plastic zone area according to rising load S in the cycle for the 1st, 2nd, 10th and the 100th cycle. The area is measured in the crack plane perpendicular to the crack front. Two planes are evaluated; the centre and the side plane. The considerable reduction of the plastic zone size can be seen after 1st cycle, but the size in the following cycles in plane strain region remains the same for each load level in upward part of the cycle. The data can be fitted by the power law, that is shown in the logarithmic scale as the straight line. During monotonic loading, it can be seen the increase of the size with the power exponent $k = 4.3$ for both plane stress and plane strain conditions. During cyclic loading, the residual plastic zone is also present at the zero load. By increasing the load, the residual plastic zone decreases and at certain load level S_{open}, the forward cyclic plastic zone starts to form. The development of the cyclic plastic zone should be described from the opening stress S_{open} when the plastically deformed material is not present. The normalized loading $(S - S_{open}) / (S_{\max} - S_{open})$ in the upward part of the cycle is therefore the unit of the x-axis and the power law will form in Eq. 1, where C and k are coefficients.

$$y = C\left(\frac{S - S_{open}}{S_{\max} - S_{open}}\right)^{k} \tag{1}$$

Applying the cyclic loading causes the cyclic plastic zone. It´s development measured in the area was described with the power exponent $k = 1.4$ for both plane stress and plane strain conditions. The exponent remains almost the same for the 2nd and the following cycles. The plastic zone size in the upward part of the cycle during cyclic loading is therefore almost the same in the studied range of cycle number. The plastic zone tends to stabilize under the plane strain conditions even near the free sides. The residual plastic zone size development according to cycle number is shown in Fig. 2 b). The size of the residual plastic zone increases with applied cycles.

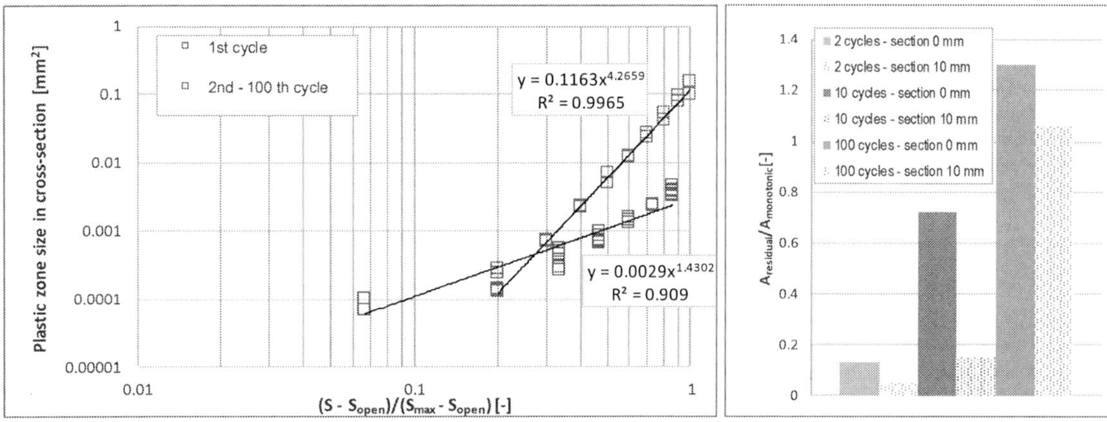

Fig.2 Area of plastic zone after 1 - 100 cycles ($B/r_{p\max}=79$) a) plastic zone size development in upward part of the cycle, b) Residual plastic zone size versus size at the maximum load level in the cycle

The plastic zone size measured as the length in the crack plane. The development of plastic zone size in the crack plane r_p was determined by the similar way as the area of plastic zone. The size r_p is the length measure of the plastic zone size commonly determined analytically according to Irwin [3] or Dugdale [4]. The plastic zone size r_p increases in the 1st cycle on the specimen sides in plane stress with the power exponent $k = 3$ and in the centre of the specimen in plane strain with the power exponent $k = 1.8$. The following cycles create the cyclic plastic zone $r_{p\,cyclic}$. Its size is smaller than the size of the monotonic zone from the x-axis rate of 0.5 and increases according to the loading with the power exponent $k = 0.7$ for both locations; the centre and the sides of the specimen.

During cyclic loading, in upward part of the cycle, the plastic zone size at the specimen's sides is larger and only at the maximum load level of the cycle the size in this location is close to the size reached in the 1st cycle. However, the size at the sides is related to the remote plastically deformed point whereas the continuous plastic zone size is smaller and it is close to the size in the centre under plane strain conditions. The development of plastic zone size for $B/r_{p\max}=79$ during loading is shown in Fig. 2 a).

The residual plastic zone size r_p has almost the same dimension in the centre under plane strain condition for all cycles but slightly increases with rising number of cycles for the sides of the specimen. In comparison with the maximum dimension at the maximum load level in the cycle, the average is around 1.1 except the size after 100 cycles causing the increase of the residual plastic zone at the sides of the specimen with the ratio of 1.6 (see Fig 3 b).

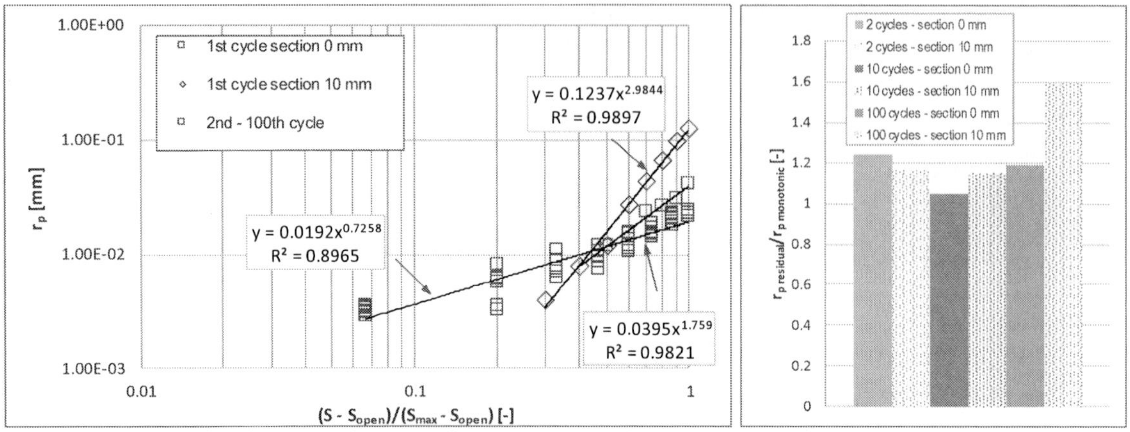

Fig.3 Plastic zone size r_p after 1 - 100 cycles ($B/r_{p\max}=79$) a) plastic zone size development in upward part of the cycle, b) Residual plastic zone size versus size at the maximum load level in the cycle

Estimation of the cyclic plastic zone size

In order to predict the size of the cyclic plastic zone $r_{p\,cyclic}$ during cyclic loading, the interpolation function is shown in Fig. 4 together with the values determined by the FE-analysis. Note that the plastic zone size in upward part of the cycle $r_{p\,cyclic}$ appears at the opening stress load level. The opening stress for analysed coefficient of cycle asymmetry $R=0$ was determined close to the load level $S/S_{\max}\in(0.2,0.3)$. This value corresponds well with the opening stress of 0.25 determined according to Newman's function for opening stress S_{op} [5]. The limit straight line shown in Fig. 4 estimates the values of $r_{p\,monotonic\,RD}/r_{p\,cyclic}$ from the left side. The subscript *RD* denotes the plane strain conditions.

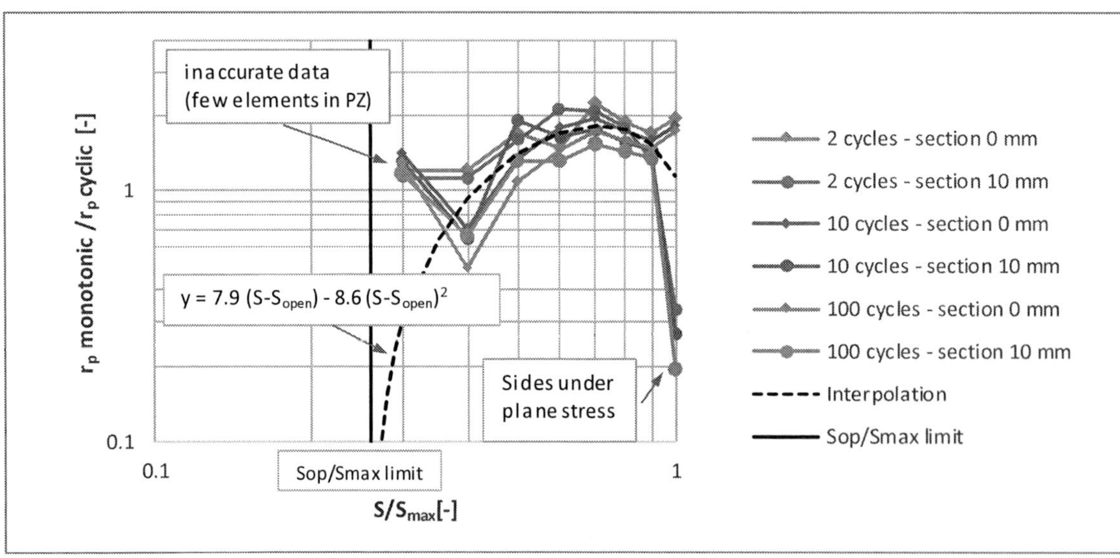

Fig.4 The ratio of $r_{p\,monotonic\,RD}/r_{p\,cyclic}$ based on the FE-analysis and the interpolation function for $B/r_{p\max}=79$

The interpolated function yielding the ratio of the monotonic plastic zone size $r_{p\,monotonic\,RD}$ under plane strain conditions versus the cyclic plastic zone size $r_{p\,cyclic}$ is proposed by Eq. 2.

The values of plastic zone size ratio close to the opening load level are influenced by the mesh size and are not assumed. The formula estimates the cyclic plastic zone size $r_{p\,cyclic}$ at each load level in upward part of the cycle for $R = 0$.

$$\frac{r_{p\,monotonic\,RD}}{r_{p\,cyclic}} = A\left(S - S_{open}\right) + B\left(S - S_{open}\right)^2 \qquad \text{for} \qquad S \in \left(S_{open}, S_{\max}\right) \tag{2}$$

The constants A = 7.9 and B = - 8.6 were determined for specimen´s characteristic dimension ratio $B / r_{p\max} = 79$. The ratio of plastic zone sizes $r_{p\,monotonic\,RD} / r_{p\,cyclic}$ is increasing along with the loading from S_{open} up to 0.7 $S_{\max}$ and then slightly decreases in the case of studied configuration with $B / r_{p\max} = 79$ and $R = 0$. The plastic zone size for $S / S_{\max} = 1$ depends on the position; i.e. at the sides or in the inner location. Evaluating the plastic zone size at the sides to the most remote plastically deformed element yields the rate of $r_{p\,monotonic\,RD} / r_{p\,cyclic} \approx 1/4$, but measuring continuous plastic zone yields $r_{p\,monotonic\,RD} / r_{p\,cyclic} \approx 3/2$.

Conclusions

The development of the plastic zone in M(T) specimen during cyclic loading is analysed in the 1st, 2nd, 10th and the 100th cycle using finite elements. The size and the shape are reported in the centre and at the sides of the specimen.

After the 1st monotonic cycle, the cyclic plastic zone is created. Its size in the centre of the specimen, where plane strain prevails, is approximately constant, but near the free sides, the zone disintegrates along with increasing number of cycles and tends to form the shape and size under plane strain conditions. The development of the plastic zone in upward part of the cycle can be described by power law. The coefficients of power law are summarized in Tab. 1 for the cross-section areas development and for the zone lengths as well.

Tab.1 Power law coefficients of plastic zone development ($B / r_{p\max} = 79$)

$y = C\left(\frac{S - S_{open}}{S_{\max} - S_{open}}\right)^k$		Monotonic plastic zone		Cyclic plastic zone	
		C [-]	k [-]	C [-]	k [-]
y = Cross – section area	Plane strain (centre of the specimen)	0.116	4.3	0.003	1.4
	Plane stress (specimens sides)				
y = Plastic zone size r_p	Plane strain (centre of the specimen)	0.040	1.8	0.019	0.7
	Plane stress (specimens sides)	0.124	3.0		

The ratio of monotonic plastic zone size versus cyclic plastic zone size $r_{p\,monotonic\,RD} / r_{p\,cyclic}$ can be described by interpolation function with the extreme between S_{open} and $S_{\max}$ and does not change significantly during cyclic loading except the maximum load level. On the sides at the maximum load level $S_{\max}$ the ratio $r_{p\,monotonic\,RD} / r_{p\,cyclic}$ is increasing due to the decreasing plastic zone size along with applied cycles. This behaviour indicates that the non-monotonic phenomenon appears. The development of plastic zone size $r_{p\,cyclic}$ according to monotonic plastic zone size contains the extreme value inside the range (S_{open}, S_{max}). At the sides, where plane stress condition prevails the ratio $r_{p\,monotonic\,RD}/r_{p\,cyclic} \approx 1/4$ was determined. As the plastic zone disintegrates after several cycles applied the ratio changes to approximately 3/2 at the load level S_{max}.

The results obtained by presented analysis are determined under several constraints and so further analyses should follow. The description of plastic zone size development during cyclic loading can be utilized for the estimation of its size based on the monotonic plastic zone size derived according to analytical formulas or FE-analysis.

Acknowledgement

This work was funded by the Ministry of Industry and Trade of the Czech Republic in the framework of the Institutional support of Research organizations.

References

[1] J. Kunz, Applied Fracture Mechanics, Czech Technique Association – CTU publishing (originally: Aplikovaná Lomová Mechanika, Česká Technika – nakladatelství ČVUT), (2005).

[2] J. Šedek, Numerical Investigation of 3-D Strain Constraint in Lab Test Specimens, Trebuňa F. et al., (eds.). EAN 2017 - 55th Conference on Experimental Stress Analysis 2017: Conference Proceedings, May 30th – June 1st, 2017, Nový Smokovec, Slovakia, TU Košice – Faculty of ME, (2017) 760 pages, pp. 174-177; ISBN 978-80-553-3167-6.

[3] G.R Irwin, Plastic Zone Near a Crack and Fracture Toughness, Proc. 7th Sagamore, (1960) conf. p. IV-63.

[4] D.S. Dugdale, Yielding of Steel Sheets Containing Slits, J. Mech. Phys. Solids 8, (1960) pp. 100 – 104.

[5] JC Newman Jr., A Crack Opening Stress Equation for Fatigue Crack Growth, International Journal of Fracture, Vol. 24, No. 3, (1984) pp. R131 - R135.

2018
56th conference on
experimental stress analysis

FEM Calculation, Measuring and Testing of Flood Protection Barriers

ŠEVČÍK Ladislav[1,a] and LUFINKA Aleš[1,b]

[1]Technical University of Liberec, Studentská 2, 46117 Liberec 1, Czech Republic

[a]ladislav.sevcik@tul.cz, [b]ales.lufinka@tul.cz

Keywords: flood protection barrier, FEM, testing, properties measuring.

Abstract. A new type of flood barrier to protect doors or windows has been developed as part of the research project VI20152018005 of the Ministry of the Interior of the Czech republic. Designing, finite element modeling and testing of the barrier prototype real properties are described in this paper.

Introduction

The newly designed flood barrier will be used to protect doors or windows against water ingress. This barrier is supposed to replace the sandbags so far used. Its advantage is quick and easy assembly and reusability. Its principle and real prototype are shown in the Fig. 1.

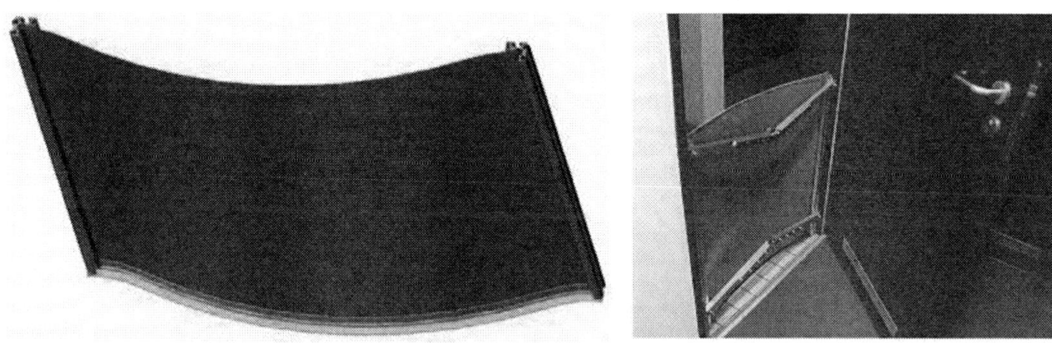

Fig.1 The flood protection barrier principle (left) and its real prototype (right)

The barrier is made up of a flexible composite plate, which is anchored on the vertical sides in the sidewalls with a rubber seal, which is also on the underside of the composite plate. The sidewalls are joined by flexible pulls. By pushing the pulls the plate is bent and the barrier can be inserted into the door frame. After loosening the bars, the bending of the plate causes a force that makes the barrier firmly anchored in the door frame. When the barrier is flooded, the hydrostatic pressure acting on the curved plate further increases the compressive force. The barrier is anchored in the door frame without any mounting. The bent shape of the board further eliminates certain inaccuracies in the width of the door frames.

CAD modeling and FEM Calculation

Composite plate properties are subject to contradictory requirements. The board must be flexible enough to be bent only by human force without the use of tools during its inserting to the door. The opposing requirement is that the bending of the plate must cause sufficient compressive force to anchor the barrier sidewalls in the door frame. The plate must also withstand the hydrostatic pressure from the retained water, and it must not collapse. The sidewalls profiles must allow the plate to be bent while ensuring watertightness. The barrier design including the seal was created using 3D CAD modelling and the parameters were optimized using FEM calculations [1, 2].

The basic problem was the decision how to solve computationally this case, whether to apply the bending theory of thin composite plates or the theory of buckling of thin composite boards for the calculation of expansion forces. The decision was mainly to place the board (boundary conditions) and the method of assembly (preload) and operation of the plate (hydrostatic pressure). Installation of flood protection is done in several steps. These steps affect to the calculation process. The assembly procedure is as follows: aluminum profiles are fitted with a rubber ring seal. Seal has the slot for the composite board (see Fig. 2).

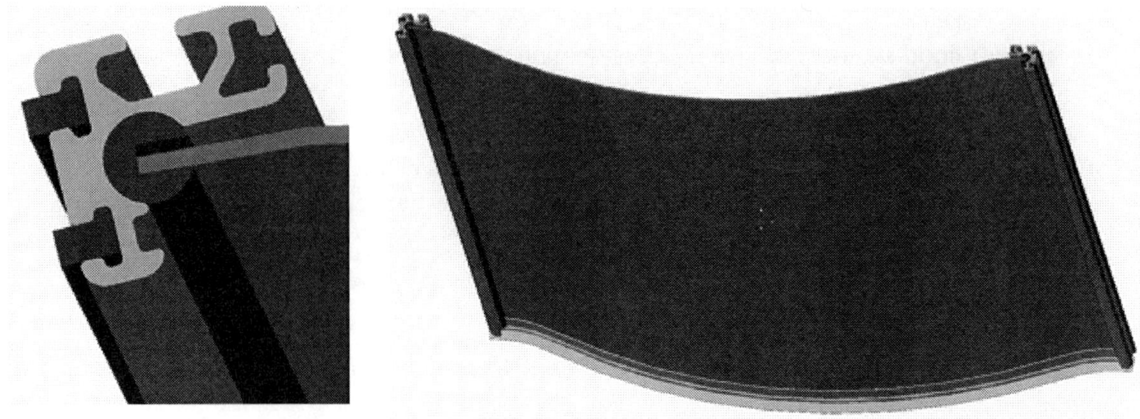

Fig. 2 Detail of the aluminium profile with rubber ring seal and composite plate (left) and the flood protection barrier (right)

The right and left aluminum profiles are connected by two jointed spacers. These spacers have joints at the junction with the profiles in the middle. The distance between the profiles maintains the spacers. Pressing the spacers together will decrease the distance of the profiles. The locations of forces are indicated by an arrow (see Fig. 3).

Fig.3 System of spacers with direction of pressing forces

In free states the composite plate is slightly longer than the pitch of the profiles including the grooves in the rubber. The board is inserted with preload force given by operator, during assembly. The plate deflects to shape near arc. The barrier holds together by action of the preloading force. This pressures force of buckling cause the frictional forces between the plate, sealing and the profile. These forces prevent the individual components of the barrier from ejecting. This folded barrier is inserted into the construction hole. The hole must be smaller than the size of assembled barrier. By computing and measuring the size of this dimension has to be verified. By pushing the spacer, the profiles are approached and the complete barrier is inserted into the door or window frame.

When the spacer's aren´t press, the plate is expanded and fixed in the hole. The prepared plate is waiting for the torrent water. This water loaded plate by hydrostatic and somewhere hydrodynamic pressure. Press of water acts against plate swelling. To some extent, this can increases the expansion force. This depends especially on the size of the construction hole and on the stiffness (thickness) of the composite board. In the case of a thin composite board, the hydrostatic pressure of the water causes its instability and thus the plate to break out. The results is that the plate has 1.5 waves in the loaded area. This wave loses its distance from the water level.

The optimization of the design is to design the thickness of the laminate plate to suit the given width of the opening and the desired level of the torque wave. Available theories are mainly solved by composite boards loaded onto buckling. In order to take into account all buckling modes (global, local and their interaction), the plate two-dimensional theory has been adopted to model the structures under analysis. Basic assumptions for thin plates were given by Kirchhoff law for the linear classical thin plate theory (CPT) and by von Kármán and Marquerre for the nonlinear CPT.

They made their assumptions for isotropic materials. In nowadays the knowledges have extended those assumptions for orthotropic or even for composite multilayer thin plates. The assumptions are as follows:

• The plate is homogeneous (for example, orthotropic homogenization is made for a fiber composite—resin matrix and fiber-reinforcement).
• The plate is thin—other dimensions (length and width) are at least 10 times higher than the plate thickness.
• The material of the plate is deformable and it is subjected to Hooke's law.
• The plane stress state is considered for the plate—the stress acting in the plate plane dominates the plate behaviour, stresses acting in the direction normal to the plate plane are assumed to be zero.
• All strains (normal and shear) in the plate plane are low compared to unity and they are linear.
• The strains normal to the plate mid-surface are neglected (the plate thickness does not change after deformation)—this assumption is made according to Kirchhoff-Love; normal to the mid-surface after deformation.
• There are no interactions in the normal direction between layers parallel to the middle surface;
• Deflections of the plate can be considered in terms of nonlinear geometrical relations.

Additionally, it is assumed that the principal axes of orthotropy do not need to be parallel to the edges of analyzed structures.
Examples of FEM modelling results are shown in the Fig. 4 and the Tab.1.

Tab.1 3-axis forces and torques of the bent plate

	Compression 20mm		Compression 40mm	
Direction	Force [N]	Torque [Nm]	Force [N]	Torque [Nm]
X	1212	1.5	1274	4.2
Y	9	52.3	52	96
Z	46	5.1	148	4.4

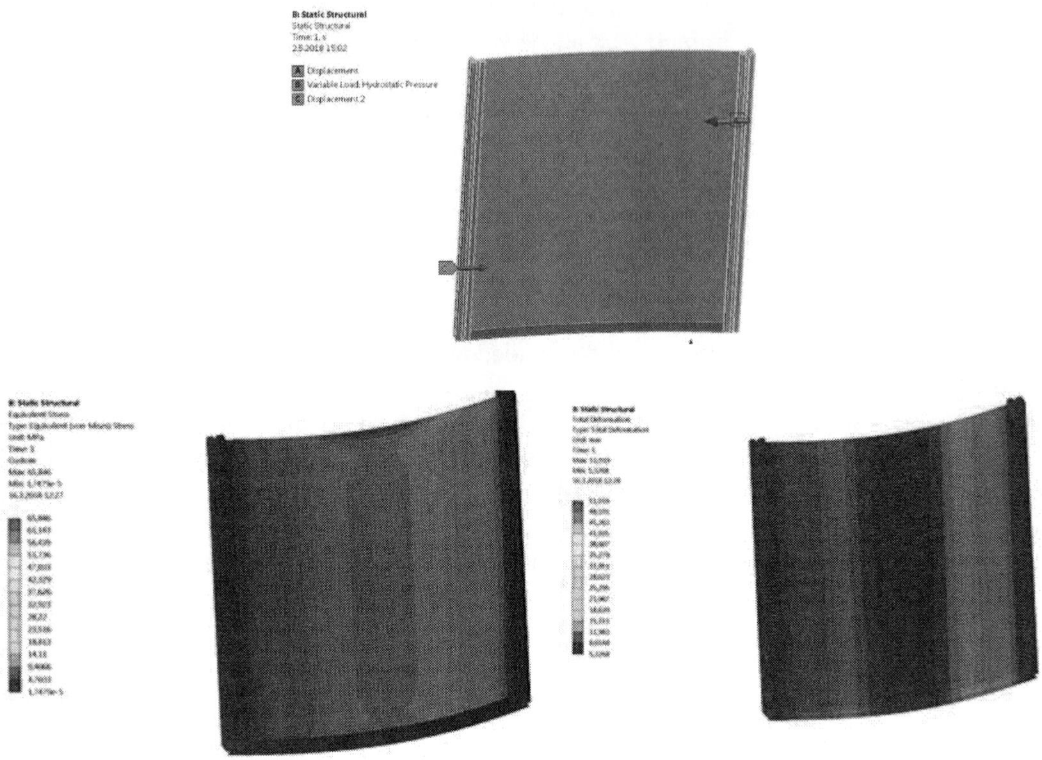

Fig. 4 Boundary condition for deformation 20 and 40 mm (up) and results of HMH stress for deformation left 20, right 40 mm

Real prototype measuring

The special test device was built to validate modelling results. It allows to set different door frame widths and to simulate the load of the barrier by hydrostatic pressure. The hydrostatic pressure of flood water is simulated using a plate driven by the linear hydraulic engine. The test device CAD design model and real test device configuration are shown in the Fig. 5.

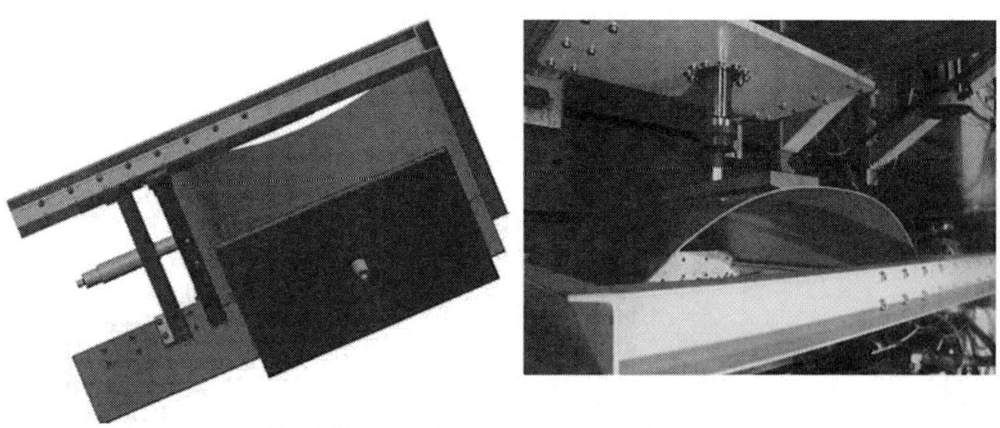

Fid. 5 Testing device and its CAD model

The real properties of the various composite plates were measured using this device. The barrier was inserted into the test device and hydrostatic pressure was simulated using the hydraulic engine (see Fig.6 up). The force acting to the barrier, its deflection and side force from the barrier to the door frame were measured. Measured results were then compared with FEM calculations (see Fig. 6 down).

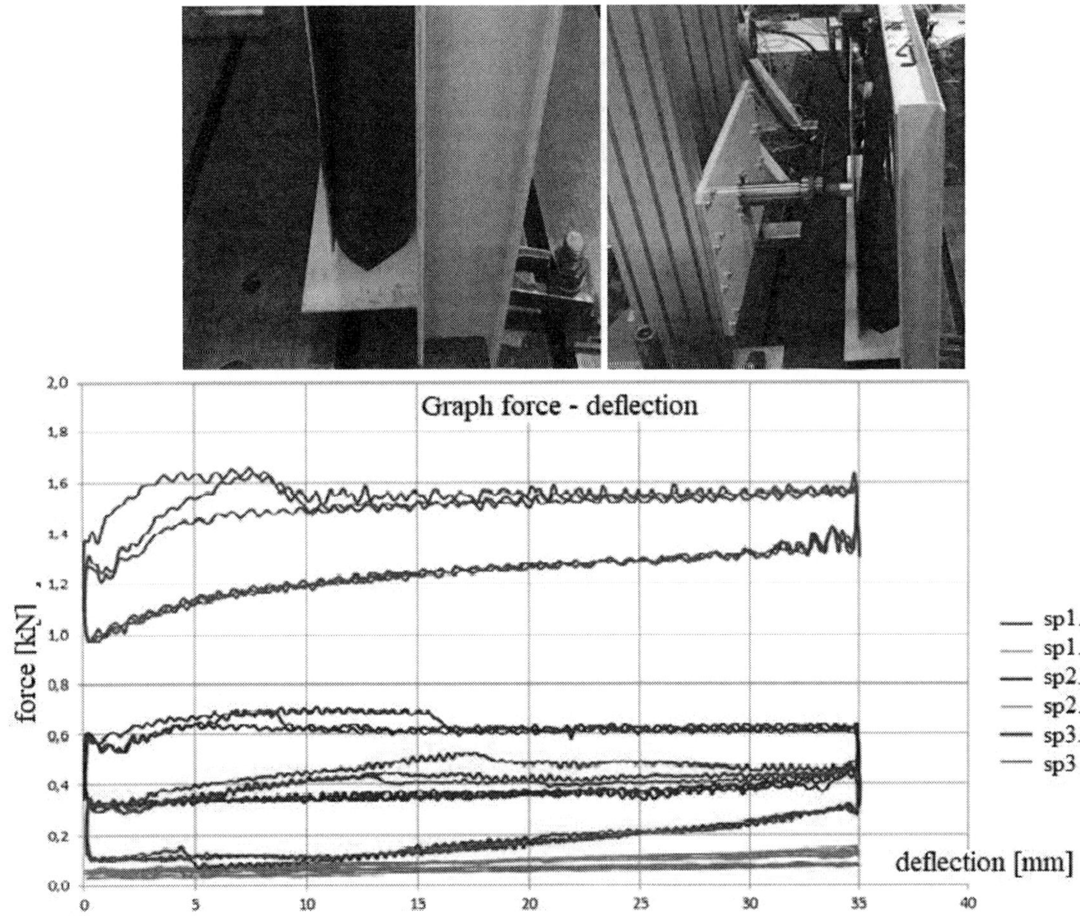

Fig. 6 Real flood barrier testing (up) and Examples of tests result(down)

Conclusions

The flood protection barrier assembly has been optimized by combining FEM modelling and measuring real properties of composite plates. The flood barrier is now ready for the final test of real water flooding. The advantage of the proposed solution is that the buckling force changes minimally in the extent of the barrier function.

The results of this project VI20152018005 were obtained with co-funding from the Ministry of the Interior of the Czech republic as part of targeted support from the project "Vývoj protipovodňových systémů pro zvýšení ochrany obyvatelstva a infrastruktury".

References

[1] Kubiak T: Static and Dynamic Buckling of Thin Walled Plate Structures, Springer 2013 ISBN 987-3-319-0D653-6,

[2] N. REDDY and A. KHDEIR. "Buckling and vibration of laminated composite plates using various plate theories", AIAA Journal, Vol. 27, No. 12 (1989), pp. 1808-1817., https://doi.org/10.2514/3.10338

EAN 2018
56th conference on experimental stress analysis

Numerical and Experimental Prediction of Lamb Wave Scattering from Horizontal Notch

ŠOFER Michal[1,a], FERFECKI Petr[1,2,b], FUSEK Martin[1,c], ŠOFER Pavel[3,d] and LIČKOVÁ Dagmar[1,e]

[1] Department of Applied Mechanics, Faculty of Mechanical Engineering, VŠB-TU Ostrava, 17. listopadu 15/2172, 708 33 Ostrava-Poruba, Czech Republic

[2] IT4Innovations National Supercomputing Center, VŠB-Technical University of Ostrava, Studentská 6231/1B, 708 33 Ostrava-Poruba, Czech Republic

[3] Department of Control Systems and Instrumentation, Faculty of Mechanical Engineering, VŠB-TU Ostrava, 17. listopadu 15/2172, 708 33 Ostrava-Poruba, Czech Republic

[a]michal.sofer@vsb.cz, [b]petr.ferfecki@vsb.cz, [c]martin.fusek@vsb.cz, [d]pavel.sofer@vsb.cz, [e]dagmar.lickova@vsb.cz

Keywords: Lamb wave, Wave scattering, Wavelet analysis, Orthogonality relation

Abstract: The presented paper deals with the topic of experimental and numerical prediction of energy transmission and reflection coefficients in the case of Lamb wave scattering on a two-millimetre deep and two-millimetre wide horizontal notch milled in a three-millimetre thick aluminium sheet. The paper contains the description of the experimental and numerical approaches, which were used for obtaining the above-mentioned coefficients. Great emphasis is also placed on the mutual comparison of results between these approaches as well as clarification of non-compliance of the results for particular Lamb wave modes.

Introduction

Lamb waves, as one of the types of guided waves, are extensively used for inspecting large structures as well as for structure health monitoring (SHM) [1]. These waves are one of the types of guided waves, which can exist only in geometries with a finite thickness such as plates, rods, or tubes [2]. Thanks to the ability to travel over long distances without much attenuation, an interaction of Lamb waves with particular types of geometric discontinuities has been an important topic of research [3-6]. In practice, the most challenging thing is in many cases the signal processing process related to the identification of the discontinuity type present in the inspected geometry [7]. Generally, the higher the number of Lamb wave modes present in the geometry, the more information related to a potentially present defect will be available. The level of post-processing complexity rapidly increases with the increasing number of present Lamb wave modes [8]. In many practical applications, we are often forced to utilize first two low-order Lamb wave modes at most. The main aim of the proposed paper is to present an experimental method, incorporation of the wavelet transform [9], for estimating energy reflection and transmission coefficients of low-order Lamb modes (A0/S0 mode), which are scattered by 2 mm wide and 2 mm deep vertical notch milled in 3 mm thick aluminum sheet. For subsequent verification of the obtained results, a hybrid method based on FEM approach and incorporation of orthogonality relation [8] is used.

Experimental Method for Scattering Coefficients Evaluation

The experiments were carried out on a three-millimetre thick aluminum sheet with the dimensions of 560 x 600 mm with a two-millimetre wide, two-millimetre deep and three hundred-millimetre long vertical notch milled on the top surface of the aluminum sheet (see Fig. 1). The coefficients were determined with respect to incident A0 Lamb mode, which has been excited by a focused air-coupled UT probe (Starmans PLF 19-012-T/F 120 kHz). The reflected and transmitted Lamb wave modes were detected by DWC B454 acoustic emission sensor, suitable for modal acoustic emission analysis. The Vallen AMSY-6 acoustic emission system served as a data acquisition unit as well as a signal generator. A five-cycle sinusoidal tone burst of 400 $V_{peak\text{-}peak}$ and the central frequency of 150 kHz (Fig. 2) have been used as a driving signal for the above-mentioned air coupled transducer.

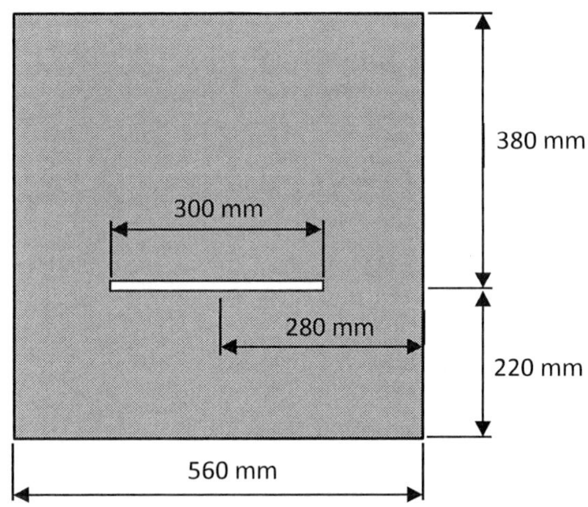

Fig.14 Schematic representation of the aluminum sheet including the position of the notch

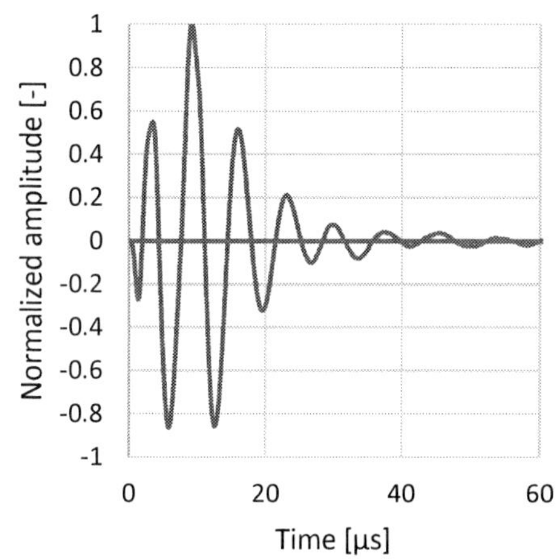

Fig.2 Driving signal for wave excitation

For proper generation of the incident A0 mode, the air coupled transducer has been placed into the positioning frame, which allows us to precisely adjust the exact position within four axes. The angle α (Fig. 3) can be derived using Snell´s law:

$$\alpha = \arcsin\frac{c_{air}}{c_{\text{phase}}}, \tag{1}$$

where c_{air} is the speed of longitudinal wave in air (suppose 0.33 mm/μs) and c_{phase} is the phase velocity of the A0 mode in the aluminium plate of the given dimensions and material properties, which arise from the defined values of longitudinal and transversal speeds. Due to the discrepancy between the frequency of the driving signal and the resonant frequency of the transmitter, there is a need to determine the frequency of the generated A0 mode in order to retrospectively modify the angle α. In the first step, we suppose that the frequency of the generated A0 mode will be around 130 kHz. The A0 phase velocity is then equal to 1.69 mm/μs and finally, the value of angle α is equal to 11.1 deg. Subsequently, we will measure the frequency of the generated A0 mode in the aluminium plate, the value of which was 140 kHz. Then, we have to repeatedly determine the phase velocity for the 140 kHz A0 mode and finally calculate the resulting value for the excitation angle, which is equal to 10.6 deg. The sensor layout depended on whether we wanted to measure the transmission or the energy reflection coefficients. The following figure summarizes the two used different configurations. The diagram in Fig. 3 to the left displays the configuration used for estimation of transmission coefficients. The transmitter is placed in front of the vertical notch, while the receiver is placed at a sufficient distance from the notch in order to ensure a proper separation of the transmitted A0 and S0 modes from each other. For estimation of energy reflection coefficients, a configuration in the figure to the right has been used. Again, the distance between the notch and the receiver had to be such that a proper separation of the reflected A0 and S0 modes is ensured.

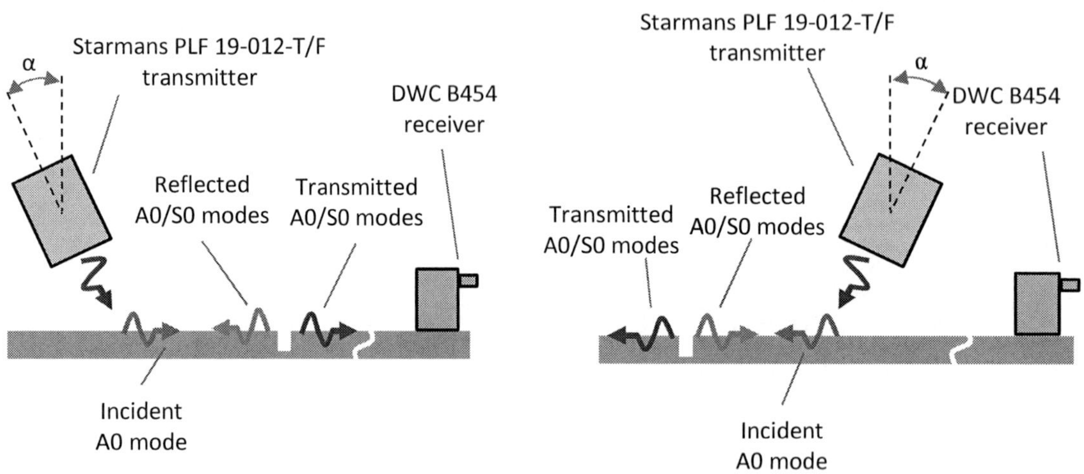

Fig.3 Left - configuration for estimation of the transmission coefficients; Right - configuration for estimation of the reflection coefficients

The calculation of the energy transmission and reflection coefficients was realized on the basis of time amplitude records of guided waves. Figures 4 and 5 display the time record in the case of the reflected and transmitted A0/S0 modes including the back-wall reflection from the plate´s free end. The time amplitude records were, with regard to a better clarity, normalized.

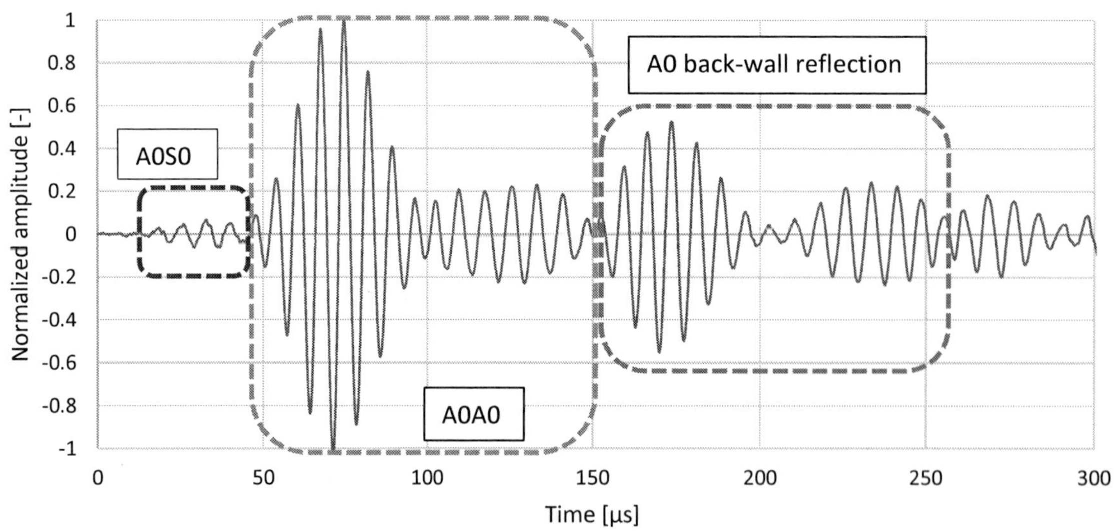

Fig.4 Time amplitude records of the reflected A0/S0 modes

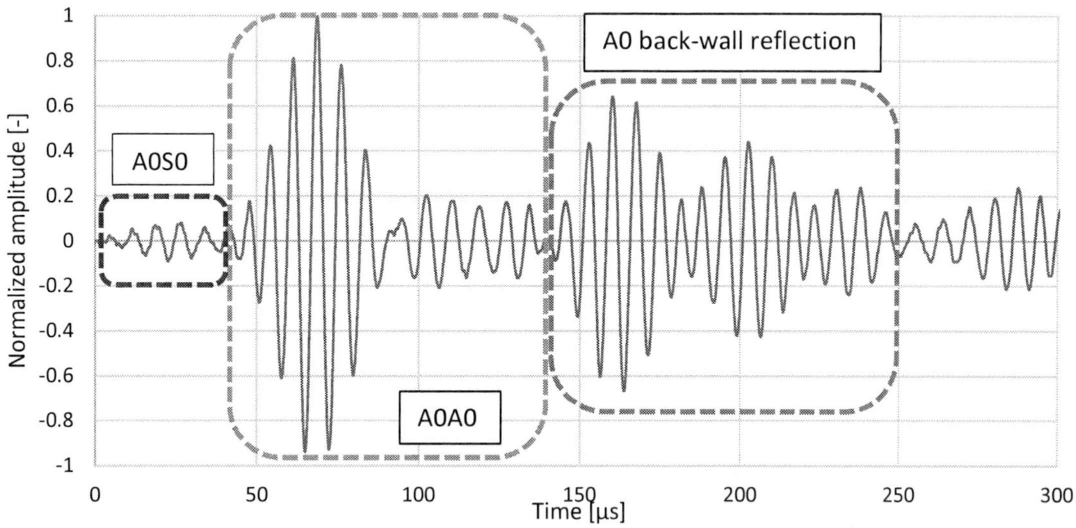

Fig.5 Time amplitude records of transmitted A0/S0 modes

The calculation process of the energy scattering coefficients includes the estimation of the energy related to the incident mode the value of the coefficient in itself is referred to. Therefore, a reference time amplitude record of the A0 mode (Fig. 6), which has propagated at the same path corresponding to the distance between the notch and the excitation point, has to be obtained.

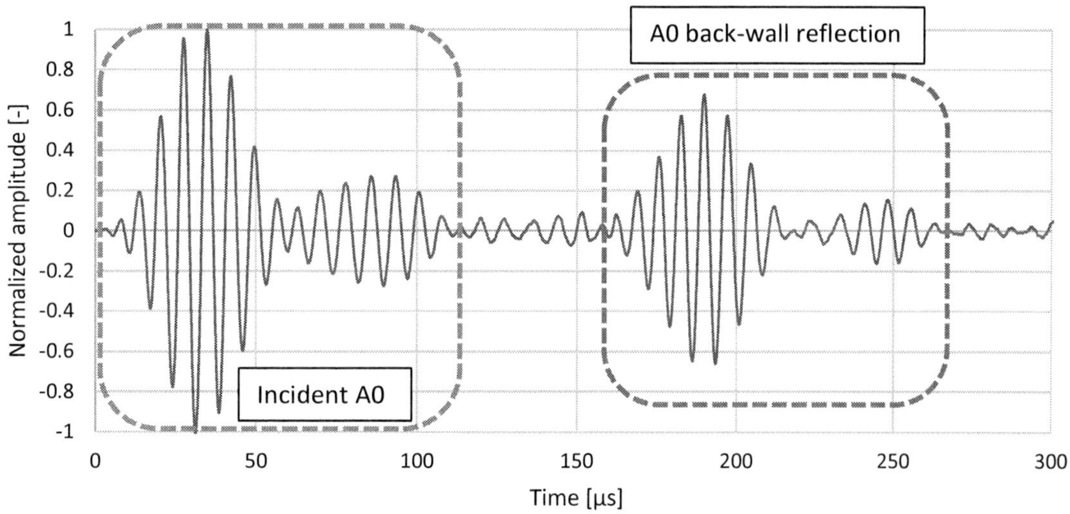

Fig.6 Time amplitude records of the reference A0 mode

After analysis and identification of the reflected and transmitted A0/S0 modes in the signal recordings (Figs. 4-6) were performed, a continuous wavelet transform using a Morlet mother wavelet function [9] was applied on the time amplitude records. In the following step, a projection of the desired part of the time-frequency spectrum belonging to the transmitted mode has to be performed into the frequency-max($c_{\mathrm{WT}}(t,f)$) plane , , where $c_{\mathrm{WT}}(t,f)$ expresses the continuous wavelet transform coefficient, which depends on time t and frequency f. The frequency-max(c_{WT}) representation can be further used for final calculation of the power of a particular mode using the Parseval's theorem [10]:

$$P = \int_{t_1}^{t_2} \left| c_{\mathrm{WT_{MAX}}}(t,f) \right|^2 \mathrm{d}t, \tag{2}$$

where $c_{\mathrm{WT_{MAX}}}(t,f)$ is the projected wavelet coefficient into the frequency-max(c_{WT}) plane and t_1 and t_2 express the time interval of occurrence of scattered Lamb wave mode, for which the time-averaged power has to be determined. It has to be noted that the wavelet coefficients displayed in Figs. 7 and 8 were normalized in order to achieve a better scale representation and has not been used for the previous computations.

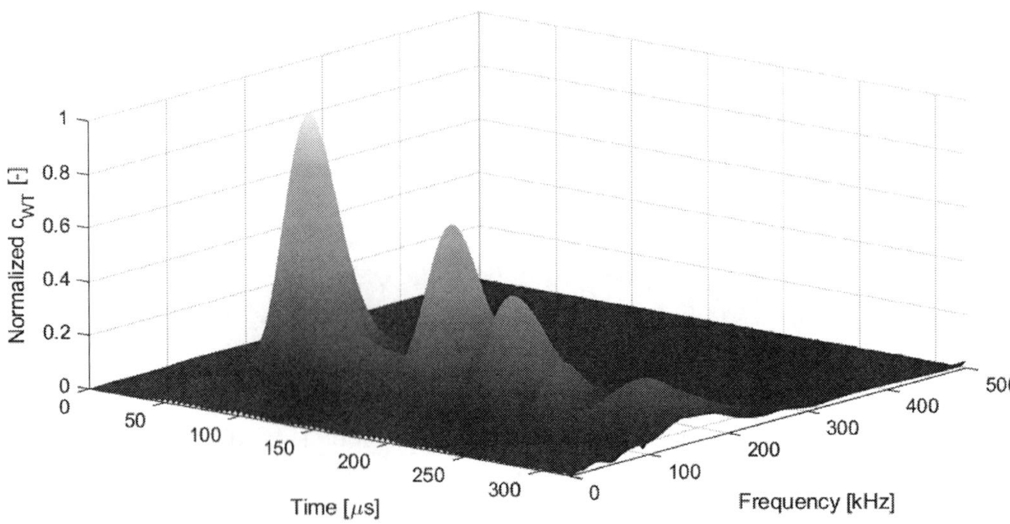

Fig.7 Calculated c_{WT} coefficients over the time-frequency domain in the case of the transmitted A0/S0 modes

The procedure of power calculation has to be carried out for each scattered mode ($S0_{ref}$, $S0_{trans}$, $A0_{ref}$, $A0_{trans}$) as well as for the incident A0 mode. Finally, the energy transmission/reflection coefficients can be computed as follows:

$$k_{\mathrm{S0ref}} = \frac{P_{\mathrm{S0ref}}}{P_{\mathrm{A0inc}}}, k_{\mathrm{A0ref}} = \frac{P_{\mathrm{A0ref}}}{P_{\mathrm{A0inc}}}, k_{\mathrm{S0trans}} = \frac{P_{\mathrm{S0trans}}}{P_{\mathrm{A0inc}}}, k_{\mathrm{A0trans}} = \frac{P_{\mathrm{A0trans}}}{P_{\mathrm{A0inc}}}, \quad (3)$$

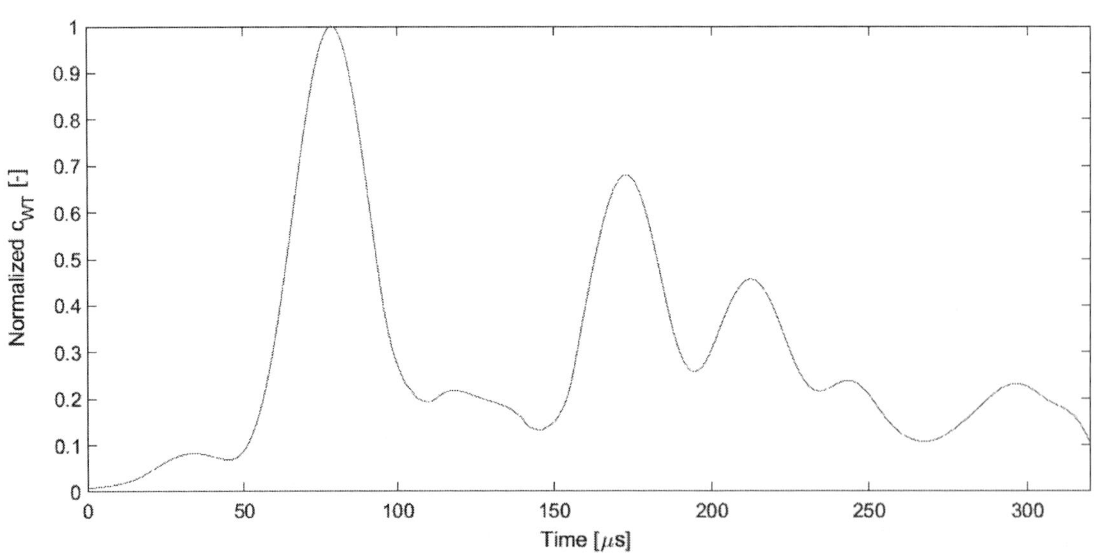

Fig.8 Normalized c_{WT_MAX} coefficients in the frequency-max(CWT coefficient) plane

where k_{S0ref} is the reflection coefficient of the reflected S0 mode, k_{A0ref} is the reflection coefficient of the reflected A0 mode, k_{A0trans} is the transmission coefficient of the transmitted A0 mode, k_{S0trans} is the transmission coefficient of the transmitted S0 mode, P_{S0ref} is the time-averaged power of the reflected S0 mode, P_{A0ref} is the time-averaged power of the reflected A0 mode, P_{A0trans} is the time-averaged power of the transmitted A0 mode, P_{S0trans} is the time-averaged power of the transmitted S0 mode and P_{A0inc} is the time-averaged power of the incident A0 mode.

Numerical Method for Scattering Coefficients Evaluation

Currently, there is a number of more or less complex methods for evaluation of scattering coefficients related to the interaction of Lamb waves with geometric discontinuities [1,6,8]. In this article, a modal extraction method based on an orthogonal property of Lamb modes has been used [8]. By means of combining the knowledge of the orthogonal property with information related to the through-thickness displacements and stress distribution, it is possible to derive the energy transmission and reflection coefficients for any present Lamb mode. A certain disadvantage lies in the need to determine the stress and displacement fields in the close vicinity of the discontinuity, for which the FEM method must be used in most cases [8]. For this case study, the stress and displacement fields were predicted using the COMSOL software. The FE simulation was realized in the frequency domain in order to obtain the above-mentioned through-thickness displacements and stress distribution in front of and above the discontinuity (vertical notch). The generation of the incident mode has been realized with the use of locally applied excitation of normal force with a Gaussian-windowed spatial distribution by the following function [8]:

$$F(x,f) = F_0 \mathrm{e}^{-ikx} \mathrm{e}^{\frac{A(\mathrm{x}-x_0)^2}{B^2}}, \tag{4}$$

where F_0 (set at 100 N) represents the amplitude of the excitation force, k is the wavenumber, x refers to the spatial coordinate, A (set at 30 mm) defines the width of the excitation zone, and B (set at 60 mm) defines its length. The wavenumber of the incident A0 mode is a function of material properties, thickness of the plate as well as a function of the frequency of the wave. The wavelet analysis showed that the frequency of the excited A0 mode propagating in unloaded aluminium plate with the thickness of 3 mm is equal to 137 kHz with respect to the above-mentioned excitation method, i.e. combination of the used transmitter and the driving signal. In the next step, the wavenumber k for the A0 mode of frequency-thickness value equal to 0.411 MHz·mm has to be determined for the material parameters explicitly expressed by means of transversal and longitudinal wave speeds. Figure 9 displays the scheme of the FE model the displacements and stress distribution were obtained from. The PML abbreviation expresses the Perfectly Matched Layer property, which is an artificial absorbing layer option available in the COMSOL software. The PML property enables us to minimize the dimensions of the model including the elimination of the back-wall reflection at the same time. After the data extraction, a method based on an orthogonal property of Lamb modes can be applied in order to evaluate the energy transmission and reflection coefficients.

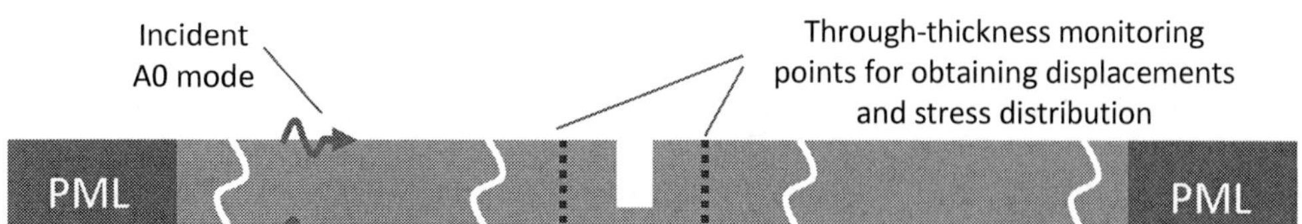

Fig.9 Scheme of the FE model for obtaining displacements and stress distribution

Results and Discussion

Two methods were used in order to evaluate the energy transmission and reflection coefficients in the case of Lamb wave scattering on a two-millimetre deep and two-millimetre wide horizontal notch milled in a three- millimetre thick aluminium sheet. As the incident wave, the A0 mode of Lamb wave was chosen especially for its sensitivity to such type of defects. The experimental method consisted of the measurement of the reflected and transmitted time amplitude records, which have been subjected to subsequent wavelet analysis in order to calculate the time-average power of the selected wave modes using Parseval´s theorem [10]. In order to prove the experimental results, a hybrid method incorporating the FE method and the modal extraction method based on an orthogonal property of Lamb modes were used. The results of both approaches are summarized in Table 1.

Tab.1 Measured and computed results of the transmission/reflection coefficients

Transmission/reflection coefficients	Experimental method incorporating wavelet transform	Hybrid method incorporating the FE method and modal extraction method
k_{S0ref} [%]	0.30	13.9
k_{A0ref} [%]	44.1	39.7
$k_{S0trans}$ [%]	0.50	16.0
$k_{A0trans}$ [%]	25.5	30.6
Energy balance [%]	70.4	100.2

It is obvious that the hybrid method exhibits perfect results in terms of energy balance criterion. The experimental approach, on the other hand, reports a loss of one-third of the energy mainly caused by almost zero values of scattering coefficients related to the S0 Lamb wave mode. The reason for such disagreement lies in the wave structure of A0 and S0 modes having the frequency·thickness parameter of 0.411 MHz·mm.

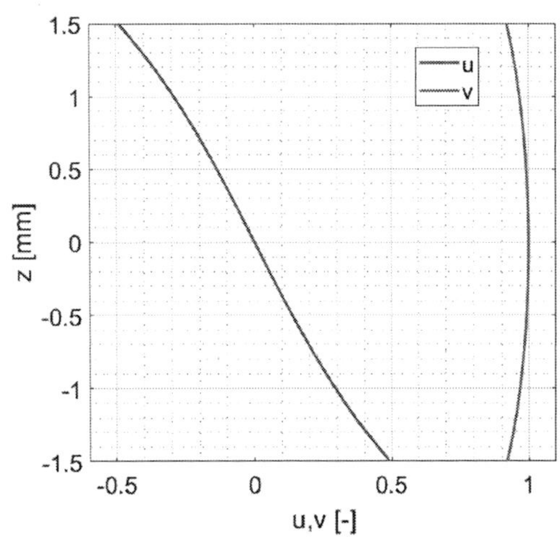

A0 mode

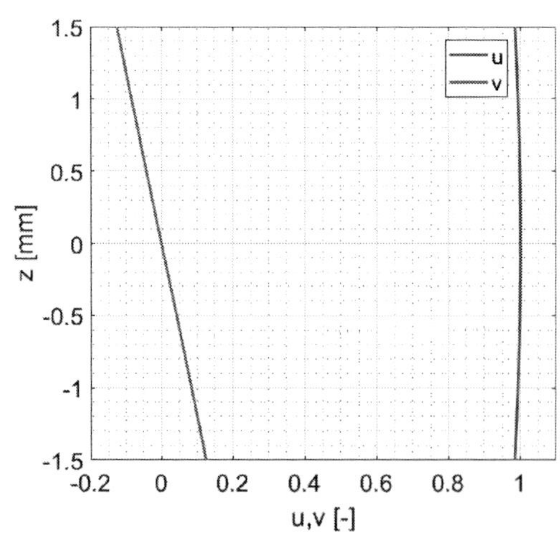

S0 mode

Fig.10 Normalized through-thickness displacements along x (u) and z axis (v)

Figure 10 displays the normalized through-thickness displacements u and v along x and z axis, respectively. The A0 mode of the frequency·thickness parameter of 0.411 MHz·mm shows relatively large values of through-thickness displacement in z direction compared to through-thickness in x direction, which is a favourable situation for standard contact piezoelectric transducers sensitive to longitudinal vibration. On the other hand, the S0 mode with the mentioned frequency-thickness parameter exhibits dominant displacements along x axis, while the displacement in z axis is more than five times smaller.

Therefore, the S0 mode will be obviously disadvantaged thanks to displacement ratio and its structure resulting in the predominant amplitudes of through-thickness displacement in x direction. This fact is, however, taken into account in the case of the modal extraction method, which incorporates both of the displacement components into its calculations. We can also notice that the values of transmission and reflection coefficients related to the A0 mode are very close to each other. The rate of the conformity between the experimental and numerical approaches therefore strongly depends on the types of Lamb modes they are working with.

Conclusion

The main aim of the proposed paper was to present the practical view on the usability of the experimental approach, which incorporates the wavelet transform as a tool for calculation of the scattering coefficients of low-order Lamb modes. The experimental approach has been tested on a simple vertical notch, and the results were then confronted with the hybrid method incorporating the FE method and the modal extraction method, which is based on an orthogonal property of Lamb modes. The mutual conformity between these two approaches depends mainly on the structure of considered wave modes the scattering coefficients are evaluated for. The experimental approach, however, proved to be a solid performer in terms of predicting the scattering coefficients for A0 mode. A deeper study of the mutual relationship between the particular low-order Lamb mode and the detectability for higher values of frequency·thickness product is also considered.

Acknowledgement

This work was supported by Specific Research (SP2018/63) and by The Ministry of Education, Youth and Sports from the National Programme of Sustainability II (LQ1602). The support is acknowledged. The authors would also like to thank the Starmans electronics Ltd. company for the provided ultrasonic probes and devices.

References

[1] B. Poddar, V. Giurgiutiu, Complex modes expansion with vector projection using power flow to simulate Lamb waves scattering from horizontal cracks and disbands, J. Acoust. Soc. Am. 140 (2016) 2123-2133.

[2] T.N. Grigsby, E.J. Tajchman, Properties of Lamb Waves Relevant to the Ultrasonic Inspection of Thin Plates, IRE Trans. Ultrason. Eng. 8 (1961) 26-33.

[3] F. Feng, J. Shen, J. Deng, Q. Wang, Analytical Solution of Lamb Wave Scattering at Plate End, Advanced Materials Research 199-200 (2011) 949-952.

[4] J.L. Rose, Ultrasonic Waves in Solid Media, Cambridge University Press, Cambridge, 2002.

[5] G. Shkerdin, C. Glorieux, Interaction of Lamb modes with an inclusion, Ultrason. 53 (2013) 130-140.

[6] F. Feng, J. Shen, S. Lin, Scattering matrices of Lamb waves at irregular surface and void defects, Ultrason. (2012) 760-766.

[7] D. N. Alleyne, P. Cawley, The Interaction of Lamb Waves with Defects, IEEE Trans. Ultrason., Ferroelectrics and Frequency Control 39 (1992) 381-397.

[8] L. Moreau, M. Castaings, B. Hosten, An orthogonality relation-based technique for post-processing finite element predictions of waves scattering in solid waveguides, J. Acoust. Soc. Am. 120 (2006) 611-620.

[9] H. Suzuki, T. Kinjo, Y. Hayashi, M. Takemoto, K. Ono, Wavelet Transform of Acoustic Emission Signals, J. Acoust. Em. 14 (1996) 69-84.

[10] L. M. Surhone, M. T. Tennoe, S. F. Henssonow, Parseval's Theorem, Betascript Publishing, 2010.

Substituting a Conventional Ram Frame with a Sandwich Structure

TANČIN Milan[1,a] and LIMBERG Luboš[1,b]

[1]University of West Bohemia, Univerzitní 22, 306 14, Plzeň, Czech Republic

[a]milax@rti.zcu.cz, [b]limbi@rti.zcu.cz

Keywords: Frame, ram, sandwich, simulation, modal analysis, static analysis

Abstract. The main goal is to determine if a sandwich structure (with steel sheets as outer plates and aluminium foam as filler) is suitable for a ram in an overhead gantry. The main expected features of the new design are fewer deformations, lower weight and natural frequencies than the current solution. A semi-circular sandwich structure in two thickness variants (V1) and (V2) will be implemented into the adjusted frame. It is also necessary to keep in mind the possibility of mounting the sandwich structure into the adjusted frame. Comparative analysis was done to determine if and how the new design is better than the current solution. The primary tools used for the comparison are modal analysis and static analysis for several states.

Introduction

Strong competition between machine tool manufacturers is the driving force behind innovations. Therefore it is necessary to develop and improve current solutions used in machine tools. One of the main component groups of machine tools are rams. The main feature of a ram is to hold the machining head which is connected to the ram by bolts on a circular flange and afterwards connected with other parts of whole machine tools. The slide is connected to the ram and performs horizontal transverse movements. Longitudinal movements are performed by a crossbeam which is placed on two supports which transfers all the loads to the base. Sleeve bearings connect the basic parts of the machine tool. Basic outer dimensions of the whole machine tool are (l x w x h) 10100 mm x 5280 mm x 7112 mm. Dimensions of the fixing desk are 6200 mm x 2670 mm.

Fig.15 Overhead gantry with ram (in yellow)

The ram performs vertical movements in a defined range. Furthermore, the ram is another part of a machine tool for which it is possible to use the advantages of sandwich materials, such as high bending stiffness, low weight and vibration damping. Several approaches for implementing sandwich materials will be shown. The approaches will be verified by simulations, which will be performed on a conventional frame and on the newly designed frame.

Material selection for designing a sandwich structure

A sandwich material composed from two steel sheets as outer plates and aluminium foam filling was selected for the design. Steel sheets were selected considering the connection with the rest of the frame by various methods including welding. The main advantages of aluminium foam are high stiffness, low density, high toughness, corrosion resistance and very good vibration damping. Aluminium foam can be used not only for sandwich structures but also for filling more complex products such as cast engine brackets.

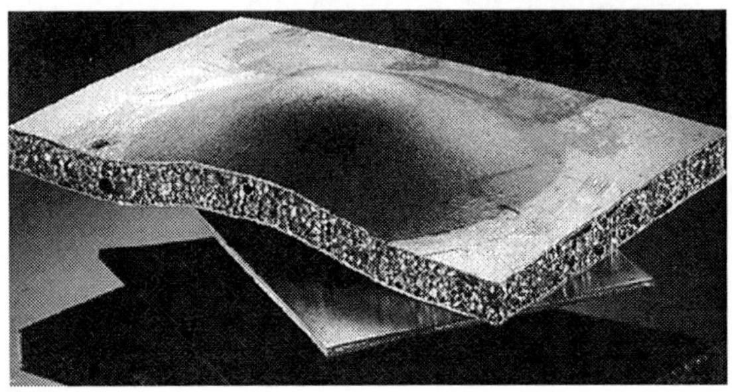

Fig.16 Sandwich material composed from two steel sheets and aluminium foam [2]

Foam can be prepared in several ways. It is a very porous material and its structure is similar to natural structures such as bone and coral. Cellular structures have relatively high stiffness. In comparison with other metal foams aluminium foam has many advantages, as mentioned above. The principle is to create bubbles in the melt and then cool the melt. There are two main types of alloys for manufacturing the foam – alloys that are suitable for casting and alloys that are suitable for forming. The first have smaller, irregular shaped pores, the walls are thin and the structure is non-uniform. Alloys suitable for forming create foams with spherical pores and the walls are thicker. Bubbles can be created as a result of melting the semiproduct, which contains a frother. Another way is to blast gas into the melt from an external gas reservoir. Thermal decomposition of added frother in the melt is also possible. When the melt has low viscosity, the resultant foam contains bigger spherical pores, which rise to the surface. The tendency is to create smaller pores. Manufacturers of aluminium foams include LKR, Cymat and Shinko Wire Company.

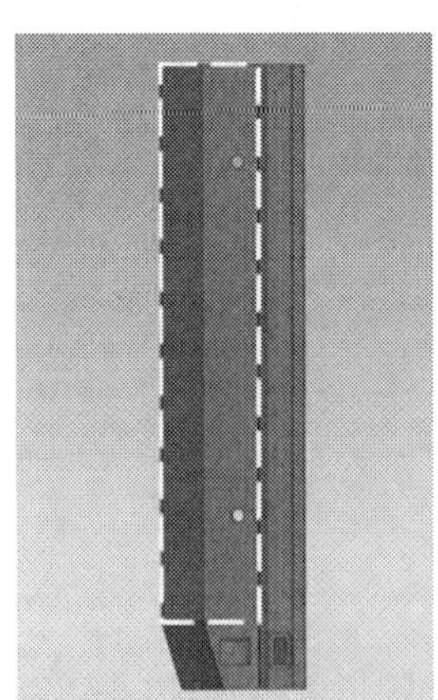

Fig.17 Conventional ram with area to be replaced by sandwich structure (white dashed line)

Principle of substitution

The objective is to use sandwich material in the area shown in **Chyba! Nenalezen zdroj odkazů.**. The area is composed from inner and outer steel sheet envelopes which are connected to each other by ribs with various thickness. The weight of the existing frame is 2640 kg. The dimensions of the frame are (l x w x h) 705 mm x 620 mm x 3000 mm. The new design of the frame should preserve these dimensions to ensure the simple replacement of the conventional frame with the new
design, which is without the struts and instead has a sandwich panel. The new design is adjusted in the modified area for placing the sandwich structure in grooves. The sandwich panel is fixed above by a steel element with the same groove. The adjusted frame, sandwich panel and upper element are connected to each other by welding. This is shown in yellow in **Chyba! Nenalezen zdroj odkazů.**.

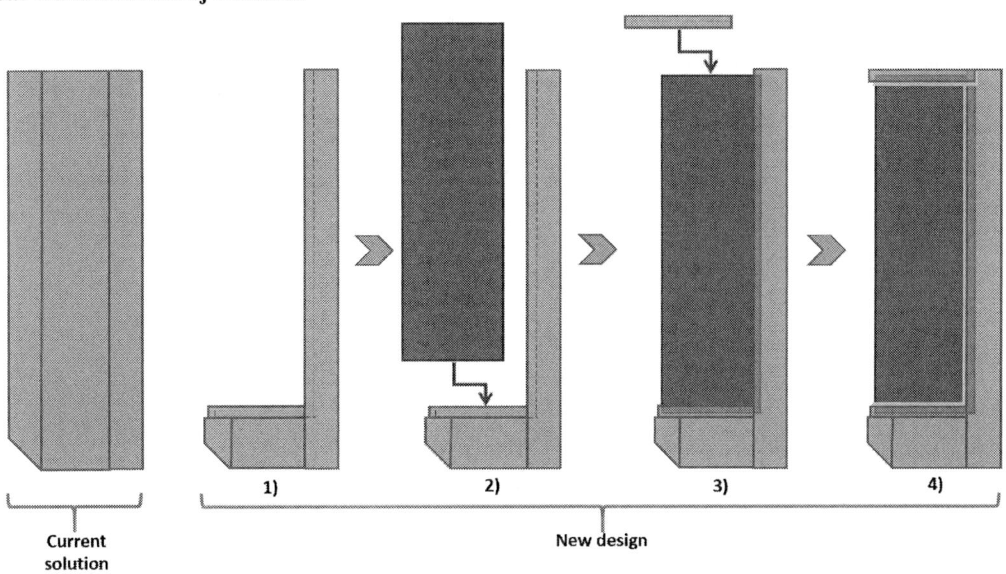

Fig.18 Diagram of current solution and new design including manufacturing steps

The sandwich panel is composed from two bent steel sheets as outer plates and filled with aluminium foam. The new design is slightly different from the current state and its profile is semi-circular. This profile is good for manufacturing because there are no other connections. The sandwich plate is completely implemented as one part. Small holes and hollows are not taken into consideration during the following simulations due to their small influence on the final results. Two versions of the new design are tested – V1 and V2. These variants have different layer thicknesses. V1 is (outer plate – filler – outer plate) 5mm – 10mm – 5mm. V2 is (outer plate – filler – outer plate) 12mm – 26mm – 12mm. Total wall thickness in the current state is 85mm, so wall thicknesses are about 76 % (V1) and 41% (V2) thinner.

Preparing the models of conventional and new frames

Conventional frame

First, modal analysis and static analysis were performed and these results are compared with the new designs. CTETRA4 with element size 25mm with the possibility of local refinement was selected as the mesh. For modal analysis, the model was not fixed and not loaded. Static simulation was carried out for minimum and maximum overhang and with a load (1000N) on two axes in the horizontal plane.

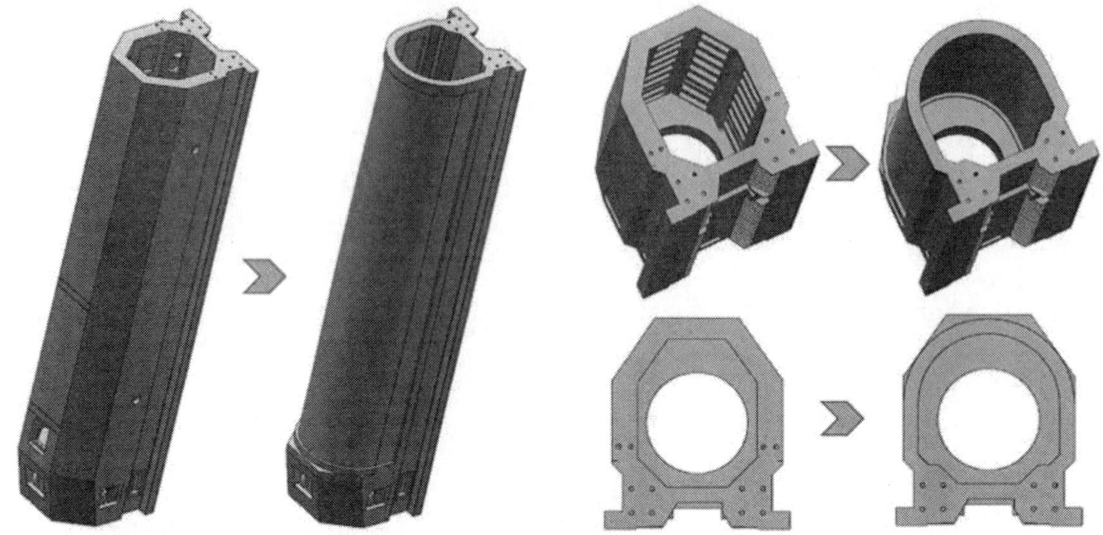

Fig.19 Comparison of current and new frame

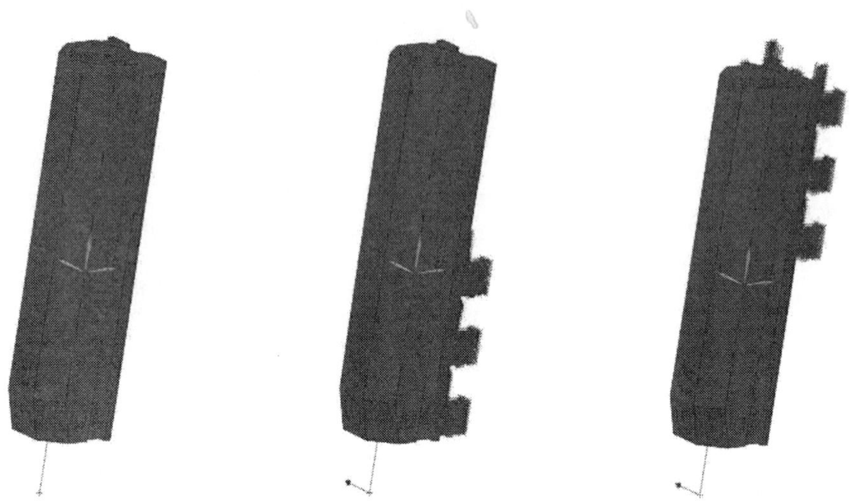

Fig.20 Computational models of conventional frame. (Left) modal analysis; (middle) static analysis - low overhang; (right) static analysis - high overhang

New design – V1 and V2

The sandwich material was defined by a laminate modeller (Siemens NX), which means that the sandwich structure was modelled as two 2D faces. First, both 2D faces ware meshed by a 2D dependent mesh with 25mm elements to ensure the defined layer composition between those two faces is filled. Thicknesses of layers were modelled in two variants – thinner (V1) and thicker (V2). V1 is composed of 5 mm steel sheets for the outer layers and 10 mm aluminium foam (Density 400 kg/m^3, Young´s Modulus 2374 MPa, Poisson´s Ratio 0.29) as filler. V2 is composed from 12 mm steel sheets for the outer layers and 36 mm of aluminium foam. The rest of the frame was adjusted by cutting and removing the conventional struts in the modified area. The grooves and the upper element, mentioned above, were modelled in this area. All these adjustments create one integral part. The outer steel sheets of the resultant sandwich panel are connected to the adjusted frame by Surface-To-Surface Gluing. This is a simplifying element which substitutes for the connection by welding.

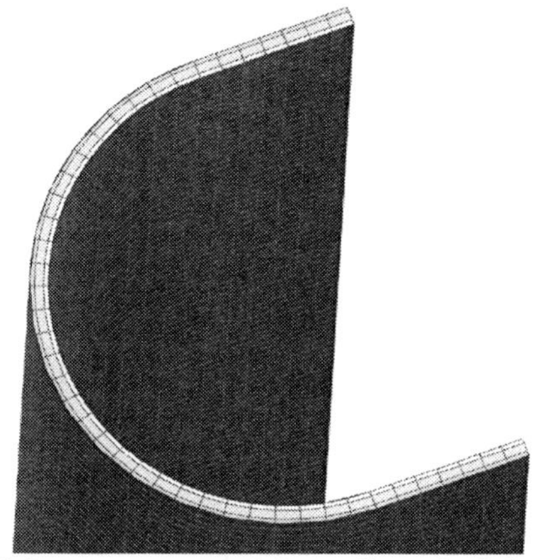

Fig.21 Layer composition of sandwich panel

Figure 22 Computational models of V1 and V2 frames. (left) modal analysis; (middle and right) static analysis - low overhang and high overhang

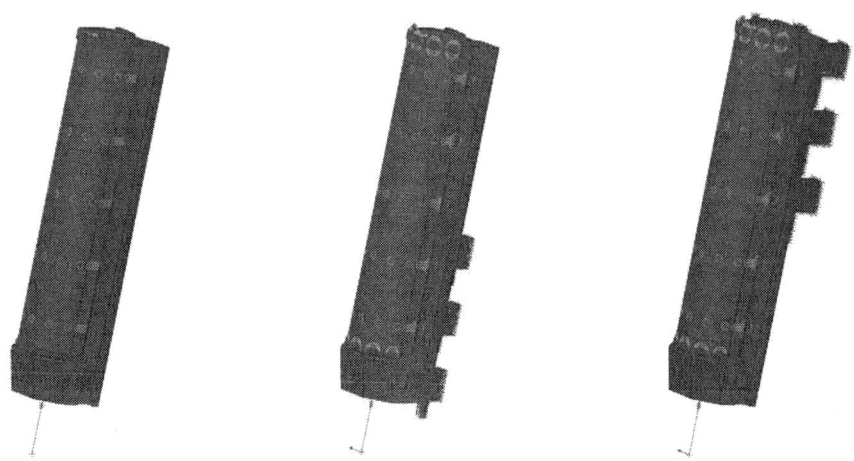

Fig.23 Computational models of V1 and V2 frames. (left) modal analysis; (middle and right) static analysis - low overhang and high overhang

Results

The results of the modal and static analysis are used as reference results for the following evaluation of the new designs. Fig.24 shows the first four natural frequencies and Fig.26 shows the results of static analysis. The same results are also shown for the modified variant V2 - Fig.25 shows natural frequencies and Fig.27 shows static analysis. Only V2 was selected because its results are more similar to the current solution. The first 10 natural frequencies for each design (current frame, V1 and V2) were simulated and four loading states were prepared for each design. In general, the resultant natural frequencies of modal analysis for V1 have lower values than the conventional frame over the whole range. The deformations of V1 are better only for a low overhang. The high overhang shows bigger differences and the conventional frame has less deformation for this state. The main advantage of V1 is that it is about 25% lighter. Variant V2 has modes of modal analysis almost the same as the conventional solution. This is a desired feature. In general, modes of modal analysis for V1 have lower values than the conventional frame over the whole range. Deformations of V1 are better for low overhang. High overhang shows bigger differences and the conventional frame has lower deformations for this state. The main advantage of V1 is that it is about 25% lighter. Variant V2 has modes of modal analysis almost the same as the conventional solution. The biggest difference is in mode 6, which is about 17%. Deformations are lower from 1 to 13% for all states. It is more than 10% lighter. In absolute terms, this means a saving of 266 kg.

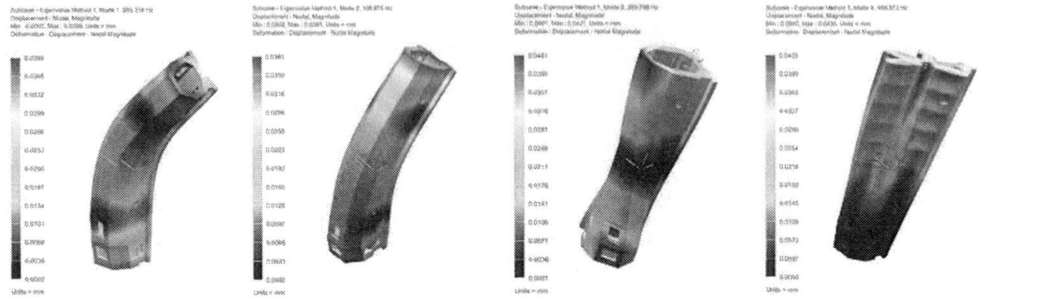

Fig.24 Results: Modal analysis of conventional frame with directions of first 4 natural frequencies

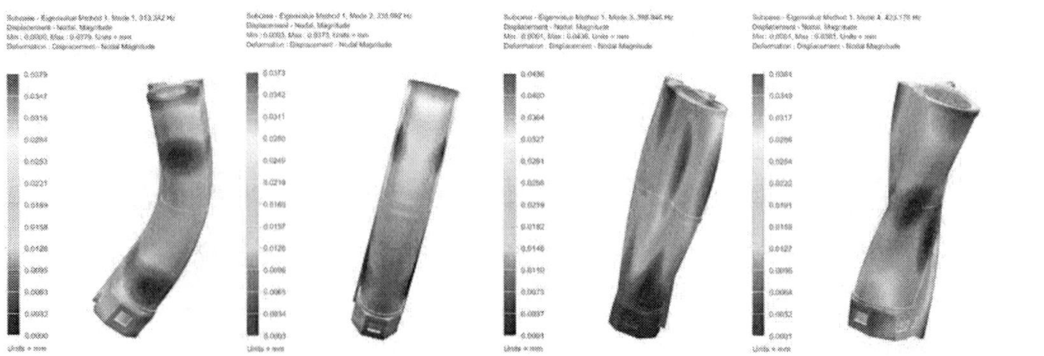

Fig.25 Results: Modal analysis of new design V2 with directions of first 4 natural frequencies

Tab.18 Comparison of results – natural frequencies

	CURRENT SOLUTION	V1		V2	
MODE	**VALUE (HZ)**	**VALUE (HZ)**	**DIFFERENCE (%) (V1 VS. CURRENT)**	**VALUE (HZ)**	**DIFFERENCE (%) (V2 VS. CURRENT)**
1	1. 303.3	2. 277.9	**3. -8.4**	4. 313.3	**5. 3.3**
2	6. 338.9	7. 311.8	**8. -8.0**	9. 338.1	**10. -0.2**
3	11. 389.8	12. 323.6	**13. -17.0**	14. 399.0	**15. 2.4**
4	16. 489.6	17. 359.8	**18. -26.5**	19. 423.2	**20. -13.6**
5	21. 535.5	22. 363.7	**23. -32.1**	24. 462.9	**25. -13.6**
6	26. 566.1	27. 407.3	**28. -28.1**	29. 468.0	**30. -17.3**
7	31. 602.5	32. 426.3	**33. -29.2**	34. 524.5	**35. -12.9**
8	36. 623.9	37. 532.9	**38. -14.6**	39. 612.5	**40. -1.8**
9	41. 652.4	42. 545.8	**43. -16.3**	44. 643.1	**45. -1.4**
10	46. 712.3	47. 664.6	**48. -6.7**	49. 671.9	**50. -5.7**

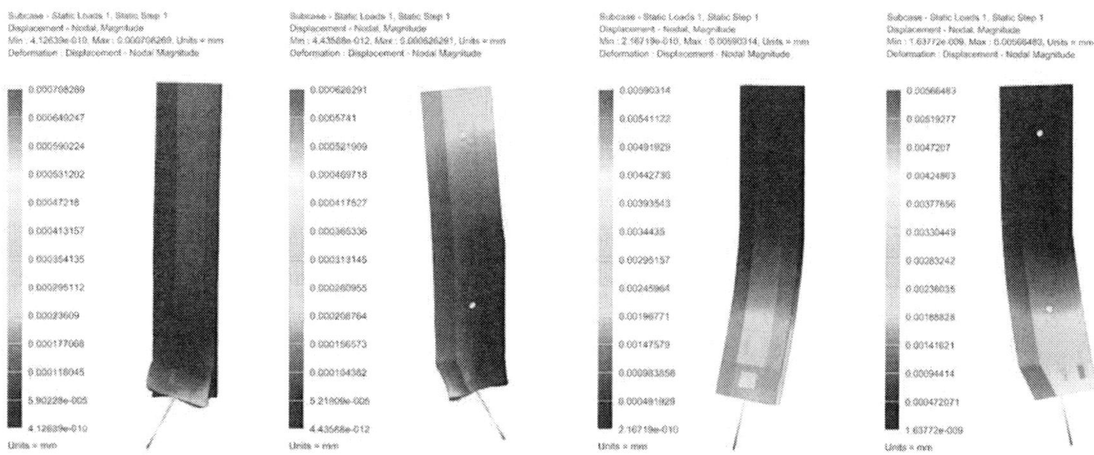

Fig.26 Results: Static analysis of conventional frame. From left: 1) low overhang, load in X direction, 2) low overhang, load in Z direction; 3) high overhang, load in X direction; 4) high overhang, load in Z direction

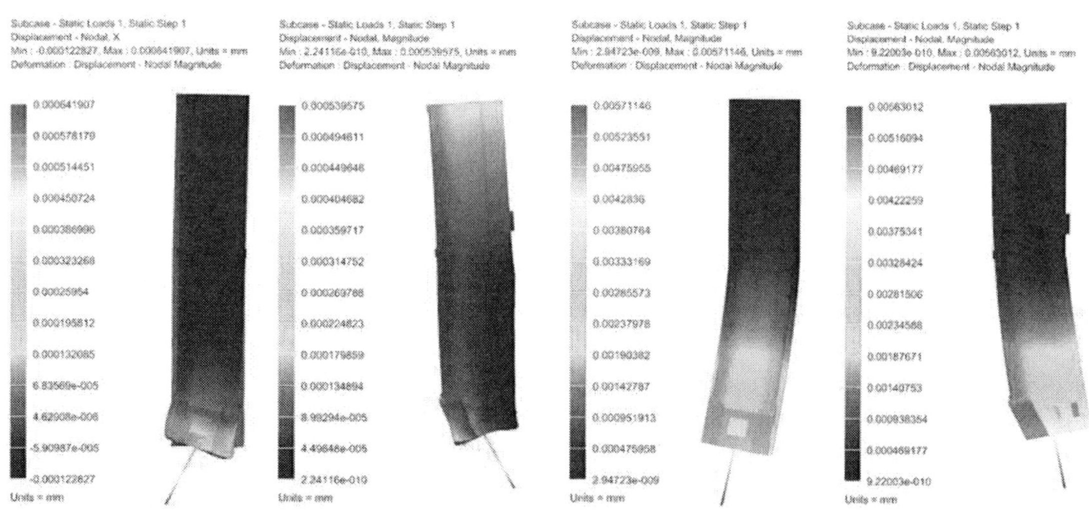

Fig.27 Results: Static analysis of new design V2. From left: 1) low overhang, load in X direction, 2) low overhang, load in Z direction; 3) high overhang, load in X direction; 4) high overhang, load in Z direction

Tab.19 Comparison of results - static analysis and weight

	MINIMUM OVERHANG (mm) Max Deformation X	MINIMUM OVERHANG (mm) Max Deformation Y	MAXIMUM OVERHANG (mm) Max Deformation X	MAXIMUM OVERHANG (mm) Max Deformation Y	Total weight (kg)
Conventional frame	0.000708	0.000626	0.005903	0.005665	2641.1
V1	0.000688	0.000622	0.006593	0.007523	1949.4
V2	0.000642	0.00054	0.005711	0.00563	2375.4
DIFFERENCE (%) Conventional vs. V1	**-2.8**	**-0.6**	**11.7**	**32.8**	**-26.19**
DIFFERENCE (%) Conventional vs. V2	**-9.3**	**-13.7**	**-3.3**	**-0.6**	**-10.06**

Conclusion

This article deals with the comparison of a conventional ram frame and an alternative construction, which is composed of sandwich material. Two variants of the new design with aluminium foam filler were simulated. Simulations confirmed that this material is useable for similar operations as the conventional ram. Substitution of the conventional design in the ram with similar natural frequencies and lower weight and deformations is possible by using variant V2. The method is also applicable for previously manufactured conventional frames which can be also adjusted as mentioned above. In general terms, the method cannot be prepared without simulations of a virtual frame. Substitution cannot be based only on estimation. There are still opportunities for optimization. The frame could be reinforced by several struts designed around the sandwich panel which would mean that the steel panels of the sandwich panel would be thinner, ensuring lower final weight. The next step is to compare other shapes of sandwich plates with the synergy of struts to reinforce the structure and investigate these influences on the results. Another way is to replace the material of the outer plates. Aluminium sheet can be used instead of steel sheet. In this case, connection by bolts or glue is better. Investigations will mainly focus on the reduction of weight, resultant deformation, preservation of current natural frequencies and manufacturing costs. The optimized ram will also be tested on a whole machine tool. Other parts of the machine tool will be the same. The results from this simulation will also be dependent on the correct design of the slide assembly. These results will be more important because the new design should be usable mainly for a complete machine tool and its cutting conditions.

Acknowledgments

We would like to thank mainly our supervisor for her valuable advice, and also our colleagues.

References

[1] Nová, Iva a Machuta, Jiří. Výrobky z hliníkové pěny. *MM Spekturm.* Inovace, 18. 12 2007, 12, str. 58.

[2] Banhart, John. Manufacture, characterisation and application ofcellular metals and metal foams. *Progress in Material Science.* 2001, 46, str. 578.

[3] *Materials engineer / Matwriálový inžinier.* místo neznámé : Peter Oslanec - Entos, 2009. ISSN 1337 - 8953.

Effect of Cement/Water Ratio on Mechanical Properties of Cement Pastes Containing Recycled Micronized Concrete Powder and Chemical Additives

TESÁREK Pavel[1,a], PROŠEK Zdeněk[1,b], RYPAROVÁ Pavla[1,c], HRŮZA Jaromír[1,d], NEŽERKA Václav[1,e] and FLÁDR Josef[1,f]

[1] Faculty of Civil Engineering, Czech Technical University in Prague, Thákurova 7, 166 29 Prague, Czech Republic

[a]tesarek@fsv.cvut.cz, [b]zdenek.prosek@fsv.cvut.cz, [c]pavla.ryparova@fsv.cvut.cz, [d]jaromir.hruza@fsv.cvut.cz, [e]vaclav.nezerka@fsv.cvut.cz, [f]josef.fladr@fsv.cvut.cz

Keywords: Micronized Recycled Concrete, Water/Cement Ratio, Recycling, Cement Pastes, Mechanical Properties.

Abstract. Cement pastes containing recycled concrete powder (RCP) from an old road drainage channel in 3/2 weight ratio were prepared and their basic and mechanical properties macroscopically tested. These were in particular composed of standard Portland cement CEM I 42.5 R (locality Radotín, Czech Republic), a micronized road drainage channel, and additives to improve workability. Namely, we used one plasticizer and two different solidification accelerators. The main goal of this research was to reduce the amount of kneading water while ensuring sufficient workability of the fresh cement pastes. The performance of individual additives was assessed in terms of selected material parameters as bulk density, being an indicator of weathering resistance and durability, and mechanical strength (compressive strength and bending strength) of hardened 28 days old specimens. To that purpose, a comparison with the reference, untreated pure cementitious paste, was done. The results indicate that additives should assist the production RCP-rich cementitious composites to avoid compromising their mechanical performance.

Introduction

Incorporation of recycled cement powder (RCP) into fresh concrete mixes leads to higher demands on kneading water and the higher water-to-cement ratio (*w/c*), resulting in deterioration of mechanical properties. However, methods to reduce the needed amount of water in fresh concrete exist. Soares et al. [1] investigated two types of aggregate to – natural and recycled of high quality. The concrete mixture containing recycled aggregate was modified by plastificators or superplastificators. It was shown that such a modified mix exhibited up to 40 % higher compressive strength, compared to a reference mix. The *w/c* resulted in the formation of a matrix having higher density, rendering the mix more durable and less susceptible to weathering. Pereira et al. [2] designed several concrete mixes with concrete recycled aggregate and outlined a self-cleaning system for such aggregates. They found that newly make the concrete with such a recycled aggregate exhibited the same mechanical compressive strength and mechanical strength as one with natural (extracted/quarry) aggregate. Çakır [3] modified properties of concrete mixes with the use of natural and recycled aggregate with other additives, such as microsilica or slag. To improve workability and reduce the amount of kneading water in the fresh mix, he used superplastificator. Some specimens containing recycled components exhibited better durability than those containing only natural aggregate. However, their mechanical strength was compromised. The same effect as plastificators or superplastificators can also have a micronized marble or rock powder. However, the upper limit of applications is 10 % of weight additions, since higher amounts have a negative effect on mechanical properties of the cementitious composites [4, 5]. These findings and outcomes of other studies were exploited when designing the cementitious mixes with micronized concrete recycled materials for applications in the construction industry in different ways, e.g., as interior or exterior plasters, masonry blocks or pastes [6, 7, 8].

Materials and Samples

Pastes made of Portland cement (C) or Portland cement with RCP mixed were prepared. The cement CEM I 42.5R was produced in Radotín, Czech Republic, while RCP was prepared by micronizing an old road drainage channel. A high-speed mill manufactured by LAVARIS Ltd. was used for micronizing. To enhance the properties of the pastes containing RCP, we used one plasticizer (P3) and two distinct solidification accelerators (U1 and U2). These additives were carefully chosen from 3 distinct plastificators / superplastificators and three distinct accelerators of hardening – these acts also as plastifying agents.

The experimental samples were denoted by the code name "XX YY," which included a type of a binder, respectively samples with and without the micronized concrete material (RCP), and type of used additives. The samples with cement CEM I 42.5R without RCP were denoted as "C" and ones with cement and RCP with a weight ratio 60/40 (cement/RCP) were denoted as "CR." These both materials were used as reference material, and they contained no additives. The water/cement ratio for samples C was 0.29 and for CR was 0.31, a higher ratio of water/cement for sample CR is due to added RCP. Other samples contained additives. The water/cement ratio for samples CU1 and CU2 was 0.29; both samples contained added additives, in this case, 1.5 wt. % of solidification accelerators from the cement of cement/RCP. The water/cement ratio for samples CRU1 and CRU2 was 0.31; both samples contained the same values od additive as CU1 and CU2. For samples, CP3 and CRP3 with plasticizer were the reduced water/cement ratio 0.21, respectively 0.25. Amount of additives (g) was 1.0 wt. %. Values of bulk density depended on added RCP and a type of used additives; the higher values were achieved for samples without RCP, namely 2054 ± 28 kg/m^3 for CP3. For other samples from this group were the values in a ranking from 1923 ± 31 to 1962 ± 22 kg/m^3. The bulk density of the sample with cement and RCP was 1749 ± 28 to 1768 ± 18 kg/m^3.

Experimental Methods and Results

The six prismatic 40×40×160 mm specimens were prepared to represent each sample. The specimens were stored at standard laboratory conditions with temperature 22 ± 1 ° C and relative humidity 50 ± 2 %. The 28-days old specimens were tested.

Bulk density was calculated from the measured dimensions and weight of the individual samples, resulting values were averaged from six specimens. Compressive strength was calculated from the contact stress on measured samples, where the load area of the press has dimensions of 40 × 40 mm. From the resulting force and contact stress on the samples, the compressive stresses in the samples were calculated. Twelve samples (12 sample halves) were tested for each set. The bending strength, three-point configuration, was calculated from the exact dimensions of each sample measured immediately before the destructive test. The strength is calculated from a linear dependence of the bending moments, which are based on the strength at the breach limit, and the section modulus, which is based on accurately measured dimensions of the samples.

Comparison of bending and compressive strengths is presented in Figure 1, including standard deviations of the measured values. From the obtained results it is obvious that the plasticizer P3 has a positive impact on bending strength when compared to the reference cement paste samples without any additives. The effect of the solidification accelerators was found similar. There was a slight increase in bending strength compared to the reference mix. On the other hand, the increase of compressive strength was observed only in the case of plastificators; the use of accelerators resulted in 50% reduction of compressive strength compared to the reference mix.

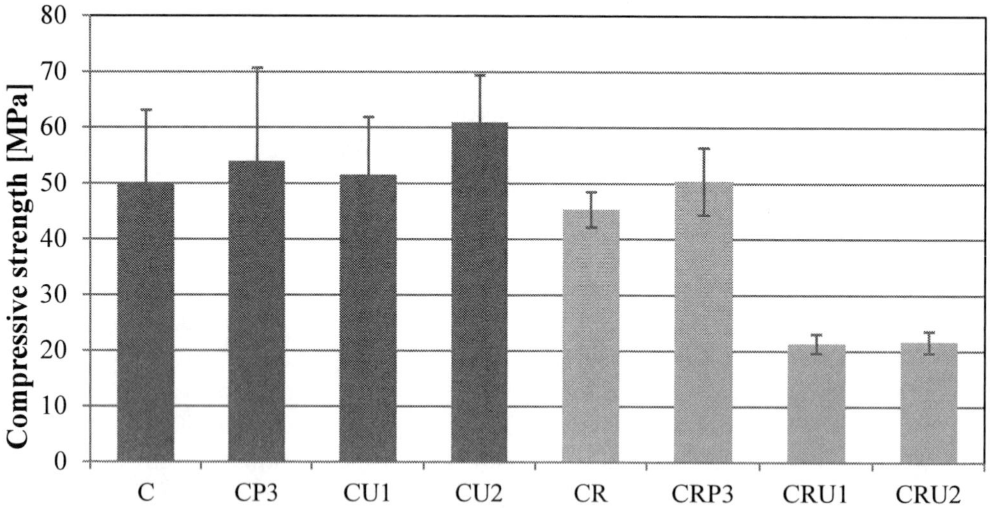

Fig.1 Comparison of bending strength (with indicators of standard deviations)

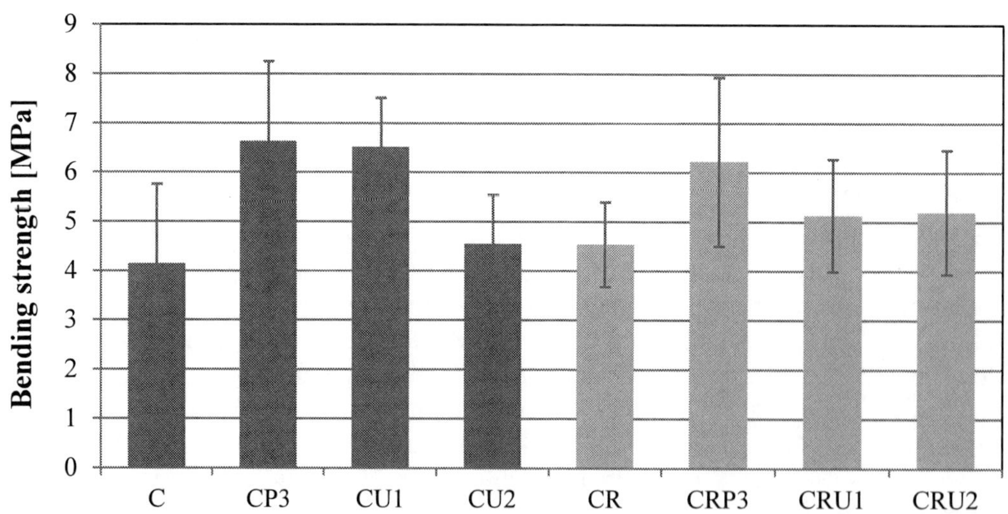

Fig.2 Comparison of bending strength (with indicators of standard deviations)

Conclusions

The results indicate the RCP can be efficiently utilized in the production of concrete, making the concrete construction more sustainable. However, agents improving the workability of the mixes must be used to minimize the needed amount of kneading water. The agents must be selected carefully. The use of suitable plastificators results in a significant reduction of *w/c*, making the cementitious matrix denser and the composite stronger. This is reflected mainly in bending strength. The following research will be focused on the study of the microstructure. To that purpose, scanning electron microscopy combined with nanoindentation will be addressed. Such approach will allow establishing the impact of micronized concrete on macroscopic properties of the cementitious composites [9, 10].

The presented results will be used for calibration of a new micromechanical model. A comprehensive study will be consisting of a microstructure investigation, nanoindentation, and micromechanical modeling was performed to identify phases and assess their impact on macroscopic properties of cement pastes containing waste marble powder. The proposed model, built on the Mori-Tanaka scheme, will be used to estimate the effective Young's modulus and compressive strength at a low computational cost. After model validation, the effects of an interfacial transition zone (ITZ) and increased porosity of marble powder rich-pastes were quantified. The study revealed a fundamental role of the ITZ formed around stiff marble powder inclusions, responsible for performance deterioration. These findings indicate that elimination of the ITZ and reduction of porosity would considerably enhance the strength of cement-based materials containing waste marble powder. As a consequence, larger amounts of waste marble could be incorporated as a cement replacement without sacrificing structural performance.

Acknowledgment

This research was supported by the GA ČR research a project under the number 17-06771S and by Czech Technical University in Prague – project under the number SGS16/201/OHK1/3T/11. The authors thank the Center for Nanotechnology in Civil Engineering at the Faculty of Civil Engineering CTU in Prague, and LAVARIS Ltd for providing the micronized concrete.

References

[1] D. Soares et al., Use of coarse recycled aggregates from precast concrete rejects: Mechanical and durability performance, Construction and Building Materials 71 (2014) 263-272.

[2] P. Pereira et al., The effect of superplasticizers on the mechanical performance of concrete made with fine recycled concrete aggregates, Cement & Concrete Composites 34 (2012) 1044-1052.

[3] Ö. Çakır, Experimental analysis of properties of recycled coarse aggregate (RCA) concrete with mineral additives, Construction and Building Materials 68 (2014) 17-25.

[4] Z. Prošek et al., Utilization of the waste from the marble industry for application in transport infrastructure: Mechanical properties of cement pastes, IOP Conference Series: Materials Science and Engineering 236 (2017) 6 p.

[5] P. Tesárek et al., Effect of a plasticizer and micronized stone powder on mechanical properties of a cement matrix. in: Experimental Stress Analysis 2016, Plzeň 2016, pp. 387-390.

[6] M. Lidmila et al., Mechanical Properties of Recycled Binder/Micro-Filler Cement-Based Material, Advanced Materials Research 2014, 1054 (2014) 234-237.

[7] M. Lidmila et al., Utilization of Recycled Fine-Ground Concrete from Railway Sleepers for Production of Cement-Based Binder, Appl. Mechanics and Materials 486 (2014) 323-326.

[8] M. Lidmila et al., Composite Material Based on Cement and PVA: Evolution of Mechanical Properties during First 28 Days, Advan. Mat. Res. 1054 (2014) 215-220.

[9] V. Nežerka et al., Investigation of crushed brick-matrix interface in lime-based ancient mortar by microscopy and nanoindentation, Cement & Concrete Composites 55 (2015) 122-128.

[10] P. Tesárek et al., Microstructural and micromechanical study of gypsum, Chemicke listy 105(S) (2011) S852-S853.

EAN 2018
56th conference on experimental stress analysis

Assessment and Increasing of Lifetime and Time of Safe Operation of Machines and Equipment by the Experimental Methods of Mechanics

TREBUŇA František[1,a], PÁSTOR Miroslav[1,b] and ŠARGA Patrik[2,c]

[1] Technical University of Košice, Faculty of Mechanical Engineering, Department of Applied Mechanics and Mechanical Engineering, Letná 9, Košice, 042 00, Slovakia

[2] Technical University of Košice, Faculty of Mechanical Engineering, Department of Mechatronic, Park Komenského 8, Košice, 042 00, Slovakia

[a]frantisek.trebuna@tuke.sk, [b]miroslav.pastor@tuke.sk, [c]patrik.sarga@tuke.sk

Keywords: experimental methods, lifetime, reliability, safe operation, residual stresses, deformation and stress analysis.

Abstract. Failures of machines and equipment are mostly caused by crossing the critical states that are defined by limit values of stresses and deformations in locations of critical loading. The paper is focused on the development of methods for the analysis of failure cases in the supporting elements of mechanical systems by the quantification of strain and stress fields using experimental and numerical methods of mechanics. The suggested treatments will be used for the reliability and residual life assessment of machines and equipment.

Introduction

Despite of fact that over the last decades' numerical methods (mainly finite element and boundary element method) have been widely spread for the solution of mechanical and mechatronic systems, computer solutions are improved on the basis of the experimentally obtained results and modelling techniques. Experimental methods of mechanics on the one hand allow to specify in more details the parameters being input into computations (e.g. time-dependent charts of load forces, influence of boundary conditions). On the other hand, application of modern experimental methods (electronic speckle interferometry (ESPI) [1], digital image correlation (DIC) [2-5], reflection photoelasticimetry (Photostress method) [6,7] and so on) allow to analyse displacements and stresses on the surface of machine parts in real conditions of loading and positioning. However, there exist areas of mechanics (e.g. area of residual stress determination), where experimental treatments are dominant and necessary for evaluation of real stress and deformation states in the supporting elements of machines and equipment [8].

The paper presents applications of methods of experimental mechanics for assessment and increasing of lifetime and safe operation of machines and equipment, where beside operational loading important roles play residual stresses created in the structure or its part due to production technology, assembling, joints, temperature influences, or as a result of their overloading. The residual stresses are superposed to the stresses resulting from actual loading of a structure and they can significantly influence the resulting stress state. This fact can have significant influence on the lifetime of mechanical parts of machines and equipment, as well as mechanical and biomechanical systems, therefore, the knowledge of residual stresses is important for the assessment of actual state of the investigated element. The authors describe methods of their measurements, including quantification and including solution of real applications of mechanical, mechatronic and biomechanical systems. The important advantage of the authors' workplace is the possibility to verify the results obtained from numerical simulations using several experimental methods (strain-gages, DIC, PhotoStress, ESPI, fatigue tests, ...). The authors' workplace is focused on the development of experimental methods of mechanics and their application to the solution of scientific and research tasks and tasks for practice. Next are given selected tasks solved for practice, the aim of which was to assess the lifespan and reliability and safe operation of the given equipment.

Analysis of probable reasons of cracks initiation on a barking drum

A barking drum serves for debarking of logs with length varying from 2.5 m to 6.0 m [9]. The apparatus consists of a drum body, driving and supporting tyres, housing and exhausting system. In the drum, wood logs up to 700 mm in diameter can be processed, maximum capacity of the equipment is 320 m^3 per hour, and angular velocity of the drum during debarking is about 5 rpm [9]. The body of the drum is designed as a cylindrical shell and consists of two parts (front and rear, Fig. 1), which are driven by a system of tyres (Fig. 2) with drives controlled by frequency changers. Rotation of the drum leads to debarking of logs inside the drum. One block of drives consists in most cases of sixteen tyres located on circumferential and longitudinal directions of the drum cylinder.

a) b)

Fig. 1 View of the barking drum a) before assembly, b) inside view

Fig. 2. Powered tyres and their connection to gearbox

During operation of the drum cracks were detected near stress concentrators in the frontal part of the drum cylindrical shell (Fig. 3). In order to establish the reasons for these failures, it was necessary to determine the levels of residual stresses near the cracks. On the basis of information gained from the customer, study of the assembly of the barking drum, as well as from the analysis of the barking drum, taking into account also time and space conditions, the authors came to the conclusion that the most suitable and relevant locations for application of strain-gages for the hole-drilling method were the places where the cracks were, i.e. where due to welding the residual stresses were high, or in places where stresses were increased by the change of thickness after welding of the rib in locations of holes for bark removal. Locations with a high number of cracks gave relevant evidence that the residual stresses were high and accordingly there were locations for application of strain-gages for the hole-drilling.

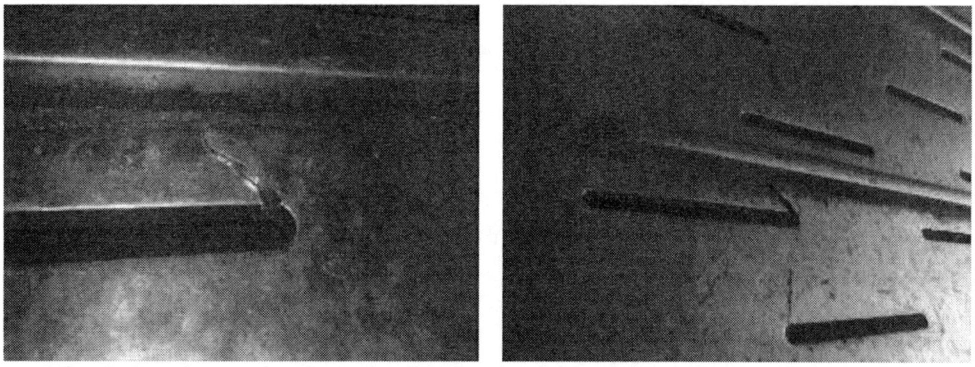

Fig. 3 Cracks on the drum beginning in the location of stress concentrator

The drilling of holes was carried out by apparatus RS 200 using strain-gage rosettes RY 21-3/120 (Fig. 4). The process of the hole drilling was performed in ten steps by 0.5 mm and the resulting hole depth 5 mm. The hole diameter was 3.2 mm. The released strains were registered in each step of measurement by strain-gage equipment P3 [10-12].

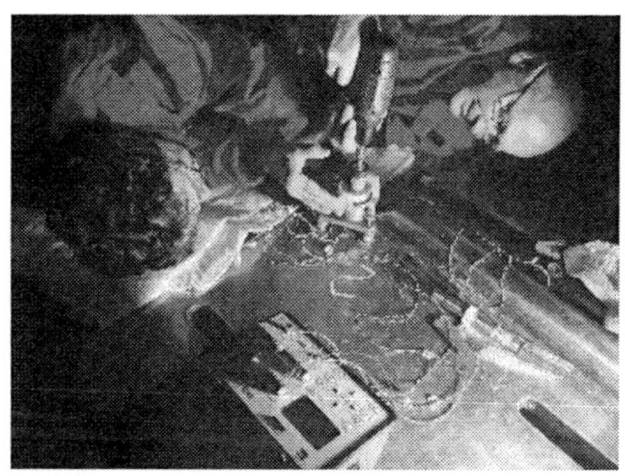

Fig. 4 Process of hole-drilling in location 2

After inspection of the internal space of the drum additional cracks were detected by authors on a number of blocks. They are marked red in Fig. 5. All cracks lie on one plane that is perpendicular to the drum axis.

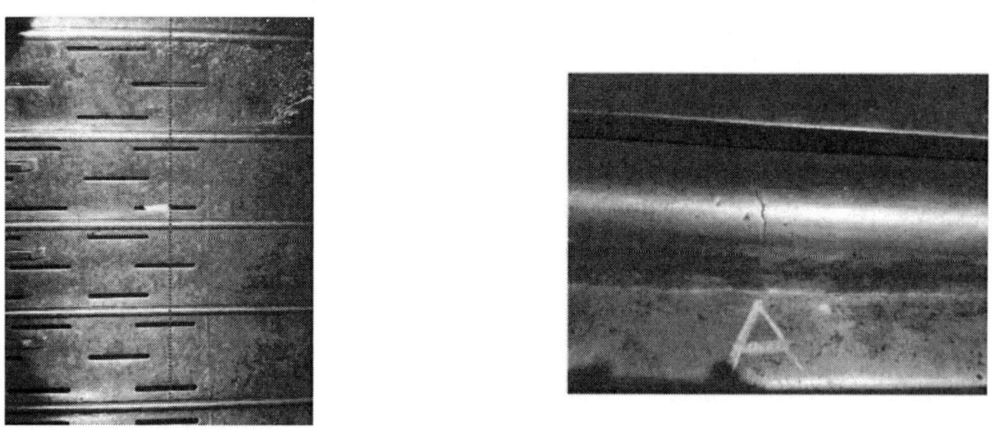

a) b)

Fig. 5 Cracks on prisms in the frontal part of the drum a) location of failures, b) detail view of the crack shape

After analysis was found out that reason of breaking of ribs in one plane can be a fact that supporting and powered tyres are do not lie on one line along the drum. Such positions lead to the additional loading that was simulated by prescribed movement of supporting point by 10 mm upwards or dawnwards. In the computational model, the loading was realized by self-weight of drum and weight of processed wood in accordance with defined power of barking drum. In order to have simple model, the holes were not modelled and accordingly the stress concentrators were removed. However, such stress concentrators are parts of drum. By the computation was determined that this movement of support can lead to stress amplitudes in blocks on the level approximately 100 MPa (Fig. 6). By such computation the authors documented possible causes of failure initiation. In one cross-section.

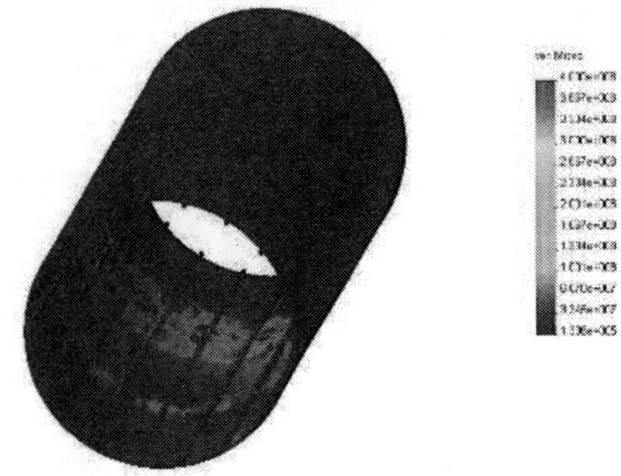

Fig. 6 Field of equivalent von Mises stresses - displacement 10 mm upward

On the basis of the thorough analysis it can be stated that:

a) Measurement of residual stresses in five locations shows that levels of residual stresses are very high. They reach 130 MPa. The stresses have character of tensile stresses and they are oriented mostly in the direction of the drum axis. High levels of residual stresses decrease stress amplitudes and these consequently invoke failures in prisms and welds.
b) The simulations confirmed that crack orientations correspond to the directions perpendicular to the directions of principal tensile normal stresses.
c) The newly detected cracks in prisms can result from inaccurate positioning of supporting and drive units.
d) The existing cracks (even if they are near the welds) cannot be considered as a result of low strength of fillet welds. It was documented by the strength of a unit length of a weld which is satisfactory and in no point the resultant force acting inside the unit length of the weld exceeds the proposed nominal value.

Analysis of crack initiation in the press frame and optimization of the frame for ensuring its further safe operation

After several years of operation of the press (Fig. 7a) used for cutting and forming of side panels of washing machines, there were detected cracks in vertical columns of the press frame (Fig. 7b). Taking into account operational conditions it was necessary to operatively analyse the causes of crack initiation in the columns of the press frame and on the basis of such analysis to propose measures in order to ensure its further failure-free operation. The paper presents the results that were reached during solution of this problem.

During press operation two operations are accomplished simultaneously in one working cycle (one ram stroke of the press), cutting from a sheet of the rectangular shape and forming the cut piece from the previous operation. From the analysis of loading of individual operations, it was found out that maximum load acting on the press ram was F_s=1800 kN during cutting and F_o=1200 kN during forming, i.e. for simultaneous execution of both operations the maximum force acting on the press ram reached 3000 kN. According to the data gained from the operator, the press force limiter was adjusted to the force magnitude in the interval from 3200 kN to 3500 kN during the whole period of its previous operation. Despite the above-mentioned fact there were detected cracks in three columns of the press approximately after ten years of operation (with approximately $8 \cdot 10^6$ working cycles).

a) b)

Fig. 7 a) General view of the press, b) cracks in the column of the press frame

The cracks were initiated in the bottom part of the press frame in the locations of transition of columns to the bottom crossbeams of the press frame. The columns had a closed rectangular cross-section created from the sheets and the cracks were spread through the whole thickness of the sheets that form the column walls. Inspection of the crack lengths showed that the carrying cross-sections of columns were decreased by 30 to 40% by cracks.

For the assessment of various reasons of crack creation on the columns of the press it was necessary to quantify the levels of residual stresses in their neighbourhood. Residual stresses occur also in the structures without loading and they can be caused by machining, assembling, but mostly by operational overloading. The residual stresses are superimposed to stresses from operational load and they can under certain circumstances (magnitude and character) cause failure of the structure. The levels of residual stresses in material can be determined by several methods. One of them is a hole-drilling method. Fig. 8 shows that the strain-gages were applied in the neighbourhood of the cracks on the inner walls of the columns with the aim to gain better knowledge about residual stress levels in the observed areas. For the measurement of residual strains, the hole-drilling equipment RS-200 was used [11-14]. For the measurement self-compensating strain-gages 1-RY-21-3/120 with radius 5.15 mm were used, and the diameter of the drilled hole was 3.21 mm. The depth of the drilled hole was 5 mm and the drilling was carried out in ten steps, each with length of 0.5 mm.

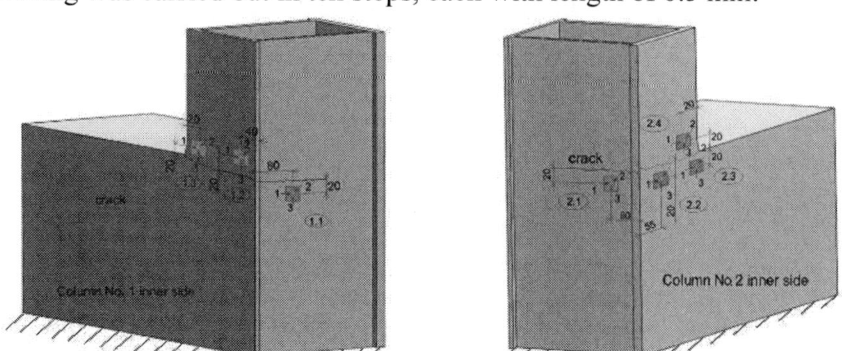

Fig . 8 Locations of applied strain-gages on columns 1 and 2 of the supporting press frame

Fig. 9a shows application of the strain-gage rosette and the location of taking material specimen for determination of mechanical properties of the material. From Fig. 9b the direction of crack propagation is obvious, it is marked red.

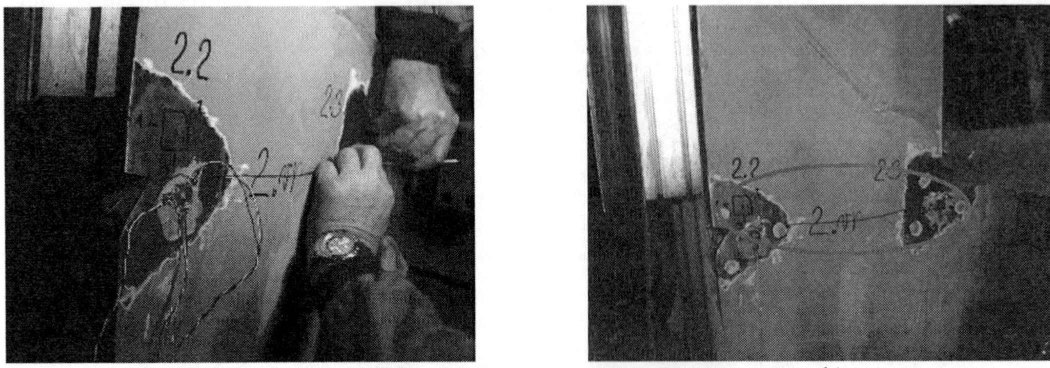

a) b)

Fig. 9 a) Application of strain-gage rosette, b) direction of crack spreading

Computed values of residual stresses determined from measured strains released by drilling on the inner sides of the column walls did not exceed 40 MPa. All values were positive, i.e. they supported growth of cracks. Their magnitudes on the sidewalls lay in the interval 20 - 40 MPa that clearly documented the fact that they were not produced during manufacturing of the frame. It can be assumed that they resulted from the loading during operation of the press. Taking into account their magnitudes they could be considered as direct reasons of cracks initiation, because during the operation of the press vanishing loading was invoked for which the middle stress was equal to half of the stress amplitude.

Beside quantification of residual stresses, the stress and deformation analysis was carried out by the finite element method. Computation was performed for a whole loading force in the ram 3500 kN (maximal allowable force determined by force limiter). This force was divided into two parts 2100 kN for cutting and 1400 kN for forming. In order to identify stress concentrators in the locations of cracks there was utilized geometrical symmetry of the frame and the computations were performed on the quarter, more loaded part of the frame. This increased precision of computation by application of smaller finite elements in the critical part and at the same time it decreased the time of computations. Boundary conditions supposed symmetrical loading of the press, which represented adverse state of loading in comparison with a real state. Stress concentration in the area of cracks is apparent from the field of equivalent stresses in Fig. 10. In this computation, a sharp corner (without rounding) was considered in the junction of the column to the crossbeam.

Fig. 10 Field of equivalent stresses in the location of the column to the bottom beam.

From the numerical computation of the press frame by the finite element method it obvious that in the junction of the frame column with the bottom crossbeam (location A in Fig. 10) the extreme stress concentration with maximal equivalent stress 273 MPa is located. Maximum stress is initiated on the inner wall side 30 mm thick in the location of a sharp (not rounded) corner. The analysis shows that cracks in the supporting structure of the press were initiated in the sharp corners of connections of the columns and the bottom beams and this was supported by unsuitable geometry and bad purity of the material used.

From the analysis of the results it follows that the cracks in the structure of press frame arose in sharp corners of the junction of the press columns and the bottom crossbeams and they were probably caused by inappropriate geometry of the press frame and low purity of the material. For the analysis of press frame strengthening, in order to ensure further operation of the press without failures, the finite element method was used. The stress analysis by the FEM showed that rounding of corners with radius 40 mm in locations of the junction of steel sheets on the columns with the bottom crossbeams (location A, Fig. 10) significantly decreased maximum value of equivalent stresses.

On the basis of the above-mentioned facts it was proposed to provide (after welding of cracks) strengthening of the press frame by welding eight plates made of material S355 to all columns (one on inner and outer side of each column). Fig. 11 shows the proposed shape of one stiffening plate.

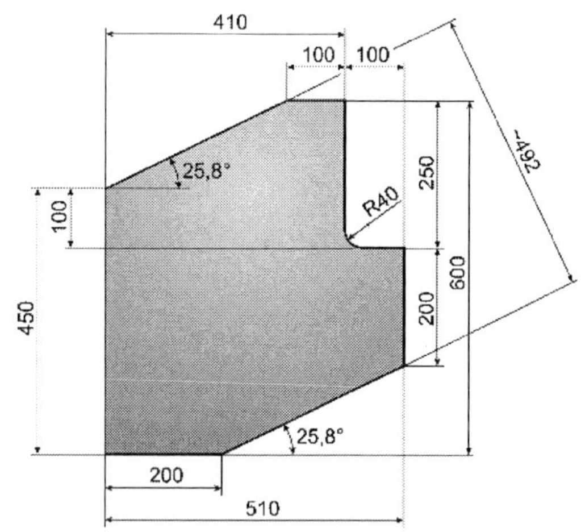

Fig. 11 Shape and dimensions of strengthening plates for inner side of columns.

Stress analysis showed (Fig. 12) that maximum equivalent stresses did not exceed 190 MPa, i.e. strengthening decreased stresses by approximately 30%. This was enough for further safe operation of the press.

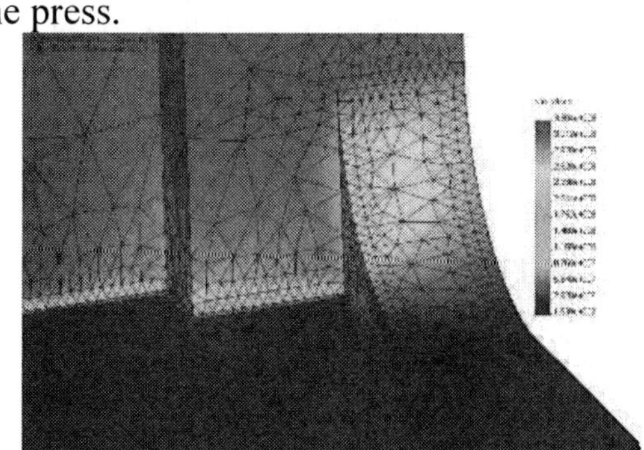

Fig. 12 Detail of field of equivalent stresses on the press frame after modifications

On the basis of numerical and experimental analysis of the press frame it can be stated:

a) Numerical analysis of the press frame showed that occurrence of sharp corners (without rounding) in the junction of the columns and the bottom crossbeams caused (in case of maximum allowable loading of the ram 3500 kN) equivalent stresses 273 MPa.
b) The residual stress levels according to ASTM E 837-01 on the sidewalls of the columns did not exceed 40 MPa and they lay in the interval from 20 to 40 MPa. Residual stresses on frontal walls of the columns did not exceed in absolute value level 62 MPa and these stresses were invoked by operation of the press.
c) Crack initiation in inner corners of the columns (location A, Fig. 10), and also in other locations could be most probably caused by overloading during operation. This did not occur during regular technological process of pressing (cutting and forming), but as a result of inappropriate position of a semi-finished product and accordingly to formation of additional forces during movement of the ram. However, this case does not have to be connected with infunctionality of a force limiter.
d) Another possible reason of cracks initiation was low purity of the press frame material that was documented by probably rolled out sulphides and consequent occurrence of lamellar splitting of the material.
e) Strengthening of the bottom parts of the columns by welded plates will result in decreasing of maximum equivalent stresses by approximately 30%, which is enough for further safe operation of press.

Stress and deformation analysis in critical joints of the bearing parts of the mobile platform using tensometry

At present, there has been a rapid advancement in the use of new materials and modern technologies in the design and technical improvement of robotic systems. Methods of numerical modelling are most commonly used in the design and development of load-bearing parts of mechanisms. When designing complicated mechanical systems such as robotic systems, it is often difficult to correctly define boundary conditions that correspond to the actual operational state using the results obtained through numerical modelling. The paper presents methods of verification of the results obtained by numerical modelling in the selected load-bearing component of a six-leg robot from the experimental measurement using tensometry with the aim to take into account the strength and stiffness parameters of the load-bearing part in order not to jeopardize safe and reliable operation of the mobile robotic platform.

The paper describes application of numerical and experimental methods of mechanics in the design of a mobile service robot of middle category (Fig. 13) which consists of two basic parts a mobile platform that enables autonomous motion in the rough terrain and ruins with payload capacity for the body weight 400 kg, max. speed 3 or 5m/s and climbing ability 45°, a robotic arm with 6 degrees of freedom and nominal load capacity 200 kg. Functional (physical) models of the above-mentioned modules were optimized by simulation computer models and tested in real conditions using experimental methods of mechanics.

Fig. 13. Mobile service robot of middle category

Robotic systems can be considered as complex mechanical systems not only in terms of kinematics but also from the point of view of the analysis of various stress and strain states due to a variety of operational loading. Multiple bearing elements were considered by numerical modelling using finite element method when designing and optimizing the bearing structure of the mobile platform. Main bearing parts of the mobile platform were designed in such a way as to ensure that their loading did not exceed the computational strength of the given material and that the deformations that occurred satisfied the required function.

To verify the results obtained using experimental methods of mechanics, a robot leg was chosen. The finite elements method was used to identify critical areas for the application of strain-gages. The robot leg was loaded at the place of the suspension of the wheel. Various combinations of the positions of individual robot legs were considered in the calculations. Fig. 14 shows the above-mentioned model of the outstretched robot leg with boundary conditions.

Fig. 14 Fields of equivalent von Mises stresses in MPa, material aluminium alloy

For the purpose of experimental measurement, a leg of the mobile robot (Fig. 15) was chosen, where strain-gages were applied. Strain-gages XY91-10/120 connected to the half-bridge were used [10]. Measuring grids of the strain-gage were positioned in the direction of the axis of the leaned leg. Compensation sensors were directed perpendicularly to the axis of the active sensors. Time records of the relative deformations in the selected areas were recorded by SPIDER8 measurement unit with Catman software.

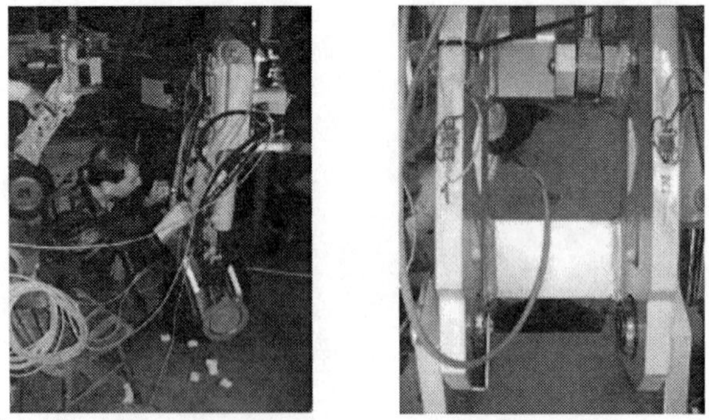

Fig. 15 Application of strain-gages on the leg of the mobile robot

The next part gives the results of the selected measurements of relative deformations and stress in the individual measured places at the selected modes of the leg loading (Fig. 16). Fig. 17 shows charts of stresses in the locations of applied strain-gages for the defined operation regimes.

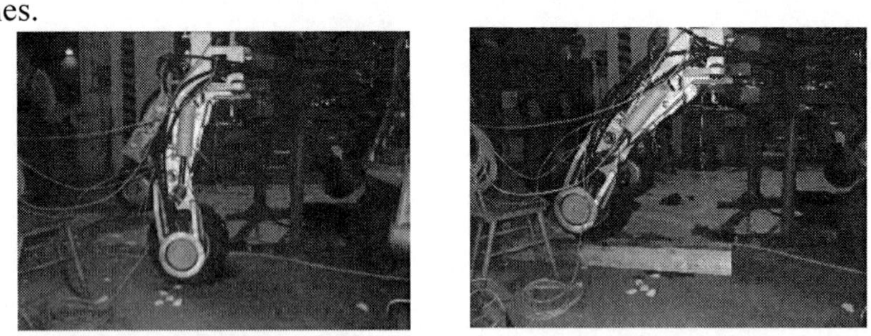

Fig. 16 Selected position of robot legs during experimental measurement.

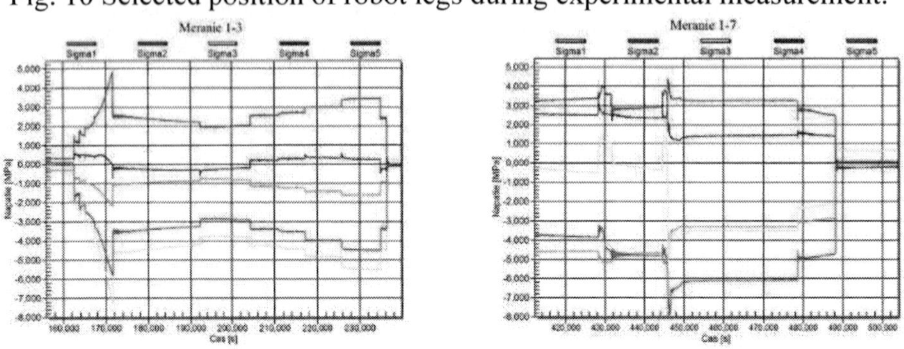

Fig. 17 Courses of time-dependent stresses in the locations of measurements under selected regimes of leg loading

In spite of the fact that nowadays great emphasis is placed on the optimization of the shape and geometry of the bearing parts of machinery and equipment in order to reduce their weight, it is necessary to state that when designing bearing parts of the mobile robotic systems of manipulators it is essential to pay close attention to the rigidity parameters of the used components. It is very important in terms of exact positioning of the end elements of the mobile robotic systems designated for application in tough conditions of rescue works, natural disasters, fires, decontamination of surfaces from toxic materials, etc., where accuracy is top priority.

Authors presents the method of verification of the results obtained using numerical modelling by experimental measurement performed in the defined areas on the robot leg of the mobile platform in different positions at the operational loading. Based on the stress values obtained by experimental measuring it can be concluded that the proposed shape, geometry and material of the robot leg in terms of hardness and strength are sufficient and the mobile platform can be safely operated without restrictions.

Conclusions

As mentioned above, experimental procedures dominate in the area of determination of the residual stress in machine parts. The hole-drilling or ring-core methods are most widely spread that are being continuously developed. The development of such methods is caused on the one hand by elaboration of the hole-drilling and ring-core method, by development of new equipment and on the other hand by increasing precision of the measured and released strains [15,16]. Significant attention is devoted to elaboration of methods for determination of residual stresses in the vicinity of weds [17,18].

The authors' workplace has been dealing with quantification of residual stresses by hole-drilling method for several decades. The experience gained not only in laboratories, but also during measurements for practice were applied in the proposal of new methodology and procedures of determination of calibration coefficients as well as processing and evaluation of measured data by the newly developed program MEZVYNA. The program was developed in the authors' workplace equipped with state-of the-art measurement systems SINT MTS-3000, RS200 and SINT MTS-3000 Ring-Core [19].

Taking into consideration the top level hardware and software equipment of the workplace in various areas of experimental methods of mechanics (strain-gage measurements, PhotoStress, DIC, ESPI), it is possible to verify the results using different methods and to create new treatments and methods that will be applied for solution of practical problems [20,21]. As the authors have extensive experience in the area of applying optical methods in the stress and deformation analysis near concentrators, they decided to design equipment for quantification of residual stresses by the hole-drilling method (strain-gage measurements) in combination with optical methods (PhotoStress, DIC) (Fig. 18).

a) b)

Fig. 18 a) model of proposed hole-drilling equipment, b) real model of hole-drilling equipment with control unit for precise positioning

An advantage of the newly developed precise positioning equipment is a possibility to verify results gained by experimental methods based on different physical principles. For the creation of methods which are used for determination of residual stresses uniformly or non-uniformly distributed along the thickness of structural supporting members with the aim to refine and improve the existing treatments, the authors proposed and built hydraulic loading equipment necessary for verification of procedures used for quantification of residual stresses in thick specimens (Fig. 19).

a) b)

Fig. 19 Hydraulic loading equipment a) computer model, b) real model including control panel and hydraulics

The authors of the paper in cooperation with other researchers from their workplace reached the results that provide a comprehensive knowledge base for development of new methods for continual evaluation of the reliability of key supporting elements of machines and equipment as well as creation of methods and programs for quantification of residual stresses using sensors for continual quantification of vibration, etc. They proposed and developed experimental procedures focused on deformation and stress analysis of supporting systems of mechanical systems with the aim to apply them for the assessment of the reliability and residual lifespan of machines and equipment. The authors analysed a broad range of problems that influence lifespan and reliability of machines and equipment (static and dynamic loading, influence of stress concentrators, problems of vibrations, residual stresses and so on) in such a way as to consider as many factors occurring during the operation as possible [22-28]. The advantage of the authors' workplace is the fact that due to the state-of-the-art equipment in various areas of experimental mechanics (SINT MTS-3000, RS 200, FL/Z-2, Q300, Q450, Quantum, Gen5i, P3, Pulse) as well as software (Ansys, Abaqus, NX, SolidWorks, Matlab, Adams, nCode) it is possible to evaluate the obtained results and propose new methodology using synergy effects of numerical and experimental methods.

Acknowledgement

This paper was supported by Project APVV 15-0435.

References

[1] M. Hagara, R. Huňady and F. Trebuňa: The Possibilities in Use of ESPI Method by Investigation of Strain Fields of Specimen with Stress Concentrator. In: American Journal of Mechanical Engineering. Vol. 4, no. 7 (2016), p. 429-434. ISSN 2328-4110

[2] M. Hagara, R. Huňady and F. Trebuňa: Stress Analysis Performed in the Near Surrounding of Small Hole by a Digital Image Correlation Method. In: Acta Mechanica Slovaca. Roč. 18, č. 3-4 (2014), s. 74-81. ISSN 1335-2393

[3] R. Huňady and P. Pavelka: Experimental modal analysis of a free suspended composite plate by using high speed DIC method. In: EAN 2017. Košice : TU, 2017 p. 223-230. - ISBN 978-80-553-3166-9

[4] R. Huňady and M. Hagara: A new procedure of modal parameter estimation for high-speed digital image correlation. In: Mechanical Systems and Signal Processing. Vol. 93 (2017), p. 66-79. ISSN 0888-3270

[5] M. Pástor, M. Hagara and J. Kostka: Stress analysis performed by photoelasticity and digital image correlation. In: Applied Mechanics and Materials. Vol. 816 (2015), p. 474-481. ISSN 1660-9336

[6] P. Frankovský, et al.: Utilisation Possibilities of PhotoStress Method in Determination of Residual Stresses. In: Applied Mechanics and Materials. Vol. 732 (2015), p. 3-8. ISSN 1660-9336

[7] P. Frankovský et al.: Experimental determination of stress fields near notches by reflection photoelasticity. In: EAN 2016. Plzeň : Zapadočeská univerzita, 2016 P. 1-10. - ISBN 978-80-261-0624-1

[8] Ľ. Gabáni et al.: Using Experimental Methods of Mechanics for Failure Prediction of Casting Pedestal. In: Acta Mechanica Slovaca. Roč. 19, č. 2 (2015), s. 42-50. ISSN 1335-2393

[9] F. Trebuňa et al.: Odhalenie príčin vzniku prasklín na odkôrňovacom bubne metódou kvantifikácie zvyškových napätí. Final Report. Košice : TU. 2013. 99 s.

[10] F. Trebuňa and F. Šimčák: Handbook of Experimental Mechanics. Typopress, Košice, 2007. ISBN 970-80-8073-816-7

[11] F. Trebuňa and F. Šimčák: Kvantifikácia zvyškových napätí tenzometrickými metódami. Grafotlač, Prešov, 2005. ISBN 80-8073-227-2

[12] Tech Note TN-503-6, Measurement of Residual Stresses by the Hole-Drilling Strain Gage Method, Measurements Group, Inc., Raleigh, NC.

[13] G. S. Schajer: Advances in Hole-Drilling Residual Stress Measurements. Experimental Mechanics, 50, 2010, p. 159-168

[14] G. S. Schajer and T. J. Rickert: Incremental Computation Technique for Residual Stress Calculations Using the Integral Method. Experimental Mechanics, 2010, doi: 10.1007/s11340-010-9408-5.

[15] Z. Barsoum and A. Lundbäck: Simplified FE welding simulation of fillet welds – 3D effects on the formation residual stresses. Engineering Failure Analysis, 16, 2009, p. 2281-2289.

[16] J. M. Costa et al.: Residual stresses analysis of ND-YAG laser welded joints. Engineering Failure Analysis, 17, 1, 2010, p. 28-37.

[17] M. Arsic et al.: Bucket Wheel failure caused by residual stresses in welded joints. Engineering Failure Analsis, 18, 2, 2011, p. 700-712.

[18] F. Trebuňa et al.: Quantification of residual stresses in the weld by the hole-drilling method . Metalurgija,2008 ,47, p. 133-137.

[19] K. Masláková, F. Trebuňa, P. Frankovský and M. Binda: Applications of the strain gauge for determination of residual stresses using Ring-core method. Procedia Engineering. 2012 ,48, p. 396-401.

[20] M. Pástor and M. Hagara: A comparison of modern and classical experimental methods of mechanics in strain investigation. In: Applied Mechanics and Materials. Vol. 611 (2014), p. 501-505. - ISSN 1660-9336

[21] M. Hagara and R. Huňady: Q-STRESS v.1.0 – a Tool for Determination of Stress Fields using Digital Image Correlation Systems. In: Procedia Engineering: Modelling of Mechanical and Mechatronic Systems MMaMS 2014. Vol. 96 (2014), p. 136-142. - ISSN 1877-7058

[22] F. Trebuňa et al.: Proposal of Methodology for Determination of Stresses around Groove by PhotoStress Method. In: American Journal of Mechanical Engineering. Vol. 5, no. 6 (2017), p. 325-328. ISSN 2328-4110

[23] P. Šarga, P. Senko and F. Trebuňa: Analysis of influence of strain gage rosette on relieved strain. In: American Journal of Mechanical Engineering. Vol. 1, no. 7 (2013), p. 309-312. ISSN 2328-4102

[24] F. Menda, F. Trebuňa and P. Šarga: New Method of Residual Stress Evaluation and its Advantages in Comparison with More Common Hole-drilling Method. In: Acta Mechanica Slovaca. Roč. 17, č. 4 (2013), s. 64-70. ISSN 1335-2393

[25] P. Šarga, F. Menda and F. Trebuňa: Experimental Verification of the Geometric Parameters in the Ring-Core Measurement. In: EAN 2015. Prague : CTU, 2015 P. 388-394. ISBN 978-80-01-05735-3

[26] F. Menda et al.: Comparison of different simulation approaches in ring-core method. In: American Journal of Mechanical Engineering. Vol. 2, no. 7 (2014), p. 258-261. ISSN 2328-4102

[27] P. Šarga, P. Senko and F. Trebuňa: Verification of equation for determining non-uniform residual stresses by FEM. In: American Journal of Mechanical Engineering. Vol. 1, no. 7 (2013), p. 169-172. ISSN 2328-4102

[28] F. Trebuňa et al.: Residual stress analysis in containers for transport of radioactive materials. In: Applied Mechanics and Materials. Vol. 732 (2015), p. 28-31. ISSN 1660-9336

EAN 2018
56th conference on experimental stress analysis

Lightweight Masonry Block made from Recycled Concrete

TREJBAL Jan[1,a], HRBEK Vladimír[1,b], ĎUREJE Jakub[1,c] and HLŮŽEK Radim[1,d]

[1]Faculty of Civil Engineering, Czech Technical University in Prague, Thákurova 7, 166 29 Prague, Czech Republic

[a]jan.trejbal@fsv.cvut.cz, [b]vladimir.hrbek@fsv.cvut.cz, [c]jakub.dureje@fsv.cvut.cz, [c]radim.hluzek@fsv.cvut.cz

Keywords: Lightweight concrete, foam concrete, recycled concrete, concrete powder.

Abstract. This work deals with design of lightweight blocks to be used as a building material for production of both non- and load bearing walls in the field of civil engineering. These blocks were based on foam concrete, containing high amount of recycled concrete at the form of high-speed ground powder. Two mixtures were designed, differed mainly in amount of used foaming additive. While in the first case, heat insulating parameters were preferred, in the second case, it was focused on mechanical properties. Prismatic samples (40×40×160 mm) were made from mixtures and then subjected to mechanical and thermo-physical testing in order to reveal their basic properties. It was shown that compressive strength of mixtures were equal to ca. 1.7 and 19.0 MPa in the case of mixtures containing high and low amount of foaming additive, respectively. Thermal conductivity was ca. 0.16 and more than 0.5 W/mK, in the same order of mixtures.

Introduction

Recycling of waste is a widely discussed topic throughout the European Union. It is obvious, that the problem of waste management strongly influences also the civil engineering. On territory of the Czech Republic, there has been produced more than 18 million tons of construction and demolition waste (C&DW) in the year of 2017. During the past decade, amount of C&DW has exceeded ca. 160 million tons in total sum; whence ca. 25 % belongs to concrete waste originated from building demolitions [1, 2]. It is worth noting that up to 50 million tons of concrete buildings are demolished in the EU per year [3]. Due to Waste Management Plan of the Czech Republic for the period 2015–2024 aiming to landfill restrictions and waste recycling acceleration [4], it is justified to find out new ways how to re-use waste concrete.

It is generally known, that recycling of concrete has been focused especially on re-use of aggregate recently. Unfortunately, a utilization of fine fractions at the form of concrete dust (containing binder and the smallest parts of aggregate) has stayed unused [5], although this material exhibits promising properties like residual binding potential of unhydrated cement grains and an ability to fill micro pores through the new concrete mixtures [6]. Moreover, in cases of autoclaved or foam lightweight materials, dust can replace conventional micro-aggregates due to similar size parameters [7]. Therefore, there is no obstacle to use waste concrete dust as so-called active micro-aggregate for the production of lightweight masonry blocks.

We made lightweight masonry blocks containing ca. 25 wt. % of recycled concrete powder that was obtained using high-speed milling. These were subjected to experiments revealing their properties from the mechanical and thermo-physical point of view.

Materials and Methods

Foam concrete composition. Two mixtures were designed, differed mainly in amount of used foaming additive. The first was determined for fabrication of masonry blocks to be used for the production of perimeter walls, where heat insulation properties plays an important role. The second mixture was designed for blocks for load-bearing walls, where mechanical parameters are the key properties. Both of them were composed from Portland cement, concrete powder, water and foaming additive Mapei. Concrete powder was obtained from structural elements of demolished building in Prague. These elements were roughly crushed by using crushing machine equipped with hydraulic jaws. After separation of steel bars, thus crushed material was finely ground using patented technology of high speed milling in company Lavaris, Ltd. (Czech Republic). Powder was composed from grains of size differed between 0.25–250 μm, as found out by granulometry. Specific surface of grains was equal to 412 m^2/kg, as measured according to Blain's method. The grain curve is shown in the Fig. 1. Prismatic samples (40×40×160 mm) from these mixtures were made and after their hardening (lasting standard 28 days) subjected to experimental tests. The whole composition of both mixtures is summarized in the Table 1.

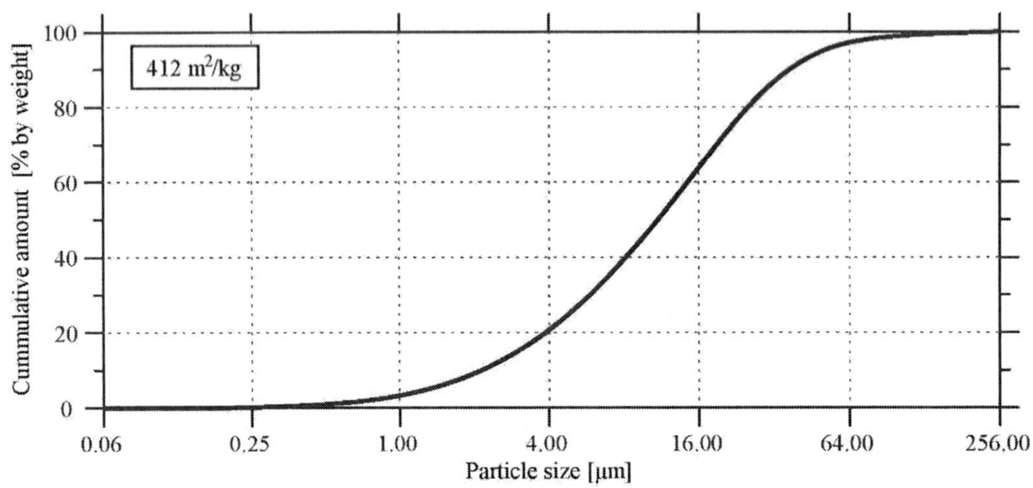

Fig.1 Grain curve of finely ground recycled concrete

Tab.1 Weight-proportional composition of tested mixtures

Mixture	CEM I 42.5 R [wt. %]	Concrete Powder	Water	Mapei Additive
M1	46.3	24.9	28.5	0.4
M2	48.1	25.9	25.9	0.2

Experimental techniques

Mechanical tests. In order to reveal pressure strength of both hardened mixtures, compression destructive test was done using loading frame Web Tiv Ravestein FP100. The experiment was displacement controlled at the constant rate of 1 mm per minute. The loading force was introduced to samples through the steel plate of dimensions equal to 40×40 mm. To obtain statistically relevant data, the measurement was repeated 6-times, then results were averaged.

Dynamic modulus of elasticity was measured by means of indirect non-destructive method. For this purpose, measuring station Brüel & Kjær (3560–B–120) was employed, equipped with piezoelectric acceleration sensor and impact hammer (both system device). The measurements was repeated 5-times on each sample, results were averaged. The process was done according to procedure described in [8].

Thermo-physical tests. Basic thermo-physical properties of samples, such as thermal conductivity and heat capacity, were measured by Isomet (model 2104, Applied Precision). The device was equipped with removable contact probe. The measurement was repeated 3-times, results were averaged. Before this experiment, all of samples were weighed and then measured to calculate their density.

Results and Discussion

Mechanical tests. It was found out from compression tests that compressive strength of M1 and M2 mixture was equal to 1.69±0.67 MPa and 18.95±2.00 MPa, respectively. It is clear that the strength of M1 does not allow blocks to be used for production of highly loaded walls. However, this strength is almost comparable to e.g. commercial aerated concrete blocks. On the other side, mixture M2 exhibited promising compressive strength, similar to lower class of plain concrete.

Modulus of elasticity measurement revealed that samples made from the mixture M1 reached to 2.8±0.8 GPa, which is according to our expectations significantly lower than in the case of samples made from M2. Those exhibited 8.9±0.4 GPa. Both of the two values overcome common aerated concretes.

Thermo-physical tests. Density of samples was 900±35 and 1390±10 kg/m^3 in the case of mixture M1 and M2 respectively. These results explained differences of mechanical parameters between the two mixtures and it is clear that they were caused by using of different amount of Mapei foaming additive (0.4 for M1 vs. 0.2 wt. % for M2).

Thermal conductivity of M1 and M2 samples was equal to 0.160±0.03, resp. 0.527±0.01 W/mK. Based on these findings, it was confirmed that masonry blocks made from M1 mixture can be used for production of perimeter walls of building because of their favorable insulation properties. The heat capacity was ca. 770, resp. 1137 J/kgK, in the same order as mentioned earlier. All of the findings obtained within this study are summarized in Table 2.

Tab.2 Summarization of tested samples basic properties

Mixture	ρ [kg/m^3]	E [GPa]	R [MPa]	λ [W/mK]	c [J/kgK]
M1	900±35	2.8±0.8	1.69±0,67	0.160±0.03	738.0–797.7
M2	1390±10	8.9±0.4	18.95±2.00	0.527±0.01	1128.6–1144.9

Conclusions

This work was focused on design of mixtures used for production of masonry blocks based on light-weighed concrete. It was targeted to use high amount of recycled concrete at the form of finely ground powder. Two mixtures were prepared, differed in amount of used foaming additive (0.4 vs. 0.2 wt. %). The first one was determined for production of masonry blocks to be used for masonry of perimeter walls, where thermal insulation properties play an important role. The second one was designed for production of blocks intended for load-bearing walls. In this case, compressive strength of them was considered as the key parameter.

Amount of used recycled concrete powder was ca. 25 wt. %. It was used as active micro aggregate. This material was obtained from structural element of demolished building in Prague. This waste concrete was roughly crushed using crushing machine equipped with hydraulic jaws. The brush was subsequently milled by means of technology of high speed milling to powder with grains of size differed between 0.25–250 μm.

It was shown that mixture designed for perimeter walls reached to comparable properties to frequently used aerated concrete in terms of heat insulation parameters (thermal conductivity ca. 0.16 W/mK) and compressive strength (ca. 1.7 MPa). On the other side, mixture intended for load-bearing blocks exhibited favorable mechanical strength properties. Their modulus of elasticity was almost 9.0 GPa, while the compressive strength was equal to ca. 18–19 MPa.

Acknowledgement

This work was financially supported by Czech Technical University in Prague – SGS project SGS16/201/OHK1/3T/11 and Czech Ministry of Industry and Trade project FV20503. The authors would also like to thank to Tereza Horová for her help with this research and to Lavaris Ltd. for their technical support.

References

[1] M. Škopán, Conditions for the Development of the Use of Recycled Building Materials in Construction, in: M. Škopán (Ed.), Proceedings of Recycling 2017, Brno, 2017, pp. 20-25.

[2] Information on https://isoh.mzp.cz/VISOH/

[3] T.C. Hansen, Recycling of Demolished Concrete and Masonry, first ed., CRC Press, London, 1992.

[4] Waste Management Plan of the Czech Republic for the Period 2015-2024, Ministry of the Environment, Prague, 2014.

[5] M. Etxeberria, E. Vázquez, A. Marí, M. Barra, Influence of amount of recycled coarse aggregates and production process on properties of recycled aggregate concrete, Cement and Concrete Research, 6 (2005) 735-742.

[6] J. Topič, Z. Prošek, Properties and Microstructure of Cement Paste Including Recycled Concrete Powder, Acta Polytechnica, 57 (2017) 49-57.

[7] S. Wei et al., Application of Organic- and Nanoparticle-Modified Foams in Foamed Concrete: Reinforcement and Stabilization Mechanisms, Cement and Concrete Research, 106 (2018) 12-22.

[8] Z. Prošek et al., Utilization of the waste from the marble industry for application in transport infrastructure: mechanical properties of cement pastes, in: P. Bílý (Ed.), Building up Efficient and Sustainable Transport Infrastructure 2017, Prague, 2017.

EAN 2018
56th conference on experimental stress analysis

Measurement of Operational Stresses of ŠKODA Trolleybus 27Tr

VÁCLAVÍK Jaroslav[1,a], NÉMET Jiří[1,b] and WEINBERG Otakar[1,c]

[1]Výzkumný a zkušební ústav Plzeň s.r.o., Tylova 1581/46, 301 00 Plzeň, Česká republika

[a]vaclavik@vzuplzen.cz, [b]nemet@vzuplzen.cz, [c]weinberg@vzuplzen.cz

Keywords: trolleybus, service life, fatigue damage, operational test, strain gauge.

Abstract. This article describes the procedures based on short and long life operational tests used for fatigue damage assessment of trolleybus ŠKODA 27Tr structure.

Introduction

ŠKODA trolleybus 27Tr is based on Solaris New Urbino skeleton. The Slovak authority, for which the articulated coaches were delivered for the first time, asked to make the strength test of the trolleybus body to ensure, that electro-equipment added masses do not influence the safety of the coach operation. On the base of Solaris acceptance and cooperation, the tests were performed by VZÚ Plzeň with the first coach prototype in Pilsen, Bratislava and Žilina, where finally all coaches are operating. The strain gauges had been installed on the raw coach welded structure before all added equipment was assembled in Solaris Company. The preliminary tests were performed in Pilsen. The main tests were realized in Bratislava. Final tests continued in Žilina, where the acceleration long time test was performed.

Trolleybus Operational tests

Test Instrumentation. Totally 62 strain gauges and several accelerometers were used for the test. The data acquisition was performed with a set of 8 HBM SPIDER8 measuring units and with on-line monitoring system EMS DV803 equipped with GSM router for data transferee and with GPS vehicle unit for tracking the trolleybus position. In parallel with strain gauges, the driving millage and derived velocity was measured with the help of two magneto-inductive sensors, tracking the rotational speed of the propeller shaft. The traction engine revolution measurement ensured to stop the data acquisition at the standing coach.

Fig. 1 Bump test

Fig. 2 Distribution of passengers loads

Performed tests. Short tests on the model road were performed in Pilsen. It concerns loading the bus, the bump test over the set of standardized bumps with the velocity of 40 km·h^{-1} (Fig. 1), riding the curve, braking and acceleration and torsion the coach with separate wheels as well as by entering the curb.

Preliminary operational tests were also made in Pilsen on several trolleybus lines for service life estimation of weak nodes (107 km).

The main operational test were performed in Bratislava on two trolleybus lines, from which the first was as mountain ride over the castle and Koliba end station and the second goes to eastern part of Bratislava with very bad pavement (232 km) (Fig. 3). Crossing over the tramlines and driving over the cubed pavement were presented during the test. The bitumen pavement was on several places of low-quality with many pot-holes. Sometimes the roadway was eng and dished with several turnings, which causes not only vertical vehicle excitation, but also coach oscillation around both vertical axes. Tests were performed with empty and loaded coach with substitution of passengers (Fig. 2).

The last part of the test was performed in Žilina. Here the two month tests with passengers at traffic lines without the operator were performed using the on-line monitoring system.

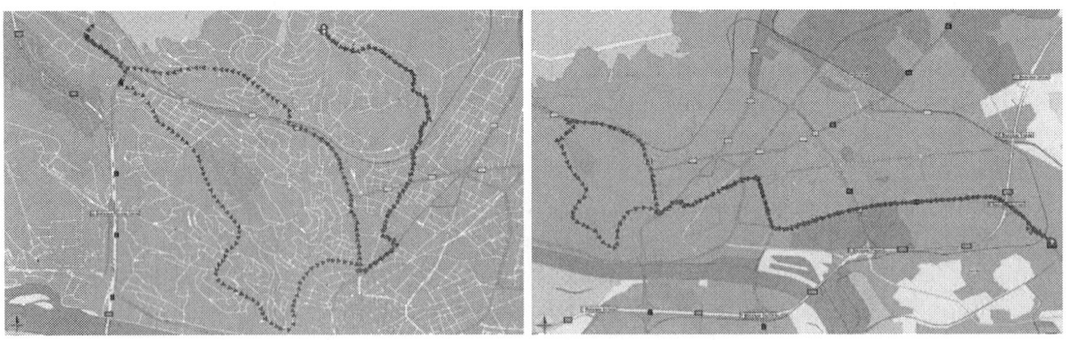

Fig. 3 Two testing loops in Bratislava

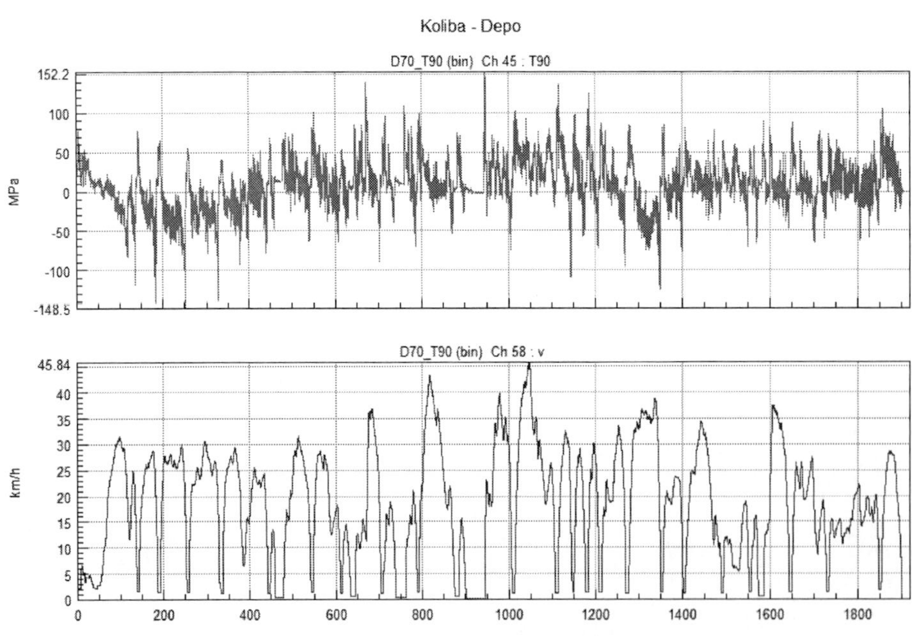

Fig. 4 Example of stress time history obtained during one testing loop and the coach velocity

Data Analysis and Results

Used Material. For the coach structure is mostly used the ferrite stainless steel 1.4003. It deals above all about welded hollow profiles, locally stiffened with added plates or ribs. Following material characteristics were taken from the test: ultimate strength R_m = 450 MPa, yield point R_e = 280 MPa. According test results of welded beam with but weld, given from the producer, the fatigue limit stress amplitude for R = 0 is 75 MPa; this value is valid for nominal stress.

S-N curves. The derivation of S-N curves went out from the recommendations of IIW [2], present standards for traffic vehicles, from the fatigue test of fillet joint from steel 1.4003 [1] and from the fatigue test results, obtained from the customer.

The existing standards classified the welded joints to classes, from which each class has the S-N curve with the same slope of m = 3, but different fatigue limits. For fillet welded joints, which are mostly presented at critical measured destinations, the best suited is the class F2. These curves are designed for the probability of survival P > 95%. Fatigue curves, based on the tests of steel 1.4003 are evaluated as for the probability P = 50% as for higher probability or only the mean curve is given.

When comparing all the standards with performed tests, the fatigue limits fits surprisingly very well.

For all welded joints was conservatively for the fatigue calculation chosen the fatigue limit σ_{ac} = 60 MPa for R = -1 for N = 2E6 and the slope m = 3. There was used only one slope of the S-N curve and the curve was cut at N = 1E7 (IIW recommendation [2]).

For the calculation of service life, the methodology and software, developed in the frame of the project EU CP 940520 Bus-Expert-System were used. During the calculation, which goes from upper given presumptions, the decomposition of stress time history to the closed stress cycles using rainflow method is used. The sensitivity to the mean stress cycle is given to the consideration with the factor M = 0.23, which was derived from the fatigue test results for pulsed (R = 0) and oscillated (R = -1) cycles.

Fatigue Estimations. The measured stresses were put to the Goodman diagrams to check the unlimited service life (Fig. 6). Here the upper envelope curve represents the fatigue limit of the basic material and the bottom of the fillet weld.

The strain gauges were installed in sufficient position from welds to be able to use the nominal approach for limited fatigue life calculation. The sensitivity for mean stress cycle was also taken in to account. The linear fatigue damage accumulation hypothesis was used for fatigue life calculation based on stress history decomposition using rainflow method according the relation

$$D = \Sigma(n_i/N_i) \qquad (1)$$

where

n_i ... number of stress cycles on i^{th} stress level with the amplitude a_i

N_i limit number of cycles on the same stress level, for the parameters of S-N curve.

The service live L in [km] was calculated according the relation

$$L = NSU\cdot(D_c/D) \qquad (2)$$

where

D_c is the critical value of the fatigue damage, which equals according Haibach hypothesis $D_c = 1$.

Estimated damage for all measured joints for design service life of 1 million km is shown in Fig. 5.

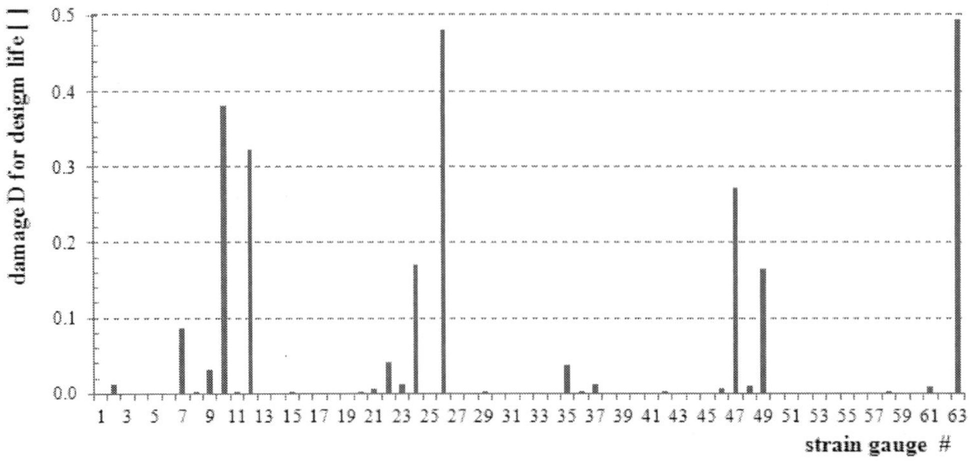

Fig. 5 Calculated damage for measured trolleybus joints for design life of 1 mil. km

Acceleration Evaluation. Measured acceleration time histories were divided to parts with constant millage and for each part the effective values were calculated and FFT spectra were averaged. The resulting spectral densities were compared with those, given in standard (Fig. 7). The goal was to ensure, that the standard values match the real ones.

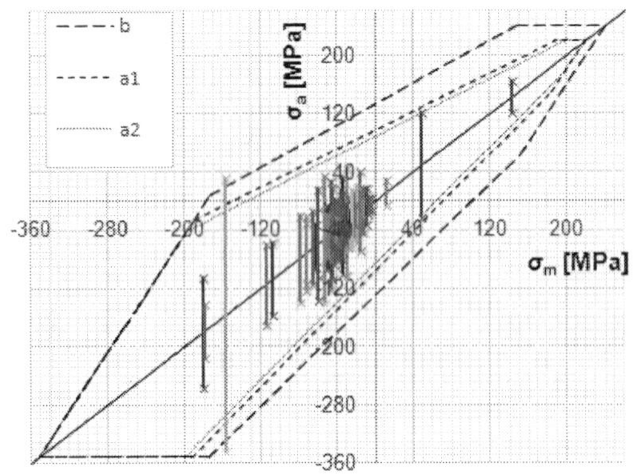

Fig. 6 Goodman diagram for tested parts

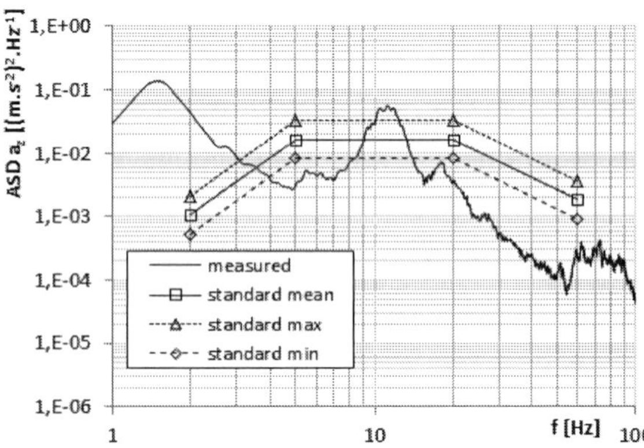

Fig. 7 Acceleration a_z spectral density

Conclusions

The estimated service life of all tested trolleybus joints exceeds the value of 1 million km. The highest damage was found on upper T-joint door pillars and the supporting bracket for the traction engine. The spectral densities of roof acceleration have different form and levels when compared with the standard EN 61373 for rail vehicles.

The article has originated in the framework of the institutional support for the long-term conceptual development of the research organization.

References

[1] K. E. Lahti, H. Hänninen, E. Niemi: Nominal stress range fatigue of stainless steel fillet welds — the effect of weld size. Elsevier, J. Const. Steel Research 54 (2000) 161–172.

[2] A.F. Hobbacher: Recommendations for Fatigue Design of Welded Joints and Components. Second edition. IIW document IIW-2259-15, Springer International Publishing Switzerland, 2016.

EAN 2018
56th conference on experimental stress analysis

Evaluation of Mechanical Behavior of 3D Printed Lattice Structure by SLM: Experiment and FEA

VRÁNA Radek[1,a], VOSYNEK Petr[1,b], KOUTNÝ Daniel[1,c], NÁVRAT Tomáš[1,d] and PALOUŠEK David[1,e]

[1]Faculty of Mechanical Engineering, Brno University of Technology, Technická 2896/2, 616 69 Brno, Czech Republic

[a]vrana@fme.vutbr.cz, [b]vosynek@fme.vutbr.cz, [c]navrat@fme.vutbr.cz, [c]koutny@fme.vutbr.cz, [c]palousek@fme.vutbr.cz

Keywords: selective laser melting, additive manufacturing, material properties, lattice structure.

Abstract. This paper deals with the prediction of the mechanical behaviour of lattice structures made by selective laser melting (SLM). The work was divided into two stages where the first was aimed on evaluation of the material model for finite element analysis (FEA) from proposed tensile test specimens. The second stage was using previously evaluated material model for more complex structures composed of periodical truss cells. It was found that the major issue was the real geometry of additively produced specimens. For evaluation, stress-strain diagrams were used. All manufactured parts were made of aluminium alloy AlSi10Mg.

Introduction

Selective laser melting (SLM) is layer by layer additive manufacturing (AM) method which allows creating objects with a complex shape that are very difficult to produce by conventional technologies. There are many great examples in nature, e.g. bone tissue composition - trabecular bone and osteons in compact bone, both oriented in the direction of principal stresses [1] or internal structure of plants such as bamboo.

The general shape such as bone or bamboo is very difficult to create in 3D software, therefore general shape is replaced by the lattice structure with periodically repeating unit shape in engineering [2]. Due to high strength to weight ratio and good energy absorption capacity, Body Centred Cubic (BCC) shape of lattice structure is very popular [3].

Mechanical properties of the SLM parts are strongly dependent on used laser process parameters of the technology. It is clearly shown in the articles [3, 4, 5], where authors present influence of three main parameters – laser speed (LS), laser power (LP) and laser hatch distance (HD) on material and mechanical properties of aluminium alloys. Many authors describe mechanical properties of a bulk material in tension, e.g. Kempen et. al. [6] describes different mechanical properties in XY and Z direction due to elongation at break which is caused by porosity close to the part surface. If we consider a lattice structure formed by thin trusses, this close to surface porosity can greatly decrease the mechanical properties of each truss in the structure. The other authors also found out that the shape of parts and then also mechanical properties in lattice structure are dependent on building direction and surface roughness [7, 8].

Based on these studies, the methodology for obtaining a material model of BCC lattice structure material for FEA analysis was created.

Materials and methods

Additive manufacturing. All samples were made of aluminum powder AlSi10Mg by the SLM 280^{HL}, the metal additive technology machine with building space 280x280x350 mm^3. The device allows producing parts from many types of metal powder materials (aluminum alloys, titanium alloys, stainless steel, etc.). The metal powder is melted by YLR-Laser with a power of 400W [9]. During the production, there is the inert atmosphere N_2 and Ar inside the build chamber. Therefore, the device can be used even for very reactive materials.

Tensile Specimens. All tensile specimens were produced with a default setting for the AlSi10Mg material and added with a cooling delay of each layer (15 seconds) before production of next layer. Samples were not further modified (heat treatment or surface finishing). To capture the effect of size (truss diameter: 0.45 mm, 1 mm) and different building angle (90°, 45°) [3], four groups of five pieces were made. Samples of different groups have the same total area of cross-section for the purpose of comparison the mechanical parameters.

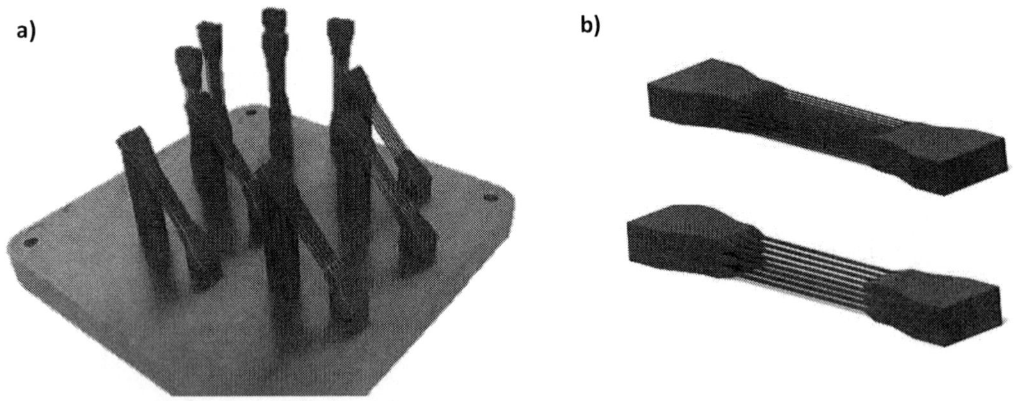

Fig. 1(a) Specimens for tensile tests on the platform after manufacturing by SLM 280HL; (b) CAD models of the tensile specimens

Compression Specimens. All compression samples were sandwich blocks 20x20x20 mm^3 of BCC lattice structure with the bottom and upper plate (t = 0.3 mm). A unit cell was composed of eight trusses with nominal diameter d = 0.5 mm and its dimensions were 4x4x4 mm^3. It results that every layer was created by 5x5 unit cells. All compression samples were also manufactured with the default setting for AlSi10Mg material and not further modified (heat treatment or surface finishing).

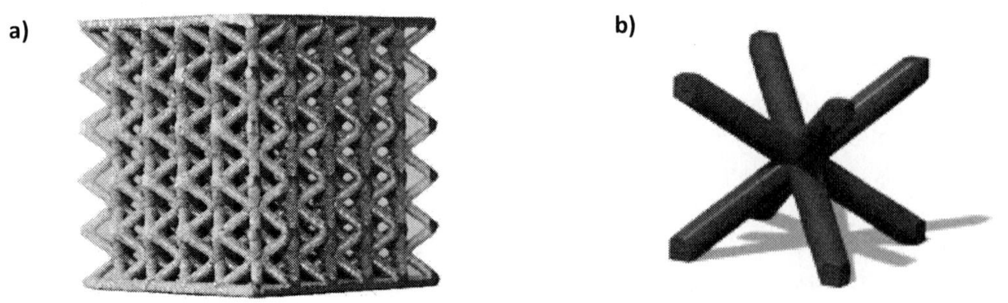

Fig. 2 (a) Specimens for compression tests manufactured by SLM 280HL; (b) CAD model of the BCC unit cell

Quality and surface roughness. The tensile samples were digitized in the as-built state on the ATOS Triple Scan. The acquired 3D geometry of the real trusses was analyzed in the GOM Inspect software, to find the difference between design and real geometry. For better quality of measurement, the surface of the samples was mated with titan powder [10].
Mechanical testing. In order to get an information about the mechanical behavior of the manufactured structures, a tensile and compressive test were performed. For the mechanical testing universal testing machine, ZWICK Z020 was used.
FEA. For the purpose of the further design of 3D printed lattice structures, the computational model was made. Ideal geometry was first considered and further corrected for the deviation of the real trusses delivered from the 3D scanner. To simulate the behavior of the tensile test in the elastic-plastic region, nonlinear model of material needs to be involved in FEA. The bilinear behavior was chosen. The parameters of the bilinear model of material were iteratively modified to ultimately match the tensile test stress-strain diagram.

Results and analysis

Tensile experiments. The tensile test results show the different deformation behavior of the produced samples (Fig. 3a), when design dimensions were used for evaluation. The results are divided into the two regions according to the diameter of the trusses. Due to the same nominal area of the cross-section, this division should not have occurred. To eliminate this effect, real diameters obtained by optical measurement were applied. There were significant differences between design and real diameter and between 45° and 90° manufacturing orientation. For the design diameter d_1 = 0.45 mm with manufacturing orientation 90°, the real diameter d_1' = 0.64 mm were evaluated. For orientation 45°, the real diameter was d_2' = 0.66 mm. The same effect with smaller deviation was observed for design diameter d_3 = 1.00 mm. The real diameter d_3' = 1.04 mm for 90° and d_4' = 1.05 mm for 45° orientation were evaluated. The results after the correction are more consistent and the differences in the linear region of the stress-strain diagram are small (Fig.3b).

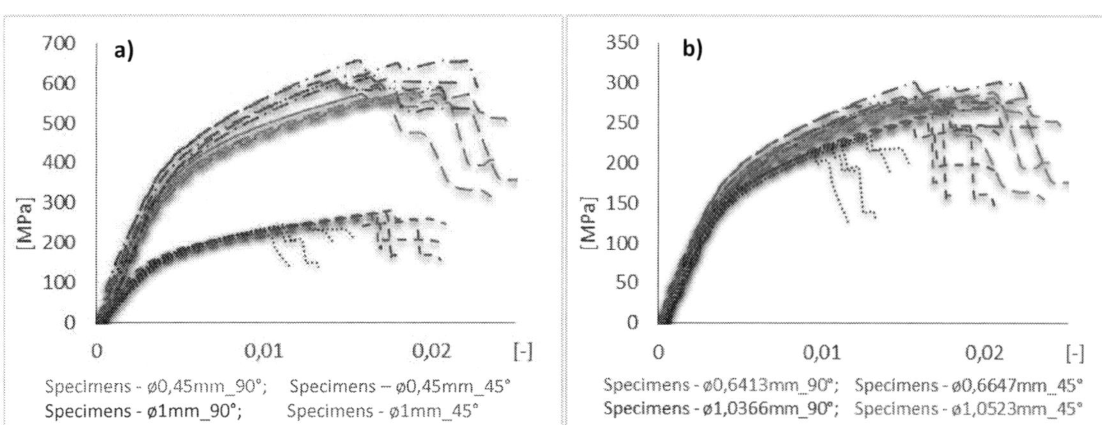

Fig. 3 (a) Tensile test evaluation with design diameters; (b) Tensile test evaluation with real diameters

Tensile simulations. Based on the experimental data, the bilinear model of material was evaluated for trusses with nominal diameter 0.45 mm and orientation 45°. Such set up is the closest to the dimensions and building orientation for BCC lattice structure which was planned to be tested afterward. Geometry was discretized by the tetrahedral quadratic elements. The model of material parameters was at this stage: Young modulus E = 49 GPa, yield strength $R_{p0,2} = 211$ MPa, tangent modulus $E_T = 12.1$ GPa (evaluated according to equations 1-4) and Poisson's ratio $\mu = 0.33$. The simulation set up was following the tensile testing with the correction on real beams geometry. After nonlinear solution, the stress-strain diagram was compared with experimental data (fig.4).

$$R_m = E_T \cdot (\varepsilon_m - \varepsilon_{el}) + R_e \tag{1}$$

$$E_T = \frac{R_m - R_e}{\varepsilon_m - \varepsilon_{el}} \tag{2}$$

$$\varepsilon_{el} = \frac{R_e}{E} = \frac{211}{49000} = 0.00431 \tag{3}$$

$$E_T = \frac{280-211}{0.01-0.00431} = 12118\, MPa \tag{4}$$

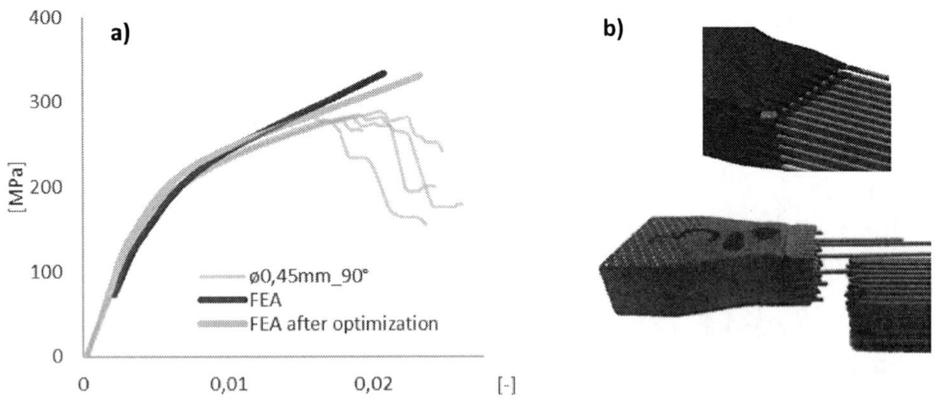

Fig. 4 (a) Stress-strain diagram after non-linear solution; (b) The results of the non-linear solution compared to real failure

The difference is probably caused by other uncertainties which are typical for the SLM technology. It is mainly high surface roughness due to the partial melting of the surrounding metal powder. This rough surface can easily initiate the crack. The second main uncertainty is the porosity and it's weakening of the cross-section area [6]. To correct the difference in stress-strain dependencies, Young modulus and tangent modulus were modified (fig.4a): E = 60 GPa, $E_T = 9.4$ GPa. All other measured mechanical properties are shown in fig. 5 (default settings for boxplots were used in Minitab software).

Compression experiments. During the compression, the individual layers are gradually deformed and densification of lattice structure material occurs. Such behavior is seen on the force-deformation dependency, where the compression load rapidly decreases and then increases after the first collapse (fig. 6a). The main objective of our study is to predict the mechanical behavior of lattice structure made of AlSi10Mg up to the first irreversible collapse, by setting up correctly the model of material in our FEA analysis.

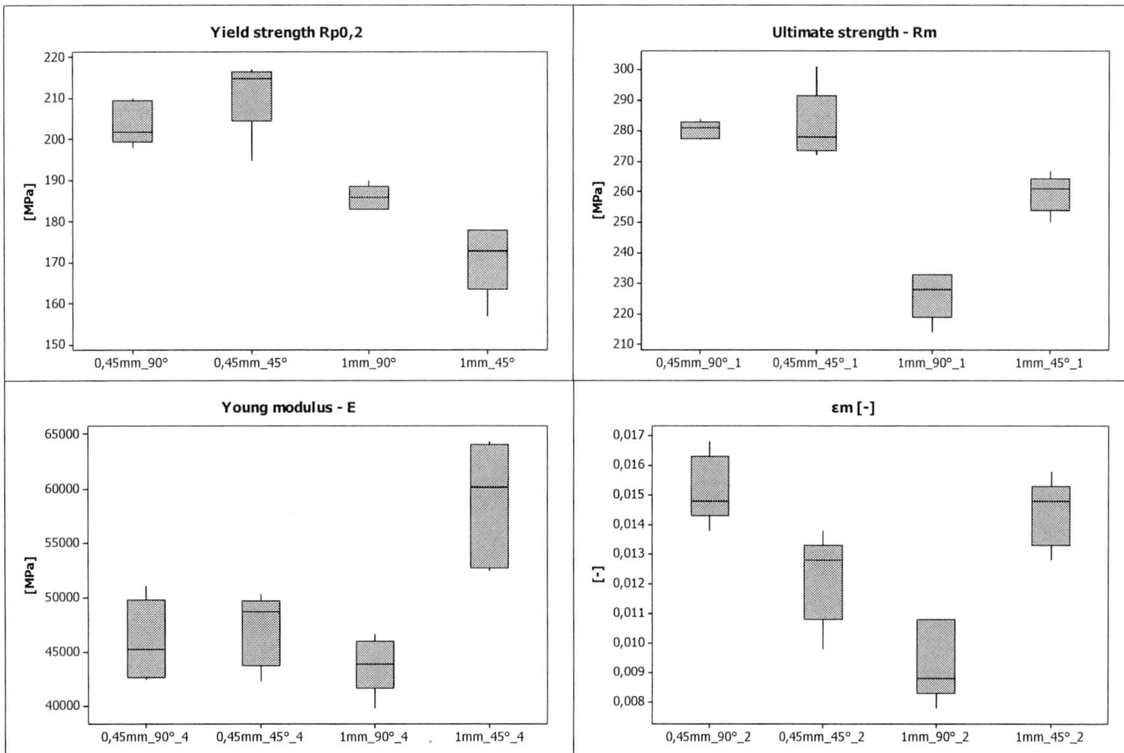

Fig. 5 Box plots of the measured mechanical parameters

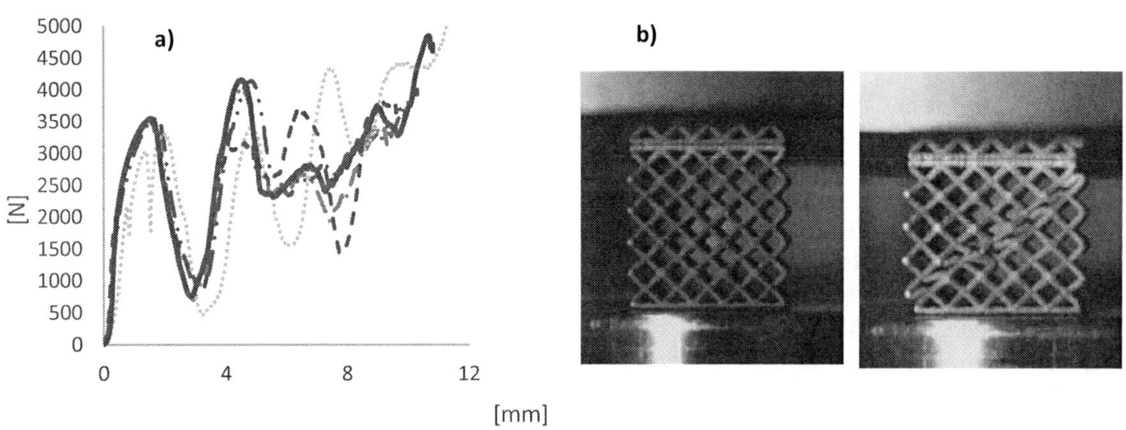

Fig. 6 (a) Force – deformation compression test data with design diameters; (b) Real deformation of the lattice structure

Compression simulations. To validate the obtained material model, a static load simulation of the lattice structure was performed. As before, in the case of tensile specimens, the surface of samples was optically digitized to obtain real dimensions. Based on dimensional analysis, the diameters were corrected from d = 0.5 mm to d′ = 0.72 mm. With these diameters, the volumetric geometry of the structure was imported to ANSYS Workbench. Geometry was discretized by the tetrahedral quadratic elements. The size of the elements has been chosen so that there are at least three elements across the cross-section area. The simulation set up was following the compression testing with the correction on real beams geometry, with the same clamping and loading. Fig 7.

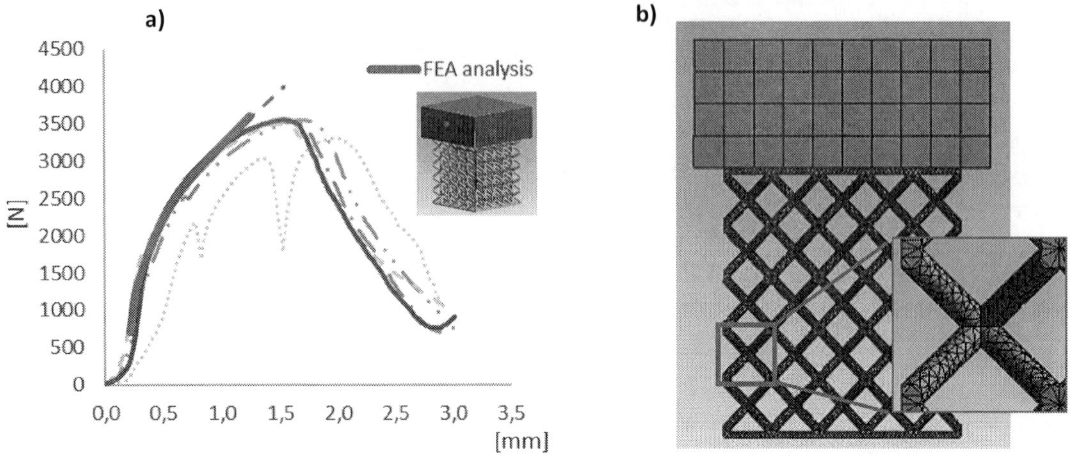

Fig. 7 (a) Force – deformation compression test data with design diameters and boundary conditions; (b) discretization of the geometry

Conclusion

Mechanical testing and computational modelling were used for the evaluation of the material behavior of lattice structures made by SLM machine SLM 280^{HT}. The main outputs of the work are: 1) material parameters (Young's modulus, strength, yielding stress, tangent modulus); 2) geometric correction for the deviation between the designed and real beams.

Acknowledgment

This work is an output of cooperation between project FSI-S-17-4144 and NETME Centre, regional R&D centre built with the financial support from the Operational Programme Research and Development for Innovations within the project NETME Centre (New Technologies for Mechanical Engineering), Reg. No. CZ.1.05/2.1.00/01.0002 and, in the follow-up sustainability stage, supported through NETME CENTRE PLUS (LO1202) by financial means from the Ministry of Education, Youth and Sports under the „National Sustainability Programme I"."

References

[1] J. Šmarda. Biologie pro psychology a pedagogy. Vyd. 2. Praha: Portál, 2007, 420 s. ISBN 978-80-7367-343-7.

[2] R. Vrána, D. Koutný, D. Paloušek. Impact Resistance of Different Types of Lattice Structures manufactured by SLM. MM Science Journal, 2016, roč. 2016, č. 6, s. 1579-1585. ISSN: 1803-1269.

[3] LI, P., Z. Wang, N. Petrinic, C.R. Siviour. Deformation behaviour of stainless steel microlattice structures by selective laser melting. Materials Science and Engineering: A [online]. 2014, 614, 116-121 [cit. 2018-04-21]. DOI: 10.1016/j.msea.2014.07.015. ISSN 09215093.

[4] D. Koutny, D. Palousek, L. Pantelejev. Ch. Hoeller, R. Pichler, L. Tesicky, J. Kaiser. Influence of Scanning Strategies on Processing of Aluminum Alloy EN AW 2618 Using

Selective Laser Melting. Materials [online]. 2018, 11(2), 298- [cit. 2018-04-21]. DOI: 10.3390/ma11020298. ISSN 1996-1944.

[5] D. Palousek, L. Pantelejev, T. Zikmund, D. Koutny. PROCESSING OF NEARLY PURE IRON USING 400W SELECTIVE LASER MELTING – INITIAL STUDY. MM Science Journal [online]. 2017, 2017(01), 1738-1743 [cit. 2018-04-21]. DOI: 10.17973/MMSJ.2017_02_2016184. ISSN 18031269.

[6] K., L. Kempen, J. Thijs, van Humbeeck, J.-P. Kruth. Mechanical Properties of AlSi10Mg Produced by Selective Laser Melting. Physics Procedia. 2012, vol. 39, s. 439-446. DOI: 10.1016/j.phpro.2012.10.059. http://linkinghub.elsevier.com/retrieve/pii/S1875389212025862

[7] M. Suard, P. Lhuissier, R. Dendievel, F. Vignat. Impact of EBM Fabrication Strategies on Geometry and Mechanical Properties of Titanium Cellular Structures. Fraunhofer Direct Digital Manufacturing Conference. 2014.

[8] D. Koutný, R. Vrána, D. Paloušek. Dimensional accuracy of single beams of AlSi10Mg alloy and 316L stainless steel manufactured by SLM. In 5th International Conference on Additive Technologies iCAT2014. Ljubljana: Interesansa, 2014. s. 142-147. ISBN: 978-961-281-579- 0

[9] Laser Beam Melting System SLM 280 HL. SLM SOLUTIONS GMBH. SLM Solutions GmbH [online]. [cit. 2014-05-16]. http://stage.slmsolutions.com/download.php?f=61008aff89ca6259285ee7f35052f942

[10] D. Palousek, M. Omasta, D. Koutny, J. Bednar, T. Koutecky, F. Dokoupil. Effect of matte coating on 3D optical measurement accuracy. Optical Materials [online]. 2015, 40, 1-9 [cit. 2018-04-21]. DOI: 10.1016/j.optmat.2014.11.020. ISSN 09253467.

EAN 2018
56th conference on experimental stress analysis

Experimental Verification of Ski Model for Finite Element Analysis

ZBONČÁK Radek[1, a]

[1]Svárovská 619, 46001 Liberec, Czech Republic

[a]radek.zboncak@vuts.cz

Keywords: Ski, Sandwich Structure, FE Model, FEA.

Abstract. The topic of this work is FE analysis as a first step to understanding of mechanical performance of a ski. At the very beginning a real part scanning has to be done because only hand drawings of the design is available. Creating of a CAD model by reverse engineering is obvious following step to prepare input CAD data for later simulation. The carbon fiber composite ski has sandwich structure with a wooden core. Material testing is an inherent part of any analysis so all represented materials are tested according to testing standards with respect to the anisotropy. In this sense there are three types of used materials – an isotropic, an orthotropic and a transversally isotropic material. Two different types of extensometers (mechanical and optical) must be used as material like wood needs special treatment. The FE model of a ski middle part is created in software Siemens NX using Laminate Composites module (NX LC). The behavior of the FE model is compared with experiment based on deflection in three point bending test. If the resultant deflections show deviation optimizing modifications are done to fit the model more precisely. Deviation of 3 % is reached for this model without any special interventions which means good experimental results. Flexural performance as a predominant loading of the ski is simulated in FEA to get the stress distribution in the ski and in each layer of sandwich structure. In this phase of analysis the exact values of stress are not so important as the way of stress distribution which the work is focused on.

Introduction

The composite materials are used in various occasions. Aerospace is an example of one of well-known hi-tech applications. But there are many other areas where high performance parts are needed or weight reduction is important aspect. The sport, where limits are moved every day, is an excellent example of application of composites. High demands on stiffness, strength and weight reduction is the main way how to reach better results.

This paper is focused on skis. First modern skis were designed by Sondre Norheim in 1870. From that time various materials and their combinations were used to get best performance. In these days, sandwich structure with wooden core is used for high performance skis. Modern cores are made of glued oriented wooden stripes and the top and bottom skins are made of laminates. This combination is proven by practice as the best. But trial and error method usually used by manufacturers is a time consuming approach which consequently costs lot of money. Despite of that fact it is still most widespread approach in ski design. That also reflects the fact that every individual skier has subjective perception of movement and ski performance which makes manufacturing of one perfect ski for everyone almost impossible. There are some efforts to use FEA and optimization methods in design process, but it still does not reach the level of required accuracy and sufficient conformity with the experiment [1]. The material properties and some additional negative effects which are inseparably connected together with manufacturing process are the biggest issue. For example, not properly fulfilled material datasheets or thermal expansion of different used materials which can cause creation of inner tension after cooling down during curing. It can be said, that mechanics of ski is very complex problematics based on understanding and creation of material respectively FE model [2] [3].

The manufacturer of the ski has not had any CAD data. The geometry of the ski is defined only by dimensions in several cross sections positioned alongside the ski length. Therefor the first step is to make a scan of the manufactured ski so the computer model can be created. The output of the scan is a cloud of surface points in *.stl* file (Fig. 1). Then the reverse engineering techniques are used to make a CAD model of the ski. Material properties of the individual constituents of the sandwich structure – wooden core, top and bottom laminate, plastic base and steel edges; are determined experimentally by testing according relevant standards.

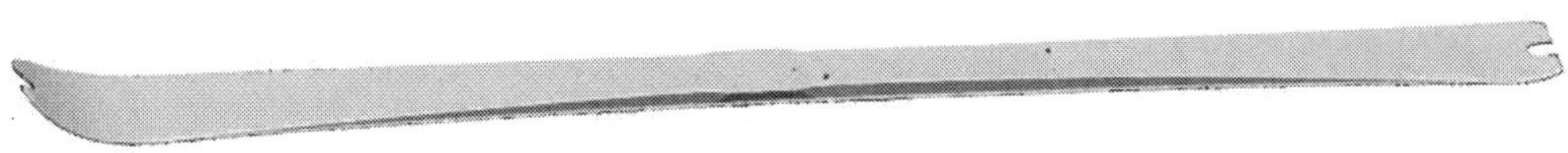

Fig.28 Scan of the ski

Experimental data are evaluated and mechanical properties of the material such as moduli and Poisson's ratios are determined by analytical calculations to define material in FE analysis. There are three types of anisotropic material used in the model – isotropic (plastic base or coating), orthotropic (laminate skins) and transversally isotropic (wooden core). To declare that wooden core is transversally isotropic material is a sort of simplification. On the other hand, the fact that the core is glued together of several oriented thin wooden stripes increases the effect of transversally isotropic behavior of the core. Setting up of the FE model is tuning of parameters so that model performs in the same way as the real part during the experiment. In our case the calculations of material properties are done mainly based on tensile tests and the flexural test of the middle part of the ski is what is used to match simulation results with experiment. The maximum acceptable deviation of simulation and experiment is about 15 % which is generally stated value of deviation tolerance.

Experiment and Calculations

Constituents' properties of laminate and properties of wooden core are identified by experiment or used from material datasheets. Sometime is needed to verify the data provided by the manufacturer or supplier of the constituents. They can differ from the reality as they are determined in laboratory conditions. In case of experiment following standards for tensile test and flexural test are used: ČSN EN ISO 527-4/1, ČSN EN ISO 527-4/10, ČSN EN ISO 527-4/50 and ČSN EN ISO 6892-1. Laminates and wood are orthotropic and transversally isotropic materials where 9 and 5 elastic constants (Young's and shear moduli, Poisson's ratio) must be determined to fully describe the material. Those which cannot be measured must be calculated or estimated by prediction methods. Different behavior in tension and compression must be taken into account as well. In case of isotropic material just 3 elastic constants must be evaluated.

Steel is used for edges to increase their endurance. Edges do not have any serious influence on the total ski stiffness or strength. The material is isotropic and Young's modulus E, shear modulus G and Poison's ratio ν are used from material datasheet because the dimension and shape of the edge does not allow the test to be performed. The relationship among these three elastic constants for mentioned isotropic material is

$$G = \frac{E}{2\cdot(1+\nu)} \quad (1)$$

In case of plastic base (coating) and wooden core, three methods of elastic constants determination are used – biaxial extensometer (Biaxial), advanced video extensometer (AVE2) and digital image correlation (DIC). Biaxial is mechanical extensometer which is clipped directly on the specimen and one can get Poisson ratio directly as the device measures longitudinal and transverse deformation at the same time. Handling with this could be a problem if the specimens do not have acceptable geometry, for example if they are too thin like the plastic base. It is realized that also soft material such as "*balsa-type*" wood can be a problem because sharp blades of the extensometer can cut in the specimen just by the gripping force. And that can influence the value of resultant deformation. Improvement in accuracy is expected from using of video extensometer (AVE2 and DIC) which is confirmed by the results. However there are still some issues with recording of deformation in transvers direction.

Plastic base is also isotropic material. Young's modulus and Poisson's ratio are determined experimentally according tensile test standard ČSN EN ISO 527-2/1. Then the shear modulus is evaluated from the equation (1).

Wooden core is made of light so called *"balsa-type"* wood, which means good mechanical properties combined with low density. The core consists of glued thin slices of wood. Generally the properties of wood vary a lot thanks to variety of imperfections in the natural structure. The consequence of such a glued structure are more balanced properties. As it is mentioned above, the wood is taken as transversally isotropic material which means that 5 elastic constants are needed to define material properties. Tensile test in longitudinal and transversal direction is performed with biaxial extensometer, which allows to construct Mohr's circle for stress and strain in case that one is able to pair two tensile tests in sense of corresponding stress and strain in longitudinal and transvers direction. Principal stresses and strains have indexes L in longitudinal and T in transverse direction.

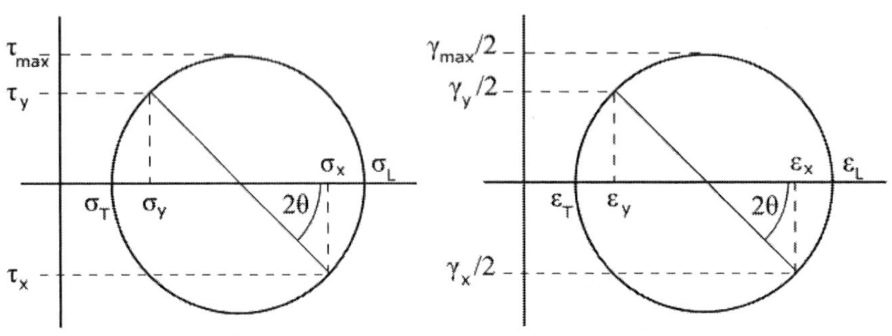

Fig.29 Mohr's stress and strain circle

The relationship between principal stresses and strains in plane is

$$\begin{Bmatrix} \varepsilon_L \\ \varepsilon_T \\ \gamma_{LT} \end{Bmatrix} = \begin{bmatrix} \frac{1}{E_L} & \frac{-\nu_{TL}}{E_T} & 0 \\ \frac{-\nu_{LT}}{E_L} & \frac{1}{E_T} & 0 \\ 0 & 0 & \frac{1}{G_{LT}} \end{bmatrix} . \begin{Bmatrix} \sigma_L \\ \sigma_T \\ \tau_{LT} \end{Bmatrix} . \tag{2}$$

One can get E_L, E_T and ν_{LT} from the first two equations and the additional one based on the matrix symmetry

$$\frac{\nu_{LT}}{E_L} = \frac{\nu_{TL}}{E_T} . \tag{3}$$

Shear modulus G_{LT} must be determined from another test. For θ equal to 45° the shear will be maximal. Applying transformation from global (*LT*) to local (*xy*) coordinate system local stresses and strains are obtained.

$$\begin{Bmatrix} \varepsilon_x \\ \varepsilon_y \\ \gamma_{xy} \end{Bmatrix} = \begin{bmatrix} \frac{1}{E_x} & \frac{-\nu_{yx}}{E_y} & \frac{\eta_{xy}}{G_{xy}} \\ \frac{-\nu_{xy}}{E_x} & \frac{1}{E_y} & \frac{\mu_{xy}}{G_{xy}} \\ \frac{\eta_x}{E_x} & \frac{\mu_y}{E_y} & \frac{1}{G_{xy}} \end{bmatrix} \cdot \begin{Bmatrix} \sigma_x \\ \sigma_y \\ \tau_{xy} \end{Bmatrix}, \tag{4}$$

where η and μ are nonconventional coupling coefficients which show that normal stress induces a distortion. To remind, the matrix in *x-y* coordinates is symmetrical. Additionally the tensile test of the coupon with different fiber orientation than 0° is made to get E_x. With knowledge of E_L, E_T and ν_{LT} from the previous tests and measured E_x the G_{LT} can be calculated.

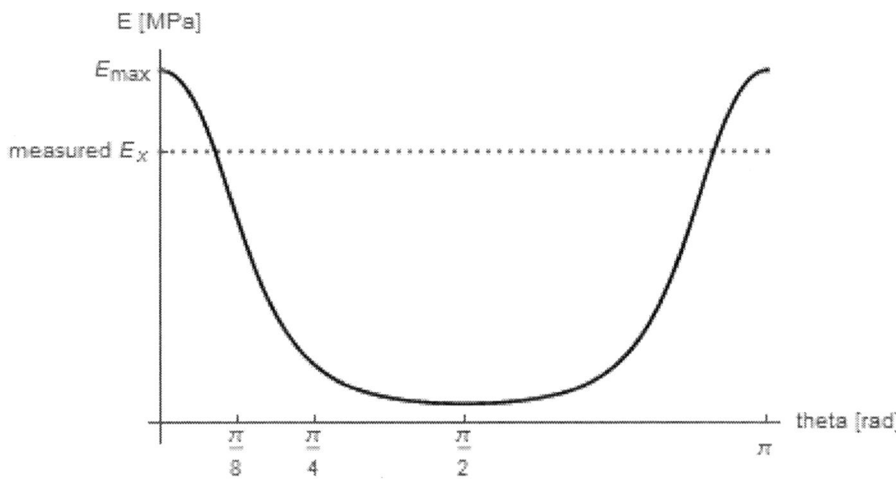

Fig.30 Transformation of modulus E_x where $E_{max} = E_L$

Laminate properties are evaluated from the theoretical values for high strength (HR) carbon fibers and epoxy resin using predictive micromechanical models [4]. This partially idealized approach is selected because of circumstances arising from impossibility to manufacture the specimens with the same hot press molding technology under the same conditions. Laminate is created in SW Siemens NX, where constituents' properties are defined and the software itself computes the laminate properties.

Finite Element Model

As mentioned above, the key is a creation of a precise model of the ski. Because of its complexity, a certain degree of simplification is always needed. And that is what this approaches differ in. Other models are often made as a simple prismatic sandwich beam which seems to be too big simplification where the idealization causes difference of simulation results and experiment. This could lead to inappropriate adjustment of material properties values. The point is to find such approach which will be simple enough to be computed and in the same time complex enough to get the results corresponding the experiment or reality with respect to part geometry.

First of all the scan of the real ski is done and CAD model is created from *.stl* file by reverse engineering techniques. The middle part of the ski is cut out and that as a solid body is used for finite element model.

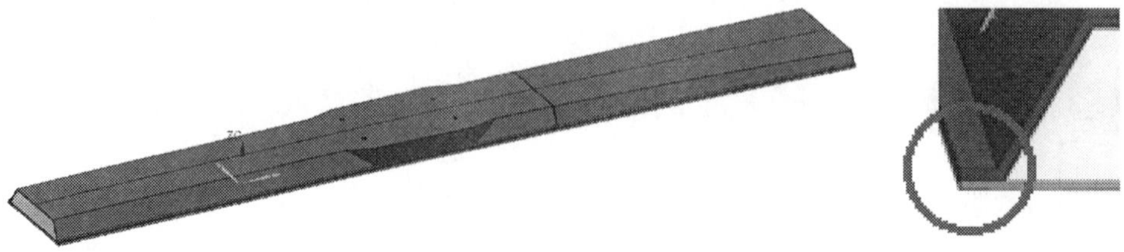

Fig.31 CAD model of the part middle part of the ski (left) and detail of the ski edge (right)

It is split into top, middle and bottom parts which stand for top laminate, core and bottom laminate of the ski. Laminates are extruded on 2D element meshes and core is meshed with 3D elements [1]. All geometrical details are preserved as shown on Fig.31.

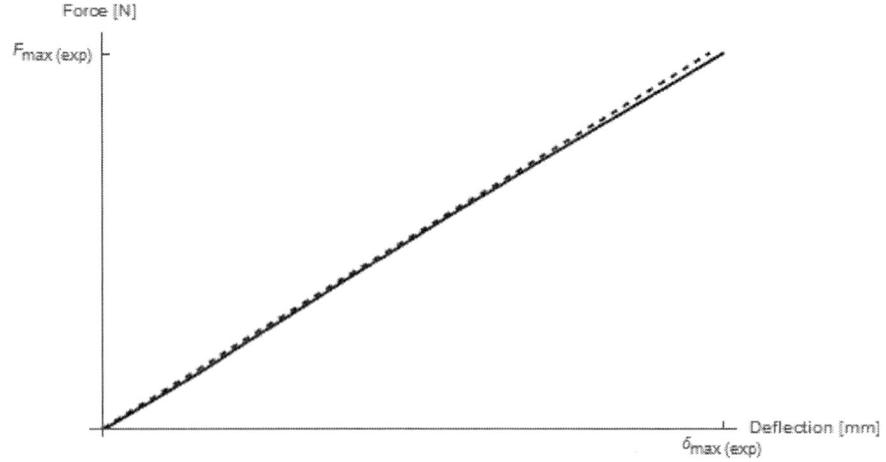

Fig.32 Deflection of the ski from the experiment (thick line) and from FEA (dashed line)

All elastic constants are determined in tensile test. But the bending as a type of major loading is simulated. That fact is taken as an advantage because it should eliminate the potential introducing an error into the result. The actual difference between experiment and FEA result is 2.3 %. The fact that FEA model has lower deflection than the real part is a consequence of geometrical idealization of a model and simplifications made during calculations. The conformity of simulation and experiment is under the acceptable deviation and so defined material can be used for simulation of entire ski.

Finite Element Analysis of Entire Ski

Based on the created model of the ski structure the FEA of the entire ski is done. As the ski is symmetrical along its length, only the half of the ski is simulated. Load and constraints are selected to correspond with typical loading state for ski which is primarily bending. The same three-point bending as for finite element model of the ski middle part is used to have the same conditions for the ski. Constraint supports are moved apart as much as possible to simulate worst loading conditions.

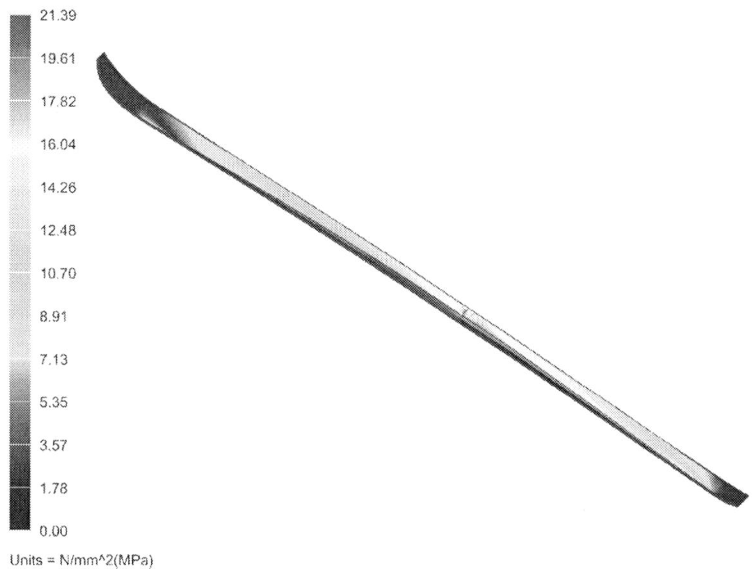

Fig.33 Simulation result

The absolute stress values are not so important in this moment. The attention is focused on the stress distribution in the ski and in each layer of sandwich structure. This information is used as input data in further calculations and FEA with the aim to optimize the structure. Optimization criteria are high stiffness and maximal weight reduction at the same time. These are activities of further work.

Conclusions

Even if the geometrical model is prepared precisely and the simulation constraints are set in an appropriate manner to correspond the real state, the deviation of the result of the complex model is still considerable. The obvious reasons are simplified calculation theories and inevitable idealization of the model. And there are some influences which still cannot be included in the simulation at all, such as some consequences of technological processes. On the other hand, partial results have sufficient precision and can be successfully used in particular steps during design process which definitely reduce labor time and cost. It is possible to design sandwich parts (facesheet and core) with sufficient precision if the model has basic geometry. Then the results can be successfully used for further computations such as optimization of laminate by certain parameters.

The experimental determination of material data is an issue itself. Especially for materials like wood which has scattering of values of its mechanical properties. Such experiments are very sensitive to size of the specimens and test conditions. Also large number of samples must be tested to get a good statistics.

Resultant deflection of the FE model differs about 3 % from the experiment. That is considered as a good result. But this is just first step to further work which is optimization process and one must remember that the results does not include for example influence of binding which causes local stiffening in the middle section of the ski.

Acknowledgement

The publication was created with the support of the project LO1213 – Excellent Engineering Research, the grantor – MŠMT, beneficiary – VÚTS, a. s., in support of the National Sustainability Program, with the financial contribution of the Ministry of Education, Youth and Sports of the Czech Republic.

References

[1] P. Boušková, Numerický model sjezdové lyže s dřevěným jádrem a vrstvou laminátu, Plzeň: Západočeská univerzita v Plzni, Fakulta aplikovaných věd, Katedra mechaniky, 2014.

[2] P. Federolf, A. Luthi and M. Roos, "Parameter study using a finite element simulation of a carving Apline ski to investigate the turn radius and its dependence on edging angle, load and snow properties," Sports Engineering, pp. 135-141, 28 March 2010.

[3] S. Holmberg, K. Persson and H. Petersson, "Nonlinear mechanical behaviour and analysis of wood and fibre materials," Computers and Structures, no. 72, pp. 459-480, 1999.

[4] R. Zbončák, Metody odhadu mechanických vlastností dlouhovlákenného kompozitu, Liberec: VÚTS, a. s., 2017.

EAN 2018
56th conference on experimental stress analysis

Application of Laser Shearography to Vibration Mode Shape Analysis of Composite Panel during Acoustic Load

ZDENĚK Pavel[1,a], KADLEC Martin[1,b] and BĚHAL Jiří[1,c]

[1]VZLU - Czech Aerospace Research Centre, Beranových 130, 199 05, Prague 9 – Letnany, Czech Republic

[a]pavel.zdenek@vzlu.cz, [b]kadlec@vzlu.cz, [c]behal@vzlu.cz

Keywords: Laser shearography, modal analysis, acoustic loading, acoustic fatigue.

Abstract. Acoustic fatigue is important topic of durability in aerospace. Progressive wave tunnel was built to carry out the tests of acoustic fatigue properties of flat composite panels. Vibrations were induced by sound wave. The mode shapes of the vibrations of panel were measured by laser shearography in a form of shearograms – fringe patterns of the out-of-plane surface displacement gradients. Finite-element modal analysis of the panel was performed. FE based displacement gradients were compared with shearograms. Excited vibration mode shapes can be qualitatively estimated on the basis of these comparisons. Few suggestions to enhance both the resolution of shearograms and FE analysis were made.

Introduction

Experience in aircraft industry has shown that the wide band high intensity noise pollution produced by jet exhaust, propellers etc. can cause structural fatigue failure in regions close to the source of the noise [1]. Even though no catastrophic failure of an airplane has occurred due to the acoustic fatigue yet, nevertheless, the economic importance of this topic is not inconsiderable. The major expenses caused by acoustic fatigue are spent on maintenance, inspection and loss of aircraft use. Designing against acoustic fatigue is therefore the key process to prolong the intervals of inspection and thus to lower operational costs.

An air inlet duct of a jet engine is a typical example of the part of an aircraft which is exposed to high intensity sound loading. Development of modernized version of the air inlet is one of main goals in project *TE02000032* (*Advanced Aerostructures Research Centre project*) of the Technology Agency of the Czech Republic. In this case, composite material is proposed to be used as a replacement of aluminum alloy, partly for the reason of low fatigue endurance of aluminum alloy. Testing of material properties, with special focus on the high frequency fatigue behavior, is therefore needed to be carried out. For this purpose, acoustic fatigue testing device, called progressive wave tunnel (PWT), was designed, optimized and constructed [2, 3].

Vibrations of a flat panel clamped on the edges are induced by progressive sound wave. The mode shapes of the vibrations of the panel determine the locations of the stress peaks, where fatigue damage is prone to occur. The displacement field of vibrations of the panel is thus desired to be measured. A potential indirect technique for the mode shape measurement is laser shearography [4, 5] with the alternatives such as the field of accelerometers or digital image correlation techniques.

Theoretical background

Acoustic fatigue is basically fatigue damage caused by the vibrational response of the structure subjected to high intensity noise. The acoustic loading is generally not very effective in inducing high amplitude of displacements or stresses in structures. The combination of very low stress amplitudes and high frequencies of vibrations is practically harmless for the duration of service life of the majority of structures thanks to the nature of S-N curve for high-cycle fatigue. The exceptions are the structures exposed to extreme sound level, such as aircraft, rocket or space shuttle, and structures such as pipeline systems, with natural frequencies close to the exciting sound frequencies. In these cases, the vibration frequency is high enough to cause fatigue damage in not negligible amount of service time.

The laboratory testing of materials is focused on three topics –vibration response of the structure subjected to acoustic loading, acoustic fatigue life and fatigue damage propagation under acoustic loading. The latter is very important in damage tolerant philosophy, for prediction of fatigue crack propagation [6].

Laser shearography is a full-field optical inspection system that uses interference of monochromatic laser light for detection of slight surface deformation due to subsurface discontinuities. It is an active non-destructive method, which measures the response of a material to an applied stress [7]. Detectable surface displacements are in the order of nanometers. Conventional application of laser shearography is the visualization of change of structural stiffness, such as disbond, delamination or change in structure. Examples of these applications are in Ref. [8, 9].

Phase stepping laser shearography works with the Michelson cube for laser interference. The specimen is illuminated with two laser beams, with focal points laterally shifted by the length of a shear vector. The process commonly consists of 4 steps, when the interfering light waves are shifted against each other by a known phase of $\frac{\pi}{2}$. The intensity $I(x,y)$ of reflected light is measured on CCD for each step:

$$I_i = I'(x,y) + I''(x,y) \cdot \cos\left[\phi(x,y) + \frac{\pi}{2} \cdot (i-1)\right], \tag{1}$$

where I', I'' are the distributions of light intensity of the two lasers, ϕ is a distribution of phase value. The phase value is calculated as

$$\phi(x,y) = \tan^{-1}\left(\frac{I_4 - I_2}{I_1 - I_3}\right). \tag{2}$$

Then, the material is stressed, and the steps are repeated, with the phase value post stress denoted as ϕ'. Relative phase difference of the pre-stressed state and post stress state is calculated as

$$\Delta = \phi - \phi'. \tag{3}$$

The relative phase difference represents the relative change of the out-of-plane displacement δw in the direction of shear vector. The out-of-plane gradients of displacement $\frac{\delta w}{\delta x}$ and $\frac{\delta w}{\delta y}$ can be measured by choosing the perpendicular orientations of shear vector for two consecutive measurements

Deformations of plate. Let us define the rectangular coordinate system with axis x and y in the plane of the panel and the z axis in the out-of-plane direction. The displacement components are denoted as *u, v, w* respectively. For simplicity, the deformation state of the tested panel can be described using Kirchhoff-Love theory of thin shells [10]. Its postulates are the negligibility of normal stresses σ_z, conservation of the thickness of the plate during a deformation, conservation of the straightness and normalness of the straight lines normal to the mid-surface after the deformation. [11]. As a consequence, the displacements of the mid-plane of the panel are zero in the *xy* plane, u = v = 0. The rotations of the deformed mid-plane are

$$\omega_x = \frac{\partial w}{\partial x}, \quad \omega_y = \frac{\partial w}{\partial y}. \tag{4}$$

Experimental setup

Device setup. The PWT device is composed of a pair of loudspeakers TVM ARA-389-00/8 with a preamplifier Dynacord S1200, transition parts, a rectangular tunnel of wooden fiberboard and a controlling PC. The tested panel is clamped on a hole on the side of the tunnel. The basic instrumentation consists of two microphones located in the tunnel and an accelerometer mounted by an adhesive wax to the center of the panel. PWT device is shown in Fig. 1

Fig. 1 Progressive wave tunnel (PWT) device

Test specimen. A flat octagonal shaped composite panel (729×729 mm) was designed with lay-up orientation of $[45/\text{-}45/0/45]_s$ and with an artificial defect. Thickness of the panel is 1,8 mm. The interlaminar defect was created by inserting a circle foil (D = 200 mm) between the middle layers at the center of the panel. The panel was manufactured in Aero Vodochody Aerospace company using Hexply 8552/AGP193-PW prepreg. The octagonal shape and the dimensions of the panel are derived from the optimization of a testing stand for high velocity impact testing [12]. The final shape of boundary of the panel is a compromise between the symmetry of circular-shape and the enforcement of prescribed clamping boundary condition of a rectangular shape. The choice of the same shape and size of specimen is driven by later possibility of acoustic fatigue testing of impact-damaged specimen.

Experimental procedures. Odd natural frequencies were estimated from the power spectral density of the response of the center of the panel to the white noise signal loading. Dantec Q-800 laser shearography system was used to measure the gradients of deformation on the estimated frequencies. The panel was illuminated by two dispersed laser beams with horizontal shear vector and excited by the sound of a chosen frequency. Horizontal gradient of panel deformations was derived from the phase difference of the reflected and interfered laser signals in the CCD sensor. The vertical gradients were obtained analogically by setting the shear vector to the vertical direction.

Numerical simulation

Finite element (FE) simulations were performed using the NX Nastran software package. The composite panel was modelled using 2D orthotropic material properties and composite lay-up with plies orientation defined by the production drawing. Literary values [13] were used as material properties. Two variants of the panel were modelled, one with the artificial defect and the other without defect. The meshes are shown in Fig. 2 and Fig. 3. The defect was modelled by two separate layers using only half of the lay-up. Both layers had the thickness of 0,9 mm (half of the thickness of intact panel). These layers were connected to the rest of the panel on circular boundary of modelled defect. Natural frequencies and mode shapes has been extracted from the Normal Modes/Eigenvalue Analysis (SOL103) using Lanczos method with the limitation to the first 60 eigenvalues.

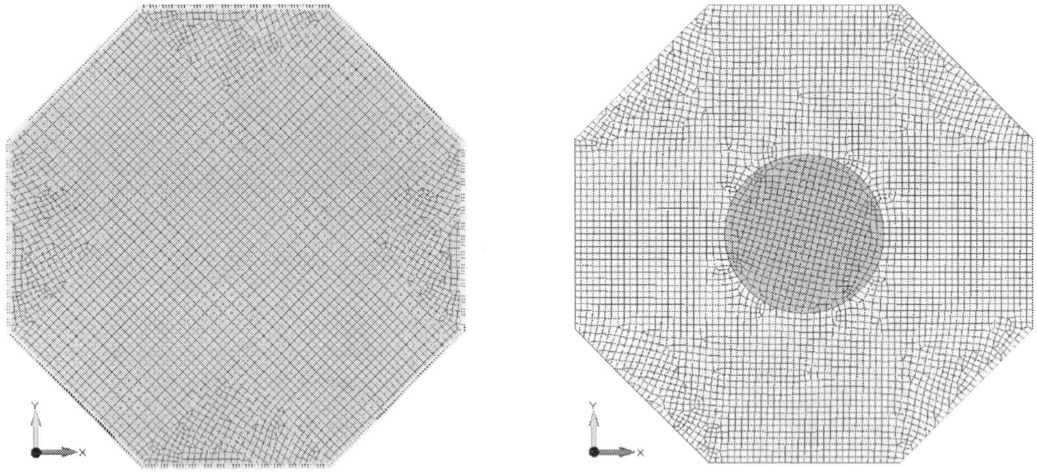

Fig. 2 Variants of the meshes of the panel. a) Panel without artificial defect. b) Panel with artificial defect. The area with defect is marked by green color.

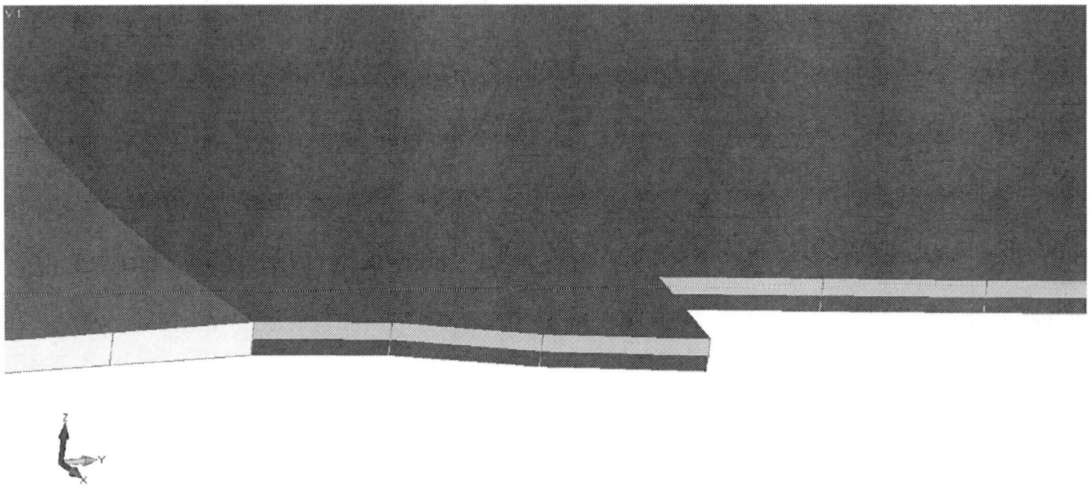

Fig. 3 Detail of the meshes of the defected panel at the edge of artificial delamination. Green and blue layers are separated

Results. As stated in the theoretical background, laser shearograms depict the gradient of out-of-plane displacement. Kirchhoff-Love theory of thin shells states, that these gradients can be easily obtained as the components of the rotation output vector from FE simulation. NX Nastran uses more sophisticated Reissner-Mindlin formulation of shell elements [14], however the differences of results of both theories are negligible for thin shells, so the assumption of the Kirchhoff-Love theory may be considered as accurate enough.

As a result, the shearograms can be directly compared with the corresponding Rotation output value of NX Nastran. The limitation of this comparison is its purely qualitative interpretation. The field of displacements cannot be determined by this method.

Horizontal and vertical gradients of deformation from the laser shearography and from FE simulation were compared. Few examples of the comparison are in Fig. 4, Fig. 5 and Fig. 6.

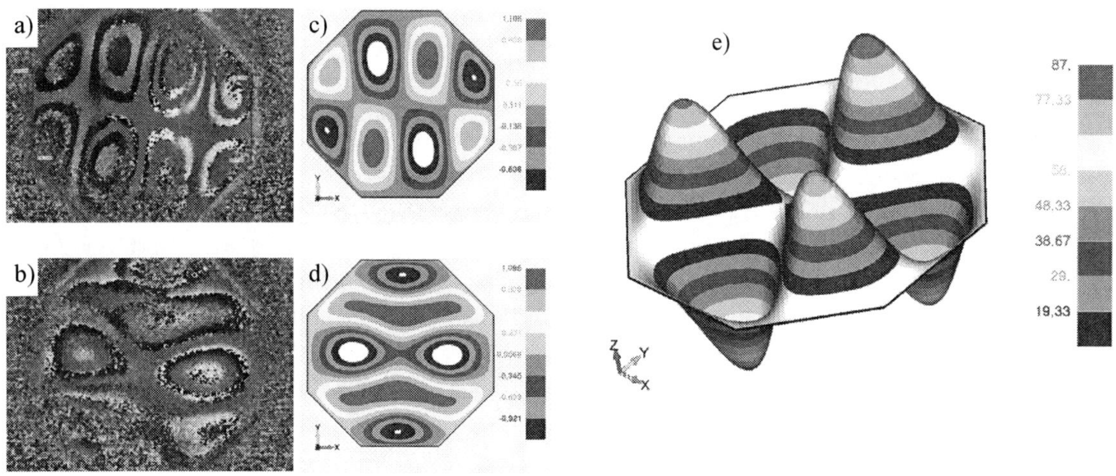

Fig. 4 Comparison of the gradients of deformation determined from laser shearography: a) shearogram –shear in x-axis direction, b) shearogram – shear in y-axis direction, both for exciting frequency 184 Hz, c) FE simulation – output vector Rotation R2, d) FE simulation – output vector Rotation R1 and e) the mode shape from FE simulation, mapped with output vector Total Translation, FE simulation output is for eigenfrequency 273 Hz.

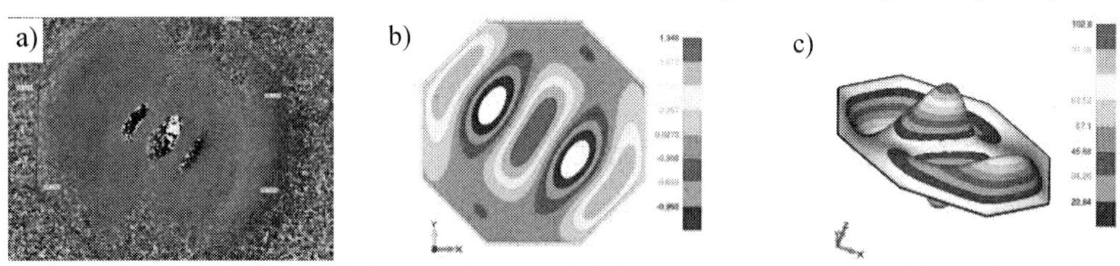

Fig. 5 Comparison of the gradients of deformation determined from laser shearography: a) shearogram – shear in x-axis direction for exciting frequency 500 Hz, b) FE simulation – output vector Rotation R2, c) the mode shape from FE simulation, mapped with output vector Total Translation, FE simulation output is for eigenfrequency 329 Hz.

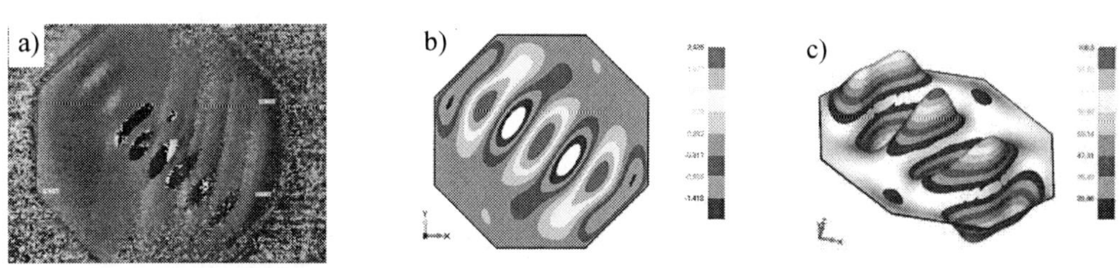

Fig. 6 Comparison of the gradients of deformation determined from laser shearography: a) shearogram –shear in x-axis for frequency 967 Hz, b) FE simulation – output vector Rotation R2, c) the mode shape from FE simulation, mapped with output vector Total Translation, FE simulation output is for eigenfrequency 650 Hz.

Discussion. The similarity between the experimental and numerical deformation gradient, represented by shearograms and FE Rotation output components is very good. Better similarity of the gradients of deformation was achieved using the model of undamaged panel. This may be a sign, that the adhesive forces of the separated surfaces might be strong enough to restrain from the total separation of these surfaces. Another but non-exclusive explanation of indicated phenomenon is that the mass forces of the oscillating plate are lower than the atmospheric pressure, which prevents from the separation and dilatation of the unbonded surfaces. However, the excited vibrations had to be very small because of the effective range of measuring technique and thus this result cannot be generalized. It is assumed that with rising amplitude of vibrations both the adhesive forces and the atmospheric pressure will be exceeded and different vibration modes may appear.

This method can be used to the qualitative determination of the mode shape, however with limitation to the cases, when linear behavior is expected. The accuracy of FE simulation is also limited, mainly due to the lack of experimental supporting data of the material properties of the panel. For example, the stiffness of the composite was described by external literary values [13], which may significantly differ from the properties of the tested specimen. There is a clear mismatch between the numerical eigenfrequencies and experimental exciting frequencies of matching shapes. Apart from the possible inaccuracy of stiffness description, numerical eigenfrequencies can be significantly affected by the boundary conditions. For example, the edges of the panel might have been not ideally clamped. Alternatively, the vibration mode shapes on Fig. 5 and Fig. 6 may have been excited by overtones of basal natural frequencies. Consequently, the differences between numerical estimation of frequency and experimental observation would lower to more acceptable range. FE frequency response analysis of the panel should yield more accurate results.

The absolute range of the displacements is needed to be determined by a different technique. Another drawback is the level of noise in the gradients of deformation from shearography, which can be effectively lowered by the use of shearographic system with more laser beams.

Conclusions

Experimentally measured shearograms can be compared with the components of Rotation output vector of the FE modal analysis. The quality of the interpretation is effectively limited to the scope of measuring displacement range of laser shearography. Other limitation factors of this approach are given by the level of noise in shearograms and by the accuracy of supporting data of FE simulation. Better results could be obtained using more advanced shearographic system, with more laser beams and higher imaging resolution and lesser influence of the noise. On the other hand, the reliability of the FE simulation could be improved by the use of experimentally measured material properties. FE based frequency response analysis of the panel is also desired to be studied.

Acknowledgement

This work was supported by the Technology Agency of the Czech Republic (TA ČR) TE02000032 / Advanced Aerostructures Research Centre project.

References

[1] Thomson, A. G. R., Acoustic fatigue design data, AGARD-AG 162, Part I, Advisory Group for Aerospace Research and Development, Engineering Sciences Data Unit LTD, London (1972), http://www.dtic.mil/dtic/tr/fulltext/u2/744210.pdf

[2] Běhal, J., Proposal of test facility for acoustic fatigue test of flat composite panels. (in Czech), VZLU technical report R-6591, 2016

[3] Běhal, J., Zděnek, P., Effect of interlaminar flaw on composite panel behavior under acoustic loading (In preparation). In Proceedings of International Conference of Engineering Against Failure, Chios Island (2018)

[4] Chiriboga, P., Finite element modeling of vibration in canvas paintings, doctoral thesis, Delft University of Technology, Faculty of Aerospace engineering, Delft (2013). https://doi.org/10.4233/uuid:a30b358e-d0de-4a81-92f3-e9f255443043.

[5] Toh, S. L., Tay, C.J., Shang, H.M. & Lin, Q.Y., Time-average shearography in vibration analysis, Optics and Laser Technology, vol. 27, no. 1 (1995), pp. 51-55.

[6] Federal Aviation Regulation: Part 25 Airworthiness Standards: Transport Category Airplanes, Sec. 25.571 - Damage-tolerance and fatigue evaluation of structure

[7] Kadlec M., Laser Shearography Course, VZLU technical report R-6785-V1 (2017)

[8] Kadlec, M., Růžek, R., A Comparison of Laser Shearography and C-Scan for Assessing a Glass/Epoxy Laminate Impact Damage, Applied Composite Materials, Vol.19, No. 3-4, 2012, pp.393-407

[9] Růžek, R., Lohonka,R., Jironč, J., Ultrasonic C-Scan and Shearography NDI Techniques evaluation of impact defects identification. NDT&E International, Vol. 39, Issue 2, March 2006, pp.132-142.

[10] Love A. E. H., On the small free vibrations and deformations of elastic shells, Philosophical trans. of the Royal Society. London (1888), Vol. serie A, N° 17 p. 491–549.

[11] Reddy, J. N., Theory and analysis of elastic plates and shells, CRC Press, Taylor and Francis. London (2007), 547 p.

[12] Doubrava, R., Raška, J., Oberthor, M: Bird strike tests of flat composite panels for validation of numerical analyses for development of air duct of L39-NG (in Czech), Transfer 29/2017, ISSN 1801 – 9315

[13] Sanchez-Saez, S., Barbero, E., Navarro, C., Analysis of the dynamic flexural behavior of composite beams at low temperature, Composite Science and Technology, 2007, vol. 67, n. 11-12, p. 2616-2632.

[14] Siemens, NX Nastran Element Library Reference. 2014

Index of Authors